DYNAMICAL MASSES OF LOCAL GROUP GALAXIES

IAU SYMPOSIUM 379

COVER ILLUSTRATION:

Original images by N. Risinger (Milky Way), 2002 R. Gendler, Photo by R. Gendler (Andromeda), ESO/VMC Survey (Large Magellanic Cloud), ESO/VMC Survey (Small Magellanic Cloud), composite image designed by K. J. Kanehisa & M. P. Júlio

IAU SYMPOSIUM PROCEEDINGS SERIES

INTERNATIONAL ASTRONOMICAL UNION

UNION ASTRONOMIQUE INTERNATIONALE

International Astronomical Union

IAU

DYNAMICAL MASSES OF LOCAL GROUP GALAXIES

PROCEEDINGS OF THE 379th SYMPOSIUM OF THE INTERNATIONAL ASTRONOMICAL UNION POTSDAM, GERMANY 20–24 AUGUST, 2023

Edited by

PIERCARLO BONIFACIO

GEPI, Observatoire de Paris, Université PSL, CNRS, France

MARIA-ROSA CIONI

Leibniz-Institut für Astrophysik Potsdam (AIP), Germany

FRANÇOIS HAMMER

GEPI, Observatoire de Paris, Université PSL, CNRS, France

MARCEL S. PAWLOWSKI

Leibniz-Institut für Astrophysik Potsdam (AIP), Germany

and

SALVATORE TAIBI

Leibniz-Institut für Astrophysik Potsdam (AIP), Germany

CAMBRIDGE
UNIVERSITY PRESS

CAMBRIDGE UNIVERSITY PRESS
University Printing House, Cambridge CB2 8BS, United Kingdom
1 Liberty Plaza, Floor 20, New York, NY 10006, USA
10 Stamford Road, Oakleigh, Melbourne 3166, Australia

First published 2025

Printed in Great Britain by Henry Ling Limited, The Dorset Press, Dorchester, DT1 1HQ

Typeset in System LaTeX 2ε

A catalogue record for this book is available from the British Library Library of Congress Cataloguing in Publication data

This journal issue has been printed on FSC™-certified paper and cover board. FSC is an independent, non-governmental, not-for-profit organization established to promote the responsible management of the world's forests. Please see www.fsc.org for information.

ISBN 9781009399111 hardback
ISSN 1743-9213

For EU product safety concerns, contact us at Calle de José Abascal, 56, 1°, 28003 Madrid, Spain, or email eugpsr@cambridge.org

Table of Contents

Session 3: Dwarf galaxies: dwarf spheroidals, ultra-faint dwarfs, dwarf irregulars

Session 4: The Magellanic System: the Clouds, the Stream, and the Leading Arm

Preface

Mass is one of the most fundamental quantities that characterise a stellar system, from star clusters to large galaxies. It determines both the internal dynamics of the system and its gravitational interaction with other systems. We know from the fluctuations in the microwave background and the standard cosmological model that mass comes in two flavours: baryonic and non-baryonic. The latter is often simply referred to as "dark matter". Contrary to baryons, which form stars that emit light, non-baryonic matter does not possess electromagnetic interaction and does not emit light. Therefore, in general we cannot simply rely on the light to determine the mass of a system. Instead, we have to use dynamics that take into account both components. Unfortunately, such determinations of masses are generally fraught by large uncertainties. However, for what concerns the galaxies in the Local Group, a real revolution in the topic has occurred with the Gaia satellite that provides six-dimensional data (three-dimensional positions and velocities) for almost 1.5 billion sources, located both in the Milky Way and in Local Group galaxies.

We therefore thought it timely to gather the community working on this topic in an IAU Symposium, to discuss and collect recent advances and provide an opportunity for researchers to meet, exchange ideas, and compare methods and hypothesis. In spite of the very focused topic the symposium addressed a very broad range of astronomical fields (from studies of variable stars for estimating distances, to dynamical modelling), of wavelengths to be investigated (from radio and optical for estimating the HI gas and stellar content and their motions, to X-rays for estimating the warm and hot ionised gas), and has major implications for fundamental questions (from the distribution and nature of dark matter to alternative models of gravity). Measuring galaxy masses at different scales in the Local Group impacts a broad range of astronomy, from stars, star clusters, to the Milky Way, distant galaxies, and Cosmology.

The meeting took place in Potsdam, on the historic Telegrafenberg, from March 20–24, 2023. We are grateful to the Leibniz-Institut für Astrophysik, Potsdam (AIP) for the support and for hosting the symposium. We also thank the International Astronomical Union for supporting the meeting and for sponsoring travel grants, and the Deutsche Forschungsgemeinschaft (DFG, German Research Foundation) for funding in support of the meeting (project number 514600258).

The topics covered where:

- Astrometry and 3D velocities: Gaia and HST
- Distances, Chemistry & Star formation in the Local Group: large surveys of resolved stars
- The Galaxy mass and size: rotation curve, halo stars, globular clusters
- Galactic halo dwarf spheroidal and ultra-faint dwarfs
- Magellanic System: the Magellanic Clouds, the Stream and the Leading Arm
- Dwarf irregular galaxies: dynamics, ram pressure effects
- Andromeda System: rotation curve, halo dwarfs
- Galaxy formation: histories of the Milky Way, of Andromeda, compared to field galaxies
- Near field cosmology: galaxy dynamics, dark matter content, cosmological models

The papers assembled in this volume capture the essence of the material presented at the symposium. We are grateful to all participants for contributing to a lively and discussion-heavy meeting, and in particular wish to thank all the authors of these proceedings for their contributions. We also attempted to capture the discussions that took place during the meeting, adding questions and answers at the end of the papers, whenever possible. We hope this volume will be a valuable reference for the next years, at least until the field will have undergone further advances by when a new Symposium may be timely.

Piercarlo Bonifacio
Maria-Rosa Cioni
François Hammer
Marcel S. Pawloswki
Salvatore Taibi

Editors

Piercarlo Bonifacio
GEPI, Observatoire de Paris, Université PSL, CNRS, France

Maria-Rosa Cioni
Leibniz-Institut für Astrophysik Potsdam (AIP), Germany

François Hammer
GEPI, Observatoire de Paris, Université PSL, CNRS, France

Marcel S. Pawlowski
Leibniz-Institut für Astrophysik Potsdam (AIP), Germany

Salvatore Taibi
Leibniz-Institut für Astrophysik Potsdam (AIP), Germany

Organizing Committee

Scientific Organizing Committee

Francois Hammer (chair), Observatoire de Paris, France
Marcel Pawlowski (chair), AIP, Germany
Beatriz Barbuy, IAG, Brazil
Maria-Rosa Cioni, AIP, Germany
Cuihua Du, UCAS, China
Anna-Christina Eilers, MIT, United States
Eva Grebel, Heidelberg University, Germany
Roger Ianjamasimanana, Instituto de Astrofísica de Andalucía (IAA), Spain
Stacy McGaugh, CWRU, United States
Angeles Perez Villegas, Instituto de Astronomia-UNAM, Mexico
Justin Read, University of Surrey, United Kingdom
Marina Rejkuba, ESO, Germany

Local Organizing Committee

Marcel Pawlowski (chair), AIP, Germany
Maria-Rosa Cioni (chair), AIP, Germany
Salvatore Taibi, AIP, Germany
Pengfei Li, AIP, Germany
Kosuke Jamie Kanehisa, AIP, Germany
Elena Sacchi, AIP, Germany
Nikolay Kacharov, AIP, Germany
Mariana Pouseiro Julio, AIP, Germany
Anna Queiroz, AIP, Germany
Stefano O. Souza, AIP, Germany
Yongjun Jiao, Observatoire de Paris, France

Section 1:
From ground and space: motions, distances, chemistry and star formation measurements

Dynamical Masses of Local Group Galaxies
Proceedings IAU Symposium No. 379, 2025
P. Bonifacio, M.-R. Cioni, F. Hammer, M. S. Pawlowski & S. Taibi, eds.
doi:10.1017/S1743921323003009

Detection of density variations and off-track features in stellar streams

A. Viswanathan, E. Starkenburg and A. F. Esselink

Kapteyn Astronomical Institute, University of Groningen, Landleven 12, 9747 AD Groningen, The Netherlands
email: viswanathan@astro.rug.nl

Abstract. Stellar streams unveil galactic histories through gravitational interactions, mergers, and tidal disruptions. To explore the complex morphology of streams, we use a ~47 million halo main sequence stars catalogue using Gaia DR3 proper motion and photometry information, the combination of which renders the reduced proper motion parameter. This sample with reliable photometric distances reaches out to ~21 kpc thereby probing much further out than would be possible using reliable Gaia parallaxes. Binned velocity moments on-sky pop up several known streams in the inner halo - particularly retrograde structures, due to the kinematic selection. We select and characterise the streams GD-1, Jhelum and Phlegethon. The faint signs of disequilibrium in the form of kinks and density variations in these thin streams will paint a more detailed picture of the existence and properties of the dark matter sub-haloes that perturb them and in turn, the mass distribution of our Galaxy.

Keywords. Galaxy evolution, Galactic halo, Stellar streams, Galactic dynamics, Dark matter

1. Introduction

Numerous smaller thin streams have been discovered in the outer halo thanks to digital sky surveys such as SDSS, 2MASS, WISE, and the Dark Energy Survey (DES) (Belokurov et al. 2006; Shipp et al. 2018). They are believed to originate from less massive systems such as globular clusters and small dwarf galaxies. The "Field of Streams" derived from SDSS data has played a crucial role in the discovery of many streams, with the Sagittarius streams being particularly prominent within this field. The discovery of the Sagittarius dwarf galaxy by Ibata et al. (1994) played a significant role in the field of galactic archaeology. The Sagittarius streams are currently undergoing a merger with the Milky Way. The advent of missions like Gaia has provided valuable data for running pattern-finding algorithms such as STREAMFINDER, developed by Malhan & Ibata (2018). This algorithm is specifically designed to identify dynamically cold and thin stream structures within Galactic surveys. STREAMFINDER employs a search algorithm that identifies stars with coherent dynamics within surveys. It is versatile and can utilize various combinations of data sets, including positional and kinematical information along with its photometry, to uncover these stream structures (Ibata et al. 2021). Gaia has also helped in the detection of substructures within stellar streams. For example, in the GD-1 stream, the presence of a gap and a streak of off-stream stars indicates the potential existence of a dark satellite in the halo (Bonaca et al. 2019a).

The study of such density variations in stellar streams is important for several reasons such as:

• probing the galactic structure - by analyzing the density variations, we can infer the properties of the progenitor systems, such as their mass, orbit, and tidal disruption processes

• constraining the galactic potential and dynamics - density variations in stellar streams can be sensitive to the potential of the Milky Way that we choose to fit them. By comparing observed density variations with theoretical models, we can refine our understanding of the Galactic potential and its components, including the distribution of dark matter. Density variations in stellar streams can provide insights into its interactions with satellite systems.

• probing the dark matter substructure - density variations in stellar streams can also reveal the presence of substructures, such as dark matter subhalos or perturbing encounters with massive objects. These density perturbations can be used to constrain the abundance and properties of substructures within the Galactic halo, providing insights into the nature of dark matter and the formation of small-scale structures in the Universe.

In summary, the study of density variations in stellar streams offers valuable insights into galactic structure, potential, stellar evolution, dynamics, and the nature of dark matter.

2. Reduced proper motion halo catalogue

For the purpose of quantifying the density variations in stellar streams robustly, we make use of main sequence halo stars selected using the reduced proper motion parameter and made available public by Viswanathan et al. (2023). This work has compiled a catalogue of inner halo dwarf stars by employing a kinematic selection technique based on reduced proper motion property (RPM) that relates photometry and proper motion, independent of the use of distances and radial velocities. This selection method follows the approach described in Koppelman & Helmi (2021) that used to make a similar catalogue for Gaia DR2. The stars in the catalogue are plotted in a diagram where the reduced proper motion (RPM) H_G is plotted against the colour $G_0 - G_{RP,0}$. This diagram resembles an absolute colour-magnitude diagram (CMD), but with the different evolutionary phases shifted vertically due to the dependence of H_G on tangential velocity. As a result, the main sequence (MS) stars appears just below the densely populated disk stars in Figure 1. To specifically select halo dwarfs, the box with higher H_G values (corressponding to high tangential velocities (200-800 km/s) is chosen in the main sequence colour regime, as it corresponds to dwarfs on halo orbits. Distances to these stars are fitted using their simple colour-absolute magnitude relation which makes it the most reliable distances (typical uncertainties down to 0.39 kpc in the colour regime where the main sequence is narrower and less likely to be influenced by sub-giant branch or the limiting magnitude of Gaia in the faint end). This catalogue is five times larger than the one made using Gaia DR2 (see title of bottom panels of Figure 1).

The use of such a main sequence halo catalogue is advantageous when studying density variations and off-track features in stellar streams for several reasons:

• Abundance: Main sequence stars are the most abundant type of stars (100 times more abundant than red giant stars), making them readily available for statistical analyses and providing a large sample size for studying the faint substructures within the stream morphology.

• Homogeneity: Main sequence stars share similar properties, such as age and chemical composition, within a given stellar population. This homogeneity simplifies the analysis and interpretation of density variations in stellar streams, but is also helpful in deriving photometric distances for them.

• Sensitivity: Density variations and off-track features in stellar streams can be more easily identified and characterized using main sequence stars. Any deviations from the

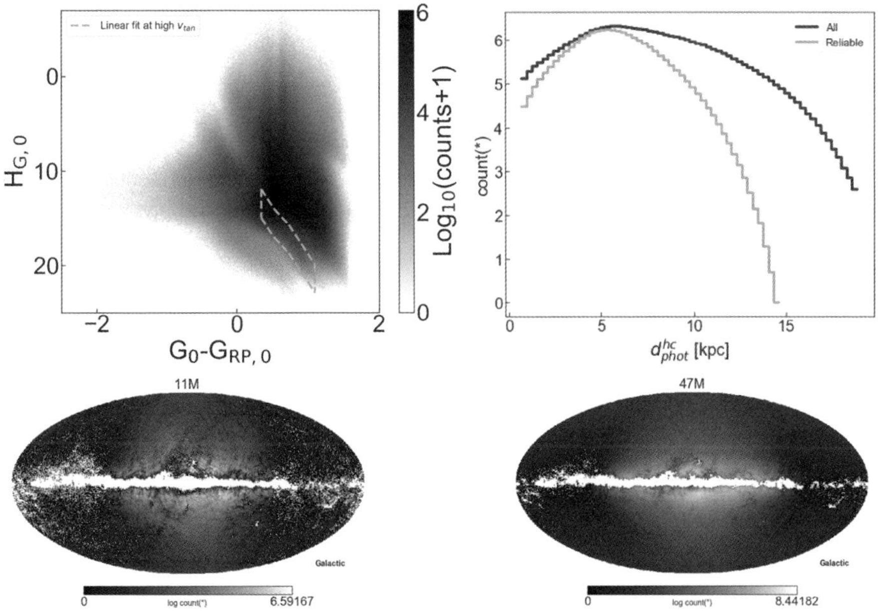

Figure 1. Top left: RPM diagram of all stars in Gaia DR3 with the high tangential velocity halo main sequence selection in yellow dashed polygon. The high density region in the plot belongs to disk stars. Top right: Photometric distance distribution for all stars in dark purple and reliable distances distribution in colour range where we know that the star is in main sequence regime in light purple. Bottom: Mollweide view of density distribution of all stars in the RPM halo selection using Gaia DR2 (left) and Gaia DR3 (right) data.

expected distribution of main sequence stars along the stream can indicate perturbations or interactions with other structures or gravitational forces.

• Simple selection: Main sequence stars can be easily selected using their proper motions and photometry (RPM diagram), which can be used to study the astrometry of stellar streams.

3. Stream selection

We select stellar streams such as GD-1, Jhelum and Phlegethon using simple cuts (similar to what's being used in literature) on-sky and in proper motion (or velocity) space. We remove contaminants by fitting proper motion trends along the stream or galactic coordinates. In the subsection below, we characterise the substructures and density variations in these streams.

3.1. *GD-1*

GD-1 stream is one of the most studied streams in terms of density variations. This stream was discovered by Grillmair & Dionatos (2006). We convert the member stars to stream coordinates using the conversion system used in Koposov et al. (2010). We fit a stream track to GD-1 that is similar to what has been defined in literature and plot the stream track variation with respect to the stream's main coordinate ϕ_1 in Figure 2. In our analysis, we observe gaps and overdensities in the stellar stream that are similar to those reported in the literature, but often with enhanced contrast, such as the Spur and the blob feature. Specifically, in the case of the GD-1 stream, we identify a density gap located around a position angle of $\phi_1 \sim 40°$. This region is a potential candidate for the location of GD-1's fully disrupted progenitor, as suggested by Webb & Bovy (2019).

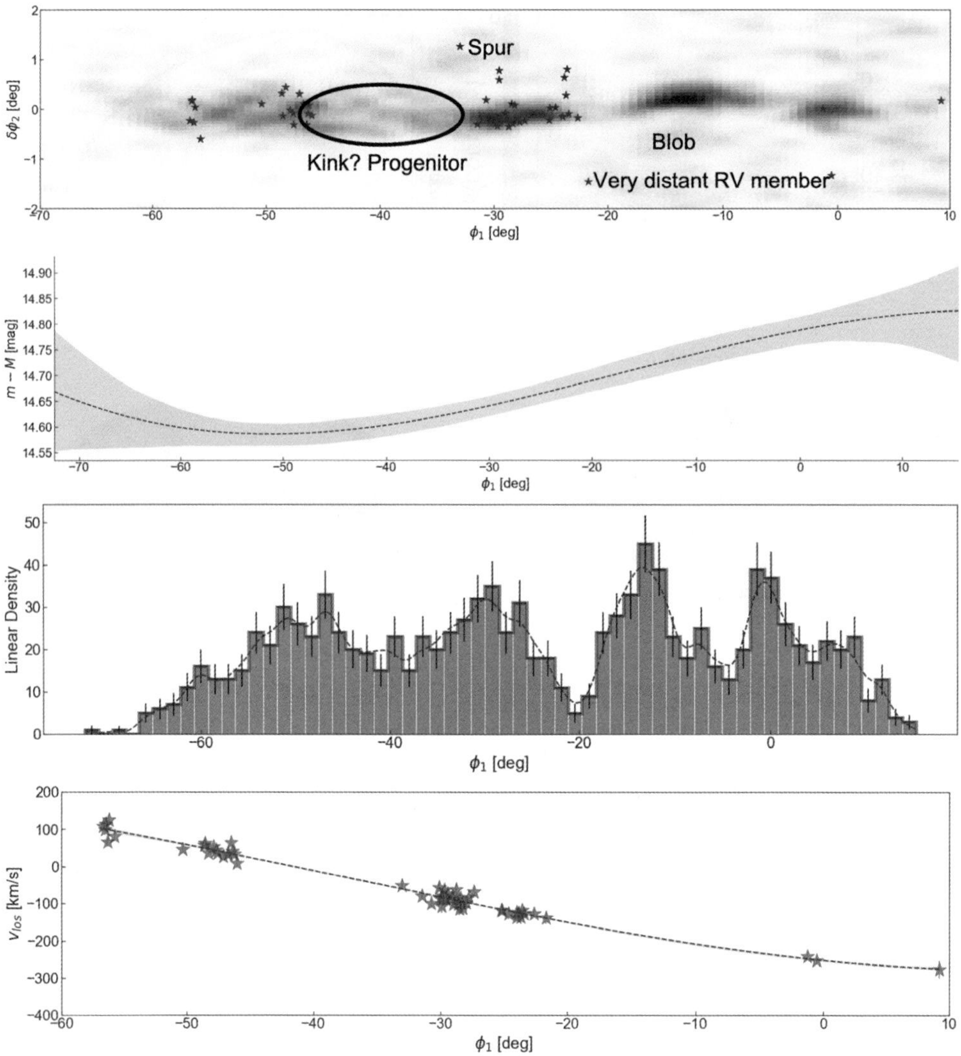

Figure 2. Stream track variations with respect to main stream coordinate ϕ_1 with the radial velocity members from SEGUE overplotted as purple stars (top), distance modulus versus ϕ_1 (middle top), density variation of the stream along the main stream coordinate ϕ_1 (middle bottom), radial velocity trend with respect to ϕ_1 axis (bottom) for GD-1 stream.

However, it is important to note that the interpretation is not conclusive. To further investigate and confirm the nature of this density gap, additional member stars and their radial velocities would be valuable for more detailed analysis and interpretation. We also find a very distant radial velocity member from SEGUE data crossmatch. We can see in Figure 2 bottom panel, the nicely fit radial velocity trend for the member stars selected using this catalogue which gives more confidence in our selection (very high success rates of ~90% of the member stars selected turn out to be true members). We are also able to quantify the distance modulus trend like how the stream has an increasing distance towards increasing ϕ_1 and all the gaps in the density distribution that could be caused by dark matter sub-haloes interactions with the stream, like the gaps at ϕ_1 ~-20° and ϕ_1 ~-5°.

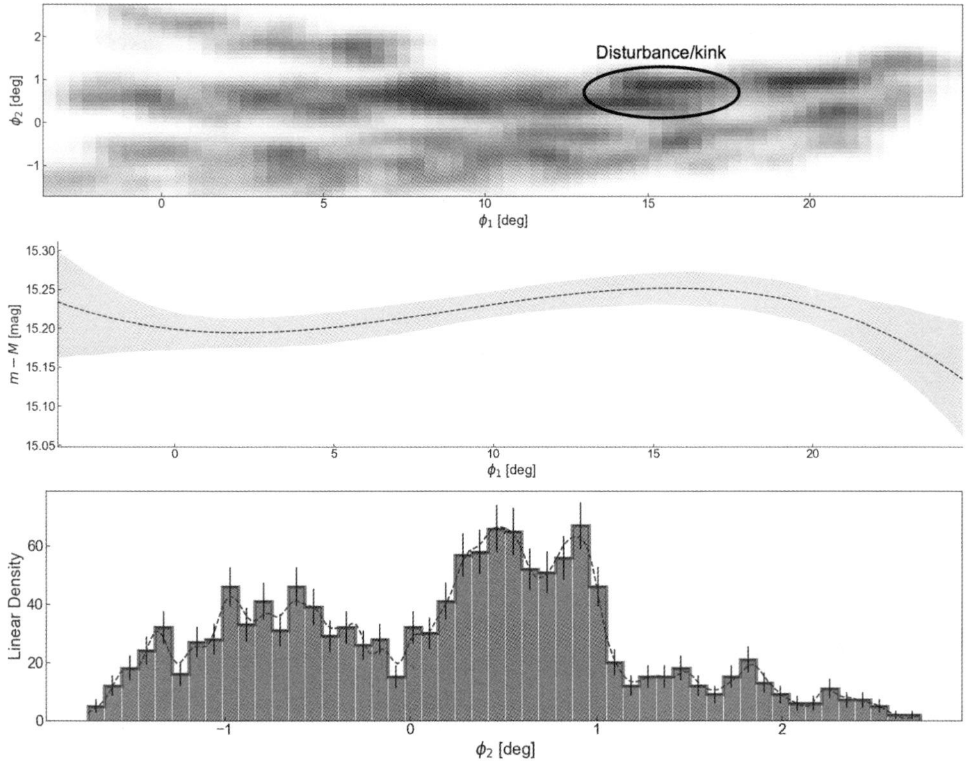

Figure 3. Density distribution in stream coordinates (top), distance modulus versus main stream coordinate ϕ_1 (middle), density variation of the stream along the stream coordinate perpendicular to the stream track ϕ_2 (bottom) for Jhelum stream.

3.2. *Jhelum*

The Jhelum stellar stream is known for its complex morphology and was discovered by Shipp et al. (2018). We convert the member stars to stream coordinates using the conversion system used in Bonaca et al. (2019b). We see three subcomponents of the Jhelum stream - the narrow and broad component and the tertiary spur feature (see top panel of Figure 3). This is also confirmed by the three peaks in the density distribution of ϕ_2. The distance modulus trend of Jhelum cannot be read at face value as the drop in distance at $\phi_1 \sim 20°$ is also influenced by Gaia's limiting magnitude. We also observe a distinct density gap located at a position angle of $\phi_1 \sim 15°$. The improved astrometry and photometry from Gaia and sophisticated selection using the RPM catalogue have allowed us to resolve this feature more effectively. Recent investigations using N-body simulations have explored the multiple components of the Jhelum stream, suggesting a potential close interaction of a globular cluster stream with the Sagittarius dwarf galaxy. More radial velocity members will be helpful to resolve the conflict about the origin of the Jhelum stream (Awad et al., in prep).

3.3. *Phlegethon*

The Phlegethon stream was initially discovered by Ibata et al. (2018) using the STREAMFINDER algorithm applied to Gaia DR2 data. This retrograde stream is believed to be a remnant of a globular cluster. The stream's intriguing characteristics include its proximity to the galactic plane and its low surface brightness. Figure 4 top

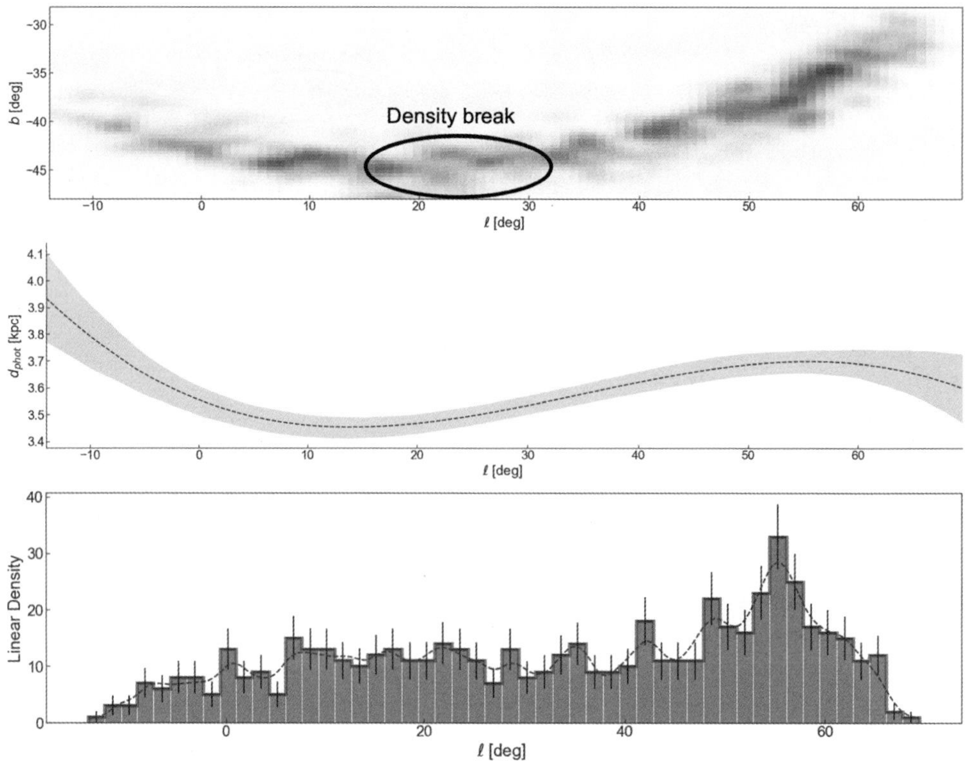

Figure 4. Density distribution in galactic coordinates (ℓ,b) (top), photometric distance versus galactic longitude ℓ (middle), density variation of the stream along the galactic longitude ℓ (bottom) for Phlegethon stream.

panel shows the density distribution of the stars in galactic coordinates, which show a clear stream-like structure. The distance distribution agree well with what was observed in literature before. The stream shows clear epicyclic density spikes along with an intriguing density break at $\ell \sim 25°$ that could be a probable progenitor location for Phlegethon. These properties are resolved for the first time using main sequence stars down to Gaia's magnitude limit.

4. Conclusion

With the release of Gaia DR3 data, we now have access to an unprecedented cartography of the Milky Way galaxy. In this proceeding, we present stellar stream selection using a catalogue of ~47 million main-sequence halo stars selected solely using Gaia DR3 proper motions and photometry. The RPM parameter allows us to pick out high tangential velocity stars in halo orbits without using line-of-sight information. Furthermore, we calculate photometric distances for these stars using a simple linear colour-magnitude relation on the main sequence, providing distance information that is more reliable and extends further than Gaia parallaxes alone. Unlike traditional methods that rely on distance and velocity cuts, our approach allows us to build a larger and more complete catalogue of inner halo stars up to a distance of approximately 21 kpc, with improved completeness beyond 5 kpc and typical uncertainties down to 7%, thanks to Gaia DR3. It is important to keep in mind the kinematic bias of this sample towards high proper motion halo stars, unreliable distances at the turn-off and Gaia's scanning pattern in the faint end. Like any other galaxy in the universe, evidence shows that the assembly

Table 1. The properties of stream members selected using the velocity space from the RPM catalogue.

Name	N	d^{hc}_{phot} [kpc]
GD-1	1155	8.82
Jhelum	1353	11.32
Phlegethon	575	3.6

of the Milky Way is irrefutably hierarchical. We study this non-linearity using stellar streams with complex morphologies that prove to be a powerful test for the nature and distribution of dark matter in the halo. We use the RPM catalogue to quantify density variations and off-track features in inner halo stellar streams such as GD-1, Jhelum and Phlegethon (see Table 1 for properties about the streams selected and number of member stars selected in the process). For these streams, we resolve the gaps, and density breaks reported in the literature more clearly, due to the added information in derived distances as well as the higher sensitivity to low surface brightness features. The use of main-sequence stars, rather than brighter giants, allows us to trace low surface brightness counterparts, pushing the substructure searches to Gaia's magnitude limits. These member stars selected can be used for high-resolution follow-up to confirm their membership and to study the origin of these density variations.

References

Belokurov V., et al., 2006, ApJ, 642, L137

Bonaca A., Hogg D. W., Price-Whelan A. M., Conroy C., 2019a, ApJ, 880, 38

Bonaca A., Conroy C., Price-Whelan A. M., Hogg D. W., 2019b, ApJ, 881, L37

Grillmair C. J., Dionatos O., 2006, ApJ, 643, L17

Ibata R. A., Gilmore G., Irwin M. J., 1994, Nature, 370, 194

Ibata R. A., Malhan K., Martin N. F., Starkenburg E., 2018, ApJ, 865, 85

Ibata R., et al., 2021, ApJ, 914, 123

Koposov S. E., Rix H.-W., Hogg D. W., 2010, ApJ, 712, 260

Koppelman H. H., Helmi A., 2021, A&A, 645, A69

Malhan K., Ibata R. A., 2018, MNRAS, 477, 4063

Shipp N., et al., 2018, ApJ, 862, 114

Viswanathan A., Starkenburg E., Koppelman H. H., Helmi A., Balbinot E., Esselink A. F., 2023, MNRAS, 521, 2087

Webb J. J., Bovy J., 2019, MNRAS, 485, 5929

Dynamical Masses of Local Group Galaxies: IAU Symposium 379
Proceedings IAU Symposium No. 379, 2025
P. Bonifacio, M.-R. Cioni, F. Hammer, M. S. Pawlowski, & S. Taibi eds.
doi:10.1017/S1743921323002107

Chemical enrichment of the thin and thicker discs of Andromeda: Oxygen to argon abundance ratios for planetary nebulae

M. Arnaboldi[1], S. Bhattacharya[2], O. Gerhard[3] and C. Kobayashi[4]

[1]European Southern Observatory, Karl-Schwarzschild-Str. 2, 85748 Garching, Germany
email: marnabol@eso.org

[2]Inter University Centre for Astronomy and Astrophysics, Ganeshkhind, Post Bag 4, Pune 411007, India

[3]Max-Planck-Institut für extraterrestrische Physik, Giessenbachstraße, 85748 Garching, Germany

[4]Centre for Astrophysics Research, Department of Physics, Astronomy and Mathematics, University of Hertfordshire, Hatfield AL10 9AB, UK

Abstract. We use oxygen and argon abundances for planetary nebulae (PNe) with low internal extinction (progenitor ages of > 4.5 Gyr) and high extinction (progenitor ages < 2.5 Gyr), as well as those of the HII regions, to constrain the chemical enrichment and star formation efficiency in the thin and thicker discs of M31. As the argon element is produced in larger fractions by Type Ia supernovae compared to oxygen, we found that the mean log(O/Ar) values of PNe as a function of their argon abundances [12 + log(Ar/H)] trace the interstellar medium (ISM) conditions at the time of birth of the M31 disc PN progenitors. Thus, the chemical enrichment and star formation effciency information encoded in the [α/Fe] versus [Fe/H] distribution of stars is also imprinted in the oxygen-to-argon abundance ratio log(O/Ar) versus argon abundance for the nebular emissions of the different stellar evolution phases. The chemical evolution model that reproduces the mean log(O/Ar) values as a function of argon abundance for the high- and low-extinction PNe requires a second infall of metal-poorer gas during a gas-rich (wet) satellite merger for the M31 disc region within 14 kpc. A strong starburst is ongoing in the intermediate radial range ($14 < R_{GC} < 18$ kpc). In the outer M31 disc ($R_{GC} > 18$ kpc), the log(O/Ar) versus argon abundance distribution of the younger high-extinction PNe indicates that they too were formed in a burst, though mostly from the metal-poorer gas. In M31, the thin disc is younger and less radially extended, formed stars at a higher star formation effciency, and had a faster chemical enrichment timescale than the more extended thicker disc. Both the thin and thicker discs in M31 reach similar high argon abundances (12 + log(Ar/H))$\sim$6.7. The chemical and structural properties of the thin and thicker discs in M31 are thus remarkably different from those determined for the Milky Way thin and thick discs.

Keywords. stars: AGB and post-AGB – planetary nebulae: general – stars: abundances – galaxies: individual: M31 – galaxies: structure

1. Introduction

The different merger histories of the Milky Way (MW) and the giant Andromeda galaxies result in their large separation in the halo metallicity versus total stellar mass diagram, where the MW and M31 are placed at opposite edges of the distribution of galaxies measured by the Galaxy Halos Outer disks Substructures Thick discs and Star Structures (GHOST) survey (Monachesi et al. 2019). In the MW, the most impactful merger took place ~ 10 Gyr ago, after which the MW disc is thought to have evolved mainly by secular evolution. M31 has experienced a more tumultuous recent history instead, as shown by the extended substructures in its inner halo (Ibata et al. 2001; McConnachie et al. 2009). Its most recent important merger is believed to have happened $\sim 2.5 - 4.5$ Gyr ago, according to the observed steep age velocity dispersion relation (Bhattacharya et al. 2019b) and merger simulations (Hammer et al. 2018; Bhattacharya et al. 2023).

Planetary nebulae (PNe) are useful tracers for constraining the kinematics and chemical abundances over a large radial range in nearby galaxies of different morphological types (e.g., Aniyan et al. 2021; Cortesi et al. 2013; Pulsoni et al. 2018; Hartke et al. 2022). Elemental abundances of PNe sample the interstellar medium (ISM) conditions at the time of formation of their parent stars. When the PN ages are also constrained, it becomes possible to map abundance variations across different epochs of star formation in galaxies.

With this aim, a large survey of M31 PNe was undertaken (Bhattacharya et al. 2019a). Bhattacharya et al. (2022) measured distinct oxygen and argon abundance distributions for the thin and thicker discs in M31. They measured flat or slightly positive oxygen and argon gradients for the older, thicker disc and a negative metallicity gradient for the younger thin disc in M31. In the simulations of Hammer et al. (2018), a gas-rich satellite is accreted onto M31 with an orbit along the giant stellar stream (GSS). The merger then heats the pre-existing M31 disc, generating the observed thicker disc. The cold gas accreted through the 'wet' merger would lead to a burst of star formation and the formation of a late, more centrally concentrated thin disc. Because of the merger-driven disc evolution, stars in the thin and thicker discs of M31 would have formed at different epochs under different chemical conditions. Those in the younger thin disc would have formed out of the pre-enriched interstellar gas in M31, mixed with metal-poorer gas brought in by the satellite.

In Arnaboldi et al. (2022), we studied whether the different chemical enrichment of the thicker and thin discs in M31 provide 'smoking gun' evidence that the secondary gas infall predicted by such a gas-rich merger indeed took place.

2. Oxygen versus argon as tracers of the enrichment history in the M31 discs

The star formation history of a galaxy leaves chemical imprints in its ISM through enrichment with different elements. The information of the ISM chemical conditions at the time of birth of a star is encoded in its element abundances. While both argon and oxygen are produced from core-collapse supernovae, argon is additionally produced by Type-Ia supernovae (Kobayashi et al. 2020a,b). Thus, even though both argon and oxygen are α-elements, they do not have lockstep behaviour and the star formation histories of parent stellar populations will imprint information on the log(O/Ar) versus argon abundance, 12 + log(Ar/H), distribution of stars with different ages.

Arnaboldi et al. (2022) studied the distribution of the log(O/Ar) average values in the 12 + log(Ar/H) bins, for the low and high–extinction PNe in M31, see Figure 1. The trend in the mean log(O/Ar) values versus 12 + log(Ar/H) for the low-extinction PNe, which trace the early evolution, clearly show a dependency of decreasing oxygen values with increasing argon abundances. Such a decrease would signal an increase in the Type Ia supernova contribution, which is indeed shown by the galactic chemical evolution (GCE) models by Kobayashi et al. (2020a). Furthermore unlike the MW thick disc, the 4.5 Gyr and older thicker disc in M31 reaches higher argon abundances than the MW thick disc and as high as those reached by the extended star formation in the solar neighbourhood.

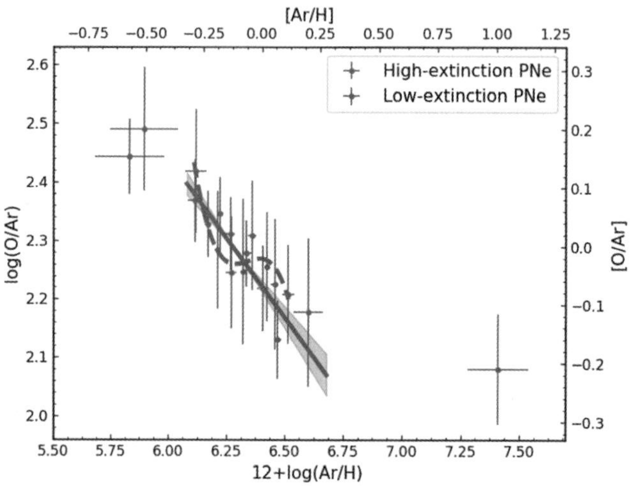

Figure 1. Log(O/Ar) average values for the low-(red) and high-(blue) extinction PNe in M31 binned as a function of their 12 + log(Ar/H) values; for each bin the mean log(O/Ar) is plotted along with the sigma-clipped error. In the figure, the linear fit is shown with uncertainty shaded for the low-extinction sample. For the high-extinction PNe, the dashed blue line shows a third-order polynomial curve that better fits the lower mean log(O/Ar) values of the high-extinction PNe in the 12 + log(Ar/H) = 6.1–6.3 range. The [Ar/H] abundances and [O/Ar] ratios with respect to the solar values are marked on the top and right axes, respectively. From Arnaboldi et al. (2022).

We now turn to the high-extinction PNe in M31, which are tracing the kinematically thin disc of M31 (B+2019b), with respect to the low-extinction older PNe. We note that high-extinction PNe occupy regions of the log(O/Ar) versus 12 + log(Ar/H) plane that, even if with some overlaps, are somewhat different from the low-extinction PNe. If both the high- and low-extinction PNe were following the same enrichment history, the high-extinction PNe, which have an age of ~ 2.5 Gyr or younger (B+2019b), should populate low log(O/Ar)- higher argon abundances region of the evolutionary track. The existence of high-extinction younger PNe with high log(O/Ar) values at relative lower argon abundances support a secondary infall event with less chemically evolved gas.

3. Constraints on chemical evolution and the formation history of the M31 disc

In Figure 2 we show the distribution of the high-extinction PNe in the radial range ($R_{GC} < 14$ kpc) which correspond to the inner disc. Considering their younger age, their distribution in this plane is consistent with the chemical evolution only if also there is a dilution of the ISM by infall of metal-poorer gas. Hence, we constructed GCE models for the inner disc regions ($R_{GC} < 14$ kpc) with two gas inflow phases for the M31 disc, using the same chemical evolution code by Kobayashi et al. (2020a). The models are constrained using the star formation history and metallicity distribution function (MDF) measured for the M31 disc within the PHAT footprint (Williams et al. 2017; $R_{GC} \leq 18$ kpc) using isochrone fitting to the observed red giant branch stars (Williams et al. 2017). A detailed illustration of these models is provided in the Appendix B of Arnaboldi et al. (2022). In these GCE models, the thicker disc older stars are formed by the first gas inflow, with the infall timescale of 1 Gyr. The star formation rate of the first burst differs in the models though, with a weaker first burst by a factor ~ 2.5 for the second model, which is still within the uncertainties of the total star formation rate versus age for the entire PHAT footprint.

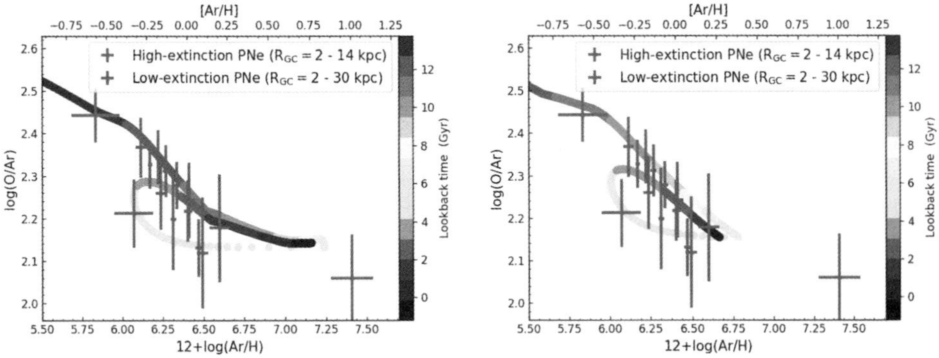

Figure 2. Mean log(O/Ar) values of the low-extinction PNe (red) over the $2-30$ kpc radial range, the high-extinction PNe (blue) within RGC < 14 kpc, and the new two-infall chemical evolution model for the M31 disc (see Sect. 4.2 for details), coloured by look-back time: the fiducial model on the left and the alternative model with a weaker first star formation on the right. The [Ar/H] abundances with respect to the solar values are marked on the top and right axes, respectively. From Arnaboldi et al. (2022).

4. Possible gas content of the merging satellite in the M31 disc

The estimated merger mass ratio from the age-velocity dispersion relation is $1:5$ (B+2019b), hence the baryonic mass of the satellite $M_{sat} = 1.4 \times 10^{10} M_{\circ}$. We then adopt the best-fit galaxy gas-fraction to galaxy stellar mass relation (Díaz-García & Knapen 2020), and the estimated gas fraction for the satellite's mass is $\sim 19\%$, which imples a total mass in gas of $M_{HI,sat} = 2.8 \times 10^9 M_{\circ}$. We can compare the estimated gas mass brought in from the satellite with the i) mass in stars formed during the last 4.5 Gyr and ii) the amount of neutral HI still detected in the M31 disc.

Williams et al. (2017) found that the total stellar mass produced over the last ~ 2 Gyr in the M31 disc is $M_{stellar} \sim 4.5 \times 10^9 M_{\circ}$, and the present-day total mass of HI gas in the M31 disc is $M_{HI} = 4.23 \times 10^9 M_{\circ}$ (Bhattacharya et al. 2019a). Thus, prior to the burst of star formation 2.5 Gyr ago (Williams et al. 2017), a gas mass of $M_{gas,2Gyr} \sim 9 \times 10^9 M_{\circ}$ must have been available in the M31 disc. This amount is much larger than the estimated amount brought in by the merging satellite. Thus, residual gas had to still be present in the pre-merger M31 disc, which was then diluted by Δ[Ar/H] ~ -0.5 dex as observed in the argon abundance distribution. The high extinction PNe were formed out of this diluted ISM within 14 kpc, as illustrated in Fig 2, and mostly from satellite gas at $R_{GC} > 18$ kpc (see Fig. 8 in Arnaboldi et al. (2022)). The chemical evidence for a "wet" merger is one of the observational results, in addition to the kinematics of the substructures in support of a major merger in M31 (Bhattacharya et al. 2023).

5. Conclusions

We have used the largest sample of PNe in the M31 disc with oxygen and argon measurements from B+2022. We have compared their distributions with chemical evolution models for the MW thick disc and the solar neighbourhood and with GCE models for the M31 discs with secondary infall phases. Our main conclusions are that the log(O/Ar) versus argon abundance distributions for the nebular emissions of the stellar evolution phases are valid alternatives to the [α/Fe] versus [Fe/H] diagrams that have been widely used to constrain the star formation effciency in the MW. By using this new approach, we determined a different chemical evolution history for the thin and thicker discs in the Andromeda galaxy, with the thin disc in M31 having been affected by a wet merger.

References

Aniyan, S. *et al.* 2021, *MNRAS* 500, 3579
Arnaboldi, M. *et al.* 2022, *A&A*, 666, 109
Bhattacharya, S. *et al.* 2023, *MNRAS*, 522, 6010
Bhattacharya, S. *et al.* 2022, *MNRAS*, 517, 2343 (B+2022)
Bhattacharya, S. *et al.* 2019b, *A&A*, 631, 56 (B+2019b)
Bhattacharya, S. *et al.* 2019a, *A&A*, 624, 132 (B+2019a)
Chemin, L. *et al.* 2009, *ApJ*, 705, 1395
Cortesi, A. *et al.* 2013, A&A 549, 115
Díaz-García, S., Knapen, J.H. 2020, *A&A* 635, 197
Hammer, F. *et al.* 2018, *MNRAS*, 475, 275
Hartke, J. *et al.* 2022, *A&A* 663,12
Ibata, R. *et al.* 2001, *Nature*, 412, 49
Kobayashi, C. et al. 2020a, *ApJ*, 895, 138
Kobayashi, C. et al. 2020b, *ApJ*, 900, 179
McConnachie, A.W. *et al.* 2009, *Nature*, 461, 66
Monachesi, A. *et al.* 2019, *MNRAS*, 485, 2589
Pulsoni, C. *et al.* 2018, *A&A* 618,94
Williams, B. *et al.* 2017, *ApJ* 846, 145

Dynamical Masses of Local Group Galaxies
Proceedings IAU Symposium No. 379, 2025
P. Bonifacio, M.-R. Cioni, F. Hammer, M. S. Pawlowski & S. Taibi, eds.
doi:10.1017/S1743921323001588

An end-to-end strategy for recovering a free-form potential from a snapshot of stellar coordinates

W. Tenachi[1], R. Ibata[1] and F. Diakogiannis[2]

[1]Université de Strasbourg, CNRS, Observatoire astronomique de Strasbourg, UMR 7550, F-67000 Strasbourg, France
email: wassim.tenachi@astro.unistra.fr

[2]Data61, CSIRO, Kensington, WA 6155, Australia

Abstract. New large observational surveys such as Gaia are leading us into an era of data abundance, offering unprecedented opportunities to discover new physical laws through the power of machine learning. Here we present an end-to-end strategy for recovering a free-form analytical potential from a mere snapshot of stellar positions and velocities. First we show how auto-differentiation can be used to capture an agnostic map of the gravitational potential and its underlying dark matter distribution in the form of a neural network. However, in the context of physics, neural networks are both a plague and a blessing as they are extremely flexible for modeling physical systems but largely consist in non-interpretable black boxes. Therefore, in addition, we show how a complementary symbolic regression approach can be used to open up this neural network into a physically meaningful expression. We demonstrate our strategy by recovering the potential of a toy isochrone system.

Keywords. Galactic Dynamics, Milky Way, Dark Matter, Unsupervised Deep Learning, Gaia, Symbolic Regression, Reinforcement Learning

1. Introduction

The Lambda cold dark matter (ΛCDM) model is very successful at reproducing large scale observations. However, it poses several challenges at the galactic scale (Bullock & Boylan-Kolchin 2017). Being able to compute a reliable and high resolution map of the dark matter distribution from observations in our galactic neighborhood would be of great value for enabling us to decide on these issues. This challenging task is being rendered feasible by the European Space Agency's Gaia mission which is measuring the distance and radial velocity of tens of millions of stars, enabling us to have access for the first time to a very large dataset of 6D (position and velocity) stellar coordinates (Gaia Collaboration 2022).

Here we present a novel strategy for mapping the galactic potential and encapsulating it in an analytical expression in an agnostic and model free manner. In Section 2 we give an auto-differentiation based framework for recovering a free form neural network model of the potential from the snapshot of frozen stellar coordinates that we are able to measure at present time. Then in Section 3, we show how our reinforcement learning based symbolic regression framework Φ-SO (Tenachi et al. 2023) can be used to distill the captured neural network potential model into an *a priori* unknown but physically meaningful functional form by searching in the space of equations.

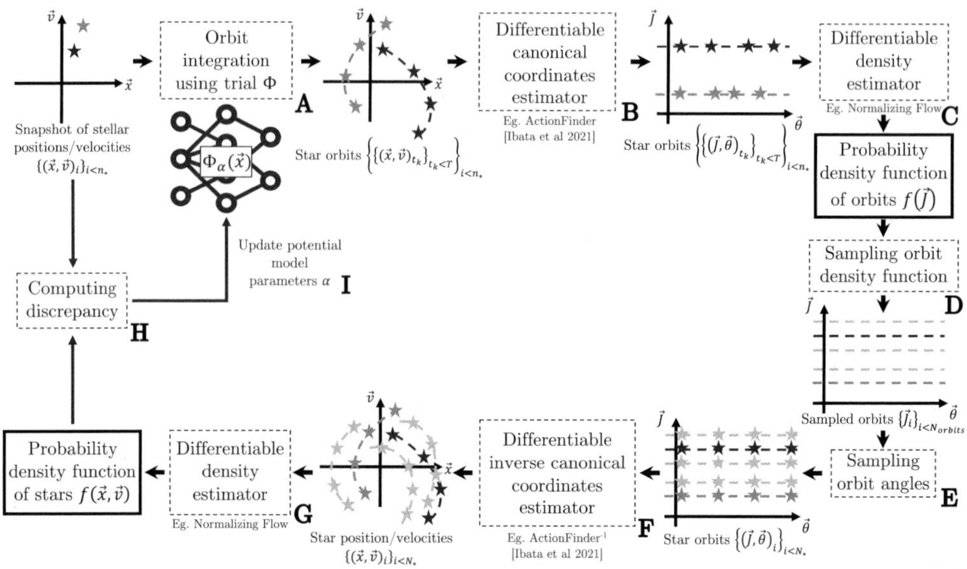

Figure 1. Proposed strategy for recovering a free form neural network potential that stabilizes a distribution of stars $\{(\vec{x},\vec{v})_i\}_{i<n_*}$ using auto-differentiation and gradient descent. See §1 of Section 2 for a full description of the workflow.

2. Learning a potential from a snapshot positions and velocities

The framework we propose for agnostically recovering the potential is shown in Figure 1. The input data of this framework consists of phase-space stellar coordinates $\{(\vec{x},\vec{v})_i\}_{i<n_*}$ obtained from catalogues such as Gaia.

(1) These stars are integrated in a trial gravitational potential Φ_α represented by a flexible free form neural network that depends on parameters α.
(2) These trajectories can then used to deduce orbits in action space using a differentiable canonical coordinates estimator. For this purpose, the neural network based `ACTIONFINDER` method (Ibata et al. (2021a)) can be used. This transformation to the space of orbits represented by three integrals of motions i.e. actions: $\vec{J} = (J_1, J_2, J_3)$ enforces physicality through the Collisionless Boltzmann equation (weak Jeans theorem) and assumes that the samples are mostly regular (i.e. non chaotic) with non-resonant frequencies (strong Jeans theorem) (Binney & Tremaine 2011), which is a reasonable assumption for the Milky Way (Michtchenko et al. 2017).
(3) A differentiable density estimator can then be employed to deduce the density function of orbits in actions $f(\vec{J})$. For this purpose, a normalizing flow (Papamakarios et al. (2021)) or a diffusion model adapted to tabular data such as `TabDDPM` (Kotelnikov et al. (2022)) can be used.
(4) Sampling this function enables us to obtain new orbits $\{\vec{J}_i\}_{i<N_{orbits}}$ in realistic proportions.
(5) These can in turn be sampled to deduce stellar coordinates in actions and angles: $\{(\vec{J},\vec{\theta})_i\}_{i<N_*}$.
(6) By applying an inverse differentiable transformation, one can obtain the Cartesian coordinates of this augmented and phased-mixed stellar population: $\{(\vec{x},\vec{v})_i\}_{i<N_*}$.
(7) These can be used to infer a smooth density function in phase-space $f(\vec{x},\vec{v})$.

(8) Finally, this density function can be compared to initial observations, using a negative log-likelihood loss function : $\sum_{i=1}^{n_*} log(f_\alpha((\vec{x}, \vec{v})_i))$.
(9) Since this final density function depends on the potential neural network's parameters α through all of the steps described above, these can be adjusted to minimize this discrepancy. This process can be repeated iteratively until convergence of Φ_α.

In essence, in our workflow, we are assuming that the system is quasi-stationary (which is well verified within a ~ 200 Myr time-scale for the Milky Way Hou & Han 2015) and computing the free form potential that stabilizes the observed stellar distribution.

It is worth noting that fitting the large number of parameters that make up neural networks is made possible through backpropagation, which involves computing gradients for each single mathematical operation performed in the workflow. While this approach is powerful, it presents challenges as it necessitates the tracking of all gradients and the utilization of differentiable operations only. We utilize `PyTorch` (Paszke et al. 2019) for this purpose.

We demonstrate the efficacy of our scheme using a toy system, substituting Gaia data with synthetic data from an isochrone whose potential is given by: $\Phi(r) = -G.M/(b+\sqrt{r^2+b^2})$, using the analytical canonical transformation to actions and angles (see Binney & Tremaine (2011)) and a simple Gaussian mixture for the density estimation.

In this toy showcase, we are able to recover the isochrone potential within a mean relative error of 0.1% showing that it is possible to use gradients to backpropagate through all of the steps necessary to recover a gravitational field from observations, including an orbit integration, a density estimation, a change of coordinates to actions/angle and a data augmentation using actions. Moreover, we note that our framework can be extended to also leverage the constraints from the numerous stellar streams recently discovered (Malhan et al. 2022; Ibata et al. 2021b; Tenachi et al. 2022; Oria et al. 2022) by requiring the potential to be such that stars from a single stream fall the same orbit $\vec{J}$.

3. Distilling a neural network into an analytical function

Although the agnostic recovery of a neural network Φ_α enclosing a potential model for the Milky Way would be of enormous value. We note that such a black box model would contrast with usual empirical laws in that it would be very difficult if not impossible to it connect with theory. Therefore, we suggest the use of symbolic regression which consists in the inference of a free-form symbolic analytical function $f:\mathbb{R}^n \longrightarrow \mathbb{R}$ that fits $y=f(\mathbf{x})$ given $(\mathbf{x}, y)$ data for distilling the potential neural network into an intelligible and interpretable analytical function. Symbolic regression is distinct from numerical parameter optimization procedures in that it consists in a search in the space of functional forms themselves by optimizing the arrangement of mathematical symbols (e.g. x, $+$, $-$, $\times$, $/$, sin, exp, log, ...).

Here we adopt the Physical Symbolic Optimization (Φ-SO) framework detailed in Tenachi et al. (2023) which was built from the ground up for physics. In this framework, the search space is reduced by leveraging physical units constraints (e.g., cos is dimensionless, velocity + length/$\Box \iff \Box$ is a time etc.) and proposing physically meaningful expressions only. As illustrated in Figure 2 Φ-SO relies on a recurrent neural network (RNN) to generate multiple trial analytical expressions. Fit quality of these expressions

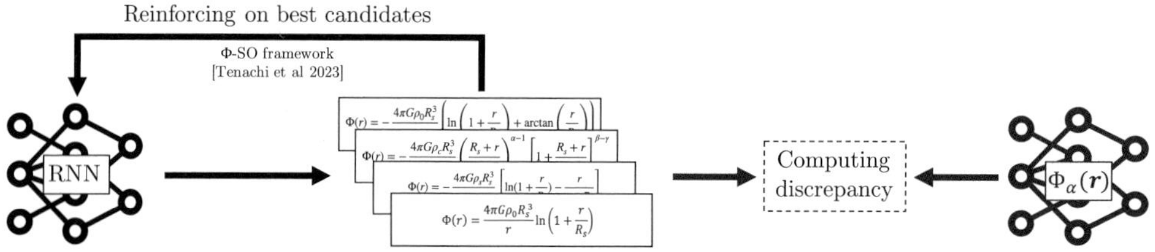

Figure 2. Distilling a neural network into a interpretable analytical expression with Φ-SO (Tenachi et al. 2023). An RNN generates trial expressions, their ability to reproduce the neural network Φ_α is assessed and the best ones are reinforced. This process is repeated iteratively until convergence of the RNN and the extraction a set of high quality expressions.

can then be assessed against data generated using the Φ_α neural network. Best expressions are then reinforced and the process is repeated until the RNN converges and a set of high quality expressions that reproduce Φ_α predictions is obtained. We note that since this framework relies on reinforcement learning, in addition to fit quality any criteria (even non-differentiable ones) can be used including the condition: $\lim_{r\to\infty} \Phi(r) = 0$. Using the Φ-SO framework, we are able to successfully recover the potential of the toy isochrone system described in Section 2.

Code availability

The documented code for the Φ-SO algorithm along with demonstration notebooks are available on GitHub github.com/WassimTenachi/PhySO.

Acknowledgments

RI acknowledges funding from the European Research Council (ERC) under the European Unions Horizon 2020 research and innovation programme (grant agreement No. 834148).

References

Binney, J., & Tremaine, S. 2011, Galactic dynamics, Vol. 13 (Princeton university press)
Bullock, J. S., & Boylan-Kolchin, M. 2017, ARAA, 55, 343
Gaia Collaboration. 2022, A&A
Hou, L. G., & Han, J. L. 2015, MNRAS, 454, 626
Ibata, R., Diakogiannis, F. I., Famaey, B., & Monari, G. 2021a, ApJ, 915, 5
Ibata, R., Malhan, K., Martin, N., et al. 2021b, ApJ, 914, 123
Kotelnikov, A., Baranchuk, D., Rubachev, I., & Babenko, A. 2022, arXiv preprint arXiv:2209.15421
Malhan, K., Ibata, R. A., Sharma, S., et al. 2022, ApJ, 926, 107
Michtchenko, T. A., Vieira, R. S. S., Barros, D. A., & Lépine, J. R. D. 2017, A&A, 597, A39
Oria, P.-A., Tenachi, W., Ibata, R., et al. 2022, ApJL, 936, L3
Papamakarios, G., Nalisnick, E., Rezende, D. J., Mohamed, S., & Lakshminarayanan, B. 2021, JMLR, 22, 2617
Paszke, A., Gross, S., Massa, F., et al. 2019, NeurIPS, 32
Tenachi, W., Ibata, R., & Diakogiannis, F. I. 2023, arXiv e-prints, arXiv:2303.03192
Tenachi, W., Oria, P.-A., Ibata, R., et al. 2022, ApJL, 935, L22

Dynamical Masses of Local Group Galaxies
Proceedings IAU Symposium No. 379, 2025
P. Bonifacio, M.-R. Cioni, F. Hammer, M. S. Pawlowski & S. Taibi, eds.
doi:10.1017/S1743921323002338

Angular momentum and orbits of Milky Way dwarf galaxies

Hefan Li[1], **Francois Hammer**[2], **Carine Babusiaux**[2,3] **and Marcel S. Pawlowski**[4]

[1]Key Laboratory of Optical Astronomy, National Astronomical Observatories, Chinese Academy of Sciences, 20A Datun Road, Beijing, China
email: lihf@bao.ac.cn

[2]GEPI, Observatoire de Paris, Université PSL, CNRS, Place Jules Janssen 92195, Meudon, France

[3]Université de Grenoble-Alpes, CNRS, IPAG, F-38000 Grenoble, France

[4]Leibniz-Institut fuer Astrophysik Potsdam (AIP), An der Sternwarte 16, D-14482 Potsdam Germany

Abstract. We calculated the positions and velocities of 47 dwarf galaxies using a Bayesian approach. Their angular momentum distribution shows that between 36% and 57% of the dwarf galaxies are located in the Vast Polar Structure (VPOS), which could conflict with expectations for a cosmological infall of primordial dwarfs. Using four gravitational potential models with mass spans from $2.8 \times 10^{11}\ M_\odot$ to $15 \times 10^{11}\ M_\odot$, we find that dwarf galaxies are over-concentrated near their pericenters compared to what Kepler's law would expect. This suggests that there may be a large number of dwarf galaxies that have not yet been discovered, or that they are not satellites of the Milky Way.

Keywords. galaxies: dwarf, galaxies: kinematics and dynamics

1. Introduction

According to the hierarchical model of galaxy formation in ΛCDM Model, small galactic building blocks formed first, and then merged into larger galaxies. Our Milky Way is the best laboratory for studying this merger process. This is attributed to the many dwarf galaxies that lie in our neighborhood. They are embedded in the halo and can reach hundreds of kiloparsecs. These galaxies are unique galaxies for which we can use the 3D positions and velocities to study the merger with the Milky Way.

The radial velocities of the member stars of the dwarf galaxy can be obtained from the spectra, and they show large radial velocity dispersion (Aaronson 1983; Kleyna *et al.* 2005). If dwarf galaxies are in dynamic equilibrium, they must be dominated by dark matter. The ratio of dark matter to baryon mass in a ultra-faint dwarfs (UFD) can be hundreds or even thousands. Radial velocity dispersion is a very effective mass estimator, even if dwarf galaxies are out of self-equilibrium due to tidal disruption (Muñoz *et al.* 2008). On the other hand, our knowledge of the proper motions of dwarf galaxies is almost exclusively limited to classical ones before the second Gaia data release (Gaia Collaboration *et al.* 2018). Now, we can get the proper motions of a large number of fainter dwarf galaxies from Gaia data. Its early third data release (Gaia Collaboration *et al.* 2021) further improves the precision of the proper motions by a factor of ~ 2.5.

Investigations of dwarf galaxies' past and future require information about their orbits. However, The orbital uncertainty comes not only from the positions and velocities of the dwarf galaxy, but also from the potential model of the Milky Way. Eilers *et al.* (2019) estimate the precise rotation velocity curve of the Milky Way based on more than 23 thousand red giant stars, ranging from 5 to 25 kpc from the Milky Way center. But at larger radii, both the uncertainty of the model itself and the differences between the models are large (Jiao *et al.* 2021). In addition to the effect of different dark matter profiles on the mass distribution of the Milky Way, they show that even the same profile can lead to a total mass difference of more than an order of magnitude.

Pawlowski & Kroupa (2013) found that 7 to 9 of the 11 most of classical dwarf galaxies have very close orbital poles. The orbital planes of these dwarf galaxies are almost perpendicular to the Milky Way disk, and these planes define a thin plane called vast polar structure (VPOS). However, they should be isotropic in a ΛCDM sub-halo system. Fritz *et al.* (2018) calculated the orbits of 39 dwarf galaxies and found that they tend to be concentrated near the pericenter. But according to Kepler's law, the object will spend very little time near the pericenter because of its high velocity. Although a double-mass dark matter halo can reduce the concentration near the pericenter, the result does not change substantially. In this work, the proper motions of 46 dwarf galaxies are determined based on Gaia EDR3. We then investigate the association of these galaxies with the VPOS and the proportion of them concentrated near the pericenter in the four different Milky Way potentials.

2. Sample of Dwarf Galaxies

We first collect the systemic proper motions and radial velocities of 46 dwarf galaxies from Li *et al.* (2021). Their systemic proper motions and errors are calculated using a Bayesian mixture model which considers both statistical and systematic errors for proper motions of member stars (Vasiliev 2019). We update the systemic proper motions of Aquarius II and Bootes II (Bruce *et al.* 2023) and obtain sky positions and distance modulus of these dwarf galaxies from Fritz *et al.* (2018) and Simon (2019). We further add a recently discovered dwarf galaxy Pegasus IV (Cerny *et al.* 2023). The final sample includes 47 dwarf galaxies.

The 3D positions (x, y, z) and velocities (v_x, v_y, v_z) are derived from Galactic coordinates (l, b), distance modulus, radial velocity and proper motions. To obtain unbiased velocities, especially tangential velocity, we adopt a Bayesian approach described in van der Marel & Guhathakurta (2008) and draw 2,000 realizations from the posterior probability for each galaxy.

3. The Prominence of the VPOS

Following the method of Fritz *et al.* (2018) and Li *et al.* (2021), and using the same definition of the VPOS, central coordinate $(l, b) = (169.3°, -2.8°)$ and opening angle $\theta_{\rm VPOS} = 36.87°$, we divide 47 dwarf galaxies into four groups based on the relationship between their orbital poles (i.e. the direction of the angular momentum) and the VPOS: (1) 20 likely members; (2) 10 uncertain members; (3) 11 unlikely members; (4) 6 impossible Members. The medians of the orbital poles of the objects in the first group are all in the VPOS. Most of them are unlikely to deviate from the VPOS by error. Only 3 dwarf galaxies in the first group are counter-orbiting. They have similar orbital planes to the rest of the group, but orbital velocities are opposite. If both Magellanic Clouds are included, the counter-orbiting fraction is only $f_{\rm counter} = 3/(20 + 2) = 14\%$.

All 9 dwarf galaxies in the second group have large errors that make it difficult to determine whether they are located in the VPOS. Even though they are intrinsically in the VPOS, most of them will deviate from the VPOS due to large errors. Although the 11 dwarf galaxies in the third group are unlikely to be members of the VPOS, 2 of them deviate only slightly from the VPOS. Small changes in the VPOS can transform them into likely members. The dwarf galaxies of the fourth group cannot align with the VPOS due to their 3D positions, regardless of their 3D velocities. Due to the similarity of our sample and the method with Li *et al.* (2021), there is only one change in our results: Pegasus IV becomes the new uncertain member.

However, the infall of primordial dwarf galaxies should be isotropic. To test how far our sample of 47 dwarf galaxies deviates from this expectation, we calculate the proportion of dwarf galaxies in different area sizes around the VPOS center $(l, b) = (169.3°, -2.8°)$. For example, Li *et al.* (2021) find 17 dwarf galaxies in about 10% of the total area around the VPOS center. This corresponds to 37% of their sample size of dwarf galaxies. The result of our sample is shown in the left panel of Figure 1. The x-axis represents the proportion of the area around the VPOS center, and the y-axis is the proportion of dwarf galaxies in this area. The blue line shows the distribution of the sample of dwarf galaxies and the shaded region represents the 1σ confidence interval calculated based on the MC realizations. The black line is the null hypothesis, i.e., the orbital poles of dwarf galaxies are uniformly distributed.

We use the Kolmogorov-Smirnov (KS) test to compare each MC realization with the null hypothesis. Note that in each comparison, the null hypothesis sample is regenerated. The p-value is $0.016^{+0.038}_{-0.014}$, which means that the orbital poles of dwarf galaxies are unlikely to be uniformly distributed. The area proportion where the KS-value is measured in 16^{+12}_{-9} percent, which gives some indication of the VPOS size. The median of this proportion is slightly greater than the 10% that Fritz *et al.* (2018) choose. However, the 84-th percentile is 28%. Such a large proportion is probably due to dwarf galaxies with larger errors. The opening angle corresponding to the proportion of total area is $46.7^{+17.2}_{-16}$ degree. There are 22^{+5}_{-5} dwarf galaxies in the area of the above proportion. Therefore, the proportion of dwarf galaxies located in the VPOS is 47^{+11}_{-11} percent. The KS-value equals to $0.32^{+0.06}_{-0.04}$ indicates an excess of 15^{+3}_{-2} dwarf galaxies compared to expectations.

4. Dwarf Galaxies are Excessively near Their Pericenter

To reinvestigate the excessive concentration of dwarf galaxies near their pericenters found by Fritz *et al.* (2018), we integrate orbits in the four potential models to obtain orbital parameters using `galpy` (Bovy 2015). These four models are consistent with the rotation velocity curve of Eilers *et al.* (2019) and have the same baryonic mass model and parameters: a Plummer (Binney & Tremaine 1987) bulge, and two Miyamoto-Nagai (Miyamoto & Nagai 1975) for the thin and thick disk. Model PNFW (Eilers *et al.* 2019) has a NFW (Navarro *et al.* 1997) dark matter halo and the total mass is $8.1 \times 10^{11}\ M_{\odot}$. Model PE_{LM} and PE_{HM} adopt an Einasto profile (Retana-Montenegro 2012) and the total mass is $2.8 \times 10^{11}\ M_{\odot}$ to $15 \times 10^{11}\ M_{\odot}$. Wang *et al.* (2022) further considers the kinematics of globular cluster to obtain model PE_{IM} with a total mass of $5.1 \times 10^{11}\ M_{\odot}$.

To measure how close a dwarf galaxy is to its pericenter, we define the orbital phase following Hammer *et al.* (2020):

$$P = \frac{t_{\rm peri}}{t_{\rm peri-apo}}, \tag{1}$$

where $t_{\rm peri}$ is the time of inward movement from the current position to the pericenter and $t_{\rm peri-apo}$ is half an orbital period. Distant and faint dwarf galaxies may not be observable due to survey limitations. To correct for this selection effect, we first apply Drlica-Wagner *et al.* (2020) method and select the absolute magnitude and half-light

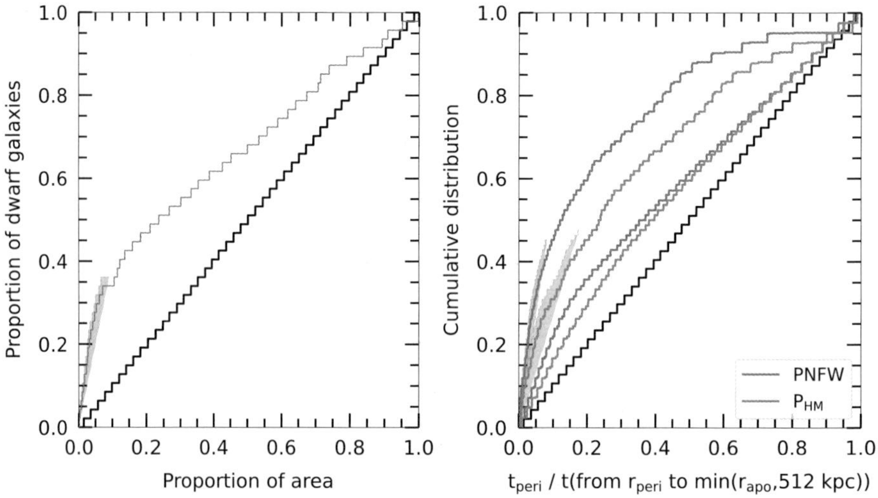

Figure 1. The area proportion of a spherical circle with the same center as the VPOS vs. the proportion of dwarf galaxies in the corresponding area (`left panel`) and the cumulative distribution of the time to the pericenter divided by the time taken from pericenter to min(apocenter, 512 kpc) of dwarf galaxies (blue and orange lines with shaded region) and satellites (black, blue and orange lines without shaded region) in different potential (`right panel`).

radius of dwarf galaxies to ensure that they can be observed at current position. Next, we assume that dwarf galaxies beyond 512 kpc are undetectable and modify the definition of orbital phase:

$$P = \frac{t_{\rm peri}}{t_{\rm peri-min(apo,512kpc)}}, \tag{2}$$

where $t_{\rm peri-min(apo,512kpc)}$ is the time spent between the pericenter and the closer one between the apocenter and 512 kpc. We know that dwarf galaxies cannot be detected at parts of their orbits, causing an additional selection effect that will be corrected later.

We only use two relatively heavy potential models (PNFW and $\rm PE_{HM}$), because most dwarf galaxies are unbound in the lighter models ($\rm PE_{LM}$ and $\rm PE_{IM}$) and cannot be satellites. The right panel of Figure 1 shows the cumulative distribution of the orbital phase. The blue line with the shaded region represents the model PNFW and the orange line with the shaded region represents the model $\rm PE_{HM}$. These shaded regions indicate the 1σ confidence interval. The expected distribution of the satellites is shown as the black line. The distribution in both models is far from this expectation. Although the satellite's orbital phases should be uniformly distributed, selection effects can cause changes in the observed distribution. For example, there would be no dwarf galaxies with large orbital phases if the dwarf galaxy could not be observed near its apocenter. We correct for this by applying the same selection effect to the expected distribution and the results are shown in the right panel of Figure 1 by the blue (PNFW) and orange ($\rm PE_{HM}$) lines without shaded region. We find that the selection effect leads to an increase in the proportion of dwarf galaxies that are closer to their pericenter, which is more pronounced in the smaller mass Milky Way. Dwarf galaxies in the more massive Milky Way have larger orbital phases at the same distance, so the loss of dwarf galaxies with larger orbital phases is smaller under the same selection effect.

We use KS-test to measure the difference in orbital phase distribution between our sample and the corrected expectation. Both of them include 2,000 MC realizations. For the median mass model PNFW, we find that p-value is $0.018^{+0.046}_{-0.015}$. We can reject the null hypothesis that dwarf galaxies are long-term satellites. The 84-th percentile p-value is 0.064, which suggests that the missing dwarf galaxies could partially alleviate the overconcentration problem near the pericenter, but not completely. The KS-value is measured at $0.24^{+0.21}_{-0.15}$, i.e. the value of the orbital phase when the difference between the sample and the null hypothesis is equal to the KS-value, meaning that dwarf galaxies are more concentrated in the quarter of orbits around their pericenter. Combined with a KS-value of $0.33^{+0.06}_{-0.05}$, we find that there are 16^{+3}_{-2} more dwarf galaxies with small orbital phases in the sample than expected.

We run the same analysis using the high mass model PE_{HM}. The p-value is $0.18^{+0.26}_{-0.14}$ and it is too large to reject. The value corresponding to the KS-value is $0.24^{+0.23}_{-0.15}$, which is almost the same as the value in model PNFW. This also indicates that dwarf galaxies are overconcentration near their pericenter, but not significantly. In summary, the selection effect is unlikely to cause dwarf galaxies are often found near their pericenter in the medium mass Milky Way, but it is likely in the massive Milky Way.

5. Conclusion

On the basis of Li *et al.* (2021), we update the proper motions of some dwarf galaxies and add a recently discovered dwarf galaxy from other literatures. The final sample includes 47 dwarf galaxies. We use a Bayesian method to propagate errors to the 3D positions and velocities and obtain unbiased results. By analyzing the angular momentum of the dwarf galaxies and the orbital characteristics of the four Milky Way potential models with total mass from $2.8 \times 10^{11}\ M_\odot$ to $15 \times 10^{11}\ M_\odot$, we find two things that could be inconsistent with expectations for a ΛCDM sub-halo system. The first one is that the orbital pole distribution of dwarf galaxies is concentrated in the VPOS. The second one is that there are too many dwarf galaxies lying near their pericenters. Selection effects are hardly the cause, but a very massive Milky Way ($15 \times 10^{11}\ M_\odot$) can mitigate the problem. Finally Hammer *et al.* (2021), Yang *et al.* (2022) and Hammer *et al.* (2023) found that because of their large orbital energy and angular momentum, it is likely that most dwarf galaxies have recently fall into the Milky Way halo, which completely change our understanding.

6. Acknowledgments

We thank especially the editor for insightful comments and suggestions, which have improved the paper significantly. This work has made use of data from the European Space Agency (ESA) mission *Gaia* (`https://www.cosmos.esa.int/gaia`), processed by the *Gaia* Data Processing and Analysis Consortium (DPAC, `https://www.cosmos.esa.int/web/gaia/dpac/consortium`). Funding for the DPAC has been provided by national institutions, in particular the institutions participating in the *Gaia* Multilateral Agreement. This work has been supported by the National Natural Foundation of China (NSFC No. 11973042 and No. 11973052). It was also supported by the Fundamental Research Funds for the Central Universities and the China Scholarship Council (CSC). We also thank for its support the International Research Program Tianguan, which is an agreement between the CNRS, NAOC and the Yunnan University. MSP thanks the Klaus Tschira Stiftung gGmbH and German Scholars Organization e.V. for support via a Klaus Tschira Boost Fund.

References

Aaronson, M. 1983, *ApJ*, 266, L11
Binney, J., & Tremaine, S. 1987, Galactic Dynamics
Bovy, J. 2015, *ApJS*, 216, 29

Bruce, J., Li, T. S., Pace, A. B., *et al.* 2023, *arxiv*: 2302.03708
Cerny, W., Simon, J. D., Li, T. S., *et al.* 2023, *ApJ*, 942, 111
Gaia Collaboration, Brown, A. G. A., Vallenari, A., *et al.* 2018, *A&A*, 616, A1
Gaia Collaboration, Brown, A. G. A., Vallenari, A., *et al.* 2021, *A&A*, 649, A1
Drlica-Wagner, A., Bechtol, K., Mau, S., *et al.* 2020, *ApJ*, 893, 47
Eilers, A.-C., Hogg, D. W., Rix, H.-W., & Ness, M. K. 2019, *ApJ*, 871, 120
Fritz, T. K., Battaglia, G., Pawlowski, M. S., et al. 2018, *A&A*, 619, A103
Hammer, F., Wang, J., Pawlowski, M. S., *et al.* 2021, *ApJ*, 922, 93
Hammer, F., Yang, Y., Arenou, F., *et al.* 2020, *ApJ*, 892, 3
Hammer, F., Li, H., Mamon, G. A., *et al.* 2023, *MNRAS*, 519, 5059
Jiao, Y., Hammer, F., Wang, J. L., & Yang, Y. B. 2021, *A&A*, 654, A25
Kleyna, J. T., Wilkinson, M. I., Evans, N. W., & Gilmore, G. 2005, *ApJ*, 630, L141
Li, H., Hammer, F., Babusiaux, C., *et al.* 2021, *ApJ*, 916, 8
Miyamoto, M., & Nagai, R. 1975, Publications of the Astronomical Society of Japan, 27, 533
Muñoz, R. R., Majewski, S. R., & Johnston, K. V. 2008, *ApJ*, 679, 346
Navarro, J. F., Frenk, C. S., & White, S. D. M. 1997, *ApJ*, 490, 493
Pawlowski, M. S., & Kroupa, P. 2013, *MNRAS*, 435, 2116
Retana-Montenegro, E., van Hese, E., Gentile, G., Baes, M., & Frutos-Alfaro, F. 2012, *A&A*, 540, A70
Simon, J. D. 2019, *ARA&A*, 57, 375
van der Marel, R. P., & Guhathakurta, P. 2008, *ApJ*, 678, 187
Vasiliev, E. 2019, *MNRAS*, 484, 2832
Wang, J., Hammer, F., & Yang, Y. 2022, *MNRAS*, 510, 2242
Yang, Y., Hammer, F., Jiao, Y., & Pawlowski, M. S. 2022, *MNRAS*, 512, 4171

Discussion

STACY MCGAUGH: Is there a possible explanation for the excessive concentration of dwarf galaxies near the pericenters?

HEFAN LI: The ram pressure is an idea that reduces the radial velocity of dwarf galaxies, causing them to move closer to the pericentric.

Dynamical Masses of Local Group Galaxies
Proceedings IAU Symposium No. 379, 2025
P. Bonifacio, M.-R. Cioni, F. Hammer, M. S. Pawlowski & S. Taibi, eds.
doi:10.1017/S1743921323001370

Equilibrium dynamical models for the Large Magellanic Cloud

N. Kacharov and M.-R. L. Cioni

Leibniz Institute for Astrophysics, An der Sternwarte 16, 14482 Potsdam, Germany
email: kacharov@aip.de

Abstract. The Large Magellanic Cloud (LMC) has a complex dynamics driven by both internal and external processes. The external forces are due to tidal interactions with the Small Magellanic Cloud and the Milky Way, while internally its dynamics mainly depends on the stellar, gas, and dark matter mass distributions. Despite the overall complexity of the system, very often simple physical models can give us important insights about the main driving factors. Here we focus on the internal forces and attempt to model the proper motions of $\sim 10^6$ stars in the LMC as measured by Gaia Data Release 3 with an axisymmetric dynamical model, based on the Jeans equations. We test both cored and cusped spherical Navarro-Frenk-White dark matter halos to fit the LMC gravitational potential. We find that this simple model is very successful at selecting a clean sample of genuine LMC member stars and predicts the geometry and orientation of the LMC with respect to the observer within the constraint of axisymmetry. Our Jeans dynamical models describe well the rotation profile and the velocity dispersion of the LMC stellar disc, however they fail to describe the motions of the LMC bar, which is a non-axisymmetric feature dominating the central region. We plan a triaxial Schwarzschild approach as a next step for the dynamical modelling of the LMC.

Keywords. Stellar dynamics, Jeans models, LMC

1. Introduction

The Large Magellanic Cloud (LMC) is the largest satellite galaxy of the Milky Way, however its total mass has been a subject of debate. With a stellar mass $\sim 2.7 \times 10^9\,\mathrm{M}_\odot$ and gas mass $\sim 0.5 \times 10^9\,\mathrm{M}_\odot$ (van der Marel et al. 2002, total luminosity of $\sim 1.3 \times 10^9\,\mathrm{L}_\odot$), it is clear that the LMC's total mass is dominated by its dark matter (DM) halo. Multiple lines of evidence (from internal kinematics, stellar streams, number of LMC satellites, effects from interactions with the Small Magellanic Cloud (SMC) and the Milky Way; see the review by Vasiliev 2023, and references therein for more details) suggest that the total mass of the LMC is in the range of $1 - 2 \times 10^{11}\,\mathrm{M}_\odot$. This implies a mass-to-light ratio in the order of $100\,\frac{\mathrm{M}_\odot}{\mathrm{L}_\odot}$, meaning that the LMC DM halo has not been stripped through the interaction with the Milky Way.

While it is indeed close to impossible to use kinematic tracers to study the dynamics of the outer regions of the LMC due to them being highly perturbed from its multiple interactions and close encounters with the Milky Way and the SMC (e.g. Choi et al. 2022), the relatively high mass density, preserved until the present day, makes equilibrium dynamical models a viable option to understand the galaxy's dynamics in its inner regions. Furthermore, any structure and details in the residuals of the equilibrium models can prove to be a valuable insight into the nature and significance of the non-equilibrium effects in the internal structure of the LMC.

We envision that a good handle of the dynamics in the innermost regions of the LMC through high quality data and careful modelling could facilitate the search for a possible massive black hole in the centre of the galaxy.

The LMC has a prominent bar structure that dominates the surface brightness and stellar kinematics in the inner region. The bar appears off-centred with respect to the centre of rotation of the H I gas, however the centre of rotation of the stellar component seems to be much better aligned with the photometric centre, close to the centre of the bar (Gaia Collaboration et al. 2021). The bar likely has a triaxial shape and is a source of non-axisymmetric perturbations in the centre of the galaxy, that may drive gas flows and trigger star formation. It was most likely formed from the mutual interactions between the LMC and the SMC (Besla et al. 2012). For the proper modelling of this structure we need to move away from the assumption of axisymmetry and use alternative methods like the Schwarzschild orbit superposition approach (van den Bosch et al. 2008; Tahmasebzadeh et al. 2022).

Here we present only preliminary equilibrium models of the LMC, based on the Jeans equations as a first step towards our goal of a complete dynamical model of the inner regions of the galaxy, using state of the art kinematic data sets. A similar analysis, based on Gaia Data Release 2 data was published by Vasiliev (2018).

2. Data and construction of the Jeans models

In this study we use Gaia Data Release 3 (DR3) proper motion (PM) data in the footprint of the VMC survey of the LMC (Cioni et al. 2011). We consider only stars that are detected in both VMC and Gaia, and have valid Gaia PM measurements. We additionally require that the Gaia DR3 PM uncertainties in the RA and DEC directions are less than 0.1 mas yr^{-1}. This leaves us with $\sim 10^6$ stars down to G = 18 mag to work with.

We assume a distance modulus to the LMC $(\mathrm{m}-\mathrm{M}) = 18.52$ mag (Crandall & Ratra 2015), which corresponds to a distance of 50.58 kpc. We fix the centre of the LMC at $\alpha_0 = 81.07$ deg and $\delta_0 = -69.41$ deg from Gaia Collaboration et al. (2021) and convert the stellar coordinates and PMs to an orthographic projection $(\alpha, \delta, \mu_\alpha.\mu_\delta) \rightarrow (\mathrm{X}, \mathrm{Y}, \mu_\mathrm{X}.\mu_\mathrm{Y})$. We also correct the PMs for perspective effects, as in van de Ven et al. (2006).

For the surface brightness distribution of the LMC, we assume an exponential profile with $\mathrm{r_e} = 98.7$ arcmin (Gallart et al. 2004) and a total luminosity of $1.31 \times 10^9\ \mathrm{L_\odot}$ from the NASA/IPAC Extragalactic Database (NED), which we represent with a Multi-Gaussian Expansion (MGE).

With the Jeans dynamical approach we can model the LMC as an axisymmetric flattened ellipsoid. The Jeans equations link the unknown gravitational potential to the stellar density and motions, which are derived from observations. In this study we make use of the Jeans Anisotropic Modelling (JAM) code by Cappellari (2008), assuming a cylindrical alignment of the velocity ellipsoid.

We model the gravitational potential with a generalised Navarro-Frenk-White (NFW) DM profile, where we can change the centre slope (γ). This profile is characterised by two parameters (scale radius – r_0 and central density – ρ_0), which are free parameters in our model. At first we consider a cuspy DM distribution by fixing $\gamma = 1$.

The rest of the free parameters in our models are related to the geometry (inclination i, rotation axis orientation θ, flattening $q = \frac{b}{a}$) and kinematical properties (anisotropy β_z and level of rotation κ) of the LMC. We also leave the bulk PMs in the x and y directions as free parameters, and one last free parameter ϵ controls the fraction of non-member stars.

We use a maximum likelihood approach on the discrete PM measurements and the Markov Chain Monte Carlo method to fit the model to the data. The total likelihood

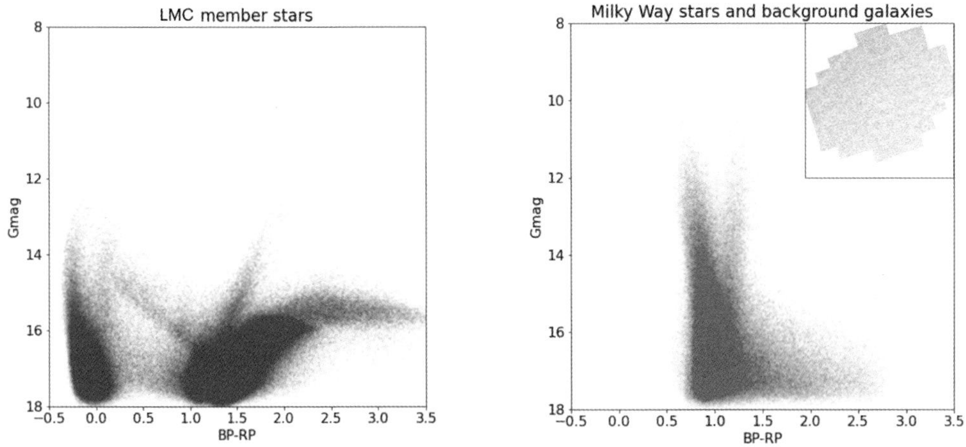

Figure 1. Left: a CMD of stars with high probability to be genuine LMC members. Right: a CMD of targets with high probability to be foreground Milky Way contaminants or background galaxies. The inlet shows their flat spatial distribution.

per star is given by the joint probability of it following the adopted exponential surface density distribution (spatial likelihood) and having kinematical properties consistent with the predictions of the Jeans equations (dynamical likelihood).

3. Results

A first outcome of our converged model fit is that we can easily compute the probability of each target to belong to either the LMC or the foreground / background population. Figure 1 shows the colour-magnitude diagram (CMD) of stars with high probability of being genuine LMC members on the left and entries with high probability of belonging to the foreground / background on the right, which also have a very flat spatial distribution throughout the VMC footprint, as expected. The left hand side CMD clearly shows the dense regions occupied by LMC red giant branch (RGB) and main sequence (MS) stars, but also the less dense sequences of the asymptotic giant branch (AGB), supergiants, red and blue He burning phases are easily identifiable.

We find a best fit inclination angle $i = 33.2$ deg in good agreement with the kinematical analysis from the Gaia Collaboration et al. (2021) and an intrinsic flattening of $q_{\rm intr} = 0.2$. Our best fit estimates for the LMC mean PMs are $\mu_\alpha^0 \cos(\delta) = 1.87\,\mathrm{mas\,yr^{-1}}$ and $\mu_\delta^0 = 0.36\,\mathrm{mas\,yr^{-1}}$. The statistical uncertainties of these parameters are very small, so they are fully dominated by systematic errors, such as choice of central coordinates, distance, surface density distribution, shape of the potential, etc. For instance changing the inner slope (γ) of the DM density profile is tightly correlated with our estimates of the LMC intrinsic flattening and inclination angle. In a dynamical model, where $\gamma = 0$ (DM core), we get a much thinner LMC disc with $q_{\rm intr} = 0.05$ and $\mathrm{i} = 32.65$ deg. On the other hand if we let γ to be a free parameter, we find best-fit $\gamma = 1.5$ and $q_{\rm intr} = 0.3$.

Figures 2 and 3 show how well our axisymmetric Jeans dynamical model fits the observed kinematics of the LMC in PM space – the galaxy's rotation and velocity dispersion profiles, respectively. The field of view in the figures is rotated such that the symmetry axis is along the vertical y direction. The rotation profile is fit generally quite well with median residuals in the order of only $1-2\%$. While the Jeans model also fits well the dispersion profile in the outer region, dominated by the LMC disc, the inner region shows strong non-axisymmetric features, which cannot be reproduced with Jeans models. This is because the kinematics in the central parts of the galaxy is dominated by

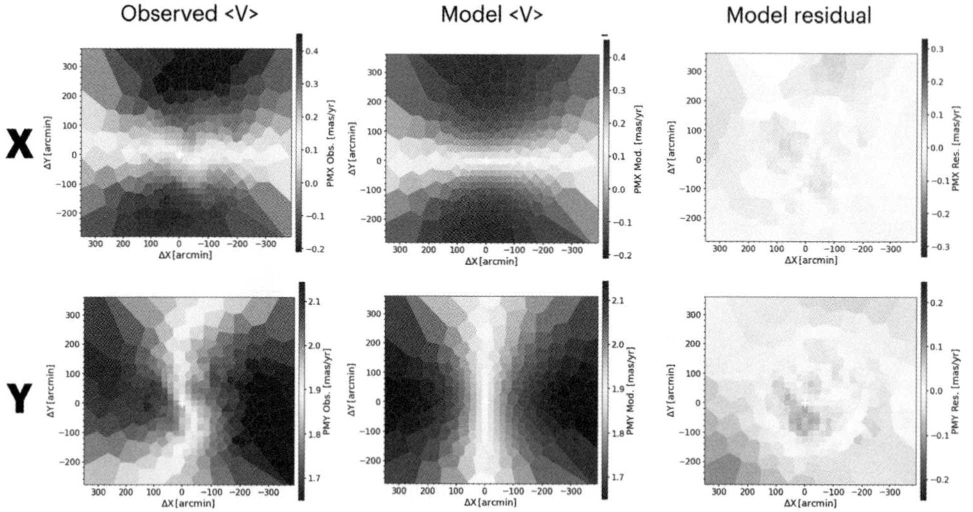

Figure 2. Observed and modelled rotation profile of the LMC. The top row shows the observed, modelled and residual PM in the direction of the horizontal axis (μ_X). Similarly, the bottom row shows the observed, modelled, and residual PM along the vertical axis (μ_Y).

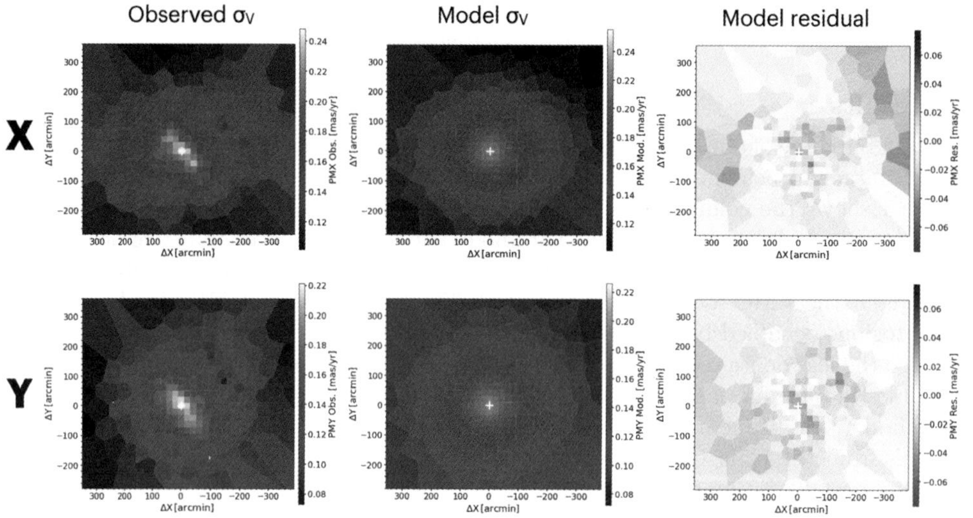

Figure 3. Observed and modelled velocity dispersion profile of the LMC. The top row shows the observed, modelled and residual PM velocity dispersion in the direction of the horizontal axis (μ_X). Similarly, the bottom row shows the observed, modelled, and residual velocity dispersion PM along the vertical axis (μ_Y).

the LMC bar. Besides not fitting well the inner region velocity dispersion spatially, the Jeans model also underestimates the central velocity dispersion. It was noted that the observed high-velocity dispersion in the centre of the LMC is artificially biased due to crowding problems with the Gaia DR3 data (Libralato et al. 2023), which could be fixed by using PM measurements from the VMC survey, which has higher spatial resolution. The non-axisymmetric effects of the bar are also visible in the rotation map, where they show as apparent wiggles of the rotation axis. Overall, we conclude that the axisymmetric Jeans model is adequate at describing the LMC disc, but it is not appropriate for the inner region of the galaxy, dominated by a non-axisymmetric bar feature.

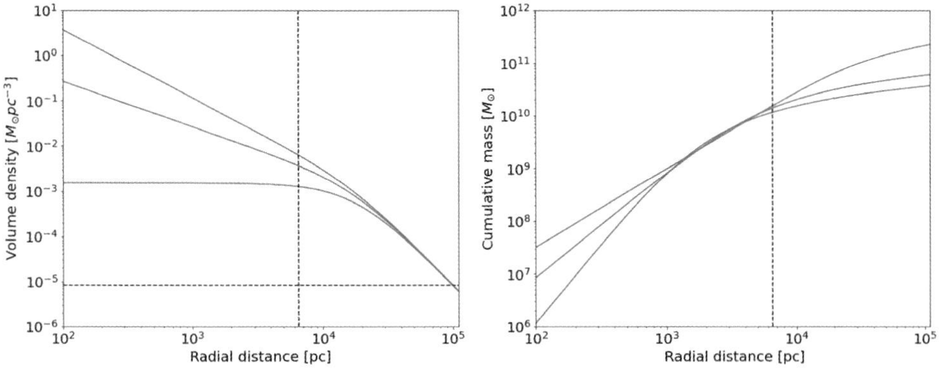

Figure 4. Left: Best fit NFW radial density profiles with central density slopes $\gamma = 0$, $\gamma = 1$, and best-fit $\gamma = 1.5$. Right: Best fit cumulative mass profiles for the three NFW density profiles with varying central density slopes. The horizontal line is at $200\rho_{\rm crit}$ level and the vertical line shows the extend of our data.

Finally, we turn to the mass estimates for the LMC from our Jeans models. Our potential tracer data extends out to 7 kpc from the centre of the galaxy. The enclosed DM mass within this radius is $\sim 10^{10}$ M$_\odot$ for all dynamical models, independent of the central density slope (γ, see Figure 1).

We can also calculate M$_{200}$ from the best-fit central density, scale radius, and central density slope of the NFW DM profiles. Our default model with $\gamma = 1$ gives a M$_{200}$ estimate of 3.3×10^{10} M$_\odot$ and the cored model ($\gamma = 0$) gives M$_{200} = 2.3 \times 10^{10}$ M$_\odot$. These mass estimates are significantly lower than independent measurements, that put the total mass of the LMC closer to 1.5×10^{11} M$_\odot$ (see Figure 1. in Vasiliev 2023). However, we mentioned above that our Jeans models generally underestimate the central velocity dispersion. This biases the best-fit NFW profile, which needs to be extrapolated to the outskirts of the galaxy and hence we obtain a lower total mass estimate. When we let γ to be a free parameter in the fit, we find best-fit $\gamma = 1.5$, which matches better the observed velocity dispersion in the inner region and we get 1.1×10^{11} M$_\odot$ in much better agreement with independent estimates.

In order to get a good handle of the total mass of the LMC we need to be able to model the stellar tracers at a wide range of radial distances, which also includes the LMC bar – a triaxial feature, that cannot be modelled with the Jeans approach.

4. Conclusion and outlook

While the axisymmetric Jeans models cannot provide a good physical representation of the dynamical state of the LMC as a whole, they could prove useful in describing its disc. The Jeans models are successful in recovering the orientation and geometry of the LMC disc under the axisymmetry assumption, fit well its kinematics in PM space, and distinguish probabilistically genuine member stars from foreground Milky Way contaminants and background galaxies.

For the inner region of the LMC, however, which is dominated by the bar structure, we need to resort to triaxial models. It is therefore interesting to explore the Schwarzschild orbit superposition approach in dynamical modelling with DYNAMITE (van den Bosch et al. 2008; Thater et al. 2022). Our goal is to obtain an orbit-based model of the LMC, where we fit the Gaia proper motions and surface density distributions of the two main components of the galaxy – an axisymmetric disc and a triaxial bar, simultaneously. This type of setup of Schwarzschild models were recently pioneered by Tahmasebzadeh et al. (2022).

It should be noted, however, that the orbit superposition method still works under the constraints of a dynamical equilibrium. The LMC is plunging into its deepest peri-centre point into the Milky Way, where maximal tidal shocks and ram pressure is expected. In that context the LMC is quite unique in having a smaller gas disc than a stellar disc. Moreover, the galaxy has recently been in collision with the SMC ~ 250 Myr ago (Choi et al. 2022). It would be interesting to explore the possibility to introduce perturbations in the dynamical models to study the deviations from equilibrium. However, a good equilibrium model could also be subtracted from the actual observations and the remaining residuals may give us valuable information about the magnitude of the complex dynamical forces that pull the LMC out of equilibrium.

Furthermore, the off-center LMC bar is likely a transient structure due to the interaction with the SMC (see e.g. the simulations in Besla et al. 2012). So it is likely that any equilibrium modelling of the bar can only capture part of its full complexity.

Acknowledgements

We thank Francois Hammer and Roeland van der Marel for the insightful discussion after the presentation. The answers to their questions and comments are incorporated in the proceeding.

References

Besla, G., Kallivayalil, N., Hernquist, L., et al. 2012, MNRAS, 421, 2109.
Cappellari, M. 2008, MNRAS, 390, 71. doi: 10.1111/j.1365-2966.2008.13754.x
Cioni, M.-R. L., Clementini, G., Girardi, L., et al. 2011, A&A, 527, A116.
Choi, Y., Olsen, K. A. G., Besla, G., et al. 2022, ApJ, 927, 153.
Crandall, S. & Ratra, B. 2015, ApJ, 815, 87.
Gaia Collaboration, Luri, X., Chemin, L., et al. 2021, A&A, 649, A7.
Gallart, C., Stetson, P. B., Hardy, E., et al. 2004, ApJL, 614, L109.
Libralato, M., Bellini, A., van der Marel, R. P., et al. 2023, arXiv:2303.00009.
Tahmasebzadeh, B., Zhu, L., Shen, J., et al. 2022, ApJ, 941, 109.
Thater, S., Jethwa, P., Tahmasebzadeh, B., et al. 2022, A&A, 667, A51.
van den Bosch, R. C. E., van de Ven, G., Verolme, E. K., et al. 2008, MNRAS, 385, 647.
van der Marel, R. P., Alves, D. R., Hardy, E., et al. 2002, AJ, 124, 2639.
van de Ven, G., van den Bosch, R. C. E., Verolme, E. K., et al. 2006, A&A, 445, 513.
Vasiliev, E. 2018, MNRAS, 481, L100.
Vasiliev, E. 2023, arXiv:2304.09136.

Discussion after the presentation

Roeland van der Marel: Congratulations on your nice work adding much needed realism to simple circular orbit models. Two comments:

(1) The velocity dispersions reported near the centre of the LMC from Gaia eDR3 by Gaia Collaboration et al. (2021) are likely very significant overestimates. See e.g. the recent proper motion data from HST and JWST and discussion in Libralato et al. (2023). So one would not expect a realistic model to reproduce them.

Nikolay Kacharov: We could cross-match the Gaia eDR3 PMs with PMs from the VMC survey to complement the data.

(2) The off-centre LMC bar is likely a transient structure due to the interaction with the SMC. See e.g. the simulations in Besla et al. (2012). So any equilibrium modelling of the bar can likely only capture part of its full complexity.

Francois Hammer: The LMC is plunging at its deepest peri-center point into the Milky Way, where maximal tidal shocks and ram pressure is expected, and indeed the one LMC peculiarity is to be quite unique in having a smaller gas disk than stellar disk. Moreover, its main stellar axis is off by $\sim 50\,\mathrm{deg}$ from its kinematics, and it has recently been in collision with the SMC $\sim 250\,\mathrm{Myr}$ ago. Would you think possible to enter some effect of this disequilibrium to verify how it can change your modelling?

Nikolay Kacharov: The orbit-superposition technique works under the assumption of dynamical equilibrium, however, examining the model residuals can inform us about the magnitude of deviation from this state.

Dynamical Masses of Local Group Galaxies
Proceedings IAU Symposium No. 379, 2025
P. Bonifacio, M.-R. Cioni, F. Hammer, M. S. Pawlowski & S. Taibi, eds.
doi:10.1017/S1743921323001709

Local Group Galaxies from an External Perspective

S. S. McGaugh

Case Western Reserve University, 10900 Euclid Ave., Cleveland, OH, USA
email: stacy.mcgaugh@case.edu

Abstract. I discuss Local Group galaxies from the perspective of external galaxies that define benchmark scaling relations. Making use of this information leads to a model for the Milky Way that includes bumps and wiggles due to spiral arms. This model reconciles the terminal velocities observed in the interstellar medium with the rotation curve derived from stars, correctly predicts the gradual decline of the outer rotation curve, and extrapolates well out to 50 kpc. Rotationally supported Local Group galaxies are in excellent agreement with the baryonic Tully-Fisher relation. Pressure supported dwarfs that are the most likely to be in dynamical equilibrium also align with this relation. Local Group galaxies thus appear to be normal members of the low redshift galaxy population. There is, however, a serious tension between the dynamical masses of the Milky Way and M31 and those expected from the stellar masshalo mass relation of abundance matching.

Keywords. Local Group (929), Milky Way Galaxy (1054), Dwarf galaxies (416), Galaxy masses (607)

1. Introduction

Local Group galaxies are accessible to detailed study with large samples of individual stars. This information is complemented by information from external galaxies, which obey strong kinematic scaling relations (McGaugh et al. 2020; Lelli 2022). These include the Baryonic Tully-Fisher Relation (BTFR: Schombert et al. 2020), the Central Density Relation (Lelli et al. 2016), and the Radial Acceleration Relation (RAR: McGaugh et al. 2016; Lelli et al. 2017). Local Group galaxies prove to be completely normal members of the general galaxy population by these standards.

2. The Milky Way

The Milky Way is of obvious interest as our home galaxy, yet our place within it is a source of constant frustration: we cannot see it from 'above' as we do external spirals. We nevertheless infer that it is a barred spiral thanks to decades of work employing a broad spectrum of methods (Bland-Hawthorn & Gerhard 2016). In external galaxies, spiral features cause bumps and wiggles in the light profile that correspond to features in the rotation curve (Sancisi 2004). This correspondence is quantified by the RAR.

The RAR can be applied to the Milky Way to build a mass model for the stellar disk (McGaugh 2016). The terminal velocities observed by McClure-Griffiths & Dickey (2007) and McClure-Griffiths & Dickey (2016) were fit in detail between $3 < R < 8$ kpc. The resulting model is shown in Fig. 1, updated to the distance scale (McGaugh 2018) indicated by the GRAVITY Collaboration (2018).

Fig. 2 shows the surface density corresponding to the rotation curve seen in Fig. 1. The model has bumps and wiggles in the profile that are reminiscent of those seen in

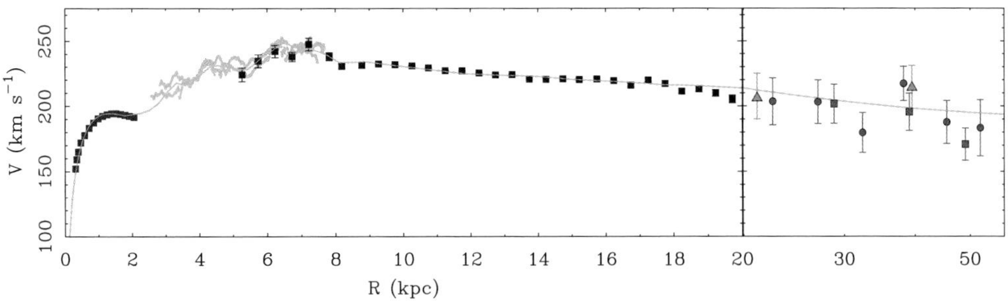

Figure 1. The rotation curve of the Milky Way. Light gray points are the terminal velocity data (McClure-Griffiths & Dickey 2007, 2016) to which the model was fit (McGaugh 2018, green line). The model was not fit to the data of Portail et al. (2017) (dark gray squares at $R < 2.2$ kpc) or Eilers et al. (2019) (black squares), but is nicely consistent with them (McGaugh 2019). It also extrapolates well out to 50 kpc as traced by K giants (red squares) and BHB stars (blue circles) (Bird et al. 2022) and by globular clusters (Watkins et al. 2019, orange triangles).

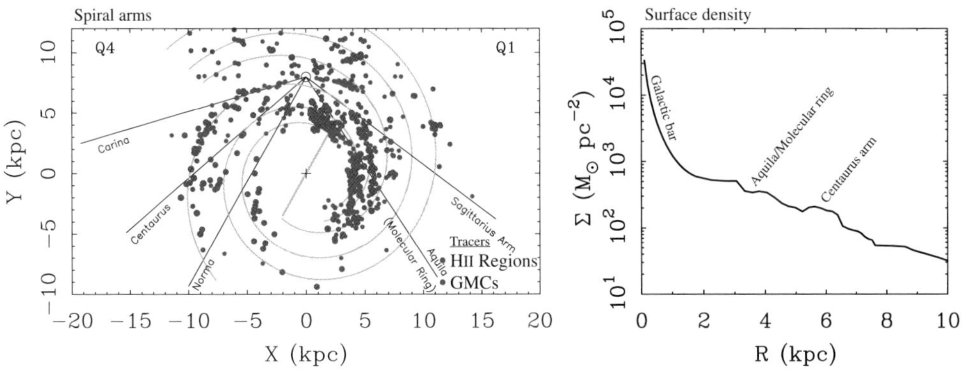

Figure 2. Spiral structure (left) and surface density profile (right) of the Milky Way. The bumps and wiggles inferred kinematically correspond well to the features known from tracers like HII regions (red points) and Giant Molecular Clouds (blue points) (McGaugh 2016).

external spirals. Indeed, these kinematically inferred features correspond well to spiral arms that are known from traditional tracers (Fig. 2).

The model (McGaugh 2018) correctly predicted the slope $dV/dR = -1.7\,\mathrm{km\,s^{-1}\,kpc^{-1}}$ of the outer rotation curve subsequently found by Eilers et al. (2019). It extrapolates well to data for globular clusters (Watkins et al. 2019) and halo stars (Bird et al. 2022) at still larger radii, and is nicely consistent with the inner mass distribution inferred by Portail et al. (2017) without that region having been specifically fit. This is a testament to how tightly the mass distribution is coupled to that of the light: the latter is predictive of the former (McGaugh 2020).

The stellar rotation curve derived by Eilers et al. (2019) is formally discrepant from the terminal velocities (McClure-Griffiths & Dickey 2007) at high significance. The model reconciles the two through numerical evaluation of the tracer term $\partial \ln \nu / \partial \ln r$ in the Jeans equation (McGaugh 2019). The apparent discrepancy is an artifact of the usual assumption of an exponential disk, which does not account for the bumps and wiggles in $\nu(r)$. Galactic data have reached a level of precision that these can no longer be neglected.

We can query the model for the halo mass it implies (Table 1). To be specific, I quote M_{200} for the standardization it provides, and the estimate I obtain is consistent with the range found by others (Wang et al. 2020). However, we should bear in mind that this mass is a rather notional quantity that involves an extrapolation to a radius that

Table 1. Mass Estimates.

Object	**Model**	M_* $10^{10}\,\mathrm{M}_\odot$	M_{200} $10^{10}\,\mathrm{M}_\odot$	c	V_{200} $\mathrm{km\,s^{-1}}$	**Ref.**
Milky Way	NFW from RAR	6.2	139	8.1	162	McGaugh (2018)
M31	NFW+SSP	13.5	160	10.1	169	Chemin et al. (2009)
M31+MW	Straight sum	19.7	299	...	...	McGaugh & van Dokkum (2021)

is not directly constrained by data using a model (Navarro et al. 1997) that persistently provides a poor depiction of observational reality.

A more robust quantity is the enclosed mass $M(R < 50\,\mathrm{kpc}) = 44 \times 10^{10}\,\mathrm{M}_\odot$. This agrees well with halo star (Bird et al. 2022) and globular cluster (Watkins et al. 2019) data. Going farther out is complicated by the perturbing influence of the LMC. Indeed, if we take seriously the downturn in the Gaia-derived rotation curve at $R \approx 19\,\mathrm{kpc}$ (Wang et al. 2023), a much lower mass is obtained (Sylos Labini et al. 2023). It is difficult to envision a model that is consistent with both this downturn and the data for halo stars.

Even in the absence of these considerations, the total mass of the Milky Way remains a fraught issue. Weak lensing observations of external galaxies (Brouwer et al. 2021) indicate that rotation curves persist in being approximately flat to hundreds of kpc, reaching beyond the expected virial radii of bright spirals. This is not consistent with an NFW profile, but it is consistent with the RAR, which is what works well above. These are not the same.

3. The Local Group

The Milky Way falls on the BTFR defined by external galaxies (Fig. 3). Indeed, all of the rotationally supported Local Group galaxies do so, within the uncertainties: there is no indication that they are abnormal in this regard. This provides both a constraint—nothing particularly peculiar is permitted—and an opportunity: if we can calibrate the dynamical masses of Local Group galaxies, we can infer those of more distant galaxies.

3.1. *Dwarf Spheroidal Galaxies and Tully-Fisher*

The BTFR may further serve as a method to unite the mass scales of pressure supported early type dwarf galaxies with their rotationally supported late type brethren. Typically we measure only a single, bulk velocity dispersion σ_* for dwarf spheroidals. The dwarf spheroidals of the Local Group fall into two populations in the BTFR plane (McGaugh & Wolf 2010): the L-σ_* data for classical dwarfs parallels the BTFR, while for the ultrafaint dwarfs σ_* is uncorrelated with L, seeming to hit a floor. As a consequence of their relatively high velocity dispersions, the latter appear to be very dense and hence safe from tidal influences (Pace et al. 2022). This may be an illusion of the assumption of dynamical equilibrium, as it seems unlikely that these tiny dwarfs can survive unmolested deep in the potential well of the Milky Way while the Milky Way itself is being strongly perturbed by the LMC. Excepting the (primarily ultrafaint) dwarfs that may be subject to tidal influence by the criterion of Bellazzini et al. (1996), the remaining dwarfs of the Local Group fall nicely on the BTFR defined by rotating galaxies given a simple transformation $M_* = \Upsilon_* L_V$ and $V_f = \beta_c \sigma_*$ with $\log \beta_c = 0.25 \log \Upsilon_* + 0.226$ (McGaugh et al. 2021). This works out to $\beta_c = 2$ for $\Upsilon_* = 2\,\mathrm{M}_\odot/\mathrm{L}_\odot$. This quantifies the circular velocity

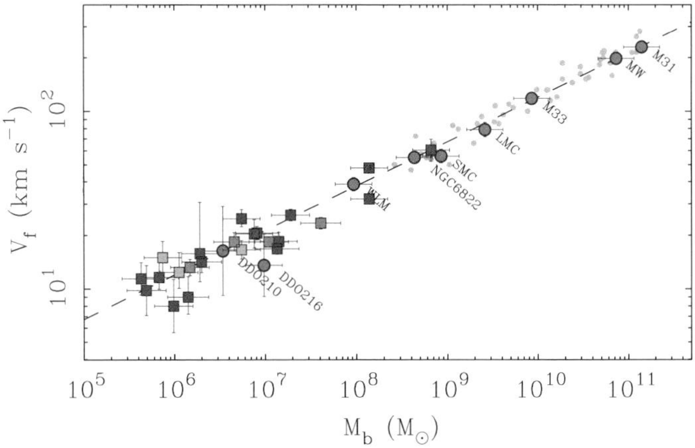

Figure 3. The baryonic TullyFisher relation of Local Group galaxies (blue circles) follows that defined by external galaxies (grey points Schombert et al. 2020). This extrapolates to pressure supported dwarfs (squares) when $V_f = 2\sigma_*$ (McGaugh et al. 2021). Red squares are satellites of M31, orange squares are satellites of the Milky Way; green squares are unaffiliated (McConnachie et al. 2021).

that is equivalent to the line of sight velocity dispersion in an average sense. The strong regularity seen in Fig. 3 is surely teaching us something about the dynamical masses of galaxies.

3.2. *Kinematic Masses and Abundance Matching*

A common method for relating the observed stellar masses of galaxies with their host dark matter halos is Abundance Matching (AM: Behroozi et al. 2013; Moster et al. 2013; Kravtsov et al. 2018; Mowla et al. 2019). Indeed, this has become an essential ingredient for understanding galaxies in ΛCDM. The Local Group provides an ideal opportunity to test these relations with independent kinematic data.

Typical AM relations are well calibrated around the knee of the luminosity function, but not to the very low masses of dwarf spheroidals (Bullock & Boylan-Kolchin 2017). The agreement between kinematic mass estimates and AM masses is reasonably good for intermediate mass galaxies like M33. The two diverge dramatically for the brightest galaxies of the Local Group, M31 and the Milky Way (Table 1 and Fig. 4)

Dynamical studies typically suggest that the total mass of the Milky Way is in the range of 0.5—2×10^{12} $M_\odot$ (Bland-Hawthorn & Gerhard 2016; Wang et al. 2020). A similar range holds for Andromeda (Chemin et al. 2009; Kafle et al. 2018). AM predicts considerably higher masses: $\sim 3 \times 10^{12}$ $M_\odot$ for the Milky Way and $> 10^{13}$ $M_\odot$ for M31. There is some room to hedge this for the Milky Way by adopting a stellar mass on the low end of the plausible range, but even then the halo mass is abnormally low. The result for Andromeda differs by an order of magnitude. There is of course some scatter in the stellar mass–halo mass relation from AM, but it is not sufficient to explain this stark discrepancy.

According to AM, a galaxy the mass of M31 should be the central object of a much larger group. Indeed, for its stellar mass, the Milky Way should be the central object of the Local Group. Andromeda should not be here. Despite its normalcy in other respects, the Local Group has far too many stars for its mass from the perspective of abundance matching.

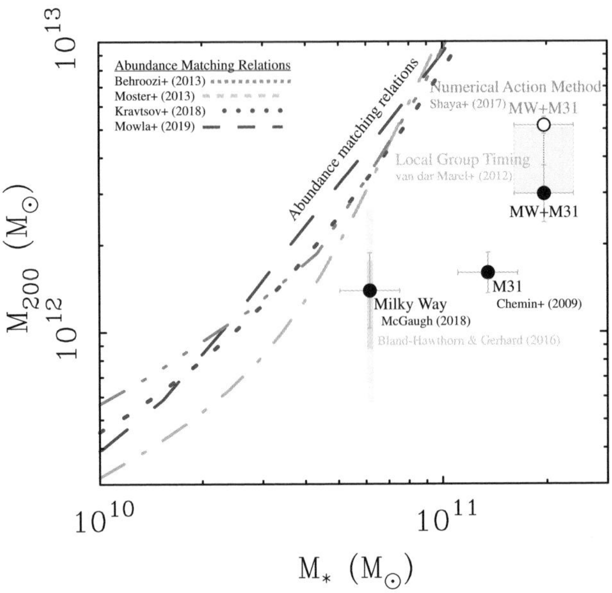

Figure 4. Stellar mass–halo mass relations in the context of the Local Group (McGaugh & van Dokkum 2021). Filled points are from rotation curve fits to the Milky Way (McGaugh 2018), M31 (Chemin et al. 2009), and their sum (Table 1). The open circle adopts for the summed total mass that estimated by Shaya et al. (2017). The blue band shows range of Milky Way mass estimates (Bland-Hawthorn & Gerhard 2016; Wang et al. 2020). Lines are the stellar masshalo mass relations from abundance matching Behroozi et al. (2013); Moster et al. (2013); Kravtsov et al. (2018); Mowla et al. (2019). The yellow box shows the timing mass of the Local Group (van der Marel et al. 2012). There is a pronounced discrepancy between dynamical and abundance matching estimates of the halo masses of the Local Group and its brightest members.

It is not clear what the resolution of this issue might be. There could be a morphological type dependence that is not accounted for in the AM relations considered here. There might be more mass at large radii than kinematic tracers have so far revealed, though this is limited by the timing argument (van der Marel et al. 2012; Peñarrubia et al. 2014; Shaya et al. 2017). The discrepancy appears to be real, and would hardly be the first indication of a problem with the dark matter paradigm in the dynamics of galaxies (Sanders & McGaugh 2002; Famaey & McGaugh 2012; McGaugh 2020; Banik & Zhao 2022).

References

Banik, I. & Zhao, H. 2022, Symmetry, 14, 1331
Behroozi, P. S., Wechsler, R. H., & Conroy, C. 2013, Astrophys. J., 762, L31
Bellazzini, M., Fusi Pecci, F., & Ferraro, F. R. 1996, Mon. Not. Royal Astron. Soc., 278, 947
Bird, S. A., Xue, X.-X., Liu, C., et al. 2022, Mon. Not. Royal Astron. Soc., 516, 731
Bland-Hawthorn, J. & Gerhard, O. 2016, Ann. Rev. Astron. & Astrophys., 54, 529
Brouwer, M. M., Oman, K. A., Valentijn, E. A., et al. 2021, A & A, 650, A113
Bullock, J. S. & Boylan-Kolchin, M. 2017, Ann. Rev. Astron. & Astrophys., 55, 343
Chemin, L., Carignan, C., & Foster, T. 2009, Astrophys. J., 705, 1395
Eilers, A.-C., Hogg, D. W., Rix, H.-W., et al. 2019, Astrophys. J., 871, 120
Famaey, B. & McGaugh, S. S. 2012, Living Reviews in Relativity, 15, 10
GRAVITY Collaboration, Abuter, R., Amorim, A., et al. 2018, A & A, 615, L15
Kafle, P. R., Sharma, S., Lewis, G. F., et al. 2018, Mon. Not. Royal Astron. Soc., 475, 4043
Kravtsov, A. V., Vikhlinin, A. A., & Meshcheryakov, A. V. 2018, Astronomy Letters, 44, 8

Lelli, F., McGaugh, S. S., Schombert, J. M., et al. 2016, Astrophys. J., 827, L19
Lelli, F., McGaugh, S. S., Schombert, J. M., et al. 2017, Astrophys. J., 836, 152
Lelli, F. 2022, Nature Astronomy, 6, 35
McClure-Griffiths, N. M. & Dickey, J. M. 2007, Astrophys. J., 671, 427
McClure-Griffiths, N. M. & Dickey, J. M. 2016, Astrophys. J., 831, 124
McConnachie, A. W., Higgs, C. R., Thomas, G. F., et al. 2021, Mon. Not. Royal Astron. Soc., 501, 2363
McGaugh, S. S. 2016, Astrophys. J., 816, 42
McGaugh, S. S. 2018, Research Notes of the American Astronomical Society, 2, 156
McGaugh, S. S. 2019, Astrophys. J., 885, 87
McGaugh, S. 2020, Galaxies, 8, 35
McGaugh, S., Lelli, F., Li, P., et al. 2020, Galactic Dynamics in the Era of Large Surveys, 353, 144
McGaugh, S. S., Lelli, F., & Schombert, J. M. 2016, Phys. Rev. Lett., 117, 201101
McGaugh, S. S., Lelli, F., Schombert, J. M., et al. 2021, Astron. J., 162, 202
McGaugh, S. S. & van Dokkum, P. 2021, Research Notes of the American Astronomical Society, 5, 23
McGaugh, S. S. & Wolf, J. 2010, Astrophys. J., 722, 248
Milgrom, M. 1983, Astrophys. J., 270, 371
Moster, B. P., Naab, T., & White, S. D. M. 2013, Mon. Not. Royal Astron. Soc., 428, 3121
Mowla, L., van der Wel, A., van Dokkum, P., et al. 2019, Astrophys. J., 872, L13
Navarro, J. F., Frenk, C. S., & White, S. D. M. 1997, Astrophys. J., 490, 493
Pace, A. B., Erkal, D., & Li, T. S. 2022, Astrophys. J., 940, 136
Peñarrubia, J., Ma, Y.-Z., Walker, M. G., et al. 2014, Mon. Not. Royal Astron. Soc., 443, 2204
Portail, M., Gerhard, O., Wegg, C., et al. 2017, Mon. Not. Royal Astron. Soc., 465, 1621
Sancisi, R. 2004, Dark Matter in Galaxies, 220, 233
Sanders, R. H. & McGaugh, S. S. 2002, Ann. Rev. Astron. & Astrophys., 40, 263
Schombert, J., McGaugh, S., & Lelli, F. 2020, Astron. J., 160, 71
Shaya, E. J., Tully, R. B., Hoffman, Y., et al. 2017, Astrophys. J., 850, 207
Sylos Labini, F., Chrobáková, Ž., Capuzzo-Dolcetta, R., et al. 2023, Astrophys. J., 945, 3
van der Marel, R. P., Fardal, M., Besla, G., et al. 2012, Astrophys. J., 753, 8
Wang, W., Han, J., Cautun, M., et al. 2020, Science China Physics, Mechanics, and Astronomy, 63, 109801
Wang, H.-F., Chrobáková, Ž., López-Corredoira, M., et al. 2023, Astrophys. J., 942, 12
Watkins, L. L., van der Marel, R. P., Sohn, S. T., et al. 2019, Astrophys. J., 873, 118

Discussion

MÜLLER: What is the reason for these strong regularities?

MCGAUGH: This behavior was predicted by MOND (Milgrom 1983).

Dynamical Masses of Local Group Galaxies
Proceedings IAU Symposium No. 379, 2025
P. Bonifacio, M.-R. Cioni, F. Hammer, M. S. Pawlowski & S. Taibi, eds.
doi:10.1017/S174392132300159X

Constraints on the dark matter distribution in dwarf spheroidal galaxies based on the 4th-order Jeans analysis

M. Dafa Wardana[1], Masashi Chiba[1] and Kohei Hayashi[2]

[1]Astronomical Institute, Tohoku University, 6-3 Aoba-ku, Sendai, Japan

[2]National Institute of Technology, Sendai College, 48 Nodayama, Natori, Japan
email: khayashi@sendai-nct.ac.jp

Abstract. The mass-anisotropy degeneracy is still one of the main issues in estimating the dark matter distribution in galaxies, especially for the commonly used 2nd-order Jeans analysis. We study the extension of spherical Jeans modeling by incorporating the 4th-order velocity moments under the assumption of dynamical equilibrium and a constant velocity anisotropy. Incorporating 4th-order moments allows stars' line-of-sight velocity distribution (LOSVD), which is sensitive to the value of β, to be flexible, covering thin-tailed to heavy-tailed distributions that are inaccessible if only 2nd-order moments are used. We test our stellar dynamical modeling using mock data that resembles Draco dSph with either central NFW cusp or Burkert core. Using 500 sample stars, our simulations show that incorporating 4th-order moments improves the results. Typically, the velocity anisotropy is constrained $\sim 2-3$ times tighter, while it is $\lesssim$30% typical bias for the constraint of the dark matter inner density slope with both parameters being recovered within 1σ uncertainties.

Keywords. dark matter - galaxies: dwarf spheroidal galaxies - galaxies: kinematics and dynamics

1. Introduction

Even though the remarkable achievements of ΛCDM model predictions range widely from large-scale structure formation, CMB, galaxy clusters, down to the general properties of galaxies (Ferreira 2021), the theory does not withstand without flaws. Along with its accurate prediction about the behaviour of the large-scale formation and evolution in the Universe, CDM theory has been facing some challenges in the sub-galactic scales, ≤ 1 Mpc. There exist some statistically significant discrepancies between properties predicted by CDM-only simulations and those inferred from observational data.

The oldest controversy between ΛCDM simulation-based predictions and observational inferences appears in the central part of dark matter halos in the mass regime roughly similar to dwarf galaxies. Generally, the central region of dark matter halos in ΛCDM simulation tends to be cuspier and denser compared to most DM-dominated galaxies whose kinematic data are enough to infer their density profile (Bullock & Boylan-Kolchin 2017).

This surviving tension emerges as ΛCDM-only simulations predict that dark halos have a universal density profile in the inner region, independent of the halo mass, initial conditions, or cosmological parameters. The universal density profile in the inner region is following $\rho(r) \propto r^{-\gamma}$ (where the inner density slope $\gamma = 1$), a cuspy density profile that rises steeply as we go to the center (Navarro, Frenk, & White 1996). Conversely, dynamical analyses using rotation curve data of dwarf spiral galaxies do not prefer this universality. The value of γ ranges to as small as $\gamma = 0$ for some galaxies, which means that they possess a flat density profile in the inner region, resembling a core (Oh *et al.* 2015).

In this work, we propose stellar dynamical modeling based on Jeans analysis for a dark matter dominated stellar system in order to estimate the dark matter distribution in its dark matter halo. Furthermore, due to their proximity so that the member stars can be resolved, extremely low baryon content, nearly spherical shape, and quite sufficient kinematics sample, dwarf spheroidal galaxies are currently among the best sites to test this modeling.

2. Dynamical modeling of dSph galaxies

We assume a spherically symmetric system that is in dynamical equilibrium with a constant value of the velocity anisotropy parameter β that is defined as

$$\beta(r) = 1 - \frac{\sigma_t^2(r)}{2\sigma_r^2(r)}, \tag{1}$$

where σ_t and σ_r are the velocity dispersion in the tangential and radial direction, respectively. For the light profile, we assume the projected Plummer profile (Plummer 1911) that can be expressed as

$$I(R) = \frac{L_{total}}{\pi a_*^2}\left(1 + \frac{R^2}{a_*^2}\right)^{-2}, \tag{2}$$

where a_* and R are the half-light radius and projected radius, respectively.

2.1. *The 2nd-order and 4th-order Jeans equations*

For a spherically symmetric system, solutions for Jeans equations after integrating along the line of sight with the boundary condition $\sigma_r \to 0$ as $r \to \infty$ for constant β can be written as

$$\sigma_{los}^2(R) = \frac{2}{I(R)} \int_R^\infty dr \left(1 - \beta\frac{R^2}{r^2}\right) \frac{\nu\sigma_r^2 r}{\sqrt{r^2 - R^2}}, \tag{3}$$

$$\overline{v_{los}^4}(R) = \frac{2}{I(R)} \int_R^\infty dr \left[1 - 2\beta\frac{R^2}{r^2} + \frac{1}{2}\beta(1+\beta)\frac{R^4}{r^4}\right] \frac{\nu\overline{v_r^4} r}{\sqrt{r^2 - R^2}}, \tag{4}$$

(Binney & Tremaine 2008; Battaglia, Helmi, & Breddels 2013) for the $2^{\rm nd}$-order and $4^{\rm th}$-order, respectively. It is convenient to express the fourth moment in the form of the line-of-sight kurtosis whose value is 3 for a Gaussian distribution (Łokas 2002),

$$\kappa_{los}(R) = \frac{\overline{v^4}_{los}(R)}{\sigma_{los}^4(R)}. \tag{5}$$

The kurtosis contains rich information about symmetric deviations of a distribution from Gaussianity. As the κ value of a distribution goes further from 3, its deviation from Gaussianity is also more prominent. Distributions with $\kappa > 3$ have stronger tails than Gaussian, and distributions with $\kappa < 3$ have thinner tails than Gaussian (see Figure 1). By invoking $4^{\rm th}$-order moments into the analysis, the LOSVD is allowed to be flexible, covering a thin-tailed up to a strong-tailed distribution which is mainly controlled by the value of β.

Table 1. Input parameters of the mock data. We use the generalized Hernquist model $\rho(r) = \rho_0(r/a_{dm})^{-\gamma}[1 + (r/a_{dm})^{\beta_{trans}}]^{-(\alpha-\gamma)/\beta_{trans}}$ for both the density profile in mock data generation and for the fitting. In this table, α, β_{trans}, γ, β, ρ_0, a_*, and a_{dm} are the outer density slope, transition density slope, inner density slope, velocity anisotropy parameter, density scale, half-light radius, and density scale-length of dark matter, respectively.

DM density slope	α	β_{trans}	γ	β	$\rho_0/(M_\odot/kpc^3)$	a_*/pc	a_{dm}/pc
Cuspy	3	1	1	0	4×10^7	214	10^3
Cored	3	1.5	0	0	5×10^8	214	500

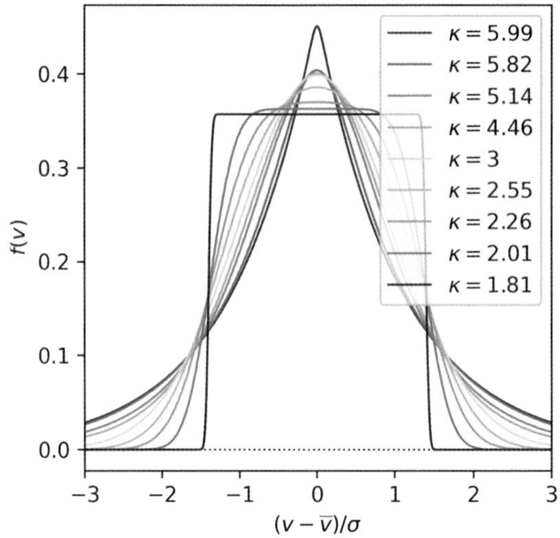

Figure 1. Velocity distribution with several kurtosis values κ–a parameter describing the 'tailedness' of a distribution. A Gaussian distribution has $\kappa = 3$ (green line). Curves in shades of red are distributions with more outliers than Gaussian ($\kappa > 3$), and curves in shades of blue are distributions with fewer outliers than Gaussian ($\kappa < 3$).

2.2. *Test using mock data sets*

In order to investigate how well our dynamical model can recover the true value of free parameters, we conduct tests using mock data. We use both a cuspy and a cored dark matter halo resembling the Draco dSph. Details on the input values are summarized in Table 1.

3. Results and discussions

Figure 2 shows the posterior distribution of β and γ for our test using mock data set for a cuspy dark matter halo. The left figure is obtained from fitting using only 2nd-order velocity moments, while its comparison in the right figure is obtained using both 2nd-order and 4th-order velocity moments. This comparison demonstrates how the error for β is shrunk by a factor of approximately $\sim 2-3$, while the bias is also reduced by a factor of ~ 2.5 after the incorporation of 4th-order moments. For γ, although the improvements are not as satisfying as it is for β, the reduction of bias in γ is not negligible. For all the MCMC samples we obtain, the bias for γ after incorporating 4th-order moments is typically $\lesssim 30\%$.

Figure 3 is the same as Figure 2 but for a cored dark matter halo. All trends are not statistically different from those mentioned in the case of cuspy dark matter halo. In general, after we include 4th-order moments, the estimation of β seems to be robust without any considerable difference among the samples. As the β estimation gets more precise,

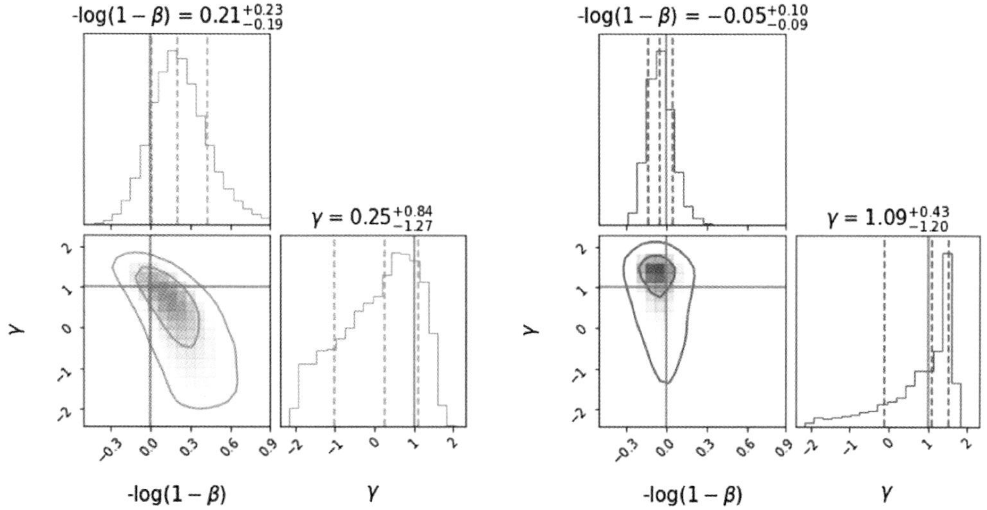

Figure 2. Posterior distributions for β and γ. The left figure shows the result for fitting using only 2nd-order moments, while the right figure shows the result using both 2nd and 4th order moments. Three dashed vertical lines mark the median and 1σ uncertainties. Blue solid lines are the input value of the mock data.

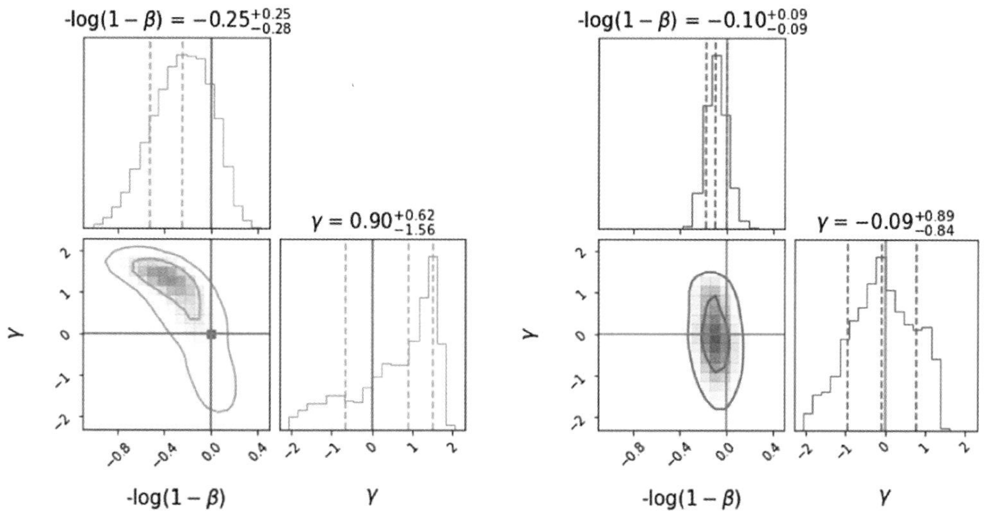

Figure 3. The same as Figure 2, but for a cored dark halo.

the $\beta - \gamma$ degeneracy is also reduced, marked by the absence of diagonally elongated features in the contour distributions. In conclusion, the dynamical analyses we conduct seem to be able to disentangle β from γ, although it still has difficulties ruling out cusp or core in a statistically significant manner. For future work, it is important to adopt a more general and flexible axisymmetric shape into the model, as it is clear that spherical and axisymmetric halos produce different velocity dispersion profiles (Hayashi & Chiba 2012; Hayashi, Chiba, & Ishiyama 2020). Also, we plan to apply these mass models to the massive spectroscopic data available from Subaru Prime Focus Spectrograph (PFS) in the coming years.

References

Battaglia, G., Helmi, A., Breddels, M. 2013, *NewAR*, 57, 52
Binney, J., Tremaine, S. 2008, Galactic Dynamics, Second Edition
Bullock, J. S., Boylan-Kolchin, M. 2017, *ARA&A*, 55, 343
Ferreira, Elisa G. M. 2021, *A&ARv*, 29, 7
Hayashi, K., Chiba, M. 2012, *ApJ*, 755, 145
Hayashi, K., Chiba, M., Ishiyama, T. 2020, *ApJ*, 904, 45
Łokas, Ewa L. 2002, *MNRAS*, 333, 697
Navarro, J. F., Frenk, C. S., White, S. D. M. 1996, *ApJ*, 462, 563
Oh, Se-Heon, *et al.* 2015, *AJ*, 149, 1800
Plummer, H. C. 1911, *MNRAS*, 71, 460

Dynamical Masses of Local Group Galaxies
Proceedings IAU Symposium No. 379, 2025
P. Bonifacio, M.-R. Cioni, F. Hammer, M. S. Pawlowski & S. Taibi, eds.
doi:10.1017/S1743921323003046

Dynamical modelling of galaxies with DYNAMITE

A. Zocchi, P. Jethwa, E. J. Lilley, S. Thater and G. van de Ven

Department of Astrophysics, University of Vienna, Türkenschanzstraße 17, 1180 Vienna
email: prashin.jethwa@univie.ac.at

Abstract. The combination of kinematic and chemical information from Galactic stars has revealed in great detail the structure, dynamics and history of our own Galaxy. In external galaxies, it is impossible to map the distribution of individual stars, but high signal-to-noise integral field unit (IFU) spectroscopy data at various wavelengths, together with sophisticated dynamical models, give us the opportunity to gather information on the structure, dynamics and formation history of these systems. The Schwarzschild method models galaxies through the superposition of stellar orbits, and is equipped to deal with very detailed kinematic measurements, allowing us to take full advantage of high-quality IFU datasets of nearby galaxies. Here we present an implementation of this method called DYNAMITE. We provide an overview of the modelling technique, introduce applications to observations and simulations, and anticipate our future plans for DYNAMITE.

Keywords. Galaxies: kinematics and dynamics, Galaxies: structure

1. Introduction

Dynamical modelling is a key instrument for studying the evolution of galaxies, because it allows us to investigate properties of these systems that are not directly observable, such as those relating to dark matter and black holes. In particular, measuring the total mass of galaxies and determining the relative contribution of luminous and dark matter is important for studying galactic evolution, and has implications for our understanding of structure formation and evolution in a cosmological context. Different approaches can be used to describe the dynamics of galaxies (e.g., see Rix & Bovy 2013), going from spherical and axisymmetric models defined by an analytical distribution function (e.g., Nagai & Miyamoto 1976; Qian et al. 1995; Magorrian et al. 1998) to Jeans modelling (e.g., Cappellari 2008), and to made-to-measure models (e.g., Syer & Tremaine 1996; Bovy, Kawata, & Hunt 2018).

Another popular approach, and the focus of this contribution, is the orbit-superposition method first developed by Schwarzschild (1979). With this technique, starting from an assumed galactic potential (that accounts for both the visible and dark matter), an orbit library is calculated, and for each orbit the relevant observables are recorded. Through a fit to the observed surface brightness and kinematic data, the particular superposition of orbits that best represents the data is constructed. This procedure makes it possible to use constraints coming from very detailed kinematic measurements, enabling a study of the intrinsic dynamical structure of galaxies in great depth. Several implementations of the triaxial Schwarzschild method exist, such as those proposed by van den Bosch et al. (2008), Vasiliev & Valluri (2020) and Neureiter et al. (2021). This modelling approach has been successfully applied to multiple galaxies, to study their internal structure (e.g., Cappellari et al. 2007; Feldmeier-Krause et al. 2017; Jin et al. 2020; Santucci et al. 2022;

Pilawa et al. 2022), to determine their dark matter content (e.g., Thomas et al. 2007; Cappellari et al. 2013; Poci, Cappellari, & McDermid 2017; Santucci et al. 2022), to measure the mass of their central super-massive black holes (e.g., van der Marel et al. 1998; Valluri, Merritt, & Emsellem 2004; Thater et al. 2019; den Brok et al. 2021; Thater et al. 2022; Merrell et al. 2022), and to identify accreted components (e.g., Zhu et al. 2020, 2022; Poci et al. 2021).

In the following we will focus on the publicly released software DYNAMITE (Jethwa et al. 2020; Thater et al. 2022), which is based on the code introduced by van den Bosch et al. (2008). We will describe the structure of the code, illustrate some of the applications for which this has been used, and outline the future enhancements that are currently planned for it.

2. The DYNAMITE code

DYNAMITE† (DYNamics, Age and Metallicity Indicators Tracing Evolution, Jethwa et al. 2020; Thater et al. 2022) is a triaxial dynamical modelling code, mainly aimed at studying the assembly history of galaxies by combining dynamical information (obtained from orbital distribution) with age and metallicity from population synthesis models. It is based on the orbit-superposition method proposed by Schwarzschild (1979).

DYNAMITE is the publically released successor of the triaxial Schwarzschild code developed by van den Bosch et al. (2008), and it is an open-source Python package, with a well-documented application programming interface. With respect to its predecessor, the DYNAMITE code has new features, such as for example the possibility to fit orbit-based models to histogrammed line-of-sight velocity distributions (LOSVDs) extracted by means of the BayesLOSVD framework (Falcón-Barroso & Martig 2021), instead of using Gauss-Hermite LOSVDs. The additions to the code and the developments planned for the future are all driven by relevant scientific questions, as illustrated in Section 3.

Figure 1 shows a scheme of the main steps needed to run DYNAMITE, which will be briefly illustrated in the following. This scheme is also a useful representation of the structure and components of the code.

Before running DYNAMITE, it is necessary to prepare the data in the correct format, i.e., to "translate" the observed kinematic data into files that are directly readable and usable by DYNAMITE. At this stage, the user can decide whether to use Gauss-Hermite moments or BayesLOSVD to characterize the kinematics; in both cases, to prepare data for use in DYNAMITE, an `astropy ecsv` file has to be created, together with some auxiliary files. At the moment, DYNAMITE accepts as input kinematic data obtained with integrated light spectroscopy, but future additions will also include the possibility to fit discrete line-of-sight velocity data and proper motions. The surface mass density and surface brightness are represented by Multi Gaussian Expansions (MGEs), also written in `astropy ecsv` format. If the assumption of a constant mass-to-light profile is valid for the stellar system under consideration, the two sets of MGEs can be identical, otherwise two different sets of MGEs need to be used, to represent light and mass respectively.

A configuration file in the `yaml` format needs to be specified in order to run DYNAMITE. This file contains two sections, where the user defines some properties of the physical system under consideration, and some settings for the modelling procedure. For example, in the first part, the user sets the starting values, step size and limiting values to be used for the parameter exploration, and can fix some parameters to specific values, if they are not fitting parameters. DYNAMITE models can contain

† The code is available on github (`https://github.com/dynamics-of-stellar-systems/dynamite`), and the documentation is also available online (`https://dynamics.univie.ac.at/dynamite_docs`).

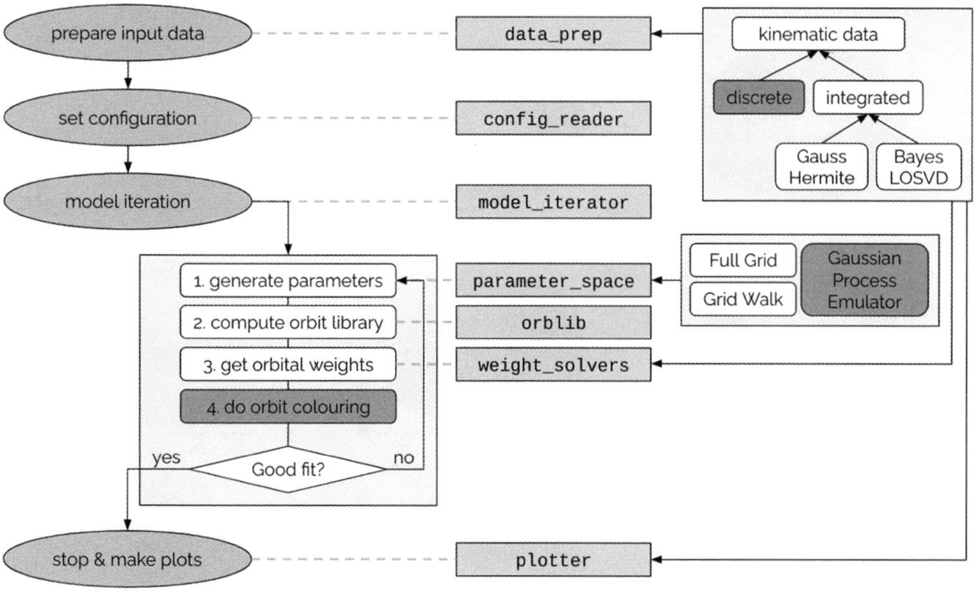

Figure 1. Scheme illustrating the basic steps of a DYNAMITE run, listed on the left in the oval green boxes. The pink boxes represent tasks that will be implemented in DYNAMITE in the future. Blue boxes indicate the relevant parts of DYNAMITE responsible for the tasks to which they are connected with blue dashed lines.

three different components: the stellar component, described by means of the parameters q, p, and u, representing the axis ratios and therefore describing the morphology of the system; a dark matter halo following a spherical Navarro-Frenk-White profile (NFW, Navarro, Frenk, & White 1996), defined by means of the concentration c and dark matter fraction f; and a central black hole, for which the mass m can be specified. In the second part of the configuration file, the user sets the dimensions of the orbit library to use, the type of weight solver and parameter generator, and multiprocessing settings.

After providing the input data and configuration file, it is possible to start a model iteration. For each choice of parameters, generated with the method specified in the configuration file, an orbit library is computed, and the corresponding orbital weights are determined, so that the superposition of orbits closely represents the data (as described in Schwarzschild 1979). This process is repeated to minimize the corresponding χ^2 function, and it is stopped when the relevant criteria set in the configuration file are met.

The best-fit model can then be used to make some plots, both to show how well the model reproduces the data and to inspect some properties of the model. As an example, Fig. 2 shows the comparison between the observed photometric and kinematic maps of galaxy NGC 4191 (panels in the left column) and the maps representing the best-fit model obtained with DYNAMITE (middle column); residuals are also shown in the right column.

3. Applications of DYNAMITE

Both DYNAMITE and the original code developed by van den Bosch et al. (2008) have been used to model the dynamics of many galaxies in the past years. A summary of the large galaxy samples that have been modelled with this code is presented in Table 1. The variety of galaxies modelled with this method and the application to data from different observing surveys reflects the plethora of scientific questions that this modelling

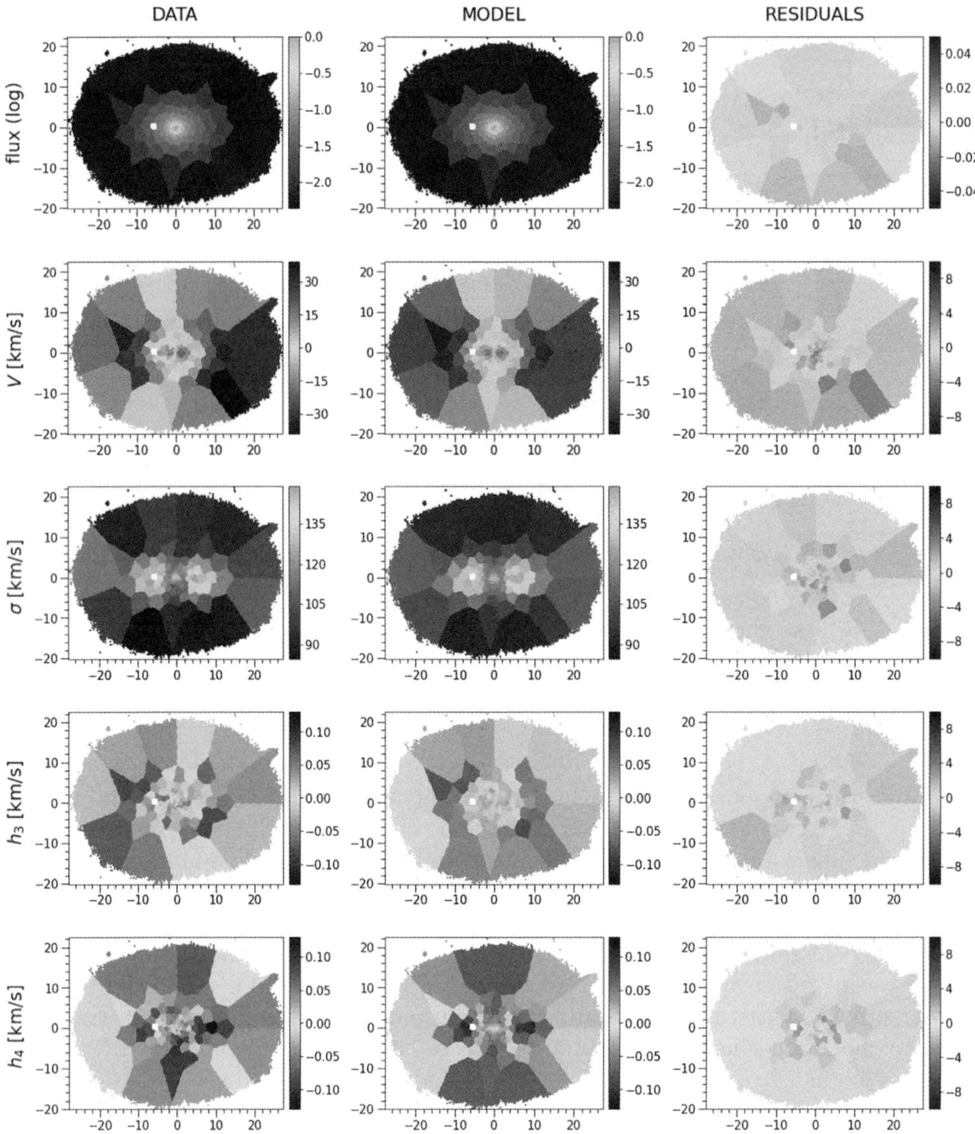

Figure 2. Comparison between the data obtained for galaxy NGC 4191 and the best-fit DYNAMITE model (Rubino et al. in prep.). The left column shows the observed photometric and kinematic maps, the middle column the ones corresponding to the best-fit DYNAMITE model, and the right column the residuals. From top to bottom, the maps represent surface brightness, mean velocity, velocity dispersion, and h_3 and h_4 Gauss-Hermite moments, indicating skewness and kurtosis deviations from a Gaussian velocity distribution.

approach can address. Some examples of what can be done with DYNAMITE, and of the upcoming further improvements to the code are presented in the following.

Thanks to the use of the BayesLOSVD framework to fit models to histogrammed LOSVDs, it is possible to describe galaxies with counter-rotating components in order to investigate their formation mechanisms. Two examples are the S0 galaxies NGC 4191 and IC 719, which are known to host two large-scale counter-rotating stellar disks (Coccato et al. 2015; Pizzella et al. 2018). The best-fit model obtained with DYNAMITE for NGC 4191 (Rubino et al. in prep.) is shown in Fig. 2, and provides a good representation of the kinematics of this galaxy.

Table 1. Overview of DYNAMITE galaxy models. For each observational survey listed in the first column, the type (second column) and number N (third column) of galaxies modelled with DYNAMITE are indicated; the corresponding references are listed in the fourth column.

Survey	Galaxy type	N	Reference
CALIFA	early-type and late-type galaxies	300	Zhu et al. (2018)
ATLAS^{3D}	early-type galaxies	$63^{(a)}$	Thater et al. (in prep)
MANGA	early-type galaxies	149	Jin et al. (2020)
SAMI	early-type galaxies	161	Santucci et al. (2022)
Fornax3D	early-type and late-type galaxies	18	Ding et al. (2023)
		(total = 691)	

Notes: $^{(a)}$ The total sample consists of 260 galaxies, whose analysis is in progress.

Efficient options for parameter search allow for a good determination of both the shapes of galaxies, and the properties of super-massive black holes (SMBHs) in relation to their host galactic nuclei. Recently, Santucci et al. (2022) applied DYNAMITE modelling to a large sample of galaxies whose data were collected in the SAMI Galaxy Survey, to recover their morphologies and inner orbit distributions. They found that the majority of galaxies in their sample are oblate, and only a small fraction appear to be prolate. Their results agree with a formation scenario proposing two main channels for the formation of slow and fast rotators, respectively characterized mainly by minor and major mergers, and show that such an investigation can provide clues on the past history of these systems. Dynamical modelling with DYNAMITE of galaxies observed within the ATLAS^{3D} survey is also in progress (Thater et al., in prep.), with the main aim of determining their intrinsic shapes. The Schwarzschild modelling approach is also useful to investigate the properties of SMBHs in the centres of galaxies, and their connection to the properties of their host nuclear star clusters. An example of this is shown in Thater et al. (2023), where DYNAMITE models are used to investigate the dynamics of FCC 47. Using this approach makes it possible to unveil the build-up of galactic centres, exploiting the fact that the evolutionary history of nuclear star clusters is imprinted into their stellar population and kinematics.

Tahmasebzadeh et al. (2022) has recently proposed an implementation of the Schwarzschild orbit-superposition method that allows the inclusion of a bar. This is currently being integrated into DYNAMITE, making it possible to fit in detail the photometry and stellar kinematics of nearby barred galaxies. Aside from a detailed insight into the orbits that built the bar, this will also yield robust measurements of the bar pattern speed, corotation radius and intrinsic bar length, which combined can answer the long-standing question whether bars are slowed down by dark matter haloes.

By inspecting the orbital distribution in a model, it is possible to identify distinct components: cold, warm and hot orbits show properties resembling thin disks, thick disks and classical bulges, respectively, as defined from cosmological simulations (Zhu et al. 2018, 2020). A colouring can be added to orbits, to reproduce metallicity and age maps on top of kinematic maps. This crucial upcoming addition to the DYNAMITE code will enable us to uncover the assembly history of galaxies, including past mergers (Zhu et al. 2022).

4. Conclusion

DYNAMITE is a user-friendly and flexible tool to construct detailed dynamical models of nearby galaxies. Based on Schwarzschild's orbit-superposition approach, it makes it possible to study morphological and dynamical properties of galaxies in detail, and to disentangle their complex build-up. DYNAMITE and its predecessor have been applied to about 700 (early-type and late-type) observed galaxies, and to several simulated galaxies, always providing a good representation of the data. The foreseen additions to the code, namely the inclusion of a barred structure and of orbit colouring, will expand the reach of this modelling tool, and will enable the users to answer many more pressing scientific questions, especially now that new and more detailed data are unveiling the complexity of galaxies and of their past evolutionary histories.

Acknowledgements

EJL & GvdV acknowledge funding by ERC under EU H2020 grant No. 724857 (ArcheoDyn).

References

Bovy J., Kawata D., Hunt J. A. S., 2018, MNRAS, 473, 2288
Cappellari M., Emsellem E., Bacon R., et al., 2007, MNRAS, 379, 418
Cappellari M., 2008, MNRAS, 390, 71
Cappellari M., Scott N., Alatalo K., et al., 2013, MNRAS, 432, 1709
Coccato L., Fabricius M., Morelli L., et al., 2015, A&A, 581, A65
den Brok M., Krajnović D., Emsellem E., Brinchmann J., Maseda M., 2021, MNRAS, 508, 4786
Ding Y., Zhu L., van de Ven G., et al., 2023, A&A, 672, A84
Falcón-Barroso J., Martig M., 2021, A&A, 646, A31
Feldmeier-Krause A., Zhu L., Neumayer N., et al., 2017, MNRAS, 466, 4040
Jethwa P., Thater S., Maindl T., van de Ven G., 2020, ascl.soft. ascl:2011.007
Jin Y., Zhu L., Long R. J., Mao S., Wang L., van de Ven G., 2020, MNRAS, 491, 1690
Magorrian J., Tremaine S., Richstone D., et al., 1998, AJ, 115, 2285
Merrell K. A., Vasiliev E., Bentz M. C., et al., 2022, arXiv, arXiv:2212.02484
Nagai R., Miyamoto M., 1976, PASJ, 28, 1
Navarro J. F., Frenk C. S., White S. D. M., 1996, ApJ, 462, 563
Neureiter B., Thomas J., Saglia R., et al., 2021, MNRAS, 500, 1437
Pilawa J. D., Liepold C. M., Delgado Andrade S. C., et al., 2022, ApJ, 928, 178
Pizzella A., Morelli L., Coccato L., et al., 2018, A&A, 616, A22
Poci A., Cappellari M., McDermid R. M., 2017, MNRAS, 467, 1397
Poci A., McDermid R. M., Lyubenova M., et al., 2021, A&A, 647, A145
Qian E. E., de Zeeuw P. T., van der Marel R. P., Hunter C., 1995, MNRAS, 274, 602
Rix H.-W., Bovy J., 2013, A&ARv, 21, 61
Santucci G., Brough S., van de Sande J., et al., 2022, ApJ, 930, 153
Schwarzschild M., 1979, ApJ, 232, 236
Syer D., Tremaine S., 1996, MNRAS, 282, 223
Tahmasebzadeh B., Zhu L., Shen J., Gerhard O., van de Ven G., 2022, ApJ, 941, 109
Thater S., Krajnović D., Cappellari M., et al., 2019, A&A, 625, A62
Thater S., Jethwa P., Tahmasebzadeh B., et al., 2022, A&A, 667, A51
Thater S., Krajnović D., Weilbacher P. M., et al., 2022, MNRAS, 509, 5416
Thater S., Lyubenova M., Fahrion K., et al., 2023, arXiv, arXiv:2304.13310
Thomas J., Saglia R. P., Bender R., et al., 2007, MNRAS, 382, 657
Valluri M., Merritt D., Emsellem E., 2004, ApJ, 602, 66
van den Bosch R. C. E., van de Ven G., Verolme E. K., et al., 2008, MNRAS, 385, 647
van der Marel R. P., Cretton N., de Zeeuw P. T., Rix H.-W., 1998, ApJ, 493, 613
Vasiliev E., Valluri M., 2020, ApJ, 889, 39
Zhu L., van den Bosch R., van de Ven G., et al., 2018, MNRAS, 473, 3000
Zhu L., van de Ven G., Méndez-Abreu J., Obreja A., 2018, MNRAS, 479, 945
Zhu L., van de Ven G., Leaman R., et al., 2020, MNRAS, 496, 1579
Zhu L., van de Ven G., Leaman R., et al., 2022, A&A, 664, A115

Dynamical Masses of Local Group Galaxies
Proceedings IAU Symposium No. 379, 2025
P. Bonifacio, M.-R. Cioni, F. Hammer, M. S. Pawlowski & S. Taibi, eds.
doi:10.1017/S1743921323001679

How 4MOST and MOONS will improve constraints on the dynamical masses of Local Group galaxies

S. Mieske[1], G. Battaglia[2,3], A. Bayo[4,5], M.-R. L. Cioni[6], O. González[7], V. Mainieri[5], A. Mucciarelli[8], L. Schmidtobreick[1], Á. Skúladóttir[9,10] and T. Szeifert[1]

[1]European Southern Observatory, Alonso de Cordova 3107, Vitacura, Chile
email: tszeifer@eso.org

[2]Instituto de Astrofísica de Canarias, 38205 La Laguna, Tenerife, Spain

[3]Universidad de La Laguna, Dpto. Astrofísica, 38206 La Laguna, Tenerife, Spain

[4]Instituto de Física y Astronomía, Universidad de Valparaíso, Chile

[5]European Southern Observatory, Karl-Schwarzschild-Strasse 2, 85748 Garching bei München, Germany

[6]Leibniz-Institut für Astrophysik Potsdam, An der Sternwarte 16, 14482 Potsdam, Germany

[7]UK Astronomy Technology Centre, Royal Observatory, Blackford Hill, Edinburgh, EH9 3HJ, UK

[8]Dipartimento di Fisica e Astronomia "Augusto Righi", Alma Mater Studiorum, Università di Bologna, Via Gobetti 93/2, I-40129 Bologna, Italy

[9]Dipartimento di Fisica e Astronomia, Universitá degli Studi di Firenze, Via G. Sansone 1, I-50019 Sesto Fiorentino, Italy

[10]INAF/Osservatorio Astrofisico di Arcetri, Largo E. Fermi 5, I-50125 Firenze, Italy

Abstract. The wide-field spectrographs 4MOST and MOONS are expected to enter operations in late 2024 at the ESO Paranal Observatory. These upcoming survey facilities will play an important role in various fields over the next decade. In particular, both will host surveys aimed at observing Local Group (LG) galaxies. We describe how their scientific performances complement Gaia and other spectroscopic surveys in the field of nearby galaxy kinematics, and provide an overview of the planned surveys that focus on LG galaxy kinematics. We outline the policies for community access to data and observing time, which is different for the two instruments. The contribution concludes with a summary of the main scientific goals of MOONS and 4MOST survey science regarding the LG galaxy masses and dynamics.

Keywords. Galaxy: halo, Galaxy: fundamental parameters, Galaxies: dwarfs, Galaxies: Local Group, Galaxies: kinematics and dynamics

---

1. Introduction - 4MOST and MOONS

The 4-metre Multi-Object Spectrograph Telescope (4MOST†) and the Multi Object Optical and Near-infrared Spectrograph (MOONS‡) are fibre-fed spectrographs that are expected to start science observations towards the end of 2024 at the European Southern

† http://4most.eu
‡ http://vltmoons.org

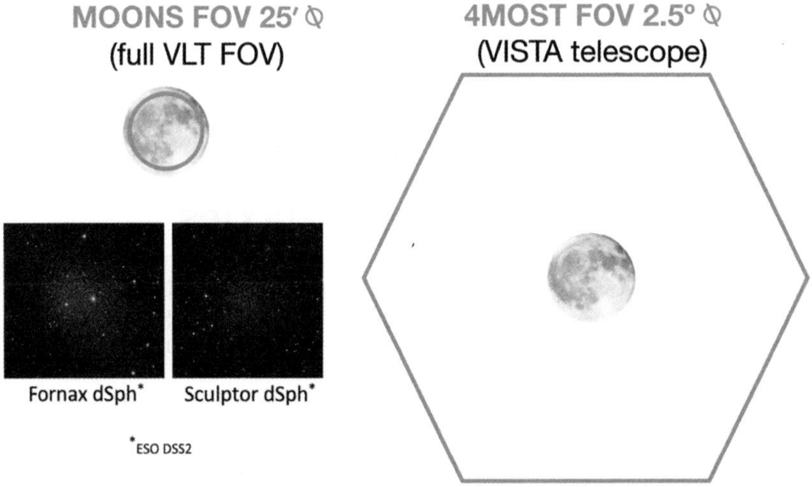

Figure 1. Comparison of the FOV of MOONS and 4MOST, including to-scale thumbnails of the two LG dSphs Fornax and Sculptor. The MOONS total FOV of 0.14 deg^2 is well matched to the dSph angular extent, whereas the 4MOST FOV (4.2 deg^2) is optimised for wide-field studies beyond individual galaxies.

Observatory (ESO)'s Paranal Observatory. 4MOST will be mounted at the Visual and Infrared Survey Telescope for Astronomy (VISTA, 4.1m main mirror diameter, covering a field-of-view of about 2.5 deg in diameter (4.2 deg^2). MOONS will be mounted at Unit Telescope 1 (UT1, 8.2m main mirror diameter) of ESO's Very Large Telescope (VLT), covering the maximum field of view of the VLT, 25 arcmin in diameter. Both MOONS and 4MOST are foreseen to physically arrive to Paranal in early 2024. In Fig. 1 we compare the field-of-view (FOV) of 4MOST and MOONS, and show in-scale thumbnails of two Local Group (LG) dwarf spheroidal galaxies (dSphs).

Details of the 4MOST design, its scientific drivers, and its survey programs are described in de Jong et al. (2019). MOONS design and science surveys are described in Cirasuolo et al. (2020). The complementarity to Gaia and Galactic Archaeology are key science drivers for both MOONS and 4MOST. The main instrument characteristics are summarised in Table 1. We also indicate the characteristics of the William Herschel Telescope Enhanced Area Velocity Explorer (WEAVE) at the WHT (Jin et al. 2023) and the Dark Energy Spectroscopic Instrument (DESI) at Kitt Peak Mayall (DESI Collaboration 2022), two contemporary multi-object spectrographs in the northern hemisphere that are optimised for wide-field surveys.

As is evidenced by the instrument characteristics, 4MOST and MOONS are very complementary: 4MOST focuses on optical wavelengths and offers unique capabilities for extremely-wide field spectroscopic surveys, ideally suited for analysis of the Milky Way halo, streams and the large LG galaxies (e.g., the Magellanic Clouds). The limiting magnitude of 4MOST spectroscopy matches the Gaia astrometric limiting magnitude of V$\sim$ 20.5 mag (Gaia collaboration 2021). MOONS has a smaller FOV but goes deeper and has a broader wavelength coverage, extending to the infrared. This enables stellar population analysis towards the Milky Way bulge, the galactic plane, and high redshift galaxy science. 4MOST radial velocity accuracy target is 1 km/s at signal-to-noise (S/N) of 10 per resolution element ($\geq$ 2.5 pixel per full width at half maximum). MOONS is expected to reach similar radial velocity accuracies, possibly down to 0.5 km/s.

An important difference between both instruments is their science operations policy.

Table 1. Instrument characteristics. LR = Low Resolution; HR = High Resolution.

	FOV	Fibres	WL coverage	Resolution	Limiting mag (LR)	Start of ops
4MOST@VISTA	2.6°	2436[1]	3600-9500 Å (LR)[3]	LR 4000-8000, HR 20000	V$\sim$ 20.5 mag	end of 2024
MOONS@UT1	25′	960/940[2]	6500-18000 Å (LR)[3]	LR 3000-7000, HR 8000-20000	R$\sim$ 22.5 mag	end of 2024
DESI@Kitt Peak	3°	5000	3600-9800 Å	2000-5000	V$\sim$ 20 mag	May 2021
WEAVE@WHT	2°	1000	3660-9590 Å (LR)[3]	LR 5000, HR 20000	V$\sim$ 20 mag	2024 for MOS

Notes: [1] 4MOST fibres are split as follows: 1624 feed the LR spectrographs, and 812 the HR spectrograph. [3] WEAVE fibres are 960 and 940, respectively, in two focal-plane plates that at night are used one after the other due to the 55 minutes configuration time. [3] The wavelength coverage in HR is less than for LR: HR for 4MOST is limited to three spectral bands between 3926 and 6790 Å, for WEAVE to three spectral bands between 4040 and 6850 Å, and for MOONS to three spectral bands between 7600 and 1640 Å.

For its first five years of operations, 4MOST will exclusively observe public surveys, which have already been selected. We highlight the most relevant 4MOST surveys for the topic of this conference in Sect. 2. All 4MOST raw data will become public immediately, and data products are provided to the community periodically during the execution of the surveys. Another important aspect of 4MOST science operations is that each pointing will usually observe science targets of *several surveys at once*. This optimisation of operations is managed by the 4MOST consortium on behalf of all surveys, including the creation of the Observing Blocks (OBs) to be executed at night at the telescope.

Instead, MOONS has the traditional VLT setup. From the start of operations open time will be mixed with guaranteed time of observations (GTO), for which the MOONS consortium was granted 300 nights over five years. OBs will have to be created by the users using the normal tools provided by ESO. MOONS data will be proprietary for the first 12 months.

2. Local Group Surveys with 4MOST

4MOST will be used in survey mode only, where 70% of the time is dedicated to 10 public surveys led by the 4MOST consortium (de Jong et al. 2019), and 30% of the time is dedicated to 15 public surveys led by the community (Mainieri et al. 2023). These surveys cover a wide range of Galactic and Extra-Galactic science.

In the following we summarise the 4MOST surveys whose scientific goals are most relevant for the topic of this Symposium. In doing so we re-use part of the text contained in the quoted references, which are ESO Messenger editions.

• **1001MC**, PI Cioni (Cioni et al. 2019), consortium survey: the *'One Thousand and One Magellanic Fields'* survey aims to measure the kinematics and elemental abundances of the stellar populations that sample the history of formation and interaction of the Magellanic Clouds (MC). This survey is motivated by the pronounced lack of spectroscopic observations across the relevant ranges of stellar populations and substructures, noting that large amounts of telescope time have been invested in imaging surveys, such as the VISTA survey of the MC system (VMC, Cioni et al. 2011). 1001MC will target about a million stars with G<20 mag. Its science goals are to: a) find and characterise kinematic and chemical patterns within the MC system; b) study links between kinematic and chemical patterns; c) establish the link of the star formation history and dynamical evolution of the MC system to these patterns; d) quantify the metallicity dependence of key distance indicators.

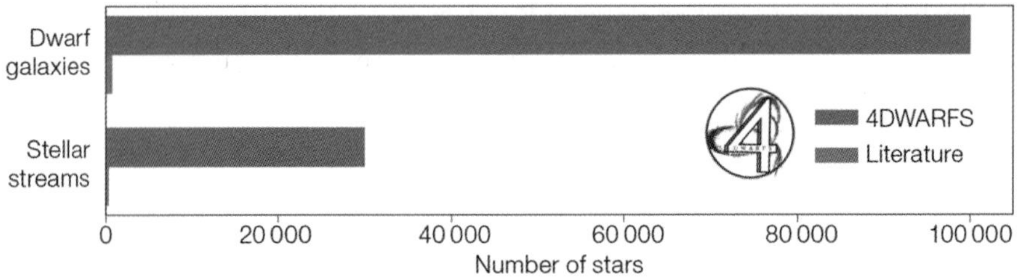

Figure 2. Sample size increase to be achieved by 4DWARFS@4MOST (Skúladóttir et al. 2023).

• **4DWARFS**, PI Skúladóttir (Skúladóttir et al. 2023), community survey: the *'4MOST Survey of Dwarf Galaxies and their Stellar Streams'* will target the dwarf galaxies and stellar streams in the 4MOST footprint (baseline survey $-70 < \text{dec} < +5$ deg) and unveil their chrono-chemo-kinematical properties. In particular, it will target Sagittarius dwarf and stream, the Fornax, Sculptor, Carina and Sextans dSphs, around 40 ultra-faint dwarfs and dozens of known stellar streams in the Milky Way halo. The survey will increase by several orders of magnitude the number of stars in those systems with detailed information on radial velocities, abundances and ages, in a homogeneous data set, see Fig. 2. The key scientific questions the survey aims to answer are: a) What are the properties of the first stars?; b) How are the chemical elements created and distributed?; c) What are the dynamical properties of dwarf galaxies?; d) What are the small-scale limits of hierarchical galaxy formation?

• **Milky Way Halo Low-Resolution Survey**, PIs Starkenburg & Worley (Helmi et al. 2019), consortium survey: the goal of this survey is to study the formation and evolution of the Milky Way halo, to deduce its assembly history and the three-dimensional (3D) distribution of mass in the Milky Way. While the community survey 4GRoundS (see above) focuses on RR Lyrae stars as tracers, the Milky Way Halo Low-Resolution Survey will produce a volume- and magnitude-limited complete sample of giant stars in the halo, covering at least 10 000 deg^2 of high galactic latitude, down to Gaia magnitude G $\sim$ 20 mag. This will allow a much more precise determination of the Milky Way mass, and importantly constrain its 3D distribution in terms of density and shape to unprecedented level. The scientific goals of this survey are to: a) determine the characteristic parameters of the dark matter halo of the Milky, including the test of alternative gravity theories such as Modified Newtonian Dynamics (MOND); b) measure pertubations induced by clumps on cold streams, leading to constraints on the mass spectrum of perturbers and the nature of dark matter; c) quantify the amount of kinematical substructure as a function of distance and location on the sky; d) characterise the metallicity and elemental abundance distribution throughout the halo and its substructures; e) characterisation of the stellar halo - thick disk interface.

• **4GRoundS**, PI Ibata (Ibata et al. 2023), community survey: the *'4MOST Gaia RR Lyrae Survey'* will measure radial velocities and metallicities of southern RR Lyrae stars, identified in Gaia Data Release 3 (DR3). It will enable studies of the orbital structure of the Milky Way's halo and outer disc, as well as stellar streams. In doing so, the mass of the Milky Way will be mapped out to 100 kpc, which will allow to test different models of the 'dark sector'. Gaia parallaxes are available until about 10 kpc distance, enabling direct distance measurements to those stars which can then be combined with proper motion and radial velocity to obtain a full set of orbital parameters. Going beyond 10 kpc requires to target tracer populations whose distance can be determined from alternative sources. To this end, RR Lyrae stars were chosen in the 4GRoundS survey due to their well calibrated use as standard candles and because Gaia DR3 provides a very clean RR Lyrae sample. The specific scientific goals of 4GRoundS are: a) Dynamical Mass Modelling of the outer Galaxy; b) Analysis of stellar streams in the outer Galaxy; c) Search for low-mass satellites; d) Analysis of the Global Halo kinematic asymmetry due to the arrival of the LMC; e) Study of the far side of the thick MW disc; f) Disentangle halo and spheroidal MW bulge; g) Detection and characterisation of spatial variations in kinematic coherence through the halo.

3. Local Group Science with MOONS

As mentioned above, MOONS will be used in the standard VLT way by providing open time access via regular call for proposals (CfPs), from the beginning of operations that is currently foreseen for early 2025. The MOONS consortium has been granted 300 nights of **GTO** over the first five years of MOONS operations (Cirasuolo et al. 2020), which comes in addition to any open time granted via the regular CfP. Of the 300 GTO nights, 100 nights will focus on Galactic Science. The two Galactic Science GTO surveys are:

- **The Milky Way Satellites**, 30 nights (PI González; González et al. 2020): this survey will target parts of the Magellanic Clouds and the Sagittarius dSph. In the LMC/SMC it will observe a mosaic of 5 x 5 / 3 x 3 pointings to study the chemical composition and kinematics of the inner region. In addition, it will observe several fields along the major axis of the LMC bar/SMC main body and in the disc/extended region. For the Sagittarius dSph it will cover 24 fields within the inner 100 arcmin, outside the tidal radius of M54, with the main goal to derive an unbiased metallicity distribution.
- **The MOONS REDened Milky WAY survey (REDWAY)**, 70 nights (PI González, González et al. 2020): this survey will sample half a million stars across the Galactic plane, tracing red clump stars in the bulge/bar and inner disc regions. It will sample the chemo-dynamics of the stellar populations in the inner 3 kpc of the Milky Way. The infrared wavelength coverage of MOONS enables this key science goal, allowing the instrument to peek through the massive extinction towards the central bulge area of the Milky Way. The survey will target: a) the nuclear region; b) the bulge and bar at low latitudes; c) the boxy/peanut bulge; d) a bulge deep field; e) inner Milky Way globular and young star clusters.

3.1. *Open time opportunities*

As discussed above, kinematical studies of LG dwarfs with MOONS-GTO are limited to the Magellanic Clouds and the Sagittarius dSph. This represents a unique opportunity for open time proposals that aim at investigating kinematical substructures of other LG dwarfs. Indeed, lab testing of the MOONS system has shown a very stable line spread function. In Fig. 3 we indicate the approximate limiting magnitude for high precision radial velocity in the low-resolution setting (1–2 km/s in 1hr integration) in the stellar CMD of the Fornax, Sculptor and Leo dSphs. In Fig. 1 we also show how the MOONS FOV is nicely matched to the angular extent of Fornax and Sculptor dSph.

4. Community data access and Summary of science questions

4.1. *Community data access*

MOONS: the MOONS instrument is foreseen to arrive to Paranal Observatory in Q1/2024. Following assembly, installation and commissioning, science verification observations may take place towards the end of 2024, and regular science observations could start in early 2025. The first Call for Proposals for regular science observations is probably for period 115, which starts April 1, 2025. The data taken for MOONS will by default have a 12-month proprietary period. A public data reduction pipeline will be available.

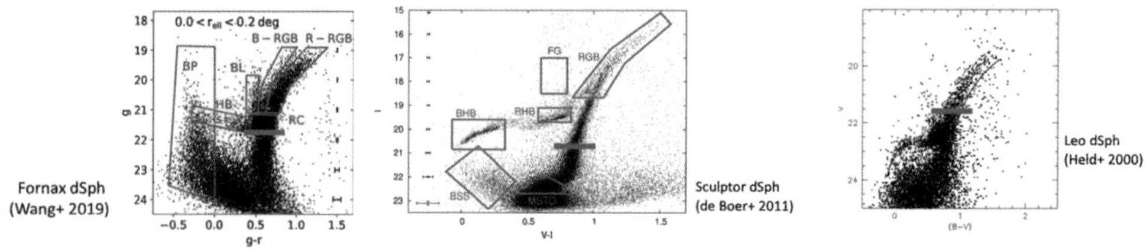

Figure 3. The approximate MOONS limiting magnitude (S/N=5, 1hr, LR mode) is indicated as a blue horizontal bar in three CMDs of LG dSphs.

4MOST: 4MOST is foreseen to arrive to Paranal Observatory in Q1/2024. Following assembly, installation and commissioning, science verification observations may take place towards the end of 2024. Regular science observations could start in early 2025. No regular call for proposals is foreseen for 4MOST during the first five years of operations, as the instrument will be exclusively dedicated to public surveys. The raw data taken for those surveys will become public immediately. However, there will not be a publicly available pipeline. Data products for each survey will be released by the 4MOST consortium to the user community on a regular basis, including reduced spectra, data products (e.g. stellar atmospheric parameters, radial velocities, chemical abundances, stellar ages), and source catalogues.

4.2. *Summary of science goals*

We conclude with a high-level summary of the key science goals related to the topic of this Symposium, that will be addressed with 4MOST and MOONS as of 2024/2025.

• Constrain the gravitational potential and dark matter halo of the Milky Way, and testing of alternative gravity theories (The 4MOST Milky Way Low-Resolution Survey, and 4GRoundS).

• Characterise properties of stellar streams in the Milky Way halo (the 4MOST Milky Way Low-Resolution Survey, 4GRoundS, and 4DWARFS).

• Identify and characterise kinematical and chemical substructures of LG dwarf galaxies (4MOST 1001MC, 4DWARFS, and MOONS GTO targeting the Magellanic Clouds and the Sagittarius dSph). There is also an opportunity for MOONS open time on other LG dwarfs, with the first Call for Proposals expected in 2024.

Acknowlegements: GB acknowledges support from the Agencia Estatal de Investigación del Ministerio de Ciencia e Innovación (AEI-MICIN) and the European Regional Development Fund (ERDF) under grant number PID2020-118778GB-I00/10.13039/501100011033 and CEX2019-000920-S.

References

DESI Collaboration, Abareshi, B., Aguilar, j., Ahlen, S. *et al.* 2022, AJ, 164, 207D
Cioni, M.R., Storm, J., Bell, C.P.M *et al.* 2019, ESO Messenger 175, Pages 54–57
Cioni, M.R., Clementini, G., Girardi, K. *et al.* 2011, A&A 527A, 116
Cirasuolo, M., Fairley, A., Rees, P. *et al.* 2020, ESO Messenger 180, Pages 10–17
de Boer, T.J.L., Tolstoy, E., Saha, A. *et al.* 2011, A&A, 528A, 119
De Jong, R., Agertz, O., Berbel, A.A. *et al.* 2019, ESO Messenger 175, Pages 3–11
Gaia collaboration, Brown A. G. A., Vallenari, A., Prusti, T. *et al.* 2021, A&A, 649, 1
Gonzalez, O., Mucciarelli, A., Origlia, L. *et al.* 2020, ESO Messenger 180, Pages 18-23
Held, E.V., Saviane, I., Momany, Y. & Carraro, G. 2000, ApJ 530L, 85
Helmi, A., Irwin, W., Deason, A. *et al.* 2019, ESO Messenger 175, Pages 23-25
Ibata, R., Battaglia, G., Bellazzini, M. *et al.* 2023, ESO Messenger 190, Pages 10–12

Jin, S., Trager, S.C., Dalton, G.B. *et al.* 2023, MNRAS tmp, 715J (arxiv:2212.03981)
Mainieri, V., Leibundgut, B., de Jong, R.S. *et al.* 2023, ESO Messenger 190, Page 3
Skúladóttir, Á., Puls, A., Amarsi, A.M. *et al.*, ESO Messenger 190, Pages 19-21
Wang, M.Y., de Boer, T., Pieres, A. *et al.* 2019, ApJ, 881, 118

Section 2:
Sizes, Masses and Formation histoires of the Milky Way and Andromeda

Dynamical Masses of Local Group Galaxies
Proceedings IAU Symposium No. 379, 2025
P. Bonifacio, M.-R. Cioni, F. Hammer, M. S. Pawlowski & S. Taibi, eds.
doi:10.1017/S1743921323001412

Dynamical properties of ancient stars in the inner Milky Way with PIGS

A. Arentsen[1], G. Monari[2], A. Queiroz[3], E. Starkenburg[4], N. F. Martin[2,5] and C. Chiappini[3]

[1]Institute of Astronomy, University of Cambridge, Madingley Road, Cambridge CB3 0HA, UK
email: anke.arentsen@ast.cam.ac.uk

[2]Université de Strasbourg, CNRS, Observatoire astronomique de Strasbourg, UMR 7550, F-67000 Strasbourg, France

[3]Leibniz-Institut für Astrophysik Potsdam (AIP), An der Sternwarte 16, D-14482 Potsdam, Germany

[4]Kapteyn Astronomical Institute, University of Groningen, Postbus 800, 9700 AV, Groningen, the Netherlands

[5]Max-Planck-Institut für Astronomie, Königstuhl 17, D-69117 Heidelberg, Germany

Abstract. We present recent results from the Pristine Inner Galaxy Survey (PIGS), which used metallicity-sensitive narrow-band CaHK photometry to identify and follow up spectroscopically thousands of ancient metal-poor candidates in the bulge. For the spectroscopic PIGS sample, we derive distances with StarHorse and compute orbital properties in a realistic potential including a bar. We find that a significant fraction of metal-poor stars is confined to the inner Galaxy (apocentre < 4 kpc), with an estimated confined fraction of 80%/50% at [Fe/H] $= -1.0/-2.0$. We also find that the very metal-poor population has a net prograde rotation, with a $v_\phi \sim 40$ km s^{-1}. It is still under discussion what the origin is of the population of very metal-poor inner Galaxy stars – it is likely a combination of *in-situ* and accreted stars. In future, spectroscopic observations from 4MOST will be crucial to complete our picture.

Keywords. Galaxy: bulge – Galaxy: halo – Galaxy: kinematics and dynamics – stars: Population II

1. Introduction

Old, metal-poor stars tell a unique story about the early history of the Milky Way. Their chemistry and kinematics are remnants from a time long gone, and teach us about early chemical evolution and the build-up of our Galaxy. The most metal-poor component of the Milky Way is the halo, and metal-poor stars have typically been searched for by and found in surveys looking up or down out of the Galactic disc – for good reasons, disc and bulge stars are metal-rich and dominate samples closer to the Galactic plane or the Galactic centre. For example the bulge red clump survey ARGOS contained only $\sim 0.1\%$ (16 out of 14 150 stars) very metal-poor (VMP) stars with [Fe/H] < -2.0 (Ness et al. 2013). This means the metal-poor inner Galaxy has remained largely unexplored, until very recently.

There are many good reasons to study the metal-poor inner Galaxy. The *oldest* metal-poor stars are expected to be very centrally concentrated (see e.g. the simulations by Tumlinson 2010; Starkenburg et al. 2017a; El-Badry et al. 2018), because the inner Galaxy is the part of the Milky Way that formed first. More observations are needed to address the nature of the metal-poor inner Galaxy, and answer open questions such as: What kind of building blocks contributed to the metal-poor inner Galaxy? Were they different from the rest of the halo? How much of the metal-poor inner Galaxy formed *in-situ*, and are there any very metal-poor stars part of the bar/bulge? Furthermore, we can gain new insights about the First Stars and galaxies from very metal-poor stars in different Galactic environments.

Recent years have seen a strong increase in interest and number of results regarding the metal-poor inner Galaxy, often using clever pre-selection methods (EMBLA/SkyMapper photometry: Howes et al. 2014, 2015, 2016, Best & Brightest IR photometry: Schlaufman & Casey 2014; Casey & Schlaufman 2015; Reggiani et al. 2020, BRAVA/RR Lyrae selection: Kunder et al. 2016, 2020; Savino et al. 2020, CaK photometry: Koch et al. 2016, 2019, COMBS/SkyMapper photometry: Lucey et al. 2019, 2021, 2022, PIGS/Pristine photometry: Arentsen et al. 2020a,b, 2021; Sestito et al. 2020, Gaia XP spectra: Rix et al. 2022). This contribution will focus on results from the Pristine survey, specifically the kinematics of (very) metal-poor stars in the inner Galaxy and what they can teach us about the ancient central Milky Way.

2. The PIGS sample with distances and orbits

The largest spectroscopic metal-poor central Milky Way survey to date is the Pristine Inner Galaxy Survey (PIGS Arentsen et al. 2020b). It uses the same technique as the main Pristine survey (Starkenburg et al. 2017b) to identify the most metal-poor stars: employing narrow-band, metallicity-sensitive *CaHK*photometry centred on the calcium H&K lines. By combining the *CaHK*photometry with broad-band photometry, one can produce a colour-colour space in which metal-poor stars separate extremely well from metal-rich stars. Candidate metal-poor stars were selected from the photometry and followed up with low/medium resolution spectroscopy with AAOmega+2dF on the AAT, 12 000 stars in total (details can be found in Arentsen et al. 2020b). Of those, 90% are metal-poor ([Fe/H] < -1.0) and the sample contains ~ 1900 VMP stars ([Fe/H] < -2.0).

In order to derive orbits for the stars in PIGS, distances are needed. Distances are challenging to determine in the inner Galaxy due to the relatively large distance (Gaia parallaxes are not good enough by themselves) and high/irregular extinction. We combine the spectroscopic stellar parameters from PIGS, photometry from *Gaia*EDR3 (Gaia Collaboration et al. 2021) and PanSTARRS (Chambers et al. 2016) and parallaxes from *Gaia*EDR3 to derive the best possible distances using the StarHorse code (as in Queiroz et al. 2023, applied to other spectroscopic surveys). We find that 90% of the stars in PIGS are within 4 kpc of the Galactic centre, see the left-hand panel of Figure 1. The distances have typical uncertainties of $\sim 15\%$.

We proceed to derive orbital properties of the stars in PIGS, adopting the StarHorse distances, the PIGS radial velocities and the *Gaia*proper motions. The orbits are integrated in a Milky Way potential with a realistic bar (Portail et al. 2017; Sormani et al. 2022). For each star we draw 50 samples from the parameters and their uncertainties for which we derive the orbits, allowing us to build a probability distribution for the orbital parameters of each star.

3. Kinematics results for the metal-poor PIGS stars

3.1. *Confined stars*

The distances show that many of the PIGS metal-poor stars are *currently* in the inner Galaxy, but do they *stay* there for most of their lives? In the right-hand panel of Figure 1 we show the fraction of stars currently within 4 kpc of the Galactic Centre that is confined to the inner Galaxy, as function of metallicity. For this analysis we have classified a star

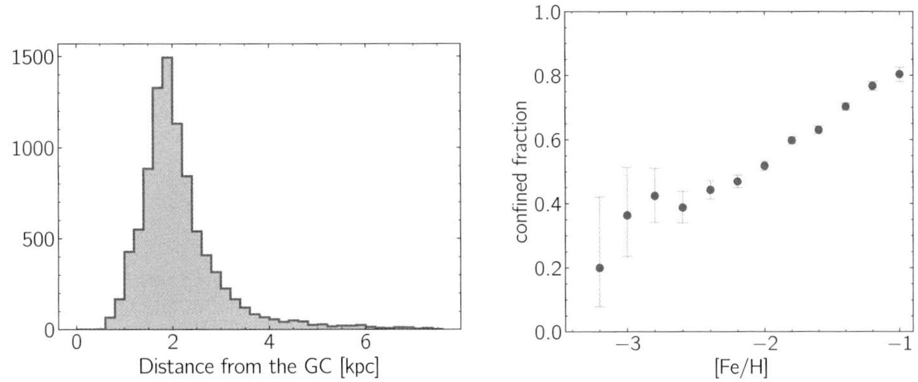

Figure 1. Left: Distribution of distances to the Galactic Centre for stars in PIGS with $[Fe/H] < -1.0$. **Right:** fraction of stars that is confined to within 4 kpc of the Galactic centre as function of metallicity.

as confined if at least 75% of the orbital samples for a star have apocentres less than 4 kpc. It can be noticed immediately that there is a strong trend with metallicity: more than 80% of the stars at $[Fe/H] = -1.0$ are confined, which steadily decreases to about 40% at $[Fe/H] = -2.5$. Below that metallicity the error bars grow, but still indicate that $20 - 40\%$ of stars are confined. The rest are halo stars just passing through the inner Galaxy. The trend of increasing number of "halo interlopers" with metallicity has been noted before (Lucey et al. 2021), based on a much smaller sample of stars, especially at the lowest metallicities. For the first time we can see that even in the very metal-poor regime there is a significant fraction of stars that is confined to the inner Galaxy (50% at $[Fe/H] = -2.0$). There appears to be a centrally concentrated metal-poor population in the inner Galaxy that has evaded previous study because typical halo surveys would not have picked up stars so close to the Galactic centre and typical bulge surveys would not have picked up enough metal-poor stars. Evidence for this was also seen by Rix et al. (2022), who used the XP spectra from *Gaia*DR3 to derive metallicities and found an over-density of metal-poor stars in the inner Milky Way.

3.2. *Rotation around the Galactic Centre*

But how does this central metal-poor population move around the Galactic centre, does it have coherent motion or is it mostly on random orbits? For stars with apocentres < 4 kpc, we projected the radial velocities to study the rotation curve as function of Galactic longitude in Figure 2. This figure is analogous to the previous PIGS rotation curves in Arentsen et al. (2020a), except that it includes 50% more data and that halo interlopers are removed. There is clearly a decrease in the rotational velocity with decreasing metallicity, but there is still some coherent rotation even for stars with $[Fe/H] < -2.0$, particularly visible at positive longitude.

We now have the full orbital properties available for the stars in PIGS instead of only the radial velocities as in Arentsen et al. (2020a). We split the sample of stars with $[Fe/H] < -1.0$ in bins of 0.25 dex and fit Gaussian distributions to the three velocity components (in spherical Galactic coordinates, v_r, v_θ, v_ϕ) using an extreme deconvolution method to take into account the effect of the uncertainties. We do this for each of the 50 orbit realisations, and only for stars with a median apocentre less than 4 kpc. For this work we only fitted one component, in the future we will test whether multiple components provide a better fit. The result is shown in Figure 3. The left-hand side presents a similar picture as Figure 2 but distilled into a mean velocity. We find that for

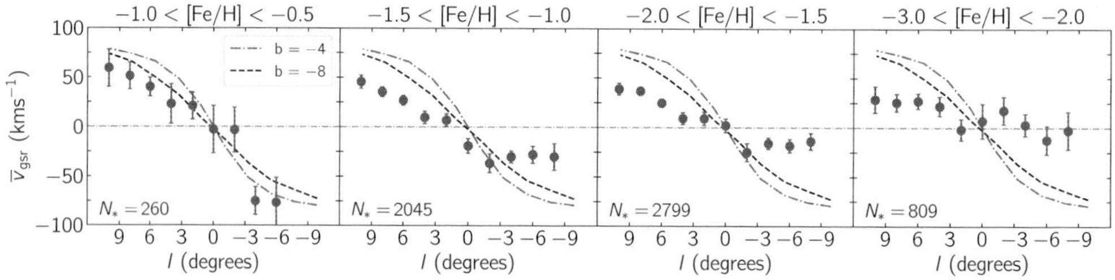

Figure 2. PIGS rotation curve as function of metallicity (only including stars with apocentres less than 4 kpc), with the number of stars included in each panel shown in the bottom left. The lines are fits to metal-rich bulge stars by Shen et al. (2010) and are the same in each panel.

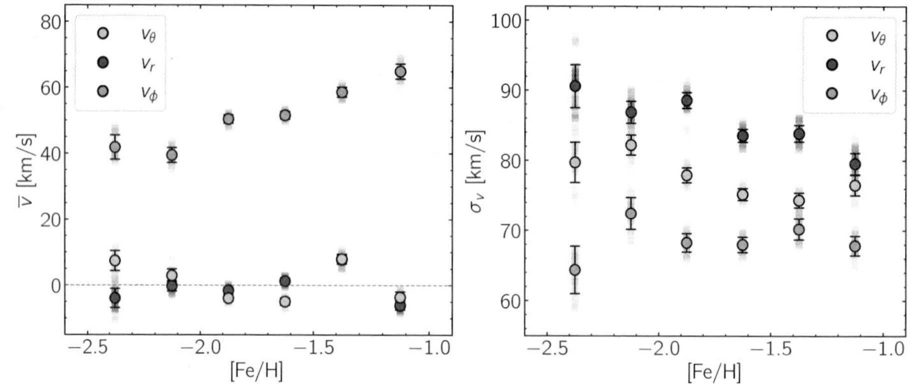

Figure 3. Mean (left) and dispersion (right) for each of the three velocity components in PIGS as a function of metallicity. The transparent squares indicate the fit for each of the 50 realisations of the PIGS orbits, the black circles and error bars the mean and standard deviation.

VMP stars, there still is an average rotational velocity of ~ 40 km s^{-1}. The right-hand panel shows the velocity dispersion. It is slightly lower for v_ϕ than for the other two velocity components, but still high (of the order of 70 km s^{-1}).

4. Discussion and conclusions

The fraction of confined stars decreases with decreasing metallicity, as well as the rotational velocity around the Galactic Centre. We do find that a large fraction of VMP stars currently in the inner Galaxy is confined to the inner Galaxy, and that this population still shows average prograde rotation. What could be the origin of this population?

Some have previously argued that there is a large *in-situ* metal-poor population in the inner Galaxy. Kunder et al. (2016, 2020) found evidence for what they called a “classical bulge”, based on the lack of net rotation of a population of RR Lyrae stars in the inner Galaxy and the confined orbits of a sub-sample of RR Lyrae. Rix et al. (2022) interpreted the spatial distribution and orbital properties of their large sample of metal-poor stars from *Gaia*XP spectra as being a remnant of the “proto-Galaxy” – the early chaotic pre-disc phase of Milky Way formation. A similar ancient pre-disc population was suggested by Belokurov & Kravtsov (2022, naming it “Aurora”), who used APOGEE spectra at slightly higher metallicity to identify *in-situ* stars based on their chemistry, which is expected to extend down to the very low-metallicity regime.

There is also evidence for accreted stars in the inner Galaxy. Kruijssen et al. (2020) and Horta et al. (2021a) suggested that there is debris from a large accreted dwarf galaxy buried in the inner Galaxy, based on the properties of globular clusters in the Milky Way or the orbital properties and chemistry of metal-poor APOGEE stars, respectively. It is unclear how many stars there would be at *very* low metallicity from such an event. From the detailed chemistry of VMP stars, there is also evidence for stars accreted from smaller dwarf galaxies (e.g. Lucey et al. 2022; Sestito et al. 2023), and there are expectations from simulations as well (e.g. Orkney et al. 2023). There are also many signs that there is a large population of disrupted globular cluster stars in the inner Galaxy based on their chemistry (Schiavon et al. 2017; Horta et al. 2021b; Lucey et al. 2019, 2022; Sestito et al. 2023), which could have come from external galaxies and/or have been born *in-situ*.

The question remains in all of these scenarios where the prograde rotation comes from for the VMP stars. It could come from the original early Milky Way build-up phase, where there was likely some net rotation after the various large building blocks coming together (e.g. suggested in the scenario of Belokurov & Kravtsov 2022). It is also possible that an originally non-rotating population gains angular momentum later on, due to the influence of the Galactic bar (e.g. Pérez-Villegas et al. 2017; Dillamore et al. 2023).

The chemical abundances of stars hold unique clues to their birth environments, and what is needed to further study the origin of metal-poor stars in the inner Galaxy, is large samples of metal-poor stars with homogeneous chemical abundances. For this we are looking forward to the upcoming multi-object spectroscopic survey 4MOST (de Jong et al. 2019), which will have dedicated metal-poor sub-surveys targeting the inner Galaxy (Chiappini et al. 2019; Bensby et al. 2019). We will provide metal-poor targets from PIGS photometry (Arentsen et al. 2020b) for the low-resolution survey, which will result in the largest sample of VMP inner Galaxy stars with multiple chemical abundances available.

References

Arentsen A., et al., 2020a, MNRAS, 491, L11
Arentsen A., et al., 2020b, MNRAS, 496, 4964
Arentsen A., et al., 2021, MNRAS, 505, 1239
Belokurov V., Kravtsov A., 2022, MNRAS, 514, 689
Bensby T., et al., 2019, The Messenger, 175, 35
Casey A. R., Schlaufman K. C., 2015, ApJ, 809, 110
Chambers K. C., et al., 2016, arXiv e-prints, p. arXiv:1612.05560
Chiappini C., et al., 2019, The Messenger, 175, 30
Dillamore A. M., Belokurov V., Evans N. W., Davies E. Y., 2023, arXiv e-prints, p. arXiv:2303.00008
El-Badry K., et al., 2018, MNRAS, 480, 652
Gaia Collaboration et al., 2021, A&A, 649, A1
Horta D., et al., 2021a, MNRAS, 500, 1385
Horta D., et al., 2021b, MNRAS, 500, 5462
Howes L. M., et al., 2014, MNRAS, 445, 4241
Howes L. M., et al., 2015, Nature, 527, 484
Howes L. M., et al., 2016, MNRAS, 460, 884
Koch A., McWilliam A., Preston G. W., Thompson I. B., 2016, A&A, 587, A124
Koch A., Reichert M., Hansen C. J., Hampel M., Stancliffe R. J., Karakas A., Arcones A., 2019, A&A, 622, A159
Kruijssen J. M. D., et al., 2020, MNRAS, 498, 2472
Kunder A., et al., 2016, ApJ, 821, L25
Kunder A., et al., 2020, AJ, 159, 270
Lucey M., et al., 2019, MNRAS, 488, 2283
Lucey M., et al., 2021, MNRAS, 501, 5981
Lucey M., et al., 2022, MNRAS, 509, 122
Ness M., et al., 2013, MNRAS, 430, 836

Orkney M. D. A., et al., 2023, arXiv e-prints, p. arXiv:2303.02147
Pérez-Villegas A., Portail M., Gerhard O., 2017, MNRAS, 464, L80
Portail M., Wegg C., Gerhard O., Ness M., 2017, MNRAS, 470, 1233
Queiroz A. B. A., et al., 2023, arXiv e-prints, p. arXiv:2303.09926
Reggiani H., Schlaufman K. C., Casey A. R., Ji A. P., 2020, AJ, 160, 173
Rix H.-W., et al., 2022, ApJ, 941, 45
Savino A., Koch A., Prudil Z., Kunder A., Smolec R., 2020, A&A, 641, A96
Schiavon R. P., et al., 2017, MNRAS, 465, 501
Schlaufman K. C., Casey A. R., 2014, ApJ, 797, 13
Sestito F., et al., 2019, MNRAS, 484, 2166
Sestito F., et al., 2020, MNRAS,
Sestito F., et al., 2023, MNRAS, 518, 4557
Shen J., Rich R. M., Kormendy J., Howard C. D., De Propris R., Kunder A., 2010, ApJ, 720, L72
Sormani M. C., Gerhard O., Portail M., Vasiliev E., Clarke J., 2022, MNRAS, 514, L1
Starkenburg E., Oman K. A., Navarro J. F., Crain R. A., Fattahi A., Frenk C. S., Sawala T., Schaye J., 2017a, MNRAS, 465, 2212
Starkenburg E., et al., 2017b, MNRAS, 471, 2587
Tumlinson J., 2010, ApJ, 708, 1398
de Jong R. S., et al., 2019, The Messenger, 175, 3

Discussion

François Hammer: I really appreciate your finding that low metallicity stars show small rotational velocities. Why do you expect no rotation at all for these stars, since almost all elliptical galaxies show even small rotation (they are generally classified as fast and slow rotators)?

Anke Arentsen: Our expectations might not have been realistic, maybe it should not be surprising that there is some net rotation indeed. But if the halo is the result of many random accretions, there should not be a large net rotation. Also the rest of the Galactic halo (at larger distances) does not appear to show a clear net rotation.

Ana Bonaca: Could there be a connection between the rotational signature of VMP stars in the inner Galaxy and the recently discovered "discy" VMP stars?

Anke Arentsen: There could be a connection indeed. In the scenario where the halo was spun up due to the Galactic bar, this could also result the local population of very metal-poor stars having net prograde rotation, looking like discy stars (see e.g. the work from Dillamore et al. 2023). There are other scenarios for the discy VMP stars discussed in e.g. Sestito et al. (2019) that can also be connected to the inner Galaxy.

Dynamical Masses of Local Group Galaxies
Proceedings IAU Symposium No. 379, 2025
P. Bonifacio, M.-R. Cioni, F. Hammer, M. S. Pawlowski & S. Taibi, eds.
doi:10.1017/S1743921323001382

The Milky Way Tomography with Subaru Hyper Suprime-Cam. I. Halo substructures

Yoshihisa Suzuki[1], **Masashi Chiba**[1], **Yutaka Komiyama**[2], **Kohei Hayashi**[3,1], **Masayuki Tanaka**[4,5], **Scott G. Carlsten**[6], **Akira Tokiwa**[7], **Tian Qiu**[7] **and Masahiro Takada**[7]

[1]Astronomical Institute, Tohoku University, Aoba-ku, Sendai 980-8578, Japan
email: yoshihisa.suzuki@astr.tohoku.ac.jp

[2]Department of Advanced Sciences, Faculty of Science and Engineering, Hosei University, 184-8584, Tokyo, Japan

[3]National Institute of Technology, Ichinoseki College, Takanashi, Hagisho, Ichinoseki, Iwate, 021-8511, Japan

[4]National Astronomical Observatory of Japan, 2-21-1, Osawa, Mitaka, Tokyo 181-8588, Japan

[5]The Graduate University for Advanced Studies, Osawa 2-21-1, Mitaka, Tokyo 181-8588, Japan

[6]Department of Astrophysical Sciences, Princeton University, Peyton Hall, Princeton NJ 08544 USA

[7]Kavli Institute for the Physics and Mathematics of the Universe, Kashiwa, Chiba, 277-8583, Japan

Abstract. We use all the Subaru/HSC-SSP photometric data to investigate spatial substructures in the outer halo of the Milky Way. In order to effectively detect them, an isochrone-filter is created for the old, metal-poor stellar systems on the color-magnitude diagram. As a result, previously discovered stellar streams (e.g. the Orphan Stream) are detected, while a new candidate substructure is discovered toward Pisces. Based on this filter, we conclude that the progenitor is an old and metal-poor system. This kind of stellar streams are expected to be found in the next-generation observation program.

Keywords. galaxies: evolution — galaxies:formation — galaxies:halos — galaxies:structure

1. Introduction

In the context of ΛCDM model, galaxies are thought to have been formed by the repeated merging and accretion of the smaller objects (e.g., White & Rees 1978). The numerical simulations predict that the past merging history should still be imprinted in the halo (Bullock & Johnston 2005; Cooper *et al.* 2010). Interestingly, such features are observationally confirmed in the halo of nearby galaxies (Martínez-Delgado *et al.* 2010; Ferguson & Mackey 2016). These evidence are key information to unravel the formation process of galaxies.

In the case of the Milky Way (MW), thanks to the significant efforts of recent wide-field surveys such as the Sloan Digital Sky Survey (SDSS) and Gaia, an important progress has been made in the studies of the inner part of the MW halo (e.g., York *et al.* 2000; Gaia Collaboration 2016). In particular, the highly precise 6-dimensional phase space information and the chemical abundance patterns provide essential insights into the formation history of the MW (Helmi *et al.* 2017; Belokurov *et al.* 2018). On the other hand, the MW outer halo is still observationally under-explored. The main reason is that the dwarf galaxies and coherent structures found there are expected to have low surface brightness (Shipp *et al.* 2022; Horta *et al.* 2023). In order to detect them, we a need wide and deep photometric survey.

Here we utilize the data obtained by the Hyper Suprime-Cam Subaru Strategic Program (HSC-SSP). By the virtue of the wide survey area ($\sim 1200\,\mathrm{deg}^2$) and deep photometry ($\lesssim 26$ mag in the i-band), these data will allow us to find the remaining substructures at the outer part of the MW halo. We report here the discovery of a candidate stellar stream at the heliocentric distance of $D_{\odot} \sim 60$ kpc. In Section 2, we introduce the data for our analysis, then describe the methodology for the detection. The result and discussion are shown in Section 3. Section 4 is devoted to our conclusion.

2. Data & Method

2.1. *HSC-SSP*

The HSC-SSP is a large-scale photometric survey that started in March 2014 and ended in January 2022, a period of about 7.5 years (e.g., Aihara *et al.* 2018). The observation was carried out using the Hyper-Suprime Camera installed on the Subaru Telescope. Finally, over the course of 330 nights, multi-band photometric data were obtained for the objects in three regions with different depths of photometry and sizes of survey areas. Of these, we make use of the imaging data obtained from the so-called Wide layer, which covers a total of ~ 1200 deg^2 in five photometric bands (g, r, i, z, and y).

In Fig. 1, we show the Wide layer field in the equatorial coordinate system. The pink, orange, and light-blue region are named Spring, Fall, and North fields, respectively. We also add the distribution of dwarf galaxies (McConnachie 2012), globular clusters (Harris 1996), and stellar streams (Mateu 2023) that have been discovered so far, marked as with symbols of plus, stars, and continuous lines, respectively. The colors range from yellow to purple, indicating increasing distance from the Sun. From this figure, it is clear that most of the currently discovered substructures are within 30 kpc of the Sun. Therefore there is a possibility that the photometric data obtained by the HSC-SSP can detect new substructures beyond 30 kpc.

2.2. *Selection of halo stars*

The HSC-SSP catalog was generated by the hscPipe 8.4, updated with an FGCM code (hscPipe 8.5.3), based on the pipeline developed for Vera C. Rubin Observatory (Ivezic *et al.* 2008; Jurić *et al.* 2017; Bosch *et al.* 2019) and calibrated against Pan-STARRS1 photometry and astrometry (Schlafly *et al.* 2012; Tonry *et al.* 2012; Magnier *et al.* 2013; Chambers *et al.* 2016). From the catalog, we first extract point-like sources with the *extendedness* parameter flagged as 0. This classification is empirical, and is assigned a value of 0 if the difference between the photometric magnitude using the point spread function (PSF) model and cmodel (a model that combines the PSF model with a typical galaxy surface brightness profile such as an exponential profile) for data collected in the same i-band is less than approximately -0.018, and 1 if the difference is grater than approximately -0.018. We then limit the data to objects brighter than 24.5 mag in the i-band. This is because the percentage of stars or galaxies that can be correctly identified as point sources is less than 50% when the magnitude is fainter than this, as verified by cross-matching the data obtained by the Hubble Telescope in the COSMOS region as the training data (Aihara *et al.* 2018).

After major contaminants are removed in this way, further hard cuts are made on a two-color diagram. In Fig. 2, we show the distribution of identified point sources brighter than 22 mag in i-band on the two-color diagram (g-r, r-i). Based on these, we also add

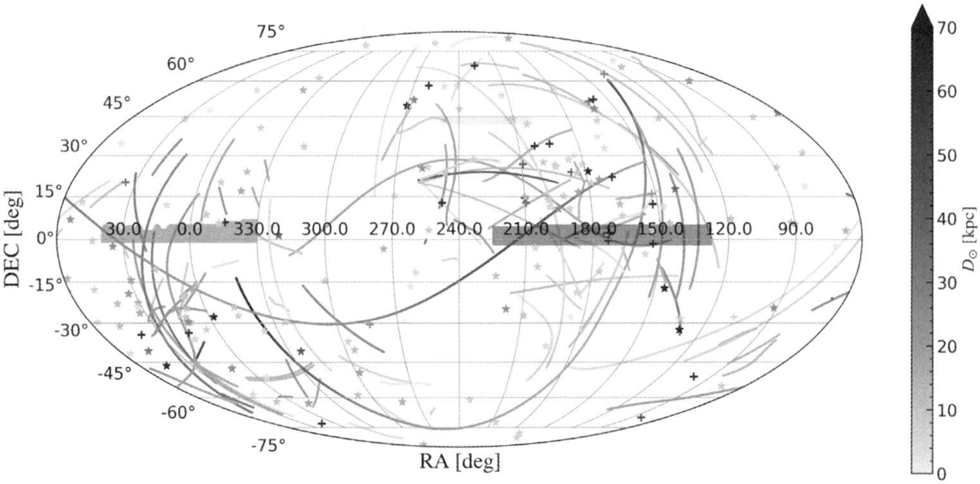

Figure 1. The Wide layer in the HSC-SSP in the equatorial coordinate. The pink, orange, and light blue regions correspond to the Wide layer field called the Spring, Fall, and North fields, respectively. The symbols represent dwarf galaxies (plus) and globular clusters (star). The continuous lines are the known stellar streams. The colors indicate the distance from the Sun, for near (yellow) and far (purple).

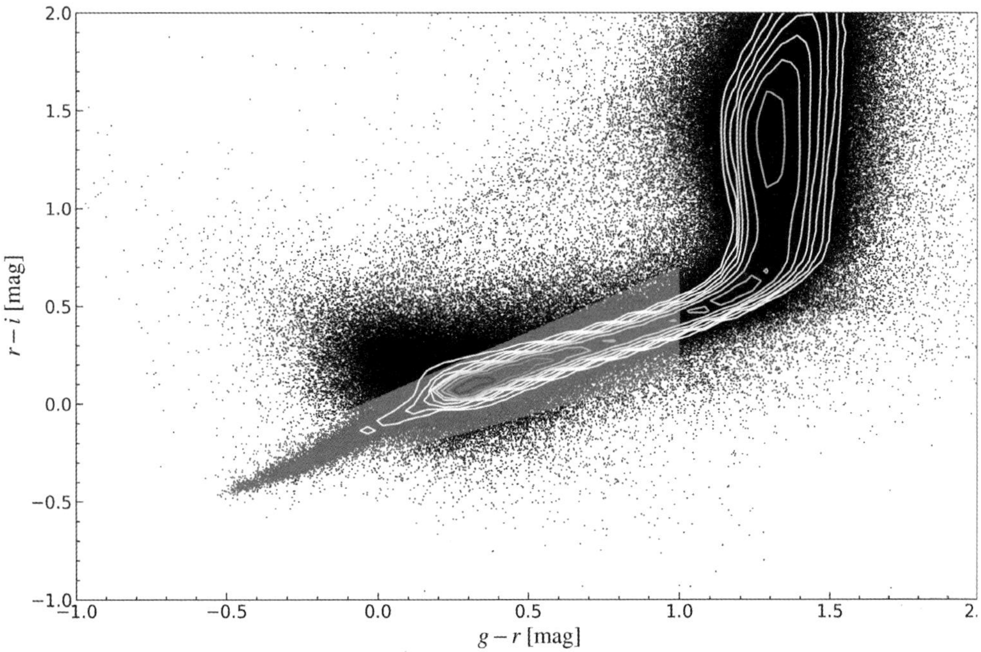

Figure 2. The distribution of point sources on the color-color diagram (g-r, r-i) for objects brighter than 22 magnitude in the i-band. The contours show the number of them displayed on a log scale from 1,000 to 100,000. In the following analysis, the point sources indicated by red dots are considered as halo stars.

contours from 1,000 to 100,000 on the log scale. In this space, we can identify a sequence of stellar concentration called the stellar locus. In particular, we can confirm the existence of a bend at around (g-r, r-i) $\sim$ (1.3, 0.5). It is known that this is caused by M and K-type stars, which are abundant in the MW thin disk. In addition, there is a component

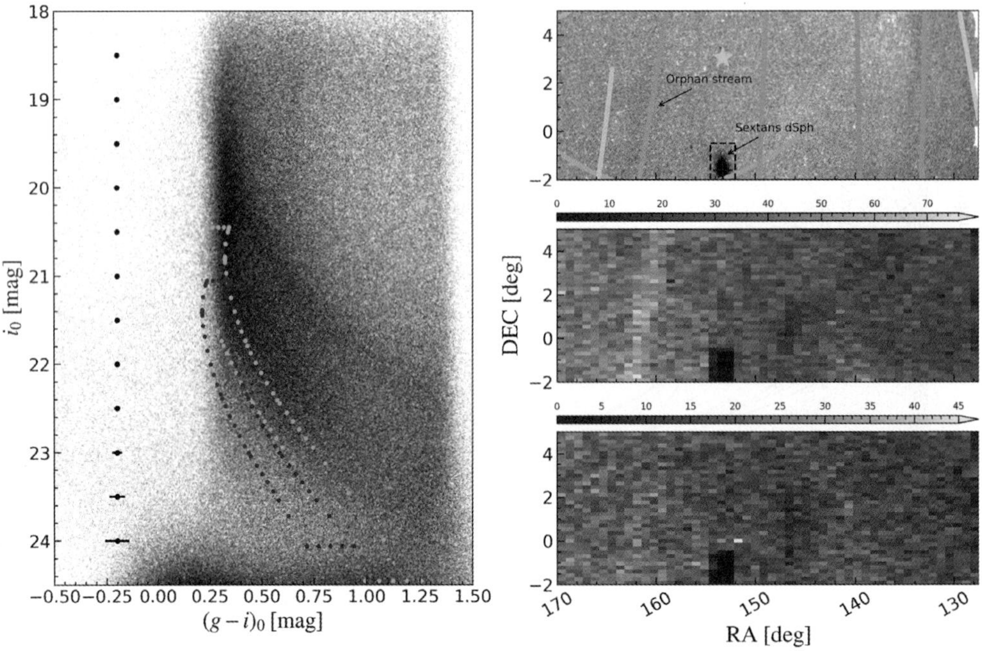

Figure 3. The effectiveness of the isochrone-filter method for the Spring region. (left) The halo stars are shown as the black dots. The typical photometric error in each i-band bin are also shown as the black dots with errorbars. (right) The top panel shows the distribution of halo stars in total as the black dots. We also add the previously discovered stellar system as the same symbols in Fig. 1. The middle and bottom panel show the number of stars within each pixel after applying the isochrone filters in the left panel shown with orange and blue dotted regions, respectively. For the star count, we remove the Sextans dSph as shown in the dashed region in the top panel.

that is not on the stellar locus at around (g-r, r-i) $\sim$ (0.0, 0.2). These are known to be quasars and star forming galaxies, as evidenced mainly by SDSS spectroscopic follow-up observations (Ahumada *et al.* 2020; Lyke *et al.* 2020). In subsequent studies, we will analyze only halo stars in the MW by removing objects with g-r greater than 1.0 as well as quasars and star forming galaxies. The resulting objects extracted as halo stars are indicated by red dots.

2.3. *Isochrone-filter method*

In order to effectively explore the substructures in the MW, we make an isochrone filter based on theoretical stellar evolution isochrones (Bressan *et al.* 2012). Assuming that most of the stellar systems in the outer halo are metal-poor and old, the theoretical curve with an age of 13.5 Gyr and [M/H] of -1.8 is adopted. Also, we confine ourselves to extract the main sequence stars to reduce the effect of foreground stars. The width of the filter is determined by the typical photometric error per bin in the i-band and intrinsic metallicity variation.

Isochrone filters made here and an example of its application to the Spring field are shown in Fig. 3. In the left panel, we show the halo stars as black points. In addition, the typical photometric errors are shown as black dots with errorbars when the photometric values in the i-band are separated by 0.5 magnitude from 18.5 to 24 and then 0.2 magnitude data is acquired around each of these magnitudes. From this figure, thanks to the small photometric errors, we can distinguish between the foreground stars and most

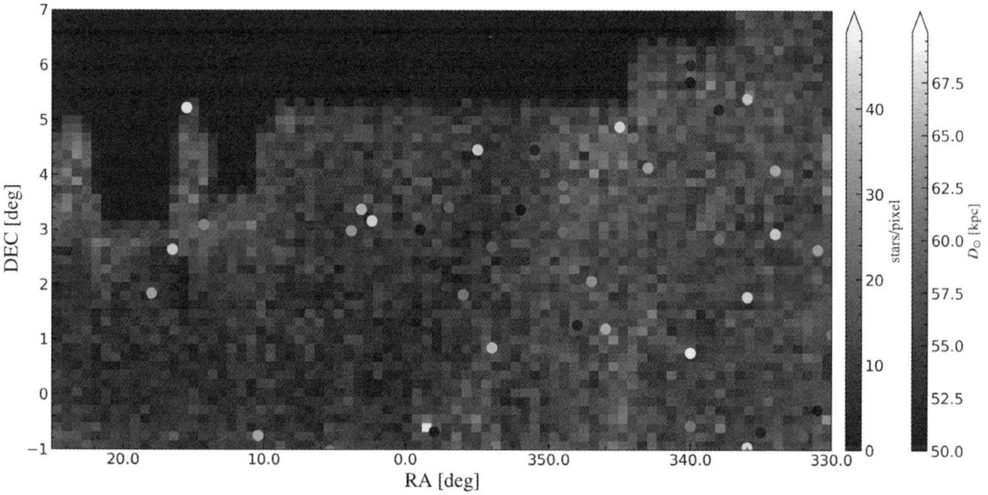

Figure 4. The candidate stellar stream and the distribution of the Blue Stragglers (BSs) in the Fall region. The dots in the figure indicate BSs, and their colors correspond to the distance from the Sun. This figure suggests that there is a correlation between the substructure discovered in this study and the distribution of BSs.

of main sequence stars in the outer halo. In the right panel, we show the distribution of halo stars in the Spring field. The top panel showed the spatial distribution of halo stars in total. The middle and the bottom panels show the number density within each pixel after applying the isochrone filters with the distance modulus (m-M) = 16.9 ($D_{\odot} \sim 24$ kpc) and 17.5 ($D_{\odot} \sim 32$ kpc), respectively, corresponding to the orange and blue dashed regions in the left panel. For the star count, a mask is made for the known dwarf galaxy, Sextans dSph. As is clear, the previously discovered stellar stream, Orphan stream is confirmed. Thus we can confidently use this isochrone-filter.

3. Result & Discussion

The result applied to to the Fall field are shown in Fig. 4. The distance modulus for the isochrone filter is 18.9 ($D_{\odot} \sim 66$ kpc). The color code is the same as used in Fig. 3. It is clearly demonstrated that we have discovered another stream-like feature.

This feature can be related to the previously discovered substructure, Pisces Overdensity (PO), which was named by Watkins *et al.* (2009) as the region where RRLyrae stars are concentrated between 60 kpc and 100 kpc from the Sun. At this time, the [Fe/H] of RRLyrae stars were thought to be about -1.5. Subsequently Nie *et al.* (2015) conducted a study using Blue Horizontal Branch (BHB) stars and showed that PO was thought to be a fairly extensive region, with a distance gradient from 60 kpc to 100 kpc. They also suggested that it may originate from small dwarf galaxies from its extent and the number of BHBs.

Based on these results, the substructure discovered in this study is almost overlapped with Pisces Overdensity. However, the stream-like feature is first discovered in this study. Moreover, this candidate substructure is extended beyond than found by Nie *et al.* (2015). Therefore, if we consider this structure to be associated with Pisces Overdensity, we can conclude that Pisces Overdensity is a part of stellar stream and is expected to be more extended structure. Moreover in Fig. 4, we overplot the candidate Blue Stragglers (BSs) identified by Fukushima et al. (in prep.) for this Fall field. BSs are thought to be abundant in old stellar systems (Piotto *et al.* 2004). Therefore, we conclude that this candidate halo substructure is considered to be a relatively old stellar system.

4. Conclusion

Based on the distribution of faint main sequence stars in the HSC-SSP footprints, we have discovered the candidate stellar stream toward Pisces. Although this candidate can be a part of the previously discovered substructure, Pisces Overdensity, this is the first time that a stream-like feature is clearly been identified. Besides this candidate is overlapped with the distribution of the BSs, thereby suggesting that the progenitor is an old and metal-poor system.

To obtain further constraints on this newly found halo structure, we are going to use the narrow-band filter $NB395$ mounted on the Subaru Telescope. This filter can potentially be useful for obtaining photometric [Fe/H] by combining the HSC-SSP g and i-band photometric data. We have already made observations along the celestial equator for the Fall field. Therefore we will be able to address the metallicity estimation in the near future.

5. Acknowledgements

This work is supported in part by JSPS Grant-in-Aid for Scientific Research (B) (No. 25287062) and MEXT Grant-in-Aid for Scientific Research (No. 16H01086, 17H01101 and 18H04334 for MC, No. 18H04359 and No. 18J00277 for KH). TM is supported by Grant-in-Aid for JSPS Fellows (No. 18J11326). The HSCcollaboration includes the astronomical communities of Japan and Taiwan, and Princeton University. The HSC instrumentation and software were developed by the National Astronomical Observatory of Japan (NAOJ), the Kavli Institute for the Physics and Mathematics of the Universe (Kavli IPMU), the University of Tokyo, the High Energy Accelerator Research Organization (KEK), the Academia Sinica Institute for Astronomy and Astrophysics in Taiwan (ASIAA), and Princeton University. Funding was contributed by the FIRST program from Japanese Cabinet Office, the MEXT, JSPS, JST, the Toray Science Foundation, NAOJ, Kavli IPMU, KEK, ASIAA, and Princeton University. This paper makes use of software developed for the Large Synoptic Survey Telescope. We thank the LSST Project for making their code freely available. The Pan-STARRS1 (PS1) Surveys have been made possible through contributions of the Institute for Astronomy, the University of Hawaii, the Pan-STARRS Project Office, the Max-Planck Society and its participating institutes, the Max Planck Institute for Astronomy and the Max Planck Institute for Extraterrestrial Physics, The Johns Hopkins University, Durham University, the University of Edinburgh, Queen's University Belfast, the Harvard-Smithsonian Center for Astrophysics, the Las Cumbres Observatory Global Telescope Network Incorporated, the National Central University of Taiwan, the Space Telescope Science Institute, the National Aeronautics and Space Administration under Grant No. NNX08AR22G issued through the Planetary Science Division of the NASA Science Mission Directorate, the National Science Foundation under Grant No.AST-1238877, the University of Maryland, and Eotvos Lorand University (ELTE).

References

Aihara, H., Arimoto, N., Armstrong, R., *et al.* 2018a, PASJ, 70, S4
Ahumada, R., Allende Prieto, C., and Almeida, A., *et al.* 2020, ApJS, 249, 3A
Belokurov, V., Erkal, D., Evans, N. W., *et al.* 2018, MNRAS, 478, 611
Bosch, J., AlSayyad, Y., Armstrong, R., *et al.* 2019, ASPC, 523, 521
Bressan, A., Marigo, P., Girardi, L., *et al.* 2012, MNRAS, 427, 127
Bullock, J. S., & Johnston, K. V., 2005, ApJ, 635, 931
Chambers K. C., Magnier, E. A., Metcalfe, N., *et al.* 2016, arXiv:1612.05560
Cooper, A. P., Cole, S., Frenk, C. S., *et al.* 2010, MNRAS, 406. 744
Ferguson, A. M. N., & Mackey, A. D., 2016, ASSL, 420, 191

Gaia Collaboration, Prusti, T., de Bruijne, J. H. J., *et al.* 2016, A&A, 595, A1
Harris, W. E. 1996, AJ, 112, 1487
Helmi, A., Veljanoski, J., Breddels, M. A., *et al.* 2017, AAP, 598, A58
Horta, D., Cunningham, E. C., Sanderson, R. E., *et al.* 2023, ApJ, 943, 158
Ivezic, Z., Axelrod, T., Brandt, W. N., *et al.* 2008, AJ, 176, 1
Jurić, M., Kantor, J., Lim, K.-T., *et al.* 2017, ASPC, 512, 279
Lyke, B. W., Higley, A. N., McLane, J. N., *et al.* 2020, ApJS, 250, 8L
Magnier, E. A., Schlafly, E., Finkbeiner, D., *et al.* 2013, ApJS, 205, 20
Mateu, C. 2023, MNRAS, 520, 5225
Martínez-Delgado, D., Gabany, R. J., Crawford, K., *et al.* 2010, AJ, 140, 962
McConnachie, A. W. 2012, AJ, 144, 4
Nie, J. D., Smith, M. C., Belokurov, V., *et al.* 2015, ApJ, 810, 153
Piotto, G., De Angeli, F., King, I. R., *et al.* 2004, ApJL, 604, L109
Schlafly, E. F., Finkbeiner, D. P., Jurić, M., *et al.* 2012, ApJ, 756, 158
Shipp, N., Panithanpaisal, N., Necib, L., *et al.* 2022, arXiv:2208.02255
Tonry, J. L., Stubbs, C. W., Lykke, K. R., *et al.* 2012, ApJ, 750, 99
Watkins, L. L., Evans, N. W., Belokurov, V., *et al.* 2009, MNRAS, 398, 1757
White, S. D. M., & Rees, M. J., 1978, MNRAS, 183, 341
York, D. G., Adelman, J., Anderson, J. E. Jr., *et al.* 2000, AJ, 120, 1579

Dynamical Masses of Local Group Galaxies
Proceedings IAU Symposium No. 379, 2025
P. Bonifacio, M.-R. Cioni, F. Hammer, M. S. Pawlowski & S. Taibi, eds.
doi:10.1017/S1743921323001278

The study of the Galactic mass distribution from Gaia DR3 RR Lyrae

Cuihua Du and Duo Li

College of Astronomy and Space Sciences, University of Chinese Academy of Sciences,
Beijing 100049, P.R. China
email: ducuihua@ucas.ac.cn

Abstract. Based on RR Lyrae with accurate proper motions and classification in *Gaia* DR3, we determine the Milky Way mass distribution from fitting dynamical models to the gravitational force field and the Galactic rotation curve. Applying Gaussian Mixture Model to the intrinsic velocity distribution, we present the result of a multi-component kinematic model of RR Lyrae in the inner regions $5 \lesssim r \lesssim 20$ kpc. Considering the early accretion history of the MW and thus the stellar halo may not be in equilibrium, we separate the halo population into an isotropic stellar halo and the radially-anisotropic population relevant to a merge event. With a Bayesian approach, we fit the potential model parameters, including the density flattening of the dark matter (DM) halo. Our best-fitting dynamical model suggests a nearly spherical spheroid shape of $q = 1.02 \pm 0.13$, a DM halo mass of $M_{200}^{\rm DM} = 0.649 \pm 0.021 \times 10^{12} {\rm M}_{\odot}$, total MW mass of $M_{200}^{\rm tot} = 0.704^{+0.017}_{-0.021} \times 10^{12} {\rm M}_{\odot}$.

Keywords. Galaxy: fundamental parameters – Galaxy: halo – Galaxy: kinematics and dynamics – Galaxy: mass

1. Introduction

According to the current ΛCDM model of cosmology, galaxies are encased in larger dark matter (DM) halos, and DM halos grow in mass and scale through both smooth accretion of diffuse matter and from mergers with other halos (Wang *et al.* 2020). The Galactic DM mass density profile and total mass are of the most importance in modern astrophysics, which could provide us some important clues about the dynamics and evolution of the Galaxy. The virial mass is critical for comparing observed properties of the Milky Way (MW) with cosmological predictions, for example, the so-called missing satellite and "too big to fail" problems (Boyan-kolchin *et al.* 2011; Bullock & Boylan-Kolchin 2017) are all required a 'lighter' MW in simulations to alleviate these problems. The total mass and the potential of our MW are also crucial for reconstructing the orbital evolution of objects and understanding dwarf galaxies dynamics and their accretion history. Having more massive MW leads to a rapid merger of our Galaxy with its brightest satellite, the Large Magellanic Cloud (LMC), and with its nearest giant neighbor, M31.

Nowadays, we are in an era of large astronomical data survey. In particular, *Gaia* data, combined with other observations, offer an exquisite data set with precise proper motion and velocity information. The mean proper motions of satellites and globular clusters and some other objects have been refined and expanded (Li *et al.* 2021; Wang *et al.* 2022). A variety of methods have been employed for determining the MW dark matter halo mass distribution by using the dynamical tracers, such as bright stars, satellite galaxies, globular clusters, HI gas clouds and tidal streams. Wang *et al.* (2020) provided a literature summary of measured virial masses using different methods in their

Figure 1. Given a reasonable model which describes their dynamics or phase-space distributions within a realistic potential profile with free parameters, one can constrain the total mass distribution of our MW.

Nevertheless, the allowed virial masses ware estimated to lie in the range between $0.5 \times 10^{12} M_\odot$ and $2 \times 10^{12} M_\odot$, and the measurements highlights that the virial mass is still uncertain to within at least a factor of two (Callingham *et al.* 2019). In previous studies, only radial velocities have been available for the tracers, which becomes increasingly difficult to perform from radial velocities alone at large distance. However, important transverse motions of the tracers could provide necessary information statistically. For example, Wang *et al.* (2022) used action-based distribution function for the dynamical model of the MW to estimate its total mass and density profile based on proper motion of Galactic globular clusters from *Gaia* EDR3 combined with disk rotation curve from *Gaia* DR2. Bird *et al.* (2022) measured the enclosed MW mass profile to Galactocentric distances of ~ 70 kpc and ~ 50 kpc using K giants and blue horizontal branch stars from LAMOST and SDSS/SEGUE and *Gaia* DR2 with accurate kinematics. A detailed review on the methods of MW mass measurement and recent study results that made use of the *Gaia* DR2 can also be found in Wang *et al.* (2020). It is expected in the following years that the *Gaia* will continue to improve precision and accuracy of the astrometry and photometry for more and more objects. With the availability of high quality proper motions information for large numbers of sources, much effort has been made to decrease the uncertainties in the MW mass estimate.

Recent studies with large spectroscopic surveys combined with *Gaia* data reveal the halo is unrelaxed or unequilibrium induced by the passage of the LMC, Sagittarius dwarf spheroidal galaxy (Sgr) (Vasiliev *et al.* 2021; Deason *et al.* 2021), and in particular, the 'Gaia-Sausage-Enceladus' (GSE) merge (Belokurov *et al.* 2018; Helmi *et al.* 2018; Naidu *et al.* 2021). These effects could affect the assumption of equilibrium and relaxed system in any dynamical modeling, which should be considered in constraining the total mass.

RR Lyrae stars (RRLs) are old low-mass, periodic pulsating variable stars, and their distances can be determined with only knowledge of the light curve. Therefore, the investigation of RR Lyrae properties is motivated by their important role both as distance indicators and tracers of old stellar populations, which are also ideal tracers that can be used to study the total mass and dark matter content of the outer halo of the MW (Prudil *et al.* 2022). The purpose of this study is to infer the mass distribution of the MW from recent *Gaia* DR3 RR Lyrae, in particular, the virial mass and shape of the DM halo.

2. Data

In this study, we adopt the rotation curve (RC) of the disk estimated by Eilers *et al.* (2019) for Galactocentric distance of $5 \leq R \leq 25$ with high precision for ~ 23000 red giants tracers. The precise distances are determined from spectrophotometric parallax combining the spectroscopic data from APOGEE DR14 and photometry from *Gaia* DR2, 2MASS, and WISE. Most measurements of circular velocity are extremely precise, with errors below 2 km/s and systematic uncertainties at the $\sim 2\% - 3\%$ level.

In order to determine the overall mass profile of the DM halo, we also use RR Lyrae (RRL) stars from *Gaia* DR3, Pan-STARRS1(PS1) and ASAS-SN as halo tracers to analyze the kinematics of DM halo. The total RRL catalogue contains 281,359 stars. The following is an overview of the criteria:

- SOS: `vari_rrlyrae` catalogue with comprehensive characteristic parameters derived from the Specific Objects Study pipeline on RRLs (Clementini *et al.* 2019, 2022), including 270 905 RRLs (175 350 RRab, 94 422 RRc)

- PS1 (56 427 stars): $\mathrm{score}_{3,ab} > 0.8$ (RRab, 44 208) and $\mathrm{score}_{3,c} > 0.75$ (RRc, 12 219)
- ASAS-SN (40 653 stars): period< 0.95d and amplitude< 2 mag (30 801 RRab, 9 852 RRc)
- SOSnotPS1: 6 016 RRLs in PS1 catalogues but not in SOS
- SOS+PS1: 275 788 RRLs with an overlap of 50 023 RRLs (the overlap is ignored for PS1 results, while adopted for SOS results)
- SOS+PS1+ASAS-SN: 281 359 RRLs with 5 571 RRLs from ASAS-SN not in SOS+PS1

To prevent possible misclassfication of RRL, we cross-match the total catalogue with the Catalina sky survey (CSS, Drake *et al.* 2017). Those stars which are not classified as RRab, RRc, RRd in CSS are excluded from the RRL sample. In order to exclude those known overdensities such as globular clusters and dwarf galaxies, we also adopt the criteria in Iorio *et al.* (2021) to obtain about 125 178 sample stars.

The application of metal dependent Period-Wesenheit (PWZ) relations for RRab and RRc in *Gaia* G-band allows us to evaluate individual distance with high precision. We adopt the linear relation proposed by Iorio *et al.* (2021), in which the metallicity relies directly on the pulsation period and Φ_{31} parameter. The adopted Wessenheit quantity centered on its photometry is defined as $\mathrm{w} = \mathrm{G} - 1.9(\mathrm{G_{BP}} - \mathrm{G_{RP}})$ in *Gaia* G-band. The application of Wesenheit quantity allows us to evaluate distance with a relative uncertainty within 5%, to which distance modulus is employed for each RRL star. The parallax method can provide more precise distance estimate for nearby stars. Adopting the Bayesian posterior analysis, we make some improvement combining the PWZ relations with parallax method to derive final distance for the same star.

Each star has a set of observable quantities such as its parallax and Wessenheit quantity in distance measurement. With the Bayesian theorem, the unnormalized posterior over the distance for a RRL star is (Muraveva *et al.* 2018),

$$P(r|\varpi, w) = P(\varpi|r, \sigma_{\varpi})P(W = w - 5\log r + 5)P(r)\,, \tag{1}$$

where $P(\varpi|r, \sigma_{\varpi})$ is the parallax likelihood as a function of distance r. The second term on the right side is the likelihood based on the PWZ relations to constrain the distance of RRL star. $P(r)$ is the prior over distance, in which $1/r^2$ is adopted to avoid bias in the distance distribution. To verify the reliability of Bayesian distance, we also compare it with distance measured by Sesar *et al.* (2017), and our results come to a good agreement with theirs. Given the heliocentric distance and the proper motion, the transverse velocity can be derived. Here, we use a right-handed Galactocentric frame and the distance from the Sun to the Galactic center $\mathrm{R}_0 = 8.122$ kpc, and the solar offset from the Galactic midplan $\mathrm{z}_{\odot} = 25$ pc.

For the majority of RRL sample stars, due to the absence of line-of-sight velocities or bias induced by periodic radial pulsation, we only use position and proper motion to restrict the velocity ellipsoid for stars in bins of distance $(R, |z|)$. According to the kinematic characteristic of stars, the sample is divided into three components: disk, halo and Gaia-Sausage. The disk component has isotropic velocity dispersion tensor $(\sigma_r \simeq \sigma_\theta \simeq \sigma_\phi)$, and high azimuthal velocity. The halo population is nearly isotropic velocities, while the GSE appears high radially anisotropic. In order to determine the kinematic properties of the DM halo, we need to isolate the isotropic halo population. The Mixture Model is used to statistically identify different populations. Of the total 72,973 RRL in our sample, 49,914 (or $\approx 68\%$) are classified as halo, 3,126 (or $\approx 4\%$) as disk; while the remaining 19,993 ($\approx 28\%$) are unclassified. and the unclassified component consist of overlapped stars between the disk and halo.

3. Galactic potential model and method

The goal of this work is to infer the mass profile of the MW, and in particular for the DM halo. Here, we use axisymmetric Galactic potential, which consist of baryon mass components and a NFW dark matter halo. We first specify the baryon mass distribution that includes five components: bulge; a thin and a thick stellar disc; an HI disc and a molecular gas disc; and a diffuse gaseous halo. The former five baryonic components follow McMillan (2017). The Galactic gaseous halo (i.e. the circumgalactic medium, hereafter CGM) contains the majority of baryonic mass at large distances from the Galactic Center, and the CGM radial density profile can be expressed as a power law of distance, i.e. $\rho_{\rm CGM} \sim r^{\beta_{\rm CGM}}$, and the power-law exponent as well as the overall density normalization are estimated. Given the fact that baryons are not the dominant component, CGM should not alter our main conclusions.

To obtain the best fitting model, we follow the Bayesian approach in which the probability of a set of model parameters values, $\theta = (M_{200}^{\rm DM}, c^{NFW}, \rho_{0,\rm bulge}, \Sigma_{0,\rm thin}, \Sigma_{0,\rm thick}, R_{\rm thin}, R_{\rm thick})$, given the data, D, is

$$p\,(\theta|D) = \frac{p\,(D|\theta)\;p\,(\theta)}{p\,(D)}, \tag{2}$$

where $p\,(D|\theta)$ is the probability of the data given the model parameters, $p\,(\theta)$ is the prior probability of parameter values, and $p\,(D)$ is a normalisation factor. We take three Gaussian priors for $(\rho_{0,\rm bulge}, R_{\rm thin}, R_{\rm thick})$, and no prior distribution for the remaining parameters. The likelihood, $p\,(D|\theta)$, is taken as the product of likelihoods associated with each circular velocity measurement. The circular velocity is calculated as $V_{\rm circ}^2 = R\ {\rm d}\Phi_{\rm tot}/{\rm dR}$, where $\Phi_{\rm tot}$ is the total gravitational potential and R is the radial distance. Then, we employ a Markov Chain Monte Carlo (MCMC) approach using the EMCEE python module (Foreman-Mackey *et al.* 2013) to fit the NFW DM halo profile.

We find that the preferred MW stellar mass is $M_{\star\ \rm total} = 5.04^{+0.43}_{-0.52} \times 10^{10} M_\odot$, mostly contributing from the thin disk and the remainder split equally between the thick disk and the bulge. The constraints on the bulge mass are mostly given by the prior since our observation data ($R > 5$ kpc) is insensitive to the mass or geometry of the bulge. Most baryonic mass within the halo contribute from the gaseous component: $1.2 \times 10^{10} M_\odot$ as HI and molecular gas, and $6.4 \times 10^{10} M_\odot$ as the CGM. We also derive the enclosed mass of the DM at various radius r. For example, our result indicates that the DM mass within the virial radius r_{200} is $M_{200} = 0.649 \pm 0.021 \times 10^{12}\ M_\odot$, which is consistent with some other research results including McMillan (2017), but slightly lower than $(9.0 \pm 1.3) \times 10^{11}\ M_\odot$ recently found by Vasiliev *et al.* (2021). The total mass of MW is $M_{200}^{\rm tot} = 0.704^{+0.017}_{-0.021} \times 10^{12} M_\odot$. Wang *et al.* (2022) listed the total MW mass measured with recent results by using *Gaia* DR2 and EDR3 in Figure 14 of their work, in which the results are grouped on the basis of the different methods used to estimate the MW total mass. Our result is at the low end of MW mass in those studies, but is in consistent with the estimate range from $5.36 \times 10^{11}\ M_\odot$ to $7.84 \times 10^{11}\ M_\odot$ in Wang *et al.* (2022).

Our inferred stellar mass is slightly smaller than the estimation in Bovy & Rix (2013), but is consistent with their result within 68% confidence regions. However, we derive a much more massive DM halo than that of Bovy & Rix (2013), and our Galactic DM halo exceeds the stellar component contribution at $R{\sim}8$ kpc, in good agreement with the theoretical predictions in Lovell *et al.* (2018). To understand various degeneracies between the model parameters, we show the posterior distribution for each pair of parameters in the DM distribution in Figure 1. In the off-diagonal panels, the red shaded regions illustrate 68 and 95% confidence regions, and the red lines in the diagonal panels show the marginalised probability of each model parameter. As shown in Figure 1, most parameters are weakly correlated but there are a few degeneracy, especially between DM halo mass

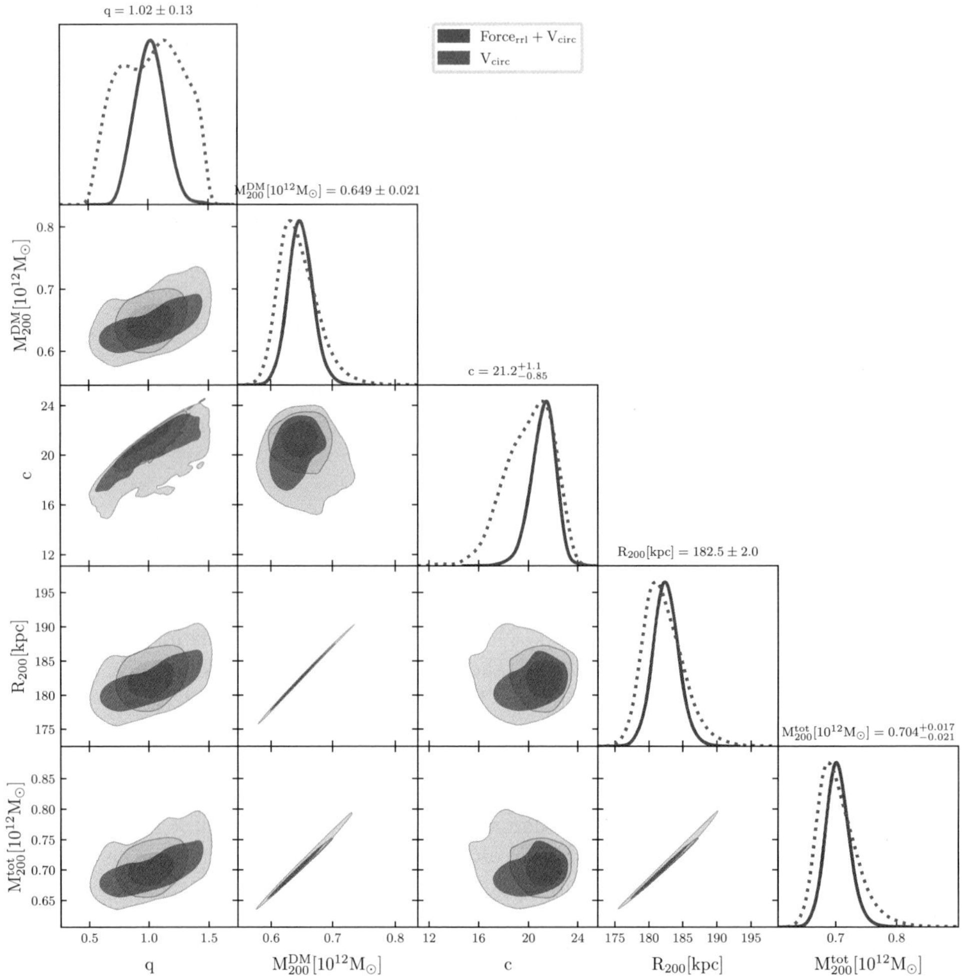

Figure 1. Posterior distributions of the parameters in the MW mass model. The red shaded regions correspond to the NFW halo model constraint by RRLs and rotation curve while the green shaded regions correspond to a NFW halo only fitted by circular velocity. The darker and lighter contours illustrate 68 and 95% confidence regions, respectively.

and concentration. Since most of the model constraints come from the inner regions, i.e. $r \lesssim 20$ kpc, the same enclosed mass can be obtained by, for example, decreasing the halo concentration in tandem with increasing the halo mass. We also notice a positive correlation between halo mass and the stellar disk stellar masses: a more massive halo prefer a more massive stellar disk. This is because a more massive halo provides a similarly good fitting if it has a lower concentration, and thus it has less mass in the inner region, which, in turn, can be compensated by a larger disk mass.

In this study, we also find that the posterior distribution is strongly peaked at $q = 1.02 \pm 0.13$, and $q > 0.963$ with 99% confidence level, which is consistent with a nearly spherical DM halo. Recently, Wegg *et al.* (2019) estimated the shape of the DM density profile by applying the axisymmetric Jeans equations for 15, 651 RR Lyrae stars at $r < 20$ kpc. They also found that the shape of the DM halo is nearly spherical, with the density flattening of $q = 1.00 \pm 0.09$. In previous studies with streams, Koposov *et al.* (2010) modelled the GD-1 stream and derived that a density flattening of $q > 0.68$ with

90% confidence level. Bovy *et al.* (2016) measured the DM density flattening to be $q = 1.3^{+0.5}_{-0.3}$ when using GD-1 stream (at $r = 14$) and $q = 0.93 \pm 0.16$ when using Pal 5 stream (at $r = 19$). By combining these two data sets, they estimated the global value $q = 1.05 \pm 0.14$. More recently, Malhan & Ibata (2019) analysed the GD-1 stream from *Gaia* DR2 to estimate $q = 0.82^{+0.25}_{-0.13}$. Although the statistical uncertainties are relatively large, all of the above-mentioned results using the GD-1 stream are consistent with a nearly spherical DM halo. In our analysis, DM's inner density slope γ is a free parameter, while the outer density slope is fixed to be $\gamma = 3$. The central 68% of the posterior distribution of γ is distributed at $1.38 < \gamma < 1.51$, which is consistent with a cusped NFW profile with $\gamma = 1.5$ (Nitschai *et al.* 2021).

4. Conclusion and summary

Based on the Bayesian posterior analysis, We combine the Period-Wesenheit (PWZ) and parallax method for about 72, 973 RR Lyrae to evaluate individual distance. With a Bayesian method, we could fit the potential model parameters, including the density flattening of the dark matter (DM) halo. Our best-fitting dynamical model suggests a nearly spherical spheroid shape of $q = 1.02 \pm 0.13$ for the NFW DM halo, a DM halo mass of $M_{200}^{\rm DM} = 0.649 \pm 0.021 \times 10^{12} {\rm M}_{\odot}$, total MW mass of $M_{200}^{\rm tot} = 0.704^{+0.017}_{-0.021} \times 10^{12} {\rm M}_{\odot}$.

We thank the editors for the insightful suggestions, which have improved the manuscript. This work was supported by the National Natural Sciences Foundation of China (NSFC No. 11973042, No. 11973052 and No. 11873053). It was also supported by the Fundamental Research Funds for the Central Universities and the National Key R&D Program of China No. 2019YFA0405501. This work has made use of data from the European Space Agency (ESA) mission *Gaia* (`https://www.cosmos.esa.int/gaia`), processed by the *Gaia* Data Processing and Analysis Consortium (DPAC, `https://www.cosmos.esa.int/web/gaia/dpac/consortium`). Funding for the DPAC has been provided by national institutions, in particular the institutions participating in the *Gaia* Multilateral Agreement.

References

Bird, S. A., Xue, X.-X., Liu, C., *et al.*, 2022, MNRAS, 516, 713
Belokurov, V., Erkal, D., Evans, N.W., Koposov, S. E., Deason, A. J., 2018, MNRAS, 478, 611
Boyan-Kolchin, M., Bullock, J. S., Kaplinghat, M., 2011, MNRAS, 415L, 40
Bovy, J. & Rix, H.-W. 2013, ApJ, 779, 115
Bovy, J., Bahmanyar, A., Fritz, T. K., Kallivayalil, N., 2016, ApJ, 833, 31
Bullock, J. S., Boyan-Kolchin, M., *ARA&A*, 2017, 55, 343
Callingham, T. M., Cautun, M., Deason, A. J., *et al.*, 2019, MNRAS, 484, 5453
Clementini, G., Ripepi, V., Molinaro, R., *et al.*, 2019, *A&A*, 622, 60
Clementini, G., Ripepi, V., Garofalo, A., *et al.*, 2022, arXiv e-prints, arXiv:2206.06278
Drake, A. J., Djorgovski, S. G., Catelan, M., *et al.*, 2017, MNRAS, 469, 3688
Deason, A. J., Erkal, D., Belokurov, V., *et al.*, 2021, MNRAS, 501, 5964
Eilers, A.C., Hogg, D.W., Rix, H.W., Ness, M. K., 2019, ApJ, 871, 120
Foreman-Mackey, D., Hogg, D.W., Lang, D. and Goodman, J., 2013, PASP, 125, 306
Helmi, A., Babusiaux, C., Koppelman, H. H., *et al.*, 2018, Nature, 563, 85
Iorio, G., Belokurov, V., 2021, MNRAS, 502, 5686
Koposov, S. E., Rix, H., Hogg, D.W., 2010, ApJ, 712, 260.
Lovell, M. R., Pillepich, A., Genel, S., *et al.*, 2018, MNRAS, 481, 1950
Li, H., Hammer, F., Babusiaux, C., Pawlowski, M. S., *et al.*, 2021, ApJ, 916, 8
Malhan, K., Ibata, R. A., 2019, MNRAS, 486, 2995
McMillan, P. J., 2017, MNRAS, 465, 76
Muraveva, T., Delgado, H.E., Clementini, G., 2018, MNRAS, 481, 1195
Naidu, R. P., Conroy, C., Bonaca, A., *et al.*, 2021, ApJ, 923, 92

Nitschai, M.S., Eiler, A.C., Neumayer, N., *et al.*, 2021, ApJ, 916, 112
Prudil, Z., Koch-Hansen, A.J., Lemasle, B., *et al.*, 2022, A&A, 664A, 148
Sesar, B., Fouesneau, M., Price-Whelan, A. M., *et al.*, 2017, ApJ, 838, 107
Vasiliev, E., Belokurov, V., Erkal, D., 2021, MNRAS, 501, 2279
Wang,W.T., Han, J.X., Cantun, M., *et al.*, 2020, SCPMA, 6309801
Wang, J., Hammer, F., Yang, Y., 2022, MNRAS, 510, 2242
Wegg, C., Gerhard, O., Bieth, M., 2019, MNRAS, 485, 3296

Dynamical Masses of Local Group Galaxies
Proceedings IAU Symposium No. 379, 2025
P. Bonifacio, M.-R. Cioni, F. Hammer, M. S. Pawlowski & S. Taibi, eds.
doi:10.1017/S1743921323003599

Milky Way Mass with K Giants and BHB Stars Using LAMOST, SDSS/SEGUE, and *Gaia*: 3D Spherical Jeans Equation and Tracer Mass Estimator

Sarah A. Bird[1,2,3,4], Xiang-Xiang Xue[3,10], Chao Liu[5,10,11], Chris Flynn[6], Juntai Shen[7,8], Jie Wang[9,10,11], Chengqun Yang[4], Meng Zhai[3,12], Ling Zhu[4], Gang Zhao[3,11] and Hai-Jun Tian[13]

[1]College of Science, China Three Gorges University, Yichang 443002, People's Republic of China
email: sarahbird@ctgu.edu.cn

[2]Center for Astronomy and Space Sciences, China Three Gorges University, Yichang 443002, People's Republic of China

[3]CAS Key Laboratory of Optical Astronomy, National Astronomical Observatories, Chinese Academy of Sciences, Beijing 100101, People's Republic of China

[4]Shanghai Astronomical Observatory, Chinese Academy of Sciences, 80 Nandan Road Shanghai 200030, People's Republic of China

[5]Key Laboratory of Space Astronomy and Technology, National Astronomical Observatories, Chinese Academy of Sciences, Beijing 100101, People's Republic of China

[6]Centre for Astrophysics and Supercomputing, Swinburne University of Technology, Post Office Box 218, Hawthorn, VIC 3122, Australia

[7]Department of Astronomy, School of Physics and Astronomy, Shanghai Jiao Tong University, 800 Dongchuan Road, Shanghai 200240, People's Republic of China

[8]Key Laboratory for Particle Astrophysics and Cosmology (MOE) / Shanghai Key Laboratory for Particle Physics and Cosmology, Shanghai 200240, People's Republic of China

[9]Key Laboratory for Computational Astrophysics, National Astronomical Observatories, Chinese Academy of Sciences, 20A Datun Road, Beijing 100101, People's Republic of China

[10]Institute for Frontiers in Astronomy and Astrophysics, Beijing Normal University, Beijing, 102206, People's Republic of China

[11]School of Astronomy and Space Science, University of Chinese Academy of Sciences, Beijing 100049, People's Republic of China

[12]Chinese Academy of Sciences South America Center for Astronomy, National Astronomical Observatories, Chinese Academy of Sciences, Beijing 100012, People's Republic of China

[13]School of Science, Hangzhou Dianzi University, Hangzhou, 310018, People's Republic of China

Abstract. We measure the enclosed Milky Way (MW) mass profile to Galactocentric distances of ~ 70 and ~ 50 kpc using the smooth, diffuse stellar halo samples of Bird et al. The samples are Large Sky Area Multi-Object Fiber Spectroscopic Telescope (LAMOST) and Sloan Digital Sky Survey/Sloan Extension for Galactic Understanding and Exploration (SDSS/SEGUE) K giants (KG) and SDSS/SEGUE blue horizontal branch (BHB) stars with accurate metallicities. The 3D kinematics are available through LAMOST and SDSS/SEGUE distances and radial velocities and *Gaia* DR2 proper motions. Two methods are used to estimate the enclosed mass: 3D spherical Jeans equation and Evans et al. tracer mass estimator (TME). We remove substructure via the Xue et al. method based on integrals of motion. We evaluate the uncertainties on our estimates due to random sampling noise, systematic distance errors, the adopted density profile, and non-virialization and non-spherical effects of the halo. The tracer density profile remains a limiting systematic in our mass estimates, although within these limits we find reasonable agreement across the different samples and the methods applied. Out to ~ 70 and ~ 50 kpc, the Jeans method yields total enclosed masses of 4.3 ± 0.95 (random) ± 0.6 (systematic) $\times 10^{11}$ $M_\odot$ and 4.1 ± 1.2 (random) ± 0.6 (systematic) $\times 10^{11}$ $M_\odot$ for the KG and BHB stars, respectively. For the KG and BHB samples we find a dark matter virial mass of $M_{200} = 0.55^{+0.15}_{-0.11}$ (random) ± 0.083 (systematic) $\times 10^{12}$ $M_\odot$ and $M_{200} = 1.00^{+0.67}_{-0.33}$ (random) ± 0.15 (systematic) $\times 10^{12}$ $M_\odot$, respectively.

Keywords. stars: individual: BHB - stars: individual: K giants - galaxies: individual: Milky Way - stars: kinematics and dynamics - Galaxy: halo - Galaxy: kinematics and dynamics - Galaxy: stellar content.

---

1. Introduction

The mass of our Milky Way (MW) is an essential parameter not only for understanding the formation and evolution and for modelling the MW but also for comparing with other galaxies and with simulations.

Although halo stars make up only one percent of the luminous mass of our MW, these stars are interesting because they are old and have large random orbits as they are a dynamically pressure supported system; thus the halo stars trace the mass distribution of the MW.

Using *Gaia* combined with large spectroscopic surveys (Large Sky Area Multi-Object Fiber Spectroscopic Telescope,LAMOST, and Sloan Digital Sky Survey/Sloan Extension for Galactic Understanding and Exploration, SDSS/SEGUE), we collect a large sample of > 13000 smooth, diffuse halo stars with 3D positions and velocities and for the first time measure the MW mass profile with such a large 3D sample, rather than just line-of-sight velocities alone, using the 3D spherical Jeans equation and the Evans et al. (2011) tracer mass estimator (TME).

2. Methods

In Bird et al. (2022) we use two methods, the 3D spherical Jeans equation and the Evans et al. (2011) TME, which assume the simplest case dynamics—a virialized system traced by a spherical non-rotating, stationary system of stars—to measure the Galactic mass profile $M(<r)$, where M is the mass enclosed within the Galactocentric radius r.

The 3D spherical Jeans equation is written as

$$M(<r) = -\frac{1}{G}(r^2 \frac{d\sigma_r^2}{dr} + r((2+\alpha)\sigma_r^2 - \sigma_\theta^2 - \sigma_\phi^2)), \tag{1}$$

where, G is the gravitational constant, the velocity dispersions in the radial and two transverse directions are given by $(\sigma_r, \sigma_\theta, \sigma_\phi)$, and α is the slope of the fit to the density fall-off of the tracers which is found to be well approximated as a power-law $\rho \propto r^\alpha$.

The second method we use is the scale-free TME of Evans et al. (2011, equation 9)†,

$$M(<r) \approx \frac{r^{0.5}(\gamma - \alpha - 2\beta)}{GN} \sum_{i=1}^{N} r_i^{0.5}(V_{r,i} - \langle V_r \rangle)^2. \tag{2}$$

† For a full derivation of the estimator, see An & Evans (2011).

Here, we have N tracer stars where the ith tracer has Galactocentric distance r_i and radial velocity $V_{r,i}$ to yield the mass $M(<r)$ of the Galaxy within the radius r of the most distant star in the sample. To build the $M(<r)$ mass profile, we apply TME to our KG and BHB stars in bins of Galactocentric radius, using the same binning scheme as we use for our application of the 3D spherical Jeans method. In this manner, γ, α, and β are functions of r. The Galactic potential is represented as a power-law with slope γ as a function of r. The tracer stars are also assumed to have a density fall-off as a power-law with slope α (as adopted for the Jeans equation). The velocity anisotropy profile along Galactocentric radius of the tracer stars is denoted as β, where

$$\beta = 1 - (\sigma_\theta^2 + \sigma_\phi^2)/(2\sigma_r^2). \tag{3}$$

To measure β, we calculate the 3D velocity dispersion profiles as functions of r, as in our application of the Jeans equation, by following Bird et al. (2021) using the `extreme-deconvolution` software (Bovy et al. 2011).

Petersen & Peñarrubia (2021) and Erkal et al. (2021) have used distant halo stars to measure the present-day reflex (or sloshing) motion of the stellar halo which is induced by the LMC. Following the recommended velocity correction of Erkal et al. (2020), we take the mean velocity into account for both the velocity dispersions (using `extreme-deconvolution`) of the 3D spherical Jeans and the velocity component in the Evans et al. (2011) TME used in this work. Applying the mean velocity corrections, we find that the changes in the mass profile are very small (of order a few percent) compared to other sources of error.

The Jeans and TME methods yield the enclosed MW mass $M(<r)$, which represents the sum of baryonic and dark components. In order to estimate the mass of the dark component, we fit the enclosed mass profiles after subtracting a model of the radial distribution of the MW baryons. Specifically, we fit the Jeans and TME enclosed mass profiles with a model of the MW baryons (consisting of a bulge and disk component Bovy & Rix 2013; Bovy 2015) and a Navarro-Frenk-White dark matter profile (NFW, Navarro et al. 1995, 1996, 1997).

We use r_{200} as the distance at which the mean density of the Galaxy is 200 times the present critical density $\rho_{\rm crit}$ of the Universe to measure the virial mass M_{200} as the dark mass enclosed within the spherical volume of radius r_{200}. Thus, by using the MW mass model to fit our enclosed mass profile measured from the data, we extrapolate to estimate the virial radius r_{200}, dark matter mass M_{200}, and total mass $M(<r_{200})$ (dark plus baryonic).

3. Data and Analysis

We use the LAMOST K-giant (KG), SDSS KG, and SDSS blue horizontal branch (BHB) halo star samples of Bird et al. (2021). These consist of line-of-sight velocities and metallicities obtained with LAMOST (Cui et al. 2012; Deng et al. 2012; Luo et al. 2012; Zhao et al. 2012; Liu et al. 2015) and SDSS/SEGUE (York et al. 2000; Yanny et al. 2009; Aihara et al. 2011b,a; Eisenstein et al. 2011; Ahn et al. 2012), star distance estimates using the method of Xue et al. (2014) and Xue et al. (2008) for KG and BHB stars (respectively), and finally proper motions from *Gaia* DR2 (Gaia Collaboration et al. 2016, 2018). The LAMOST/SDSS KG and BHB samples initially contain 12728, 5248, and 3982 stars (Table 2 Bird et al. 2021), respectively.

As described in detail in Bird et al. (2021), our samples include the propagation of (1) uncertainties in the distances of the stars, (2) line-of-sight radial velocity uncertainties, (3) two *Gaia* DR2 proper motion error estimates, and (4) the *Gaia* DR2 proper motion covariances. These four uncertainties are propagated via 1000 Monte Carlo runs per star

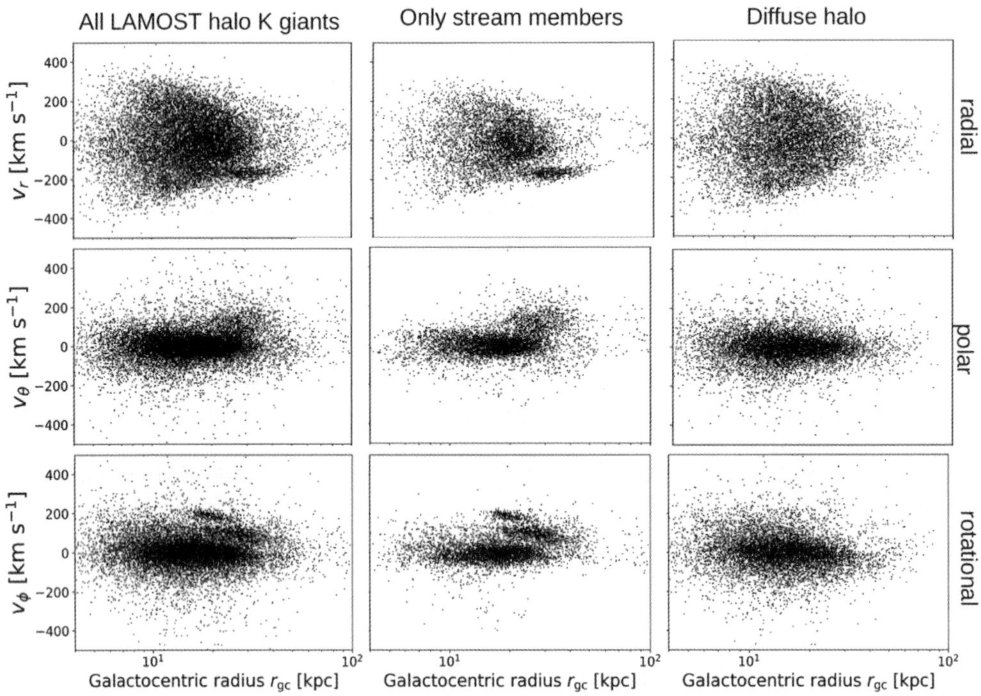

Figure 1. Substructure removal.

with the `python` package `pyia` (Price-Whelan 2018) to derive the uncertainties on the 3D kinematics of each star in the Galactocentric Cartesian and spherical coordinate systems.

Both the 3D spherical Jeans equation and TME assume the population tracing the Galactic potential is a relaxed and virialized system; we, thus, use the smooth, diffuse halo samples from which streams and substructure have been removed using the method of Xue et al. (2023, in preparation) who identify substructure through clustering in integral-of-motion space. We show the efficiency of the substructure removal process in Fig. 1.

The final sample sizes are 10762 KG and 2526 BHB stars covering a distance range from $3 < r < 105$ kpc and $4 < r < 60$ kpc, respectively (upper left two panels of Fig. 2).

We measure the enclosed mass profiles and estimate the virial mass using our KG and BHB samples separately. We have sufficiently large samples of both KG and BHB stars to estimate the 3D velocity dispersion and enclosed mass profiles separately.

We note that our KG and BHB smooth diffuse stellar halo samples peak at different metallicities as seen in the far right upper panel of Fig. 2. We have shown previously that the metallicities and velocity anisotropy β show a relation such that the more metal rich stars have more radial orbits compared to the more metal poor stars (Bird et al. 2019; Bird et al. 2021). Since our smooth, diffuse KG sample peaks at higher metallicities than our BHB sample, we expect the velocity dispersions of the samples to also differ, which we see in the lower panel of Fig. 2 where the dispersion is slightly larger in radial velocity for KG stars and in tangential velocities for BHB stars, leading to larger β values for KG stars. The different velocity distributions will influence our Jeans and TME mass estimations. Differences in the tracer density profiles will correspond to differences in the velocity dispersion profiles, since the tracer stars occupy a common gravitational potential. As the uncertainties are large for our estimated density profiles, we prefer to measure the mass from the KG and BHB samples separately and use the density profiles

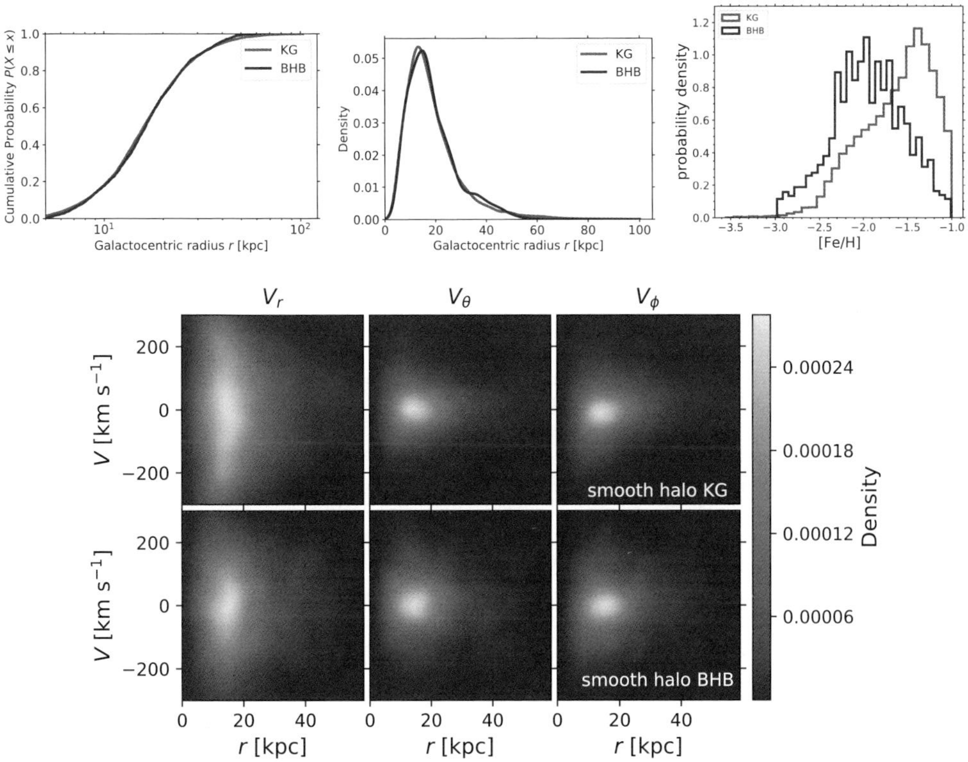

Figure 2. Cumulative probability, one dimensional kernel density estimation, and probability density distribution of [Fe/H] (upper panels from left to right, respectively) using KG (red) and BHB (blue) smooth, diffuse halo stars. Lower panel: Two dimensional kernel density estimation for the 3D spherical Galactocentric velocity components.

Table 1. Enclosed total mass estimates.

Method (star type)	<r[1] kpc	M(<r) $\times 10^{11}$ $M_\odot$
Jeans (KG)	73	4.3 ± 1.0
Jeans (BHB)	52	4.1 ± 1.2
TME (KG)	73	4.3 ± 1.1
TME (BHB)	52	3.6 ± 1.6

Notes: [1]Median Galactocentric distance of stars in the last radial bin.

from the literature measured separately for KG and BHB samples. In this way, differences in the Galactic mass profile measured between the KG and BHB tracers will give us an estimation of the mass uncertainties.

4. MW Mass

We present our mass and circular velocity profiles in Fig. 3 and summarize our mass estimate results in Tables 1–2, which give $M(< r)$ out to our furthest radial bins and our M_{200} fits for dark matter, respectively.

For our two star types and their corresponding mass profiles derived from the two methods, we find a weighted average dark matter virial mass of $M_{200} = 0.53 \pm 0.05 \times 10^{12}$ $M_\odot$ with weighted average virial radius and concentration parameter $r_{200} = 172 \pm 5$ kpc and $c = 20 \pm 3$.

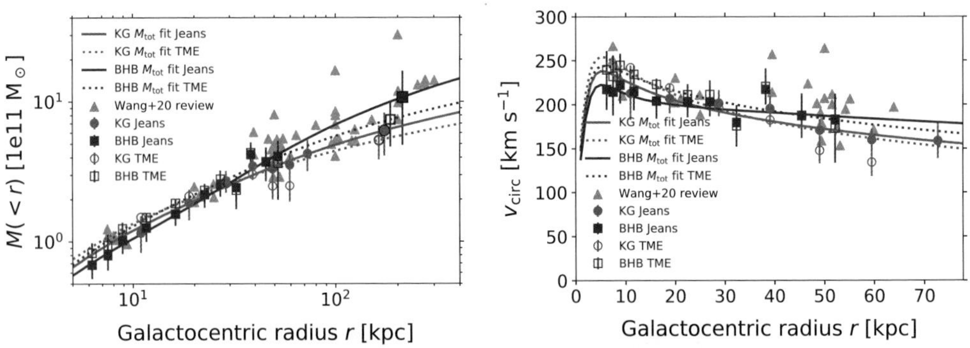

Figure 3. Profiles for MW enclosed mass $M(<r)$ (upper panel) and circular velocity $v_{\rm circ}$ (lower panel) using KG (red) and BHB (blue) stars. Jeans (filled markers) and TME (open markers) $M(<r)$ and $v_{\rm circ}$ are plot for our KG (circles) and BHB (squares) samples. Each marker represents the median radius of the stars within our selected radial bins. In the upper panel, the most distant markers of each KG (circles) and BHB (squares) mass profile are the corresponding total mass $M_{\rm tot}$ estimates within the virial radius r_{200} from the model fits. Gray triangle markers are the $M(<r)$ and $v_{\rm circ}$ estimates from the literature using various methods and are summarized in Wang et al. (2020).

Table 2. Best fit parameters for an NFW dark matter profile.

Method (star type)	M_{200} $\times 10^{12}$ M$_\odot$	r_{200} kpc	c
Jeans (KG)	$0.55^{+0.15}_{-0.11}$	173^{+15}_{-12}	$18.2^{+7.1}_{-6.8}$
Jeans (BHB)	$1.00^{+0.67}_{-0.33}$	211^{+39}_{-26}	$9.4^{+4.4}_{-3.2}$
TME (KG)	$0.46^{+0.043}_{-0.040}$	$162^{+4.9}_{-4.9}$	$28.5^{+2.4}_{-2.2}$
TME (BHB)	$0.67^{+0.21}_{-0.14}$	185^{+17}_{-14}	$19.0^{+4.7}_{-4.1}$

5. Discussion and Suggestions for Future MW Mass Measurements

In Fig. 4, we compare our dark matter M_{200} results to previous Jeans and TME MW mass estimates as well as to the recent estimates from the literature using a wide range of methods (timing, comparison to simulations, rotational velocity, streams, high velocity stars, distribution function, combination of distribution function and simulations). Taking the weighted mean of these recent mass estimates shown in Fig. 4, we find a virial mass of $\sim 0.85 \times 10^{12}$ M$_\odot$ with a scatter of 33 percent.

Despite the great efforts put in to determine the MW mass, the existing mass estimates are scattered by a factor of two; this scatter remains even after estimating the mass using the precise data from Gaia and HST.

We provide some topics for discussion and consideration for future efforts related to MW mass.

• We find $M_{\rm vir} = 0.85 \times 10^{12}M_\odot$ as the weighted average with a scatter of 33 percent of recent mass estimates shown in Fig. 4, but is there a method that is intrinsically better than others?

• What questions can be addressed using $M_{\rm vir}$ with 33 percent scatter? What questions do we still need quite better accuracy?

• What current steps can we take to more accurately measure the MW mass while we wait for the upcoming SuperGaia surveys (e.g., Vera C. Rubin Observatory Legacy Survey of Space and Time, Ivezić et al. 2019)?

• How do the results using other density models compare to using NFW, such as Einasto (Einasto 1965)?

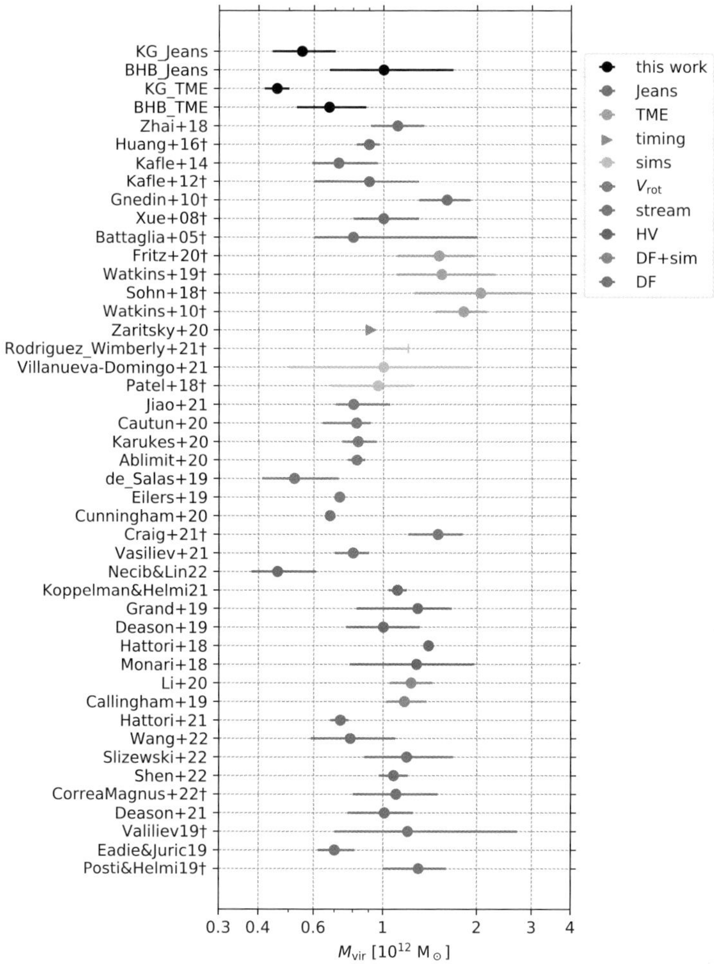

Figure 4. Comparison of our KG and BHB dark matter virial mass M_{200} estimates with virial mass $M_{\rm vir}$ estimates from the literature. We plot those measured using similar Jeans and TME methods involving tracer samples, as well as the recent virial mass estimates using various different methods. Colors represent the method as specified in the legend. The † marks virial mass estimates using definitions other than M_{200}. Further details are summarized in Bird et al. (2022).

References

Ahn, C. P., Alexandroff, R., Allende Prieto, C., et al. 2012, ApJS, 203, 21

Aihara, H., Allende Prieto, C., An, D., et al. 2011a, ApJS, 195, 26

—. 2011b, ApJS, 193, 29

An, J. H., & Evans, N. W. 2011, MNRAS, 413, 1744

Bird, S. A., Xue, X.-X., Liu, C., et al. 2019, AJ, 157, 104

Bird, S. A., Xue, X.-X., Liu, C., et al. 2021, ApJ, 919, 66

—. 2022, MNRAS, 516, 731

Bovy, J. 2015, ApJS, 216, 29

Bovy, J., Hogg, D. W., & Roweis, S. T. 2011, Annals of Applied Statistics, 5, 1657

Bovy, J., & Rix, H.-W. 2013, ApJ, 779, 115

Cui, X.-Q., Zhao, Y.-H., Chu, Y.-Q., et al. 2012, Research in Astronomy and Astrophysics, 12, 1197

Das, P., Williams, A., & Binney, J. 2016, MNRAS, 463, 3169

Deng, L.-C., Newberg, H. J., Liu, C., et al. 2012, Research in Astronomy and Astrophysics, 12, 735
Einasto, J. 1965, Trudy Astrofizicheskogo Instituta Alma-Ata, 5, 87
Eisenstein, D. J., Weinberg, D. H., Agol, E., et al. 2011, AJ, 142, 72
Erkal, D., Belokurov, V. A., & Parkin, D. L. 2020, MNRAS, 498, 5574
Erkal, D., Deason, A. J., Belokurov, V., et al. 2021, MNRAS, 506, 2677
Evans, N. W., An, J., & Deason, A. J. 2011, ApJ, 730, L26
Gaia Collaboration, Prusti, T., de Bruijne, J. H. J., et al. 2016, A&A, 595, A1
Gaia Collaboration, Brown, A. G. A., Vallenari, A., et al. 2018, A&A, 616, A1
Ivezić, Ž., Kahn, S. M., Tyson, J. A., et al. 2019, ApJ, 873, 111
Liu, X.-W., Zhao, G., & Hou, J.-L. 2015, Research in Astronomy and Astrophysics, 15, 1089
Luo, A.-L., Zhang, H.-T., Zhao, Y.-H., et al. 2012, Research in Astronomy and Astrophysics, 12, 1243
Navarro, J. F., Frenk, C. S., & White, S. D. M. 1995, MNRAS, 275, 56
—. 1996, ApJ, 462, 563
—. 1997, ApJ, 490, 493
Petersen, M. S., & Peñarrubia, J. 2021, Nature Astronomy, 5, 251
Price-Whelan, A. 2018, Adrn/Pyia: V0.2, Zenodo, doi:10.5281/zenodo.1228136
Wang, W., Han, J., Cautun, M., Li, Z., & Ishigaki, M. N. 2020, Science China Physics, Mechanics, and Astronomy, 63, 109801
Xu, Y., Liu, C., Xue, X.-X., et al. 2018, MNRAS, 473, 1244
Xue, X.-X., Rix, H., Hogg, et al. 2023, in preparation
Xue, X. X., Rix, H. W., Zhao, G., et al. 2008, ApJ, 684, 1143
Xue, X.-X., Ma, Z., Rix, H.-W., et al. 2014, ApJ, 784, 170
Yanny, B., Rockosi, C., Newberg, H. J., et al. 2009, AJ, 137, 4377
York, D. G., Adelman, J., Anderson, Jr., J. E., et al. 2000, AJ, 120, 1579
Zhao, G., Zhao, Y.-H., Chu, Y.-Q., Jing, Y.-P., & Deng, L.-C. 2012, Research in Astronomy and Astrophysics, 12, 723

Discussion

MAGDA ARNABOLDI: What is the largest most important bias in your determination of the MW M_{200}?

SARAH A. BIRD: Now with the combination of LAMOST/SDSS and *Gaia*, the velocity distributions are known with great accuracy. The largest uncertainty in our estimation is the tracer density profile. Varying the slope of the density profile power-law index by ± 1, for example, changes the estimated mass by ~ 20 percent.

YOSHIHISA SUZUKI (ONLINE): Can you use your sample to measure the density profile in different quadrants of the MW?

SARAH A. BIRD: Yes, our sample can be used for this, although careful corrections for the selection function are needed to measure the density profile directly from the data. In this work, we use the density profile models of Das et al. (2016) and Xu et al. (2018) and do not attempt to take into account variations from different quadrants of the MW.

FEDERICO LELLI: You have shown the MW circular velocity curve out to 70 kpc, but your tracers extend out to ~ 200 kpc. Does the circular velocity curve remain flat or does it show a clear Keplerian decline?

SARAH A. BIRD: Our data go out to ~ 70 kpc, although when we fit our enclosed mass profiles with a MW mass model including dark matter, we plot the fitting result of the total mass at the virial radius as the furthest point. To the extent of our tracers the MW circular velocity curve remains flat.

ELENA D'ONGHIA: The bar is strong in the MW and $M_{\rm bar} = 5 \times 10^{10}$ $M_{\odot}$, so are we accounting for enough mass in the MW? It seems to be a small mass, it may be larger.

SARAH A. BIRD: Our dark matter mass of $M_{200} = 5.3 \times 10^{11}$ $M_{\odot}$ is low. Thanks for noting that it is important to consider whether such a low mass can support a strong and massive bar.

N. WYN EVANS: With this large dataset, you could consider using an axisymmetric density model to fit the data and use this density profile within the 3D spherical Jeans mass model.

SARAH A. BIRD: Thanks for this suggestion; this is a great next step.

ANA BONACA (SLACK FORUM): How does the selection of KG and BHB stars compare between the sample used to measure their density profile and the mass modeling?

SARAH A. BIRD: Several differences exist between the samples. We do not expect these to have a large effect on the density models as the authors carefully account for completeness and selection biases. The Xu et al. (2018) KG model is based on LAMOST DR3 and the Das et al. (2016) BHB model is based on SDSS/SEGUE II. We are able to attain larger samples using the KG stars from SDSS SEGUE I and II and the more recent LAMOST DR5, and BHB stars from SDSS SEGUE I and II. For this study we prefer to use the published models as the selection functions will be quite complex for our more full datasets, and we do not expect completeness or selection biases to affect the kinematics.

Dynamical Masses of Local Group Galaxies
Proceedings IAU Symposium No. 379, 2025
P. Bonifacio, M.-R. Cioni, F. Hammer, M. S. Pawlowski & S. Taibi, eds.
doi:10.1017/S1743921323001503

The orbital poles of the Milky Way streams: testing the MW-M31 flyby scenario

Elena Asencio[1] and **Pavel Kroupa**[1,2]

[1]Helmholtz-Institut für Strahlen und Kernphysik (HISKP), University of Bonn, Nussallee 14-16, D-53115 Bonn, Germany
email: s6elena@uni-bonn.de

[2]Astronomical Institute, Faculty of Mathematics and Physics, Charles University, V Holešovičkách 2, CZ-180 00 Praha 8, Czech Republic

Abstract. Understanding the origin of the stellar streams around the Milky Way can be of great relevance to learn about the history of the Milky Way and the formation of its substructures. A previous study on the Milky Way streams (Pawlowski et al. 2012) showed that many of these (7 out of 14) present a similar orientation to that of the disk of satellite galaxies (DoS) and the young globular clusters of the Milky Way. This suggests that the DoS, the young globular clusters and a large fraction of the Milky Way streams have a correlated origin. The authors proposed that these substructures could have formed as a result of a past interaction between the Milky Way and the Andromeda galaxy. In this work, we revise the distribution of the orbital poles of the Milky Way streams in light of the latest stream dataset, which includes a total of 97 streams.

Keywords. Galaxy: structure – galaxies: interactions – galaxies: kinematics and dynamics – galaxies: formation – Local Group

1. Introduction

Stellar streams are the remnants resulting from the disruption of dwarf galaxies and star clusters. The stars in these streams are expected to follow a similar orbit to that of the original object (Odenkirchen et al. 2003), which implies that the orbital poles of stellar streams can be used to infer their origin.

Previous studies have analysed the orbital pole orientation of several Milky Way (MW) substructures. Kroupa et al. (2005) and Metz & Kroupa (2007) found that the satellite galaxies near the MW could be fitted to a disk (DoS), thus being in tension with the expected spheroidal distribution of satellite galaxies in the standard (ΛCDM) model. Pawlowski et al. (2012) also showed that the orbital poles of 7 out of 14 MW streams align with the DoS, as well as with most of the young population of MW globular clusters. The authors named this group of substructures with similar orbital poles the "Vast POlar Structure" (VPOS). The coincidence in the orbital poles of the VPOS substructures indicates that they have a correlated origin. This could be explained if they originated from a tidal tail resulting from an interaction of the MW and Andromeda (M31) (Pawlowski et al. 2012; Bílek et al. 2018; Banik et al. 2022). While this scenario is in contradiction with the standard model expectations – as the MW and M31 would have merged after interacting due to the dynamical friction of their dark matter halos – it is still the most plausible explanation for the highly phase-space correlated DoS (Pawlowski et al. 2011; Fouquet et al. 2012; Yang et al. 2014) and the mutually correlated DoS of the MW and M31 (Pawlowski et al. 2013).

Riley & Strigari (2020) repeated the analysis on the orbital poles of the Milky Way streams with a more recent and larger data set (64 streams). They concluded that there is no preferential clustering of the orbital pole orientation around the VPOS direction.

Here, we revise the orbital pole distribution of these streams by using the latest Gaia data in the galstreams catalogue (Mateu 2023) – with 97 streams. Mainly, we aim to assess the feasibility of the MW-M31 flyby hypothesis by analysing the new data while also considering the effect of gravitational precession over the poles. In contrast with the previous studies, we do not check whether the poles are aligned with the VPOS to test this hypothesis. We use the results of the MW-M31 interaction simulations by Banik et al. (2022) in order to compare the resulting pole distribution predicted by this simulation with the observed distribution.

In Section 2 we describe the observational data present in the galstreams catalogue (Mateu 2023) and the method used for inferring the orbital poles of the streams. We present our results in Section 3 and we discuss them in comparison to the results of the hydrodynamical simulations of Banik et al. (2022). Finally, we conclude in Section 4.

2. Data

To obtain the orientation of the orbital poles of the Milky Way streams, we use the data on the recently updated galstreams catalogue (Mateu 2023). Currently, this catalogue contains information on the tracks of 97 stellar streams of the Milky Way. Specifically, on the stream length, end points, mean pole, stream's coordinate frame, polygon footprint, and pole and angular momentum tracks. The stream pole, which is the main quantity that we are interested in, is obtained in the following way: two points along the stream (in Galactocentric coordinates) are chosen, then the stream-normal vector n to the plane is obtained as

$$n = \frac{s' \times e'}{\| s' \times e' \|}, \tag{1}$$

where s' and e' are the two chosen points and $\| s' \times e' \|$ is the norm of their cross product. Due to the effect of gravitational torques, the stream track is not exactly shaped as an arc, which means that the orbital poles obtained for different points on the stream track are not the same. Therefore, to assign an orbital pole for the stream, we choose the mode of the different pole values as the nominal pole, and the pole values within the 1σ confidence level – assuming a gaussian distribution – as the uncertainty.

3. Results

The distribution of the orbital poles obtained for the 93 stellar streams in the catalogue is consistent with the results of Riley & Strigari (2020): there is no preferential clustering around the VPOS direction. However, in this work, we also considered the potential effects of precession. That is, the change in the stream's orbital pole direction caused by the gravitational torque of the Milky Way. Given the strong dependency of the precession rate with the distance to the galactic centre (Erkal et al. 2016), we excluded all streams which were at a distance < 15 kpc from the Milky Way centre. The Galactic longitude and latitude of the remaining stream poles are shown in Fig. 1. This figure shows that the poles seem to be mostly concentrated along the $20° - 40°$ Galactic latitude direction and the $130° - 170°$ Galactic longitude direction.

In order to assess whether the observed pole distribution is consistent with the expected distribution for the MW-M31 encounter scenario, we compared our results with the simulations of the MW-M31 encounter. A hydrodynamical simulation of the Local Group

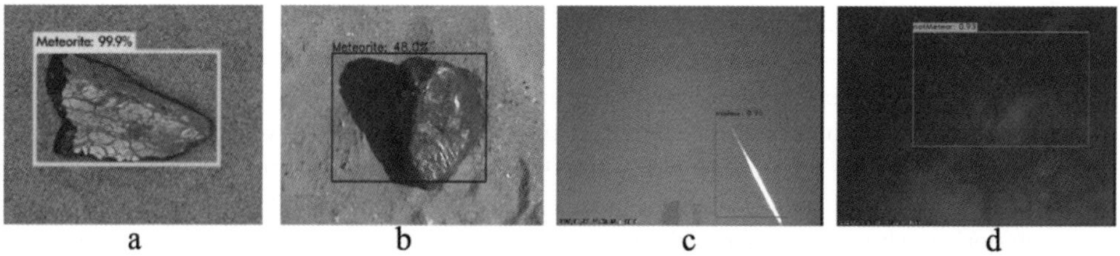

Figure 1. Distribution of the orbital poles, in galactic coordinates, for the streams in the galstreams catalogue which are at a distance > 15 kpc from the Milky Way centre. The smaller green circle corresponds to a cone with aperture angle 30° centered around the observed orbital pole of the disk of satellites. The larger green circle corresponds to a cone with aperture angle 49° centered around the preferred orbital pole in the Banik et al. (2022) simulation. This represents the estimated orbital pole dispersion of the simulated material.

was computed using the Milgromian† N-body code PHANTOM OF RAMSES (Lüghausen et al. 2015), which is adapted from the Newtonian N-body code RAMSES (Teyssier 2002). Their figure 7 shows the orbital pole distribution of material resulting from a MW-M31 interaction 8.2 Gyr ago. The model orbital poles are mostly concentrated between -20° and 40° in the Galactic latitude direction, and between 150° and 220° in the Galactic longitude direction.

The orbital poles of the selected streams (Fig. 1) are somewhat displaced with respect to the expected distribution from the simulations, but otherwise show a comparable distribution.

4. Discussion and conclusions

There is a clustering of the orbital poles, but not in the exact same direction predicted by the simulation. This could be due to several reasons:

(1) the configuration of the MW-M31 interaction could have been different from the one considered in the simulation,
(2) precession could still have a strong effect on the orbital pole of the streams even on streams at > 15 kpc from the Milky Way centre,
(3) the effect of the infalling Large Magellanic Cloud on the orbital poles of the MW streams also needs to be taken into account,
(4) many streams could have a common origin (as suggested by the observed clustering of the poles) but they do not originate from a MW-M31 interaction.

Given that the simulations of Banik et al. (2022) do not take into account the evolution of the orbital pole distribution due to precession, and that additional effects (e.g. infall of the Large Magellanic cloud, the precession rate for each stream assuming a Milgromian potential, etc.) need to be considered in order to make an accurate assessment of the MW-M31 flyby scenario, we conclude that this hypothesis can not be ruled out and requires further future exploration.

References

Banik I., Thies I., Truelove R., Candlish G., Famaey B., Pawlowski M. S., Ibata R., Kroupa P., 2022, MNRAS, 513, 129

† As we mentioned in Section 1, the MW-M31 fly-by scenario cannot be understood in the context of the standard model. Therefore, this scenario can only take place within a paradigm capable of explaining galactic dynamics without dark matter, e.g. Milgromian dynamics (MOND; Milgrom 1983)

Bílek M., Thies I., Kroupa P., Famaey B., 2018, A&A, 614, A59
Erkal D., Sanders J. L., Belokurov V., 2016, MNRAS, 461, 1590
Fouquet S., Hammer F., Yang Y., Puech M., Flores H., 2012, MNRAS, 427, 1769
Kroupa P., Theis C., Boily C. M., 2005, A&A, 431, 517
Lüghausen F., Famaey B., Kroupa P., 2015, Canadian Journal of Physics, 93, 232
Mateu C., 2023, MNRAS, 520, 5225
Metz M., Kroupa P., 2007, MNRAS, 376, 387
Milgrom M., 1983, ApJ, 270, 365
Odenkirchen M., et al., 2003, The Astronomical Journal, 126, 2385
Pawlowski M. S., Kroupa P., de Boer K. S., 2011, A&A, 532, A118
Pawlowski M. S., Pflamm-Altenburg J., Kroupa P., 2012, MNRAS, 423, 1109
Pawlowski M. S., Kroupa P., Jerjen H., 2013, MNRAS, 435, 1928
Riley A. H., Strigari L. E., 2020, MNRAS, 494, 983
Teyssier R., 2002, A&A, 385, 337
Yang Y., Hammer F., Fouquet S., Flores H., Puech M., Pawlowski M. S., Kroupa P., 2014, MNRAS, 442, 2419

Dynamical Masses of Local Group Galaxies
Proceedings IAU Symposium No. 379, 2025
P. Bonifacio, M.-R. Cioni, F. Hammer, M. S. Pawlowski & S. Taibi, eds.
doi:10.1017/S1743921323001291

Weighing Andromeda: Mass estimates of the M 31 galaxy

Souradeep Bhattacharya

Inter University Centre for Astronomy and Astrophysics, Ganeshkhind,
Post Bag 4, Pune 411007, India
email: souradeep@iucaa.in

Abstract. Andromeda (M 31) is the nearest giant spiral galaxy to our Milky Way, and, over the past few decades, has been dubbed the most massive member of the Local Group. I explore the evolution of the measured mass of M 31 over the past ~ 80 years, reviewing the different observational and modelling techniques that have developed over time to measure its mass. I discuss the best present-day constraints of the mass of M 31 and the consistency of different techniques.

Keywords. galaxies: individual: M 31

1. Introduction

The Andromeda galaxy (M 31) is the nearest giant spiral galaxy to our Milky Way (MW) and has long been considered its sister galaxy. Together they account for $\sim 90\%$ of the mass of our Local Group (LG; Penarrubia *et al.* 2016). Given its large size and proximity to the MW, M 31 has long been the subject of many path-breaking studies in astronomy, the earliest of which was the first identification of "island universes" beyond the MW (Hubble 1925, 1929). Measured from Cepheid variables to be at a distance of 285 kpc (Hubble 1925), M 31 (along with the Triangulum galaxy, M 33) was the first of many galaxies to be studied, effectively kickstarting the field of Extragalactic Astrophysics.

M 31 was also one of the first galaxies whose mass determination was attempted. Babcock (1939) constructed the rotation curve (RC) of M 31 out to $\sim 100'$ from its center by measuring the radial velocities of its HII regions from spectroscopic observations. Assuming M 31 to be at a distance of 210 kpc† and modelling the disc of M 31 as rotating concentric flattened spheroids, Babcock (1939) measured the mass of M 31 as 10.2×10^{10} $M_{\odot}$. Babcock (1939) also noted that the nearly constant angular velocity measured at the outer parts of M 31 was opposite to "planetary" type rotation, a problem that would later come to be known as the "missing mass" problem and nearly 30 years later would start the development of ΛCDM cosmology (Ostriker & Peebles 1973).

In this paper, we will go through a near-exhaustive list of mass measurements of M 31. A number of galaxy mass measurement techniques, many of which find their first application to M 31, will be discussed in Section 2. We will then compare the mass estimate of M 31 from different techniques, discuss possible reasons of discrepancies if any, and finally obtain the present-day best mass estimate of M 31 in Section 3.

† Babcock (1939) attributes this assumed distance to Hubble (1929). However, the distance for M 31 quoted by Hubble (1929) is 275 kpc.

2. Mass measurements of the Andromeda galaxy

The various mass measurements of M 31 discussed in this paper have been noted in Table 1 along with the original source of the measurement, the technique employed, the distance to M 31 assumed by the original source as well as the enclosed radius within which the mass has been measured (r_{enc}; calculated assuming the present-day best known distance to M 31 of 776 kpc; Savino 2022).The various mass measurements are discussed in the ensuing subsections.

2.1. *The disc rotation curve and early mass measurements of M 31*

Following the first measurement of the M 31 disc RC by Babcock (1939), the same data were modelled as a thin disc by Wyse & Mayall (1942) leading to a mass estimate of 9.5×10^{10} $M_\odot$ for M 31. These authors had assumed a variable mass-to-light ratio (M/L) as a function of radius in the M 31 disc. Assuming a constant M/L and thin disc for the M 31 RC from Babcock (1939) and additionally also from Mayall (1950), Schwarzschild (1954) measured a mass of 1.4×10^{11} $M_\odot$ for M 31. At that time, there was an ongoing tension in the measured distance of M 31, summarized in Hubble & Sandage (1953), and thus Schwarzschild (1954) had assumed a distance of 460 kpc for M 31, an average of the contentious distance estimates. With the same RC data and distance but with a slightly different disc mass model, Lohmann (1954) measured a mass of 3.3×10^{11} $M_\odot$ for M 31. Soon after, van de Hulst, Raimond & van Woerden (1957) reported the first measurement of the HI gas RC of M 31 from their radio observations. The mass of M 31 based on the HI gas RC was reported by Schmidt (1957) to be 3.38×10^{11} $M_\odot$, similar to that of Lohmann (1954). Brandt (1960) simultaneous modelled the M 31 disc RC as a thin disc from both HII regions (Babcock 1939; Mayall 1950) and the HI gas (van de Hulst, Raimond & van Woerden 1957) to obtain a similar mass of 3.7×10^{11} $M_\odot$ for M 31. Both Schmidt (1957) and Brandt (1960) utilised the distance modulus measured for M 31 from cepheid variables by Baade & Swope (1955) but made additional different corrections for extinction (see Table 1 for the values). Roberts (1966) obtained the HI gas RC out to $\sim$2 deg in M 31 and assuming a distance of 690 kpc (Baade & Swope 1955) and similar model as Brandt (1960), they obtained a mass of 3.1×10^{11} $M_\odot$ for the galaxy.

Rubin & Ford (1970) utilised an efficient image-tube spectrograph to measure the radial velocities of individual HII regions in M 31 in reasonable time (1–1.5 hr per HII region). They constructed the RC of M 31 from these HII regions and determined a mass of $1.85 \pm 0.1 \times 10^{11}$ $M_\odot$ for M 31. While previous authors had generally termed their determined mass as the total (or near total) mass of M 31, Rubin & Ford (1970) had noted that their determined mass was the enclosed mass and that there should be more missing mass in the outskirts of the galaxy, which later became clearer as they and their contemporaries applied their efficient observation techniques to other galaxies (see review by Faber & Gallagher 1979). However, Einasto & Rummel (1970) simultaneously modelled the observed luminosity distribution of stars in M 31 and the RC from HII regions and HI gas to obtain a similar mass of 2×10^{11} $M_\odot$ for M 31, but showing that no missing mass is required at least within the enclosed radius to reconcile the RC with the luminosity distribution of the galaxy. Similar values of the M 31 mass were computed by Gottesman & Davies (1970), Roberts & Whitehurst (1975) and Emerson (1976), and later by Braun (1991) and even Carignan *et al.* (2006) from their respective HI gas RC measurements, and by Deharving & Pellet (1975) for their RC constructed from HII regions (see Table 1). A distance of 690 kpc (Baade & Swope 1955) to M 31 was generally accepted. These authors all quoted their mass obtained as being the enclosed mass out to their probed radius (see r_{enc} in Table 1).

Table 1. Mass measurements of M 31

Publication	Method	$d_{M\ 31}$ [kpc]	r_{enc} [kpc]	M_{tot} [10^{12} $M_\odot$]
Babcock (1939)	RC – HII regions	210	22.8	0.102
Wyse & Mayall (1942)	RC – HII regions	210	22.8	0.095
Schwarzschild (1954)	RC – HII regions	460	22.8	0.14
Lohmann (1954)	RC – HII regions	460	27.3	0.33
Schmidt (1957)	RC – HI gas	630	27.3	0.338
Brandt (1960)	RC – HII regions & HI gas	600	34.2	0.37
Roberts (1966)	RC – HI gas	690	27.3	0.31
Rubin & Ford (1970)	RC – HII regions	690	29.6	0.185 ± 0.01
Einasto & Rummel (1970)	RC – HII regions & HI gas	692	34.2	0.2
Gottesman & Davies (1970)	RC – HI gas	690	34.2	0.22
Hartwick & Sargent (1974)	GCs	667	19.3	0.34 ± 0.14
Deharving & Pellet (1975)	RC – HII regions	690	22.8	0.163 ± 0.015
Roberts & Whitehurst (1975)	RC – HI gas	690	34.2	0.37
Gunn (1975)	LG kinematics	690	–	4
Hodge (1975)	Satellites	690	–	6
Emerson (1976)	RC – HI gas	690	31.2	0.218
Rood (1979)	Satellites	667	–	0.338
Bahcall & Tremaine (1981)	Satellites	667	100	1
van den Bergh (1981)	GCs	667	22.8	0.24 ± 0.12
	Satellites	667	–	0.75 ± 0.39
Braun (1991)	RC – HI gas	690	31.3	0.2 ± 0.01
Courteau & van den Bergh (1999)	LG kinematics	758	–	1.33 ± 0.18
Evans & Wilkinson (2000)	Satellites, GCs & PNe	770	–	$1.23^{+1.8}_{-0.6}$
Cote *et al.* (2000)	Satellites	780	–	0.79 ± 0.05
Klypin, Zhao & Somerville (2002)	Mass model	770	–	1.6
Evans *et al.* (2003)	GCs	770	100	1.2
	Satellites	770	–	1.1
Ibata *et al.* (2004)	Substructure	780	125	$0.75^{+0.25}_{-0.13}$
Carignan *et al.* (2006)	RC – HI gas	780	35	0.34
Galleti *et al.* (2006)	GCs	784	–	2.4
Fardal *et al.* (2006)	Substructure	784	125	0.74 ± 0.12
Geehan *et al.* (2006)	Mass model	784	–	0.71
Seigar, Barth, & Bullock (2008)	Mass model	784	–	0.82
Lee *et al.* (2008)	GCs	780	100	$1.92^{+0.14}_{-0.13}$
Chemin, Carignan & Foster (2009)	Mass model	785	–	1
Corbelli *et al.* (2010)	Mass model	785	–	1.3
Watkins, Evans & An (2010)	Satellites	785	300	1.4 ± 0.4
van der Marel *et al.* (2012)	LG kinematics	770	–	1.54 ± 0.39
Tamm *et al.* (2012)	Mass model	785	–	0.95 ± 0.15
Tollerud *et al.* (2012)	Simulation	785	–	$1.2^{+0.9}_{-0.7}$
Fardal *et al.* (2013)	Substructure	780	–	$1.99^{+0.52}_{-0.41}$
Veljanoski *et al.* (2013)	GCs	780	200	1.35 ± 0.35
Diaz *et al.* (2014)	LG kinematics	780	–	1.7 ± 0.3
Hayashi & Chiba (2014)	Mass model	785	200	$1.82^{+0.49}_{-0.39}$
Penarrubia *et al.* (2014)	LG kinematics	783	–	1.5 ± 0.3
Sofue (2015)	RC	770	31.3	1.99 ± 0.39
Penarrubia *et al.* (2016)	LG kinematics	783	–	$1.33^{+0.39}_{-0.33}$
Kafle *et al.* (2018)	PNe	780	–	0.8 ± 0.1
Zhai *et al.* (2020)	Simulation	787	–	$2.5^{+1.3}_{-1.1}$
Villanueva-Domingo *et al.* (2021)	Simulation	–	–	$2.2^{+1.3}_{-0.8}$
Escala *et al.* (2022)	Substructure	785	36.6	0.926 ± 0.278
Patel & Mandel (2022)	Simulation	776	–	$3.02^{+1.3}_{-0.69}$
Dey (2023)	Substructure	785	125	$0.63^{+0.2}_{-0.13}$

Notes: Column 1: Source publication; Column 2: Method utilised for mass measurement; Column 3: Adopted distance to M 31 in source publication; Column 4: Radius within which enclosed mass has been measured, assuming present day best measured distance of 776 kpc (Savino 2022). The radius is not noted if the method, either directly or indirectly, measures the total mass of M 31.; Column 5: Mass of M 31, as reported.

At this time, bolstered by the suggestion of a massive dark halo required to support disc galaxies (Ostriker & Peebles 1973), alternative methods to measure the mass of galaxies started becoming popular. One of the earliest applied to M 31 was by Hartwick & Sargent (1974) who applied the viral theorem to the globular cluster (GC) population of M 31 (identified by van den Bergh 1969). The GCs only enclosed a small radius in M 31 and they measured a similar mass as other authors did from the RCs (see Table 1). They had assumed a slightly lower distance of M 31 from Sandage & Tammann (1971). During this time, Gunn (1975) utilised the timing argument (Kahn & Woltjer 1959) to simultaneously measure the mass of the MW and M 31, reporting a very high but uncertain mass of 4×10^{12} $M_\odot$ for M 31. Hodge (1975) used the radial velocities of two dwarf satellites of M 31, NGC 247 and NGC 185, and applied tidal radius equations to obtain a similarly high and uncertain mass of 6×10^{12} $M_\odot$ for M 31. Rood (1979) applied the virial theorem to the radial velocities of the six most luminous satellite galaxies of M 31 and mass of 3.38×10^{11} $M_\odot$ for M 31, an uncertain value but one that is in-line with that measured in the central regions for M 31 from RCs.

Such mass determination methods of M 31 and their development to present days are discussed individually in the ensuing subsections.

2.2. *Satellite radial velocity measurements*

Given that M 31 satellites occur at larger distances from the galaxy, they are thought to trace the mass of M 31 out to larger radii. The projected mass estimator method introduced by Bahcall & Tremaine (1981) brought an improvement to the viral mass technique used by Rood (1979), determining the mass of M 31 as $\sim 10^{12}$ $M_\odot$. The same projected mass estimator method was applied using an additional M 31 satellite by van den Bergh (1981) to find slightly lower masses (see Table 1). Once more M 31 satellites were identified with measured radial velocities and the distance to M 31 was updated to 770 kpc (close to its present day value; Holland 1998), updated measurements of the mass of M 31 utilising variations of the projected mass estimator were reported (see Table 1) by Evans & Wilkinson (2000) and Cote *et al.* (2000). Notably, Evans & Wilkinson (2000) also utilised GCs and Planetary Nebulae (PNe) in M 31 to trace its mass. Evans *et al.* (2003) developed the tracer mass estimator method and applied it to the satellites of M 31 to measure a mass of 1.1×10^{12} $M_\odot$ for the M 31 galaxy. The latest iteration of this technique was applied by Watkins, Evans & An (2010) to obtain a mass of $\sim 1.4 \times 10^{12}$ $M_\odot$ for M 31 out to 300 kpc (well beyond their estimated virial radius of M 31).

2.3. *Globular cluster radial velocity measurements*

Similar to satellite galaxies, the projected mass estimator method (Bahcall & Tremaine 1981) was applied to GCs to estimate the mass of M 31 enclosed by these tracers. van den Bergh (1981) measured a mass of $2.4 \pm 0.12 \times 10^{11}$ $M_\odot$ for M 31 within ~ 22.8 kpc, similar to that from RCs at this radius. Evans *et al.* (2003) utilised the tracer mass estimator technique for GCs but now extrapolated out to 100 kpc to find that M 31 has a mass of 1.2×10^{12} $M_\odot$, consistent with that from applying the same technique to satellite galaxies. With an updated sample of GCs and slightly larger M 31 distance of 780 kpc (McConnachie *et al.* 2005), the same technique led to a larger measured mass of 2.4×10^{12} $M_\odot$ for M 31 by Galleti *et al.* (2006), with a slightly lower value from Lee *et al.* (2008). The latest application of the tracer mass estimator technique to GCs by Veljanoski *et al.* (2013) fetched a slightly lower mass of $1.35 \pm 0.35 \times 10^{12}$ $M_\odot$ with the authors stating that the determined mass is highly dependent on the chosen model and assumptions within.

2.4. *Mass models of M 31*

Similar in spirit to the technique from Einasto & Rummel (1970) but now including a bulge, an adiabatic disc and a cuspy halo component described by a NFW (Navarro, Frenk & White 1996) halo profile, Klypin, Zhao & Somerville (2002) simultaneously fitted the surface brightness profile of M 31 and its RC measurements from CO and HI gas within ~35 kpc. From their mass models, they thus obtained a mass of 1.6×10^{12} $M_{\odot}$ for M 31. Similar models by Geehan *et al.* (2006) and Seigar, Barth, & Bullock (2008), however, led to lower modelled mass values of M 31 (see Table 1), nearly half that of Klypin, Zhao & Somerville (2002). Applying similar mass models to their measured HI RCs out to ~ 40 kpc from the center of M 31, both Chemin, Carignan & Foster (2009) and Corbelli *et al.* (2010) find slightly larger masses for M 31 of 10^{12} $M_{\odot}$ and 1.3×10^{12} $M_{\odot}$ respectively. The value is consistent (see Table 1) with that of Tamm *et al.* (2012) who utilised stellar population models to obtain a stellar mass map of M 31 using multi-wavelength photometry of M 31, and modelled the total mass of M 31 similar to Klypin, Zhao & Somerville (2002). Hayashi & Chiba (2014) modelled a prolate halo of M 31 to find a total mass of $\sim 1.82 \times 10^{12}$ $M_{\odot}$ for the galaxy. They however noted that the choice of assumed halo density profile (NFW or otherwise) has strong bearing on the measured mass of M 31.

2.5. *Substructure radial velocity measurements*

From improvements in wide-field imaging techniques and instrumentation, a number of surveys conducted in the outskirts of M 31 led to the identification of a plethora of substructures in the halo of M 31 (see McConnachie *et al.* 2018, and references therein). The most prominent among these is the Giant Stream (GS). From their galaxy mass models, Ibata *et al.* (2004) found that the kinematics of the GS require a total mass inside 125 kpc of $\sim 7.5 \times 10^{11}$ $M_{\odot}$. Similar values were found by both Fardal *et al.* (2006) and Dey (2023) using a similar technique and GS kinematics out to ~125 kpc. Accounting for updated kinematic measurements of the GS out to larger radii and updated surface brightness measurements as constraints for mass models, Fardal *et al.* (2013) found a virial mass of $\sim 2 \times 10^{12}$ $M_{\odot}$ for M 31. The higher mass is also supported by the mass within ~ 37 kpc determined by Escala *et al.* (2022) from kinematics of the NE-Shelf substructure of M 31 (see Table 1).

2.6. *Local Group kinematics*

The relative motion of galaxies in the LG around its barycenter were utilised by Courteau & van den Bergh (1999) to obtain the mass of the LG. They had assumed M 31 to be 1.5 times as massive as the MW and found that M 31 had a mass of $\sim 1.3 \times 10^{12}$ $M_{\odot}$. With an estimate of the M 31 transverse and radial velocities towards the MW, van der Marel *et al.* (2012) applied the timing argument (Kahn & Woltjer 1959) to determine a mass of $\sim 1.5 \times 10^{12}$ $M_{\odot}$ for M 31. A similar value was later determined by Penarrubia *et al.* (2014) with a slightly lower value when the LMC kinematics (Penarrubia *et al.* 2016) are considered (see Table 1). By balancing mass and momentum in the LG, Diaz *et al.* (2014) simultaneously determined the total mass of the LG as the mass ratio of the MW and M 31. They also found a similar mass of $\sim 1.7 \times 10^{12}$ $M_{\odot}$ for M 31.

2.7. *Cosmological simulations*

Over the past decade, a number of large scale cosmological simulations have developed, that are well-constrained to observed scaling relations for massive galaxies (see review by

Vogelsberger *et al.* 2020). One of the earliest use of cosmological simulations to measure the mass of M 31 was from Tollerud *et al.* (2012), who actually used the virial theorem to determine the M 31 mass from satellite radial velocities but calculated a correction factor by applying the same technique to a simulated galaxy system and checking the discrepancy. Over time, techniques have developed to identify observed galaxy analogues in such cosmological simulations based on observed properties and thereby inferring unobserved properties from the simulated analogues. Zhai *et al.* (2020) identified LG analogues in the Millenium simulations based on observed properties of the two galaxies and their mutual orbits. Their M 31 analogues thus identified with a radial in-fall into the MW had a mean mass of $\sim 2.5 \times 10^{12}$ $M_\odot$ which the report as the mass of M 31. Villanueva-Domingo *et al.* (2021) carry out a similar exercise but for the SIMBA and TNG simulations identifying the masses of M 31 analogues given the input positions of the observed M 31 and its three most massive satellites (M 33, M 32, M 110) as well as their radial velocities (for M 31, tangential velocities are also used as a constrain). A similar mean mass of $\sim 2.2 \times 10^{12}$ $M_\odot$ is obtained for the analogues and attributed to M 31. Patel & Mandel (2022) also use a similar approach but identify M 31 analogues in Illustris TNG-dark simulations constrained by the radial and tangential motions of four of its satellites – M 33, NGC 185, NGC 147 & IC 10. They find a mean mass of $\sim 3 \times 10^{12}$ $M_\odot$ for the analogues and attribute it to M 31.

2.8. *Other techniques*

Sofue (2015) combined the disc rotation velocities of M 31 with the radial velocities of its satellites and GCs to construct the grand rotation curve of M 31. They then used mass models for the disc, bulge, and an NFW halo to determine the mass of M 31 to be $\sim 2 \times 10^{12}$ $M_\odot$. Kafle *et al.* (2018) used high-velocity PNe to measure the escape velocity of PNe as a function of radius, while also simultaneously constraining the virial mass and viral radius of M 31. They find a mass of $\sim 8 \times 10^{11}$ $M_\odot$ for M 31.

3. Discussion

Figure 1 shows the reported mass of M 31 as a function of publication year. From the lowest measured mass of M 31 from Wyse & Mayall (1942) to the highest value from Patel & Mandel (2022), the measured mass of M 31 has increased by almost 30 times over the span of the past ~80 years. The initial mass measurements were actually restricted to that of the luminous disc and inner regions of the galaxy. Probing greater and greater enclosed radius of M 31 (see Table 1) using the variety of techniques discussed in Section 2, coupled with improvements in its measured distance, has resulted in a steady increase of the measured mass of M 31 (see Figure 1). Most measurements of the M 31 total mass over the past ~20 years do place its mass within 1–2 $\times 10^{12}$ $M_\odot$.

Benisty *et al.* (2022) find a LG mass of $3.4^{+1.4}_{-1.1} \times 10^{11}$ $M_\odot$ from LG analogues in the Illustris TNG simulations using particularly the M 31 tangential velocities as an important criteria in choosing said analogues. The same simulation suite is used by Patel & Mandel (2022) and Villanueva-Domingo *et al.* (2021) but using different conditions for choosing M 31 analogues to find M 31 masses of ~ 3 & $\sim 2.2 \times 10^{12}$ $M_\odot$ respectively, which are then nearly ~88% &~66% of the mass of the LG. It is thus clear that the mass of M 31 determined using cosmological simulations depends heavily on the choice of priors. These large-scale cosmological simulations have their mean galaxy masses matched to the observed stellar-mass halo-mass relation. This observed relation would require a typical galaxy with the stellar mass of M 31 to have a halo mass of $\sim 10^{13}$ $M_\odot$ (McGaugh 2023). Thus it is unsurprising that, despite including some priors, M 31 analogues in cosmological simulations tend to be on the more massive end (as the typical M 31 mass galaxies in

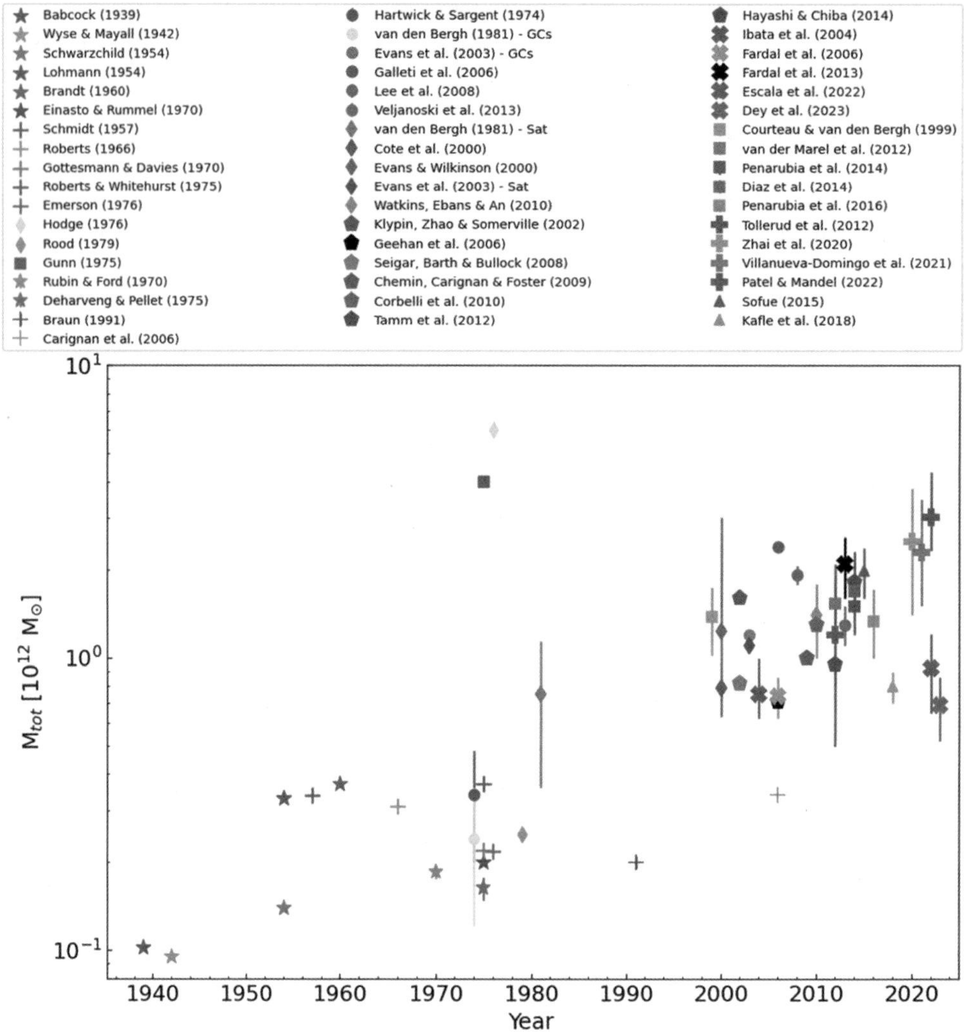

Figure 1. The measured mass of M 31 (in log scale) as a function of publication year. Different symbols refer to different mass measurement techniques that are described in Section 2. Star: RC – HII regions; Narrow plus: RC – HI gas; Diamond: Satellites; Circle – GCs; Square – LG kinematics; Pentagon – Mass models; Cross – Substructures; Broad plus – simulations; Triangles – others.

these simulations lie in more massive haloes). Stronger priors may be necessary to obtain a more accurate measurement of the M 31 mass from their simulated analogues. This is especially true given that there is observational evidence that M 31 has undergone a recent major merger (Bhattacharya *et al.* 2019a,b, 2021, 2022, 2023; Arnaboldi *et al.* 2022) consistent with 2–3 Gyr old 1:4 major merger simulations (Hammer 2018). This is in sharp contrast to that of the MW (see Bhattacharya *et al.* 2023, for a summary). Considering the merger history in choosing the simulated analogues of M 31 may yield more accurate mass measurements.

The most accurate total mass measurements of M 31 for the time being come from different techniques discussed in Section 2. To reiterate, Watkins, Evans & An (2010) find a near-total mass of M 31 as $\sim 1.4 \times 10^{12}$ M$_\odot$ from applying the tracer mass estimator technique to the radial velocities of M 31 satellites (see Sect 2.2). The same technique applied to GCs (Veljanoski *et al.* 2013) also yields a similar M 31 mass (see Sect 2.3). The mass models of M 31 constrained to the latest HI RC in the disc and the surface brightness profile also yield a similar mass (Corbelli *et al.* 2010; see Sect 2.4). A mass of $\sim 2 \times 10^{12}$ $M_{\odot}$ is determined from M 31 using models that are consistent with the GS kinematics (Fardal *et al.* 2013; see Sect 2.5). From the timing argument, Penarrubia *et al.* (2016) find the mass of M 31 to be $\sim 1.33 \times 10^{12}$ $M_{\odot}$ (see Sect 2.6). Finally, from the grand RC, Sofue (2015) measure a mass of $\sim 2 \times 10^{12}$ $M_{\odot}$ for M 31.

Considering the mean of the mass of M 31 determined from these works using the aforementioned different techniques, we find that M 31 has a mass of $\sim 1.56 \times 10^{12}$ $M_{\odot}$. There is still a significant margin to improve the mass measurement of M 31 as understood from the previous discussion.

References

Arnaboldi, M., Bhattacharya, S., Gerhard, O., *et al.* 2022, *A&A*, 666, A109
Baade, W. & Swope, H. H. 1955, *AJ*, 60, 151
Babcock, H. W. 1939, Lick Observatory Bulletin, 498, 41
Bahcall, J. N. & Tremaine, S. 1981, *ApJ*, 244, 805
Benisty, D., Vasiliev, E., Evans, N. W., *et al.* 2022, *ApJ*, 928, L5
Bhattacharya, S., Arnaboldi, M., Hartke, J., *et al.* 2019, *A&A*, 624, A132
Bhattacharya, S., Arnaboldi, M., Caldwell, N., *et al.* 2019, *A&A*, 631, A56
Bhattacharya, S., Arnaboldi, M., Gerhard, O., *et al.* 2021, *A&A*, 647, A130
Bhattacharya, S., Arnaboldi, M., Caldwell, N., *et al.* 2022, *MNRAS*, 517, 2343
Bhattacharya, S., Arnaboldi, M., Hammer, F., *et al.* 2023, arXiv:2304.14151
Brandt, J. C. 1960, *AJ*, 131, 293
Braun, R. 1991, *ApJ*, 372, 54
Carignan, C., Chemin, L., Huchtmeier, W. K., *et al.* 2006, *ApJ*, 641, L109
Corbelli, E., Lorenzoni, S., Walterbos, R., *et al.* 2010, *A&A*, 511, A89
Cote, P., Mateo, M., Sargent, W. L. W., *et al.* 2000, *ApJ*, 537, 91
Courteau, S. & van den Bergh, S. 1999, *AJ*, 118, 337
Chemin, L., Carignan, C. & Foster, T. 2009, *ApJ*, 705, 1395
Deharveng, J. & Pellet, A. 1975, *A&A*, 38, 15
Dey A., Najita, J. R., Koposov, S. E., *et al.* 2023,*ApJ*, 944, 1
Diaz, J. D., Koposov, S. E., Irwin, M., *et al.* 2014, *MNRAS*, 443, 1688
Einasto J., & Rummel U. 1970, in Becker W., Kontopoulos G. I., eds, Proc. IAU Symp. 38, The Spiral Structure of Our Galaxy. Reidel, Dordrecht, p. 51
Emerson, D. T. 1976, *MNRAS*, 176, 321
Escala, I., Gilbert, K. M., Fardal, M. A., *et al.* 2022, *AJ*, 164, 20
Evans, N. W. & Wilkinson, M. I. 2000, *MNRAS*, 316, 929
Evans, N. W., Wilkinson, M. I., Perrett, K. M., *et al.* 2003, *ApJ*, 583, 752
Faber, S. M. & Gallagher, J. S. 1979, *AR&AA*, 17, 135
Fardal, M. A., Babul, A., Geehan, J. J. *et al.* 2006, *MNRAS*, 366, 1012
Fardal, M. A., Weinberg, M. D., Babul, A., *et al.* 2013, *MNRAS*, 434, 2779
Galleti, S., Federici, L., Bellazzini, M. *et al.* 2006, *A&A*, 456, 985
Geehan, J. J., Fardal, M. A., Babul, A., *et al.* 2006, *MNRAS*, 366, 996
Gottesman, S. & Davies, R. 1970, *MNRAS*, 149, 263
Gunn, J. 1975, Comments Ap. Sp. Phys., 6, 7
Hammer F., Yang Y. B., Wang J. L., *et al.* 2018, *MNRAS*, 475, 2754
Hartwick, F. & Sargent, W. 1974, *ApJ*, 190, 283
Hayashi, K. & Chiba, M. 2014, *ApJ*, 789, 62
Hodge, P. W. 1975, Bulletin of the American Astronomical Society, 7, 506
Hodge P. W., ed. 1992, Astrophysics and Space Science Library Vol. 176. Kluwer, Dordrecht
Holland, S. 1998, *AJ*, 115, 1916
Hubble, E. P. 1925, The Observatory, 48, 139

Hubble, E. P. 1929, *ApJ*, 69, 103
Hubble, E. P. & Sandage, A. 1953, *ApJ*, 118, 353
Ibata, R., Chapman, M., Ferguson, A. M. N., *et al.* 2004, *MNRAS*, 351, 117
Kafle, P. R., Sharma, S., Lewis, G. F., *et al.* 2018, *MNRAS*, 475, 4043
Kahn, F. D. & Woltjer, L. 1959, *ApJ*, 130, 705
Klypin, A., Zhao H. & Somerville R. S. 2010, *ApJ*, 573, 597
Lee, M. G., Hwang, H. S., Kim, S. C. *et al.* 2008, *ApJ*, 674, 886
Lohmann, W. 1954, Zeitschrift für Astrophysik, 35, 159
Mateo, M. 1998, *ARA&A*, 36, 435
Mayall, N. U. 1950, Publication of the Observatory of the University of Michigan, 10, 19
McConnachie, A. W., Irwin, M. J., Ferguson, A. M. N., *et al.* 2005, *MNRAS*, 356, 979
McConnachie, A. W., Ibata, R., Martin, N., *et al.* 2018, *ApJ*, 868, 55
McGaugh, S. 2023, arXiv:2305.00858
Navarro, J. F., Frenk, C. S., & White, S. D. M. 1996, *ApJ*, 462, 563
Ostriker, J. P., & Peebles, P. J. E. 1973, *ApJ*, 186, 467
Patel, E. & Mandel, K. S. 2022, arXiv:2211.15928
Penarrubia, J., Ma, Y. Z., Walker, M. G., *et al.* 2014, *MNRAS*, 443, 2204
Penarrubia, J., Gomez, F. A., Besla, G., *et al.* 2016, *MNRAS*, 456, L54
Roberts, M. S. 1966, *ApJ*, 144, 639
Roberts, M. & Whitehurst, R. 1975, *ApJ*, 201, 327
Rood, H. 1979, *ApJ*, 232, 699
Rubin, V. & Ford, W. 1970, *ApJ*, 159, 379
Sandage, A. & Tammann, G. A. 1971, *ApJ*, 167, 293
Savino, A., Weisz, D. R., Skillman, E. D., *et al.* 2022, *ApJ*, 938, 101
Schwarzschild, M. 1954, *AJ*, 59, 273
Schmidt, M. 1957, Bulletin of the Astronomical Institutes of the Netherlands, 14, 17
Seigar, M. S., Barth, A. J., & Bullock, J. S. 2008, *MNRAS*, 389, 1911
Sofue, Y. 2015,*PASJ*, 67, 65
Tamm, A., Tempel, E., Tenjes, P., *et al.* 2012, *A&A*, 536, A4
Tollerud, E. J., Beaton, R. L., Geha, M. C., *et al.* 2012, *ApJ*, 752, 45
van de Hulst, H. C., Raimond, E. & van Woerden, H. 1957, Bulletin of the Astronomical Institutes of the Netherlands, 14, 1
van den Bergh, S. 1969, *ApJS*, 19, 145
van den Bergh, S. 1981, *PASP*, 93, 428
van der Marel, R. P., Fardal, M., Besla, G., *et al.* 2012, *ApJ*, 753, 14
Veljanoski, J., Ferguson, A. M. N., Mackey, A. D., *et al.* 2013, *ApJL*, 768, 33
Villanueva-Domingo, P., Villaescusa-Navarro, F., Genel, S., *et al.* 2021, arXiv:2111.14874
Vogelsberger, M., Marinacci, F., Torrey, P., *et al.* 2020, Nature Reviews Physics, 2, 42
Watkins, L. L., Evans, N. W. & An, J. H. 2010, *MNRAS*, 406, 264
Wyse, A. B. & Mayall, N. U. 1942, *ApJ*, 95, 24
Zhai, M., Guo, Q., Zhao, G., *et al.* 2020, *ApJ*, 890, 27

Dynamical Masses of Local Group Galaxies
Proceedings IAU Symposium No. 379, 2025
P. Bonifacio, M.-R. Cioni, F. Hammer, M. S. Pawlowski & S. Taibi, eds.
doi:10.1017/S1743921323001515

Proper Motions of M31 Satellite Galaxies

S. Tony Sohn[1], Mark Fardal[2], Ekta Patel[3], Gurtina Besla[4], Nitya Kallivayalil[5], Marcel Pawlowski[6], Jenna Samuel[7], Erik Tollerud[1], Laura Watkins[1], Daniel Weisz[8], Andrew Wetzel[9], Andres del Pino[10] and Roeland van der Marel[1]

[1]Space Telescope Science Institute, 3700 San Martin Drive, Baltimore, MD 21218, USA
email: tsohn@stsci.edu

[2]Eureka Scientific, Inc., 2452 Delmer Street, Oakland, CA 94602, USA

[3]Astronomy Dept., University of California at Berkeley, Berkeley, CA 94720, USA

[4]University of Arizona, 933 N. Cherry Avenue, Tucson, AZ 85721, USA

[5]Dept. of Astronomy, The University of Virginia, 530 McCormick Road, Charlottesville, VA 22904, USA

[6]Leibniz-Institut für Astrophysik Potsdam, An der Sternwarte 16, D-14482 Potsdam, Germany

[7]Dept. of Astronomy, The University of Texas at Austin, 2515 Speedway, Austin, TX 78712, USA

[8]Dept. of Astronomy, University of California, Berkeley, CA 94720, USA

[9]Dept. of Physics and Astronomy, University of California, Davis, CA 95616, USA

[10]Centro de Estudios de Física del Cosmos de Aragón (CEFCA), Unidad Asociada al CSIC, Plaza San Juan 1, 44001 Teruel, Spain

Abstract. The field of Galactic archaeology in the Milky Way (MW) has been significantly transformed by the advent of proper motions (PMs), but PM studies are still in their early stages in the M31 system. Measuring PMs in the M31 system poses a challenge for *Gaia*, limiting *HST* and *JWST* as the only reliable options. PM measurements of M31 satellites enable the estimation of M31's total mass, examination of the dynamical stability of the Great Plane of Andromeda, and establishing connections to M31 substructures. We have successfully measured the PMs of NGC 147 and NGC 185 using multiepoch *HST* imaging data. Recently, we have obtained second-epoch HST imaging data for seven additional satellites through the *HST* program GO-16273. We present preliminary results from the PM measurements of these galaxies along with the implications. Finally, we discuss the future prospects of measuring PMs at the distance of M31.

Keywords. Astrometry, proper motions, galaxies: dwarf, galaxies: kinematics and dynamics

1. Introduction

Our understanding of the Milky Way (MW) has grown explosively in the past decade thanks to accurate proper motion (PM) measurements for satellites and substructures of the MW. By extending the available phase-space information to the full six dimensions, PM measurements from *HST* and *Gaia* have revolutionized the field of Galactic archaeology. However, PM studies in and around our neighbor spiral galaxy M31 are still in their early stages. *Gaia* can only detect some of the brightest stars in star forming regions at the distance of M31, so *HST* and *JWST* remain the only reliable options for measuring PMs of M31 satellite galaxies dominated by old stellar populations (e.g., dwarf spheroidals) for now.

1.1. *Historical Background*

The first record of (making an attempt at) measuring PMs of M31 satellites was by van Maanen (1923). van Maanen compared photographic plates of M33 obtained 12 years apart and concluded that M33 has measurable internal rotation as well as systemic bulk motion across the sky. However, as M33's distance estimate became available through the discovery of Cepheids by Hubble (1926), it became obvious that van Maanen's measurements were erroneous and likely suffering from significant systematics. A nice review about how van Maanen's alleged PM measurements of spiral galaxies including M33 *"form a fascinating episode in the history of astronomy"* can be found in Hetherington (1972).

The first *successful* PM measurements of M31 satellites were made by Brunthaler *et al.* (2005) for M33 and Bruntaler *et al.* (2007) for IC 10, both using Very Long Baseline Array (VLBA) observations of H_2O masers against background quasars as reference sources. While VLBA provides extremely accurate positional measurements when sources are available, most of the M31 satellite galaxies are devoid of H_2O masers that are only found in actively star-forming regions. It has only been a few years since we started measuring PMs of M31 satellites using *stars*. Both *Gaia* (van der Marel *et al.* 2019, for M33) and *HST* (Sohn *et al.* 2020; Warfield *et al.* 2023, for NGC 147/185 and And VII, respectively) made this possible, and so far there are about a handful of M31 satellites with PM measurements available in the literature.

1.2. *Ongoing Projects by the HSTPROMO Group*

The *High-resolution Space Telescope Proper Motion* (HSTPROMO†) Collaboration aims at improving our dynamical understanding of stellar systems in the nearby Universe through measurement and interpretation of their PMs. We have several ongoing projects for measuring PMs of the M31 system.

Resolved Proper Motions of M31 and the M31-M32 Interaction (Sohn 2018): We are in the process of measuring the PM of M31 stars to unprecedented precision using multi-epoch HST data of seven ACS/WFC fields. Our goals for this program are threefold. Our first goal is to measure M31's mean tangential motion with an accuracy of $\sim 20\,km\,s^{-1}$. The better-determined center-of-mass (COM) motion of M31 will be essential for refining models of the local Universe, the orbits of objects in M31's halo, the total dark matter halo masses of the Milky Way (MW) and M31, and the evolution and fate of the MW. Second goal is to clearly resolve the internal kinematics of M31 with PMs by obtaining useful precision for multiple structural components and even the brighter individual stars. We will measure the bulk flows between cold components such as M31's disk and giant stream, and resolve the tangential velocity dispersion in M31's hot halo. Our third goal is to obtain the PM of M32 for the first time. This will provide crucial information about M32's orbit, evolution, and interactions with M31. In particular, we will test scenarios where it is the stripped remnant of a major merger, responsible for shaping many of M31's properties.

Resolved Proper Motion of M33 (Sohn 2020a): As the only low-mass disk galaxy in the Local Group, M33 is one of the most important and best-studied galaxies on the sky. However, with the bulge being almost non-existent, it is unclear what makes it stable to formation of a strong bar. One stabilizing factor may be the existence of dynamically

† `https://www.stsci.edu/~marel/hstpromo.html`

hot components such as a thick disk or halo, but these have long been controversial and hard to detect. Recent stellar spectroscopy gives clear evidence of a halo component, but its structural properties, its origin, and its dynamical effects are not yet clear. Our goal for this program is to measure PMs of stars in five fields located at the edge of the M33 disk. We are probing the relative rotational lag of the hotter components (the halo and a possible thick disk) versus the young, thin disk, and measuring the velocity dispersions in the planar and vertical directions separately, using dispersions measured directly from PMs as well as the correlation between the PMs and LOS velocities from spectroscopy. We are checking the halo for non-equilibrium flows indicating a possible origin in a recent merger. We expect to also provide an independent and more precise measurement of M33's COM PM.

Andromeda and the Seven Dwarfs (Sohn 2020b): Through this program, we are in the process of measuring PMs of 7 M31 satellites: And I, II, III, VII, XV, XVI, XXVIII. We are using the PM-based 6D phase space information of satellites to constrain the total M31 mass with uncertainties at the 30% level. We are also testing the dynamical stability of the Great Plane of Andromeda (GPoA) using our PM results for And I, III, and XVI, which will provide clues to the origin of satellite planes observed around M31 as well as the MW. The orbits derived for individual satellites will be used to disentangle the processes that have contributed to the quenching of star formation, to connect our target galaxies to M31 tidal structures, and to test the merger scenario of And II.

In this proceedings contribution, we describe how our PM measurement technique works, and present preliminary results of measuring PMs of M31 satellite galaxies.

2. Proper Motion Measurements

Sohn *et al.* (2012) developed a technique for measuring the PMs of stellar systems in the Local Group (LG) using multi-epoch *HST* data. Their method relies on using compact background galaxies as references for positions. Background galaxies are abundant and evenly distributed in *HST* images, making them suitable local positional references. By accurately measuring the positions of these galaxies, the motions of stars can be directly measured relative to them. The key is to precisely determine the positions of background galaxies. To achieve this, a template is created for each galaxy from stacked images obtained through multiple observations. This template represents the expected distribution of the galaxy's flux among the pixels, based on an assumed center. By fitting the template to the observed pixels using trial offsets, similar to the way an effective point spread function (ePSF) is used for star positions, the galaxy's position can be determined. These templates, referred to as galaxy spread functions (GSFs), are analogous to ePSFs.

The GSF-fitting technique has been successfully used to measure PMs of a variety of stellar systems including satellite dwarf galaxies (Sohn *et al.* 2013, 2017), globular clusters (Sohn *et al.* 2018), and even individual stars along stellar streams in the MW halo (Sohn *et al.* 2015, 2016). More recently, we started using this technique to measure PMs of M31 satellite galaxies. Section 3.1 illustrates one example of using this technique for our PM measurements of two satellites galaxies orbiting around M31, and we will continue applying the same technique to other M31 satellites as presented in Section 1.2.

3. Proper Motions of M31 Satellite Galaxies

Extending the PM measurement technique to distances of M31 and beyond opens up excellent opportunities. In this section, we first summarize our PM results of NGC 147 and NGC 185 Sohn *et al.* (2020) to demonstrate how the PM measurements can improve our dynamical understanding of M31 satellites. We then present preliminary results from

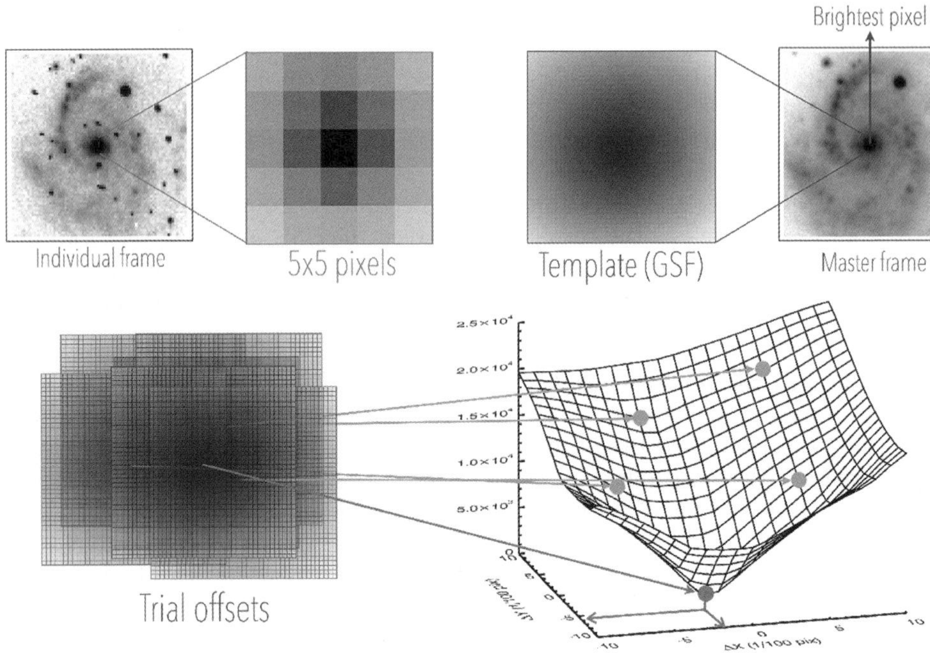

Figure 1. Illustration of the GSF-fitting method. *Upper left*: 5×5 extraction of the individual exposure centered on the brightest pixel within the galaxy. *Upper right*: The galaxy spread function (GSF) corresponding to the 5×5 extraction of the individual exposure, extracted from the stacked image created using multiple dithered observations. *Lower left*: GSF-fitting is done by moving the template in all directions and comparing it with the individual exposure. *Lower right*: The χ^2-like surface plot of the difference between the trial GSF and individual exposure.

our ongoing *HST* program GO-16273. Finally, we discuss future prospects of measuring PMs for other M31 satellites.

3.1. *Proper Motions of NGC 147 and NGC 185*

NGC 147 and NGC 185 are located in the outer halo of M31 and are the best observed of any dwarf elliptical (dE) galaxies. The two galaxies are only separated by ~ 1 deg on the sky, and it has often been assumed that they are a bound pair due to the similarities in their mass, size, and stellar metallicity (Geha *et al.* 2010). However, NGC 147 and NGC 185 are *physically* located ~ 90 kpc apart (Savino *et al.* 2022) and their star formation histories show stark differences (Geha *et al.* 2015), both hinting that the two dwarf galaxies may not share a common origin. Also, NGC 147 exhibits 50 kpc-long tidal tails pointing nearly radially toward and away from M31, while NGC 185 appears relatively undisturbed suggesting the two galaxies may have had different orbital trajectories.

In Sohn *et al.* (2020), we measured the PMs of NGC 147 and NGC 185 for the first time using multiepoch *HST* ACS/WFC and WFC3/UVIS imaging data with time baselines of ~ 8 yr. The PM results of the two galaxies are shown in Figure 2. The PM measurements based on the two different detectors (ACS/WFC and WFC3/UVIS) for each galaxy agree to within their uncertainties in each coordinate, and the PM uncertainties are 40–50 $km\,s^{-1}$ at the distance of the dwarfs. By utilizing the measured proper motion (PM) values of NGC 147 and NGC 185, we demonstrated their on-sky movements relative to M31. Our analysis revealed that the motion direction of NGC 147 aligns with its tidal tails, as depicted in Figure 3 (Left). Additionally, we observed that the motions of both

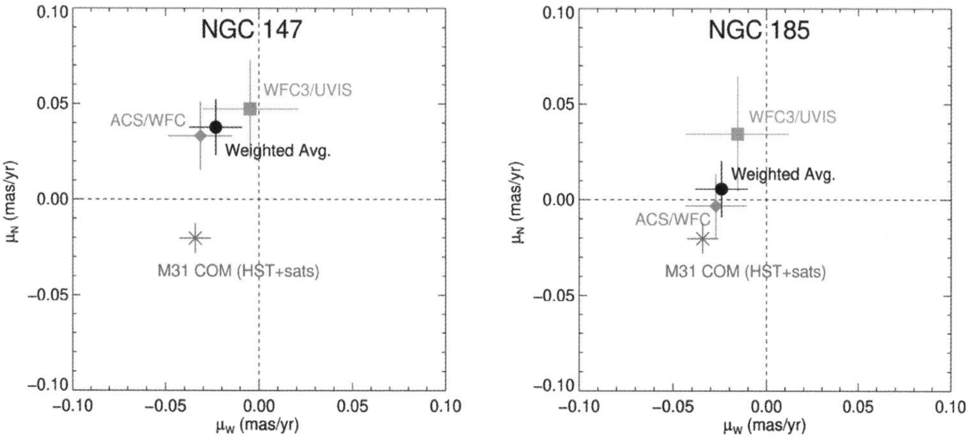

Figure 2. PM diagrams for NGC 147 (left) and NGC 185 (right). Red and blue symbols are for the ACS/WFC and WFC3/UVIS measurements, respectively. The weighted averages are shown as the black circle. For comparison, we also plot the M31 center of mass motion.

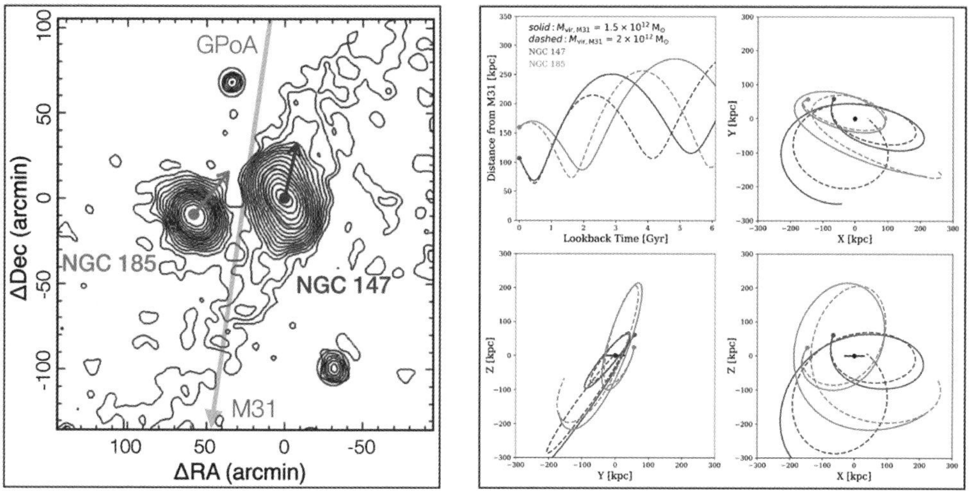

Figure 3. *Left*: On-sky motions of NGC 147 (purple) and NGC 185 (orange) with respect to M31 plotted over the surface density variation of RGB stars. The relative sizes of arrows were chosen to represent the comparative on-sky motions of the two galaxies. The green line that bisects the plot represents the the Great Plane of Andromeda.
Right: Orbits of NGC 147 (purple) and NGC 185 (orange) in the past 6 Gyr for a low-mass ($M_{\rm vir} = 1.5 \times 10^{12} {\rm M}_{\odot}$) M31 model (solid lines) and a high-mass ($M_{\rm vir} = 2 \times 10^{12} {\rm M}_{\odot}$) M31 model (dashed lines).

galaxies are in accordance with rotation along the Great Plane of Andromeda (GPoA; Ibata *et al.* 2013; Pawlowski *et al.* 2013).

We transformed the three-dimensional positions and velocities of both galaxies into a common coordinate system centered on M31 to study the detailed orbital histories of the combined M31-NGC 147-NGC 185 system using numerical orbit integration (Figure 3, Right). We found that NGC 147 (NGC 185) had its closest passage to M31 0.3–0.5 Gyr (~ 1.6 Gyr) within the past 6 Gyr at distances of ~ 70 kpc (70–260 kpc), and the pericentric times of NGC 147/NGC 185 correlate qualitatively well with the presence or absence of tidal tails seen around the galaxies.

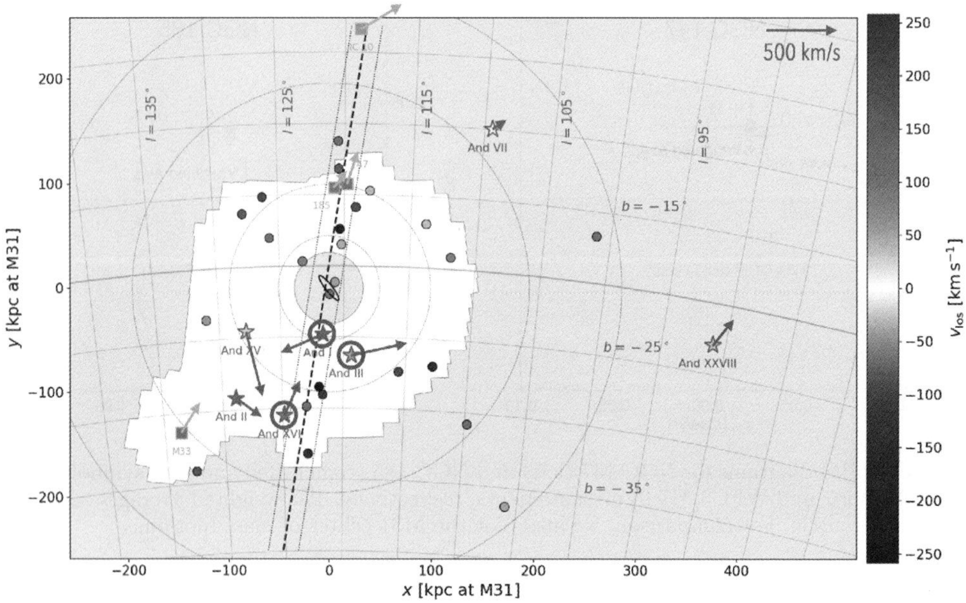

Figure 4. Spatial distribution of the M31 satellite system in Galactic (Δl, Δb) coordinates in units of kpc at the M31's distance centered on M31 (ellipse). Projected distances at 50, 100, 200, and 300 kpc are shown in dashed concentric circles around M31. The target galaxies of *HST* program GO-16273 are labeled and plotted in stars, those with existing PM measurements are plotted in squares, and all other satellites are plotted in circles. The arrows indicate two-dimensional motions on the plane of the sky with respect to M31 based on their PM measurements. The length of these arrows are drawn in relative scale with an on-sky velocity of 500 $km\,s^{-1}$ shown on the top right. The color of each satellite galaxy indicates the measured $v_{\rm LOS}$. The thick dash-line represents the best fit to the Great Plane of Andromeda (GPoA) discovered by Ibata *et al.* (2013), and the thin dotted-lines define the rms width of it. Among the targets of GO-16273, the ones identified as members of the GPoA are marked with purple circles.

3.2. *Proper Motions of Seven Dwarf Spheroidal Galaxies Orbiting around M31*

Through the *HST* program GO-16273, we obtained second-epoch imaging of seven M31 satellites (And I, II, III, VII, XV, XVI, and XXVIII), targeting the same fields with earlier deep *HST* observations to measure PMs. We have completed preliminary PM measurements of all seven dwarf galaxies using the GSF-fitting technique presented in Section 2, and the PM uncertainties are in the range 50–100 $km\,s^{-1}$ at the distances of the dwarfs. We examine the on-sky motions of the seven dwarf galaxies by subtracting the contribution of M31's motion, based on the most recent M31 PM measurement using *Gaia* EDR3 data Salomon *et al.* (2021), at the position of each galaxy based on their distances, and then plotting the corresponding vector as shown in Figure 4. Our sample includes three satellites (And I, III, and XVI) that are members of the GPoA as identified by Ibata *et al.* (2013) based on their location on the sky and $v_{\rm LOS}$ measurements. One important question about the GPoA (or any plane of satellites for that matter) is whether it is a dynamically stable structure or a simple transient feature. PM measurements are required to judge if a satellite is rotating along the plane, and both NGC 147 and NGC 185 that lie in the Northern half of M31 have been claimed to do so. Among the three satellites observed that lie in the Southern half of M31, only And XVI appear to have motions consistent with rotating along the plane, whereas And I and III are moving in a direction almost perpendicular to the GPoA. There are a total of five satellites on the plane that have PM measurements, and three of them seem

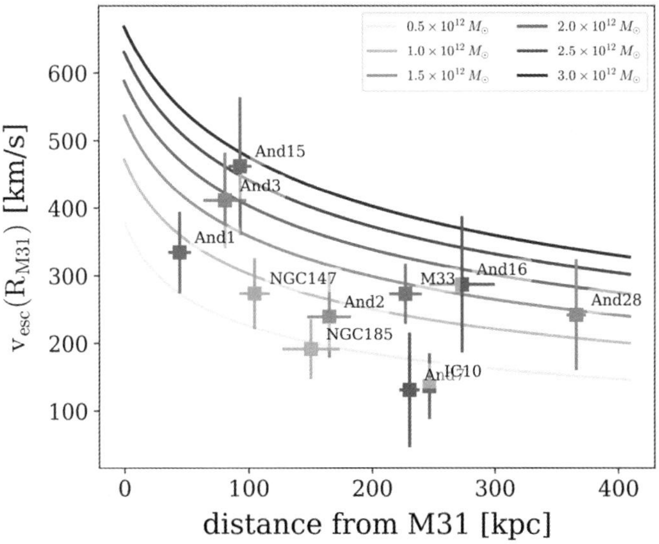

Figure 5. Total velocities as a function of M31-centric distance for all M31 satellite galaxies with proper motion (PM) measurements available. For the velocity zero point of each satellite, we adopted the M31 PM measurement by Salomon *et al.* (2021). PM measurement for M33 is from van der Marel *et al.* (2019), for IC10 is from Bruntaler *et al.* (2007), and for NGC 147 and NGC 185 are from Sohn *et al.* (2020). PMs for all other M31 satellites in this plot are from the preliminary measurements of program GO-16273. We considered six different M31 virial masses ranging from 0.5 to $3 \times 10^{12} M_{\odot}$. The escape velocity curves are calculated assuming an NFW profile for M31's halo, with the concentration parameter determined using Equation (12) from Patel, Besla, & Sohn (2017).

to be dynamically linked to the plane. Due to the limited sample size, it is challenging to draw definitive conclusions about the plane's dynamical state. However, as the sample size grows, it will undoubtedly bring fresh insights to the subject.

Satellite galaxies serve as tracers for the dark matter halo of their host galaxies, making measurements of their three-dimensional positions and velocities valuable for estimating the total mass of the hosts. Under the assumption that a satellite galaxy is gravitationally bound to its host, the total velocity ($v_{\rm tot}$) must be lower than the escape velocity $v_{\rm esc} = \sqrt{-2\phi(r)}$, where r is the distance from the center of the host galaxy to the satellite and $\phi(r)$ is the halo's gravitational potential. Thus, the escape velocity at a specific satellite distance provides a lower limit for the total mass of the host galaxy required to retain the satellite. In Figure 5, we compare the observed total velocities of all M31 satellites with existing PM measurements, including the newly-measured ones from the GO-16273 study, to the escape velocities as a function of the M31-centric distance. Figure 5 indicates that to keep And XV bound, M31's virial mass would have to be higher than $2.5 \times 10^{12} M_{\odot}$. However, considering only And XV lies above the $2.5 \times 10^{12} M_{\odot}$ curve, it may be more reasonable to deduce that the minimum virial mass of M31 is closer to $2 \times 10^{12} M_{\odot}$ based on the total velocities of three satellites at a wide range of M31-centric distance: And I, XVI, and XXVIII. Note that the escape velocity only provides insight into the *minimum* virial mass, so the results from this plot is still consistent with the recent study of Patel & Mandel (2023) which found that M31's virial mass may be as high as $\sim 3 \times 10^{12} M_{\odot}$.

3.3. *Future Prospects*

With the remarkable astrometric capabilities of the *HST*, we anticipate obtaining PM measurements for over ten satellite galaxies of M31 in the coming years, once the projects listed in Section 1.2 reach completion. Achieving accurate PM measurements for any M31 satellite galaxy necessitates an optimal set of first-epoch data. Through the *HST* program GO-15902 (Weisz 2019), we have acquired deep and well-dithered images using ACS/WFC and WFC3/UVIS for 23 M31 satellites. Our plan involves conducting follow-up imaging of these galaxies in a few years using either *HST* or *JWST* to measure their PMs, which will bring the total number of satellites with PM measurements available comparable to that of the MW.

Moreover, we are engaged in a dedicated *JWST* GTO Program (1305) specifically designed for measuring PMs of four M31 satellites: And I, III, XIV, and XVII. PM measurements for And I and III have already been obtained through the *HST* GO-16273 (Section 3.2), and our data will contribute to establishing the GSF-fitting method and controlling systematics for *JWST*. As for And XIV and XVII, their PM measurements will be acquired for the first time when we obtain the second-epoch images in a few years. The larger mirror size of *JWST* provides a significant advantage over *HST* by requiring considerably less exposure time for PM measurements. Additionally, the near-infrared domain offers substantially brighter background galaxies on average, and thereby expected to enhance the reliability of GSF-fitting based PM measurements with *JWST* compared to *HST*. Looking ahead, the Nancy Grace Roman Space Telescope, scheduled for launch in 2027, will possess a field of view 100 times larger than *HST* and astrometric capabilities akin to both *HST* and *JWST*, making it an exceptional instrument for PM measurements (WFIRST Astrometry Working Group *et al.* 2019). Considering all these factors, the future looks promising for measuring PMs of galaxies in the M31 system.

References

Anderson, J., & King, I. R. 2000, *PASP*, 112, 1360
Arias, V., Guglielmo, M., Fernando, N., *et al.* 2016, *MNRAS*, 456, 1654
Brunthaler, A., Reid, M. J., Falcke, H., Greenhill, L. J., & Henkel, C. 2005, *Science*, 307, 1440
Brunthaler, A., Reid, M. J., Falcke, H., Henkel, C., & Menten, K. M. 2007, *A&A*, 462, 101
Geha, M., van der Marel, R. P., Guhathakurta, *et al.* 2010, *ApJ*, 711, 361
Geha, M., Weisz, D., Grocholski, A., *et al.* 2015, *ApJ*, 811, 114
Hetherington, N. S. 1972, *QJRAS*, 13, 25
Hubble, E. P. 1926, *ApJ*, 63, 236
Ibata, R. A., Lewis, G. F., Conn, A. R., *et al.* 2013, *Nature*, 493, 62
Pawlowski, M. S., Kroupa, P., & Jerjen, H. 2013, *MNRAS*, 435, 1928
Patel, E., Besla, G., & Sohn, S. T. 2017, *MNRAS*, 464, 3825
Patel, E., & Mandel, K. S. 2023, *ApJ*, 948, 104
Salomon, J. -B., Ibata, R., Reylé, C., *et al.* 2021, *MNRAS*, 507, 2592
Savino, A., Weis, D. R., Skillman, E. D., *et al.* 2022, *ApJ*, 938 101
Sohn, S. T., Anderson, J., van der Marel, R. P. 2012, *ApJ*, 753, 15
Sohn, S. T., Besla, G., van der Marel, R. P., *et al.* 2013, *ApJ*, 768, 139
Sohn, S. T., van der Marel, R. P., Carlin, J. L., *et al.* 2015, *ApJ*, 803, 56
Sohn, S. T., van der Marel, R. P., Kallivayalil, N., *et al.* 2016, *ApJ*, 833, 235
Sohn, S. T., Patel, E., Besla, G., *et al.*2017, *ApJ*, 849, 93
Sohn, S. T. 2018, *HST* Proposal. Cycle 26, ID. #15658
Sohn, S. T., Watkins, L. L., Fardal, M. A., *et al.* 2018, *ApJ*, 862, 52
Sohn, S. T. 2020a, *HST* Proposal. Cycle 28, ID. #16274
Sohn, S. T. 2020b, *HST* Proposal. Cycle 28, ID. #16273
Sohn, S. T., Patel, E., Fardal, M. A., *et al.* 2020, *ApJ*, 901, 17
van der Marel, R. P., Fardal, M. A., Sohn, S. T., *et al.* 2019, *ApJ*, 872, 24
van Maanen, A. 1923, *ApJ*, 57, 264
Warfield, J. T., Kallivayalil, N., Zivick, P., *et al.* 2023, *MNRAS*, 519, 1189
Weisz, D. R. 2019, *HST* Proposal. Cycle 27, ID. #15902

WFIRST Astrometry Working Group, Sanderson, R. E., Bellini, A., *et al.* 2019, *Journal of Astronomical Telescopes, Instruments, and Systems*, 5, 044005

Discussion

OLIVER MÜLLER: Is the Giant Stellar Stream (GSS) proper motion aligned with the Great Plane of Andromeda (GPoA)?

TONY SOHN: The implied direction on the sky based on our preliminary measurements is aligned with the GSS itself, so yes the proper motion is fairly aligned with the GPoA.

Dynamical Masses of Local Group Galaxies
Proceedings IAU Symposium No. 379, 2025
P. Bonifacio, M.-R. Cioni, F. Hammer, M. S. Pawlowski & S. Taibi, eds.
doi:10.1017/S1743921323003071

Formation of the Andromeda Giant Stream and the 10 kpc Ring Structure in the Andromeda Galaxy

Masao Mori[1], Ayami Hotta[2] and Koki Otaki[2]

[1]Center for Computational Sciences, University of Tsukuba,
1-1-1 Tennodai, Tsukuba, Ibaraki, Japan
email: mmori@ccs.tsukuba.ac.jp

[2]Graduate School of Pure and Applied Sciences, University of Tsukuba,
1-1-1 Tennodai, Tsukuba, Ibaraki, Japan

Abstract. We investigate the possible link between the Andromeda Giant Stream (AGS) and the 10 kpc ring structure using N-body/SPH simulations of a minor merger between the M31 and a satellite galaxy with a mass of $10^{10} M_{\odot}$. The simulation result successfully matches the observed features of the AGS and the 10 kpc ring concurrently. The simulation reproduced the observations, showing that the stars are smoothly distributed in the galactic disk, and the gas is shaped in a ring-like structure. In addition, we demonstrate the spatial metallicity distribution of the merger remnant, assuming the metallicity gradient of the progenitor galaxy. The result remarkably captures the observed features in the AGS exhibiting non-uniform metallicity distribution perpendicular to the AGS axis. These results indicate that a minor merger with a massive dwarf galaxy is capable of simultaneously forming the AGS and the 10 kpc ring.

Keywords. Andromeda Galaxy, Galaxy Collision, Hydrodynamics

---

1. Introduction

Recent observations around M31 have revealed many traces of past interactions with satellite galaxies. In particular, the Andromeda Giant Stream (AGS) in the halo and the double ring structure in the disc have been drawing attention. The AGS is a giant structure extending more than 100 kpc from the centre of M31 and is thought to have been formed in a collision with a satellite galaxy about one billion years ago (Fardal et al. (2007); Mori & Rich 2008). On the other hand, Block et al. (2006) found the double ring structure in the M31 disc made up of gas and dust and argued that the structure was formed by a head-on collision of a satellite galaxy, M32, around two billion years ago. They conclude that the mass of M32 at the time of the collision was about ten times the present-day mass of M32 $10^{11} M_{\odot}$. Moreover, a model has been proposed by Hammer et al. (2018) in which the AGS and the double ring structure are formed simultaneously by a major merger with a mass of more than $10^{11} M_{\odot}$ for the first passage at the large pericentric distance. These situations motivate us to investigate the possible link between the Andromeda Giant Stream (AGS) and the 10kpc ring structure using N-body/SPH simulations of a minor merger between M31 and a satellite galaxy with a mass of $10^{10} M_{\odot}$.

In the following the numerical method is presented in section 2. The results of the simulations are then presented in Section 3 and summarised in Section 4.

2. Numerical method

We adopt the hierarchical tree algorithm for self-gravity. The acceleration of the particle at a position $\boldsymbol{r}_i$ in a gravitational field consisting of N particles is obtained as

$$\frac{\mathrm{d}\boldsymbol{v}_i}{\mathrm{d}t} = -\sum_{j=1}^{N} \frac{Gm_j(\boldsymbol{r}_i - \boldsymbol{r}_j)}{(r_{ij}^2 + \epsilon^2)^{3/2}}, \tag{1}$$

where $r_{ij} = |\boldsymbol{r}_i - \boldsymbol{r}_j|$ and G is the gravitational constant. $\boldsymbol{v}_j$ and m_j are the velocity vector and a particle mass at $\boldsymbol{r}_j$, respectively. To avoid numerical divergence, the softening parameter ϵ is introduced. The three-dimensional smoothed particle hydrodynamics (SPH: Lucy 1977; Gingold & Monaghan 1977) method for gas dynamics. In the SPH formulation, the gas density of one particle at a position $\boldsymbol{r}_i$ is given by

$$\rho_i = \sum_{j=1}^{N_{\mathrm{neigh}}} m_j W(r_{ij}, h_i), \tag{2}$$

where $W(r,h)$ is the smoothing kernel, h is the smoothing length, and $N_{\mathrm{neigh}} = 200$ is the number of neighbour particles. We adopt the (Wendland 1995) C^4 function,

$$W(r,h) = \frac{495}{32\pi h^3} \begin{cases} (1-q)^6(1+6q+\frac{35}{3}q^2), & q \leq 1, \\ 0, & q > 1, \end{cases} \tag{3}$$

where $q = r/h$, to avoid the clumping instability (Dehnen 2012). We basically followed the formulation of SPH introduced by Springel & Hernquist (2002). The smoothing length h_i of each particle is determined by $4\pi h_i^3 \rho_i / 3 = \overline{m} N_{\mathrm{neigh}}$, where $\overline{m}$ is an average mass of gas particles. These equations (2) and h_i need to be solved implicitly for ρ_i and h_i. However, the minimum value of smoothing length h is set to gravitational softening ϵ in order to match the spatial resolution.

The momentum equation for a gas particle is given by

$$\frac{\mathrm{d}\boldsymbol{v}_i}{\mathrm{d}t} = -\sum_j m_j \left[f_i \frac{p_i}{\rho_i^2} \nabla_i W(r_{ij}, h_i) + f_j \frac{p_j}{\rho_j^2} \nabla_i W(r_{ij}, h_j) + \Pi_{ij} \nabla_i \overline{W}_{ij} \right] \tag{4}$$

where p_i is pressure of gas particle, f_i is defined by

$$f_i = \left(1 + \frac{h_i}{3\rho_i} \frac{\partial \rho_i}{\partial h_i}\right)^{-1}, \tag{5}$$

and $\overline{W}_{ij}$ is a symmetrised kernel,

$$\overline{W}_{ij} = \frac{1}{2} \left[W(r_{ij}, h_i) + W(r_{ij}, h_j) \right]. \tag{6}$$

Artificial viscosity Π_{ij} is necessary for the proper handling of shocks. We adopt Monaghan (1997)'s artificial viscosity $\widetilde{\Pi}_{ij}$ combined with Balsara (1995)'s switch F_i. The artificial viscosity is expressed as

$$\Pi_{ij} = \frac{F_i + F_j}{2} \widetilde{\Pi}_{ij}, \tag{7}$$

where

$$\widetilde{\Pi}_{ij} = \begin{cases} -\alpha \dfrac{v_{ij}^{\rm sig} w_{ij}}{\rho_i + \rho_j} & \boldsymbol{v}_{ij} \cdot \boldsymbol{r}_{ij} < 0, \\ 0 & \boldsymbol{v}_{ij} \cdot \boldsymbol{r}_{ij} \geq 0, \end{cases} \tag{8}$$

$$v_{ij}^{\rm sig} = c_{{\rm s},i} + c_{{\rm s},j} - 3w_{ij}, \tag{9}$$

$$w_{ij} = \boldsymbol{v}_{ij} \cdot \boldsymbol{r}_{ij} / |\boldsymbol{r}_{ij}|, \tag{10}$$

and

$$F_i = \frac{|\nabla_i \cdot \boldsymbol{v}_i|}{|\nabla \cdot \boldsymbol{v}_i| + |\nabla_i \times \boldsymbol{v}_i| + 0.0001 c_{{\rm s},i}/h_i}. \tag{11}$$

In order to handle strong shock waves generated by the galaxy collisions, we put the parameter $\alpha = 5$, which adjusts the strength of the artificial viscosity. In this study, an isothermal process is assumed.

3. Simulation

In our simulation, M31 is assumed to consist of four components: a dark matter halo(DMH), a bulge, a stellar disc and a gas disc. The total mass of the stellar and gas discs is assumed to be $1.88 \times 10^{10} M_{\odot}$. The infalling satellite is also assumed to have four components, a DMH, a bulge, a stellar disc and a gas disc. The total mass of the stellar and gas discs is assumed to be $1.88 \times 10^{10} M_{\odot}$. The density distribution of the DMH is assumed to be the Navarro-Frenk-White profile (Navarro, Frenk & White 1996), where the mass of DMH is M_{200} and the masses of the bulge, the stellar disc, and the gas disc are $0.01M_{200}$, $0.45M_{200}$ and $0.45M_{200}$ respectively. The number of particles in the bulge, the gas disc, and the stellar disc are $3.24 \times 10^5 M_{\odot}$, $1.88 \times 10^5 M_{\odot}$ and $1.88 \times 10^5 M_{\odot}$, respectively. The mass of all M31 particles is set to be $10^5 M_{\odot}$. The DMH of the M31 is treated as a fixed potential. The mass of the DMH particles in the progenitor is $10^5 M_{\odot}$, and the masses of the stellar and gas particles are $10^4 M_{\odot}$, with the number of particles varying with the mass of the progenitor. The Toomre's Q-value in the stellar disc of each galaxy is set to 1.0. Initial conditions for each galaxy are created by the MAGI (Miki & Umemura 2018), and the simulation code is developed by Otaki & Mori (2022, 2023). The thermal evolution of the gas is assumed to be an isothermal process with a temperature of 10^4 K.

As mentioned before, the M31 disc contains the 10 kpc ring composed of dust, which is thought to have been formed by past galactic collisions, and the colliding galaxies are believed to be M32, a satellite galaxy of M31. Here, we examine that collisions of the AGS progenitor may have formed the 10 kpc ring. So far, in the study of these formation processes, the mass of the colliding galaxies was thought to be around $10^{11} M_{\odot}$. The results presented in our previous studies indicate that the minor merger hypothesis is favoured for the orbitals forming the AGS. Therefore, in this study, we perform to verify whether the 10 kpc ring is formed in the minor merger.

The results of our simulation show that for galaxies with a mass of $10^9 M_{\odot}$, the effect of the impact on the M31 disc is negligible, and a too-faint ring-like structure is formed. Figure 1 shows the result of the simulation for the collisions of galaxies with a mass of around $10^{10} M_{\odot}$. In this case, the galaxy collision is able to form clear ring-like structures in the gas disc while maintaining the smooth distribution of the M31 stellar disc. At the same time, the simulation results show that the AGS and the two shell structures are also well reproduced. Thus, it was shown that it is possible to form the AGS and the 10 kpc ring simultaneously, even under the conditions of the minor merger hypothesis.

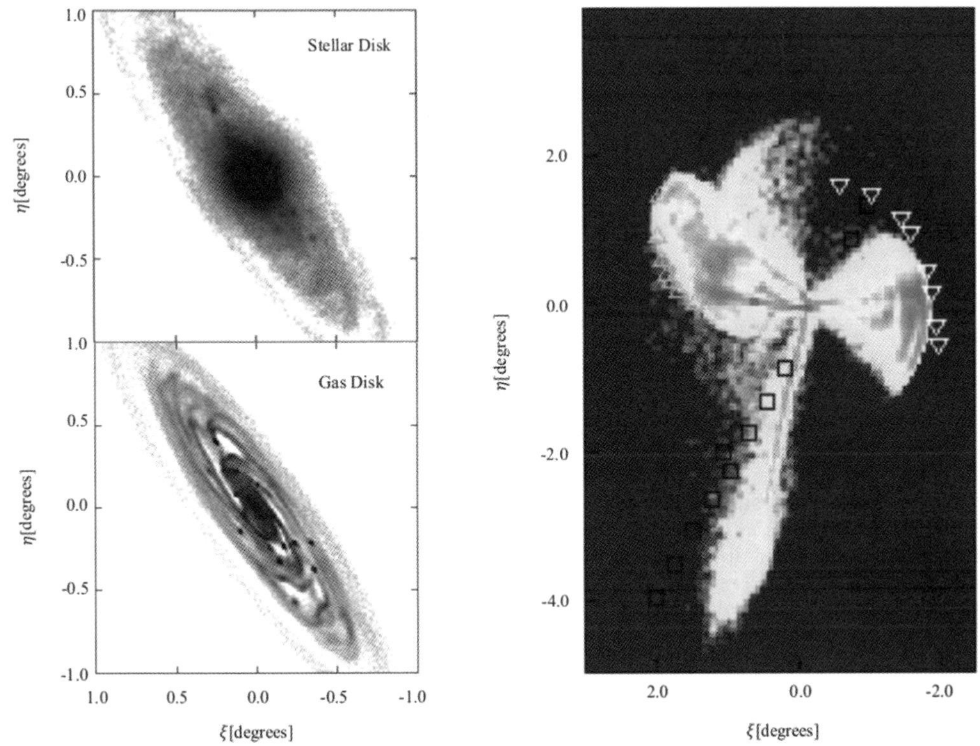

Figure 1. Result of N-body/SPH simulation: density distribution of the star (upper-left panel) and the gas (lower-left panel) in the M31 disc. The right panel shows the distribution of stars in the disc and bulge originally associated with satellite galaxies.

However, in such a case, the progenitor would need to have a mass of $10^{10} M_\odot$ instead of $10^9 M_\odot$, as considered in previous studies.

Recent observations have yielded metal abundance distributions for the halo of M31. According to Preston et al. (2021), the average metal abundance of the AGS is obtained as $[\mathrm{Fe/H}] = -0.5 \pm 0.03$. Furthermore, they showed that metal-rich stellar components dominate the stellar component of the AGS. Therefore, we investigate whether the simulation results of this study can reproduce the metal abundance information. We assume that the distribution of metal abundances within the progenitor is determined by

$$[\mathrm{Fe/H}](R) = a \times \log\left(\frac{R}{R_d}\right) + b, \tag{12}$$

and depends on the radius. In this equation, a and b are free parameters. These two free parameters were determined subject to the metal content of the AGS given by Preston et al. (2021) and the mass-metallicity relation of Dekel & Woo (2003). Figure 2 shows the results for combinations satisfying the above conditions with $a = -1.5$ and $b = -0.9$. It is clearly shown that the spatial distribution of stellar metal abundances in accreting satellite galaxies spread over the halo of M31. The left panel shows the spatial distribution of metal-poor stars with $[\mathrm{Fe/H}] < -1.0$, and the right panel shows that of metal-rich stars with $[\mathrm{Fe/H}] > -1.0$, the simulation performed in this study reproduces the observation of a predominance of a stellar component with high metal content in the AGS.

So far, the metal-rich stellar components were considered to be distributed mainly in the NE shell. However, the current simulation shows that they are distributed mainly

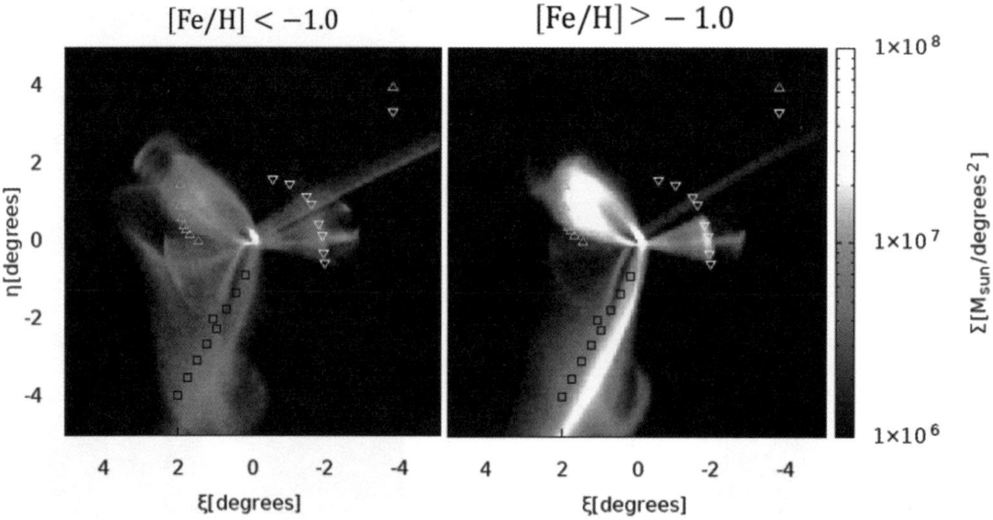

Figure 2. Spatial distribution of stellar metal abundances in accreting satellite galaxies spread over the halo of M31. Left panel: spatial distribution of metal-poor stars with [Fe/H] < -1.0. Right panel: that of metal-rich stars with [Fe/H] > -1.0.

around the AGS and NE shell in the present study. Thus, the latest observations can be reproduced by our simulation, considering past collisions.

4. Summary and Discussion

We investigated the potential correlation between the AGS and the 10 kpc ring structure by utilizing N-body/SPH simulations. These simulations were designed to model a minor merger event between M31 and a satellite galaxy possessing a mass of $10^{10} M_\odot$. The results of these simulations successfully reproduced both the observed features of the AGS and the 10 kpc ring. While the stars in the galactic disc were smoothly distributed, the gas ring was present that accurately recreated the observations. Additionally, we examined the spatial metallicity distribution of the merger remnants by assuming the metallicity gradient of the progenitor galaxy. The outcome of this analysis notably replicated the observed characteristics of the AGS, which display a non-uniform metallicity distribution that is perpendicular to its axis. These findings suggest that the minor merger involving the massive dwarf galaxy has the potential to generate both the AGS and the 10 kpc ring structure concurrently.

The present study conducted hydrodynamic calculations assuming an isothermal approximation. However, future calculations should include radiative cooling processes for a more realistic representation. Recent observations indicate that star formation is actively transpiring in the high-density zone of the 10 kpc ring. Nevertheless, as observed, the star formation timescale exceeds the dynamical time of the galactic disc. This implies that the 10 kpc ring is stationary in its position and does not move dynamically, which is challenging to maintain. Further investigations concerning multiple galaxy collisions and the effects of non-axisymmetric central bulges on dynamics are required to achieve more accurate results.

This work was supported by JSPS KAKENHI Grant Numbers JP20K04022. Numerical computations were performed with computational resources provided by the Multidisciplinary Cooperative Research Program in the Center for Computational Sciences, University of Tsukuba.

References

D. L. Block, F. Bournaud, F. Combes, R. Groess, P. Barmby, M. Ashby, G. Fazio, M. Pahre, and S. Willner. 2006, *Nature*, 443, 832

Dehnen, W. and Aly, H. 2012, *MNRAS*, 425, 1068

Dekel, A., and Woo, J. 2003, *MNRAS*, 344, 1131

Fardal, M. A., Guhathakurta, P., Babul, A. and McConnachie, A. W. 2007, *MNRAS*, 380, 15

Gingold, R. A. and Monaghan, J. J. 1977, *MNRAS*, 181, 375

Hammer, F., Yang, Y. B., Wang, J. L., Ibata, R., Flores, H. and Puech, M., 2018, *MNRAS*, 475, 275

Lucy, L. B. 1977, *AJ*, 82, 1013

Miki, Y. and Umemura, M. 2018, *MNRAS*, 475, 2269

Monaghan J. J. 1997, *Journal of Computational Physics*, 136, 298

Balsara, D. S. 1995, *Journal of Computational Physics*, 121, 357

Mori, M. and Rich, R. M. 2008, *MNRAS*, 674, 77

Navarro, J. F., Frenk, C. S. and White, S. D. M. 1966, *ApJ*, 462, 563

Otaki, K. and Mori, M. 2022, *LNCS* 13378, 373

Otaki, K. and Mori, M. 2023, *submitted to MNRAS*

Preston, J., Collins, M., Rich, R. M., Ibata, R., Martin, N. F. and Fardal, M. 2021, *MNRAS*, 504, 3098

Springel, V. and Hernquist, L. 2002, *MNRAS*, 333, 6495

Wendland, H. 1995, *Adv Comput Math*, 4, 389

Dynamical Masses of Local Group Galaxies
Proceedings IAU Symposium No. 379, 2025
P. Bonifacio, M.-R. Cioni, F. Hammer, M. S. Pawlowski & S. Taibi, eds.
doi:10.1017/S1743921323001692

Revisiting mass estimates of the Milky Way

Yongjun Jiao[1], François Hammer[1], Haifeng Wang[2], Jianling Wang[1,3] and Yanbin Yang[1]

[1]GEPI, Observatoire de Paris, Université PSL, CNRS, Place Jules Janssen, 92195 Meudon, France
email: yongjun.jiao@obspm.fr

[2]CREF, Centro Ricerche Enrico Fermi, Via Panisperna 89A, I-00184 Roma, Italy

[3]CAS Key Laboratory of Optical Astronomy, National Astronomical Observatories, Beijing 100101, China

Abstract. We use the rotation curve from *Gaia* data release (DR) 3 to estimate the mass of the Milky Way. We consider an Einasto density profile to model the dark matter component. We extrapolate and obtain a dynamical mass $M = 2.75^{+3.11}_{-0.48} \times 10^{11} M_{\odot}$ at 112 kpc. This lower-mass Milky Way is consistent with the significant declining rotation curve, and can provide new insights into our Galaxy and halo inhabitants.

Keywords. Galaxy: kinematics and dynamics – Galaxy:structure – dark matter – methods: numerical

1. Introduction

The rotation curve (RC) is an essential tool for estimating the enclosed mass at different radii. The Milky Way (MW) disc is in relatively equilibrium because there was no major merger since 9 to 10 Gyr ago (Hammer et al. 2007; Haywood et al. 2018). Then, its dynamical mass can be well established from its RC. Previous RC studies based on the second *Gaia* data release (DR2, Gaia Collaboration et al. 2018) provided a declining MW RC, with a slope of $\beta = -(1.4 \pm 0.1)$ km s^{-1} kpc^{-1} and $-(1.7 \pm 0.1)$ km s^{-1} kpc^{-1} for Mróz et al. (2019) and Eilers et al. (2019), respectively. Such declining RCs have led to estimates of the total MW mass near or well below $10^{12} M_{\odot}$ (de Salas et al. 2019; Eilers et al. 2019; Karukes et al. 2020; Jiao et al. 2021).

The *Gaia* DR3 (Gaia Collaboration et al. 2022) provides a wider sample of stars with new determinations of spectra, radial velocity, chemical abundance, etc. Recently, Wang et al. (2023) have applied a statistical inversion method introduced by Lucy (1974) to reduce the errors in the distance determination in the Gaia DR3 data-set and provided the MW rotation curve $v_C(R)$ up to about 27.5 kpc. Their RC is in reasonably good agreement with the RC measurement based on *Gaia* DR2 (Mróz et al. 2019; Eilers et al. 2019) and DR3 (Ou et al. 2023). However, Zhou et al. (2023) found a systematically higher RC based on *Gaia* DR3 when it is compared to Wang et al. (2023) and Ou et al. (2023). Ou et al. (2023) argue that this systematic differences are caused by different methods of measuring distance. In particular, Wang et al. (2023) found that a steeper slope of RC, with $\beta = -(2.3 \pm 0.2)$ km s^{-1} kpc^{-1}, which is smaller than previous studies. A more rapidly declining RC provides a more stringent constraint, leading to a lower mass MW. Jiao et al. (2021) have shown that a NavarroFrenkWhite (NFW) profile could not be reconciled with a low mass MW and that an Einasto profile gives a larger mass range. Here we focus on the Einasto profile for the fitting of the MW RC.

2. Method

The baryonic model of this study is the same model that we used in Jiao et al. (2021, model I), which was obtained from Pouliasis et al. (2017) and corresponded to a mass of $\sim 9 \times 10^{10} M_{\odot}$. Some studies argue that this model overestimates the mass of baryons towards the outer galactic radii (de Salas et al. 2019; Ou et al. 2023). We have also tested some different baryonic models with smaller mass. Though it has some minor impact on the estimated dark matter (DM) mass, it does not affect the results of dynamical mass (total mass). So the choices of baryonic mass do not change our main results.

We apply the χ^2 method to fit the RC and calculate its associated probability, for which we have tested an extremely large parameter space. The χ^2 is calculated by the sum at each disk radius R_i:

$$\chi^2 = \sum_{i}^{N} \frac{(v_{\mathrm{mod},i} - v_{\mathrm{obs},i})^2}{\sigma_i^2} \tag{1}$$

where v_{mod} is the modeled circular velocity for the cumulative baryons + DM profiles, v_{obs} is the observed circular velocity and σ_{stat} is the statistical uncertainty of the measurement so that $\sigma_{\mathrm{stat},i} = (\sigma^{+}_{v_{\mathrm{obs},i}} + \sigma^{-}_{v_{\mathrm{obs},i}})/2$ to which we have added the systematic uncertainty $\sigma_{\mathrm{sys},i}$ to calculate σ_i. Hence the χ^2 probability can be expressed as :

$$\mathrm{Prob}\left(\frac{\chi^2}{2}, \frac{N-\nu}{2}\right) = \frac{\gamma\left(\frac{N-\nu}{2}, \frac{\chi^2}{2}\right)}{\Gamma\left(\frac{N-\nu}{2}\right)} \tag{2}$$

where N is the number of independent observed velocity points in the RC and ν is the number of degrees of freedom.

3. Result and conclusion

Jiao et al. (2021) explained in detail that the NFW profile may lead to a possible methodological bias particularly against low MW masses. With the RC from *Gaia* DR3 (Wang et al. 2023), we find that the Einasto profile provides better fitting results than the NFW model. In Figure 1 we present the RC from Wang et al. (2023), 3 models (low mass fit, best fit and high mass fit) of Einasto profile and best-fit model of NFW profile. The reduced χ^2 of the best fit model is ~ 1.5 (for comparison, the reduced χ^2 of best fit NFW profile is ~ 3.7). Ou et al. (2023) and Sylos Labini et al. (2023) have also found a similar result. In short, the three-parameter Einasto profile provides a much better fit to the declining RC.

In Fig. 2. we compare the results of Einasto profile using *Gaia* DR2 and DR3. We keep the same parameter space as in Jiao et al. (2021) but the estimated MW mass range is much narrower. The best-fit dynamical mass remains consistent, from $2.77 \times 10^{11}\ M_{\odot}$ to $2.75 \times 10^{11}\ M_{\odot}$, indicating that the Einasto profile gives a consistent estimated dynamical mass.

The low mass fits of RC are mostly constrained by the data at large radii (> 17 kpc). Wang et al. (2023) find a consistent slope beginning at ~ 13 kpc. The last five points seem contradictory to each other. For now we think that inconsistencies may be due to the lack of a strict systematic uncertainty analysis, like the neglected cross term in Jeans equation and the uncertainty of scale length (see also Eilers et al. 2019). We will present this analysis in an upcoming work, but preliminary results show that the systematic does not lead to significant differences in our main results, including the choices of DM profile, estimated mass range and best fit model.

The Magellanic Clouds (MCs), Globular clusters (GCs) and dwarf galaxies are also widely used as tracers to estimate the enclosed mass of the MW at large radii. However,

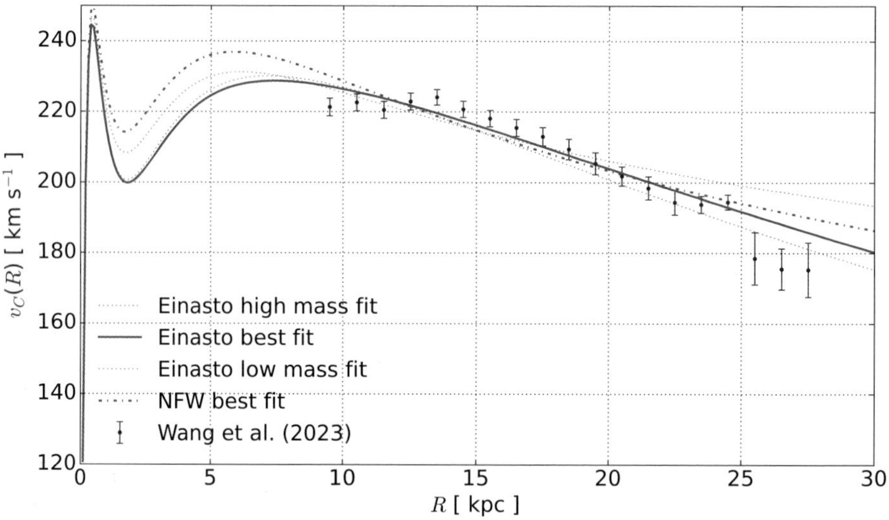

Figure 1. Rotation curve from *Gaia* DR3 (points and error bars Wang et al. 2023), Einasto profile fits (solid and dotted lines) and NFW best fit (dash-dotted line)

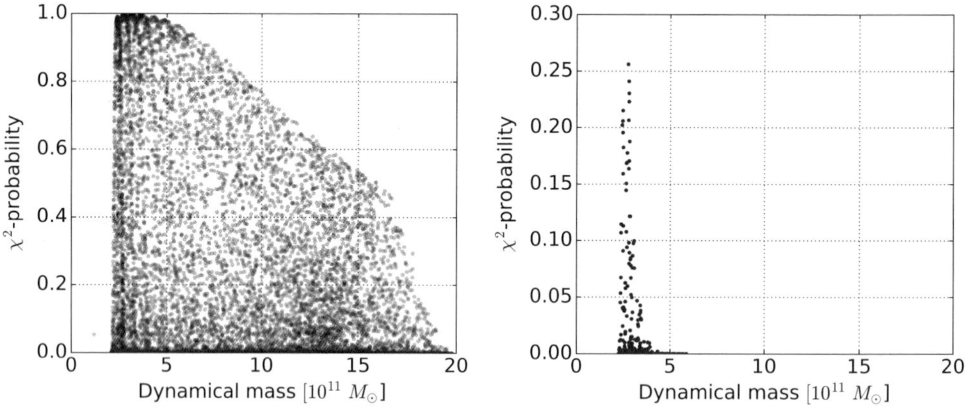

Figure 2. Estimated dynamical mass and its associated χ^2 probability of Einasto profile fits to RC from Eilers et al. (2019)(the left panel) and Wang et al. (2023)(the right panel).

these methods can easily lead to prior selection bias. Wang et al. (2022) found that by excluding two GCs with large orbit energy, Crater and Pyxis, their estimated MW mass decreased from $5.73^{+0.76}_{-0.58} \times 10^{11} M_\odot$ to $5.36^{+0.81}_{-0.68} \times 10^{11} M_\odot$ with Einasto profile. In fact, as relatively old halo inhabitants (Hammer et al. 2023), GCs behave consistently with a large MW mass range. We have tested orbits of all GCs with the MW mass from 2 to 15 $\times 10^{11} M_\odot$, and find that almost all GCs are gravitationally bound. However, about half of the dwarf galaxies are not bound with a low mass MW ($\sim 2 \times 10^{11} M_\odot$). If most dwarfs came late in the MW halo ($<$ 3 Gyr, Hammer et al. 2023), it is unrealistic to use them as tracers for estimating MW mass. For example, it is likely that the consideration of Leo I as a bound satellite would lead to a significant overestimate of the MW mass. On the other hand, orbital studies of other halo inhabitants are sensitive to the choice of MW mass and need more rigorous analysis.

References

de Salas, P. F., Malhan, K., Freese, K., et al. 2019, JCAP, 2019, 037.
Eilers, A.-C., Hogg, D. W., Rix, H.-W., et al. 2019, ApJ, 871,120
Gaia Collaboration, Brown, A. G. A., Vallenari, A., et al. 2018, A&A, 616, A1.
Gaia Collaboration, Vallenari, A., Brown, A. G. A., et al. 2022, arXiv:2208.00211.
Hammer, F., Puech, M., Chemin, L., et al. 2007, ApJ, 662, 322.
Hammer, F., Li, H., Mamon, G. A., et al. 2023, MNRAS, 519, 5059.
Haywood, M., Di Matteo, P., Lehnert, M. D., et al. 2018, ApJ, 863, 113.
Jiao, Y., Hammer, F., Wang, J. L., et al. 2021, A&A, 654, A25.
Karukes, E. V., Benito, M., Iocco, F., et al. 2020, JCAP, 2020, 033.
Lucy, L. B. 1974, AJ, 79, 745.
Mróz, P., Udalski, A., Skowron, D. M., et al. 2019, ApJL, 870, L10.
Ou, X., Eilers, A.-C., Necib, L., et al. 2023, arXiv:2303.12838.
Pouliasis, E., Di Matteo, P., & Haywood, M. 2017, A&A, 598, A66.
Sylos Labini, F., Chrobáková, Ž., Capuzzo-Dolcetta, R., et al. 2023, ApJ, 945, 3.
Wang, J., Hammer, F., & Yang, Y. 2022, MNRAS, 510, 2242.
Wang, H.-F., Chrobáková, Ž., López-Corredoira, M., et al. 2023, ApJ, 942, 12.
Zhou, Y., Li, X., Huang, Y., et al. 2023, ApJ, 946, 73.

Discussion

MANUEL BAYER: 1. Do I understand correctly that for the modeling of the Eilers et al. (2019) data-set you just take into account the uncertainties for the circular velocity?

2. Did you just use the binned data of Eilers et al. (2019)?

3. What stars do Wang et al. (2023) employ to trace the rotation curve?

YONGJUN JIAO: 1. We consider the neglected term in their Jeans equation, which is a cross-term made by the vertical density gradient of the product of the radial and vertical velocities and add an additional systematic uncertainty of 2% on the velocity scale due to the effect of changing the distance of the Sun to the Galactic center and the scale length, the proper motion of the latter, etc.

2. We directly used their binned data (Eilers et al. 2019, Table 1) with overestimated error bars (see above).

HAIFENG WANG: 3. Wang et al. (2023) use different types of stars in combination with Gaia DR3 line-of-sight velocity, which is based on the Lucy method and mainly in the galactic anticenter direction. The random errors and systematics are acceptable at least for the moment.

Dynamical Masses of Local Group Galaxies
Proceedings IAU Symposium No. 379, 2025
P. Bonifacio, M.-R. Cioni, F. Hammer, M. S. Pawlowski & S. Taibi, eds.
doi:10.1017/S1743921323001497

Bar-driven Shock Features in M31

Zi-Xuan Feng[1,2], Zhi Li[3], Juntai Shen[3,4], Ortwin Gerhard[5], R. P. Saglia[5,6] and Matias Blaña[5,7]

[1]Shanghai Astronomical Observatory, Chinese Academy of Sciences, 80 Nandan Road, Shanghai 200030, P.R. China

[2]School of Astronomy and Space Sciences, University of Chinese Academy of Sciences, 19A Yuquan Road, Beijing 100049, P. R. China

[3]Department of Astronomy, School of Physics and Astronomy, Shanghai Jiao Tong University, 800 Dongchuan Road, Shanghai 200240, P. R. China;
email: jtshen@sjtu.edu.cn

[4]Key Laboratory for Particle Astrophysics and Cosmology (MOE) / Shanghai Key Laboratory for Particle Physics and Cosmology, Shanghai 200240, P. R. China

[5]Max-Planck-Institut für Extraterrestrische Physik, Giessenbachstrasse, D-85748 Garching, Germany

[6]Universitäts-Sternwarte München, Scheinerstr. 1, 81679 Munich, Germany

[7]Institute of Astrophysics, Pontificia Universidad Católica de Chile, Av. Vicuña Mackenna 4860, 7820436 Macul, Santiago, Chile

Abstract. We have identified the velocity jump (shock) features in the inner $20' \times 10'$ region of M31 using the data of [O III] and H I. The identified shock features are found primarily on the leading side of the potential bar, displaying a typical pattern of bar-driven gas inflow. The shock features provide independent evidence for M31 being a barred galaxy. Our preliminary gas simulations with a barred potential of M31 can reproduce these shock features.

Keywords. interstellar medium; hydrodynamical simulation; barred galaxy; Andromeda galaxy

---

1. Introduction

M31, the closest massive spiral galaxy to us, still raises uncertainty regarding the presence of a bar due to the large inclination of its stellar disk. In the central region of M31, the stellar photometry indicates twisted isophotes, which were initially noticed by Lindblad (1956) and considered suggestive of a bar. Athanassoula & Beaton (2006) first employed N-body simulations to reproduce these isophotal twists and concluded that both a classical bulge and a boxy/peanut bulge are required to reproduce such features. More recently, Blaña Díaz, Matias *et al.* (2017) successfully reproduced the photometric details using a barred model. In addition to evidence from stellar photometry, the gas kinematics also provide clues regarding the presence of a bar in M31. In the absence of a bar, the gas within the stellar disk would exhibit predominantly circular motions, resulting in line-of-sight velocities that are lower than or similar to the rotation curve. However, Brinks & Burton (1984) observed high-velocity features exceeding 300 km/s in the HI observations, as identified in the position-velocity diagrams. The more recent IFU study of M31 (Opitsch, M. *et al* 2018) found twisted zero-velocity line of [O III] in the bulge region of M31. Such twisted zero-velocity line has been recognized in many barred galaxies (Leroy, Adam K. *et al.* 2021). It is worth noting that alternative explanations, such as a triaxial bulge or a recent merger event, could potentially produce these features. Consequently, we aimed to identify clearer evidence in the gas kinematics supporting the presence of a bar in M31.

2. Method

A significant feature of barred galaxies is the pair of shock related dust lanes on the leading side of the bar. If we position a pseudo-slit perpendicular to the dust lane, it is expected to find sharp velocity jump features near the dust lane. We position a series of pseudo-slits at different positions of the disk of M31, and search for such shock features in the data of [O III] (Opitsch, M. *et al* 2018) and H I (Brinks & Burton 1984). If the shock features really correspond to the expectation of bar-driven shock features (Athanassoula 1992), it will be an independent evidence for M31 having a bar.

3. Result

We found that some shock features have larger amplitude, while some shock features have smaller amplitude. Therefore we classify them by the velocity jumps into three classes of shock features. In the central region, some shock features show up in the [O III] data but not in the H I data, maybe due to the lack of neutral gas there. Farther out, the positions of [O III] and H I shock features are roughly similar. The clearest shock features are located mainly on the leading side of the possible bar in M31. The H I shock features are found even beyond the bulge region. If we look at the shock features in more details, the clearest [O III] shock features show velocity jumps as large as 200 $\mathrm{km\,s^{-1}}$. Although the shock features in H I are more clumpy and less clear to identify, the profiles of these shock features are similar to the ones in [O III].

We noticed that shock features on the far side of M31 are clearer than the ones in the other side. There are several possible explanations for the less clear shock features on the near side: 1. the warp in the gas disk of M31 becomes a foreground on the near side, and a background on the far side. The gas and dust on the warp could blur the central gas velocities. 2. the shock features in [O III] are very close to the boundary of the observational coverage. These shock features are possibly not well recognized.

Based on the barred gravitational potential from the well-constrained made-to-measure model of M31 in Blaña Díaz, Matias *et al.* (2017), we construct a series of gas models and compare the shock features in the models with those in the observations. The best-fitting gas model used a low pattern speed of $\Omega_b = 20\,\mathrm{km\,s^{-1}\,kpc^{-1}}$ and a high sound speed of $c_s = 30\,\mathrm{km\,s^{-1}}$. The bar is deviated from the disk major axis by $\phi_b = 54.7^\circ$ in the face-on view. After projection with the inclination of $i = 77^\circ$, the shock features in the model are found at positions similar to those of observations. The overall gas kinematic profiles on the position-velocity diagrams in the model and observation are similar, especially near the shock regions. We also tried a non-rotating bar model to mimic a static triaxial bulge, the gas features of which do not show any large velocity jump.

4. Conclusion

We have identified shock features with [O III] and H I data in the central bulge region of M31. The shock features are mainly located on the leading side of the possible bar of M31. The shock features in [O III] and H I are roughly similar, and their patterns correspond to the expectations of bar-driven gas inflows. These shock features provide an independent evidence for M31 having a bar. Our preliminary gas models with a barred potential can reproduce the observed shock features.

References

Athanassoula, E. 1992, *MNRAS*, 259, 345
Athanassoula, E. and Beaton, R. 2006, *MNRAS*, 370, 1499
Brinks, E. and Burton, W. B. 1984, *A&A*, 141, 195
Blaña Díaz, Matias *et al.* 2017, *MNRAS*, 466, 4279
Chemin, Laurent *et al.* 2009, *ApJ*, 705, 1395
Lindblad, Bertil 1956, Stockholms Observatoriums Annaler, 19, 2
Leroy, Adam K. *et al.* 2021, *ApJS*, 257, 43
Opitsch, M. *et al* 2018, *A&A*, 611, 380

Dynamical Masses of Local Group Galaxies
Proceedings IAU Symposium No. 379, 2025
P. Bonifacio, M.-R. Cioni, F. Hammer, M. S. Pawlowski & S. Taibi, eds.
doi:10.1017/S1743921323001448

First Results from the Triangulum Extended Survey: Insights into the Dynamical History of M33

K. M. Gilbert[1,2], A. C. N. Quirk[3], L. R. Cullinane[2], A. Smercina[4] and P. Guhathakurta[5]

[1]Space Telescope Science Institute, 3700 San Martin Dr., Baltimore, MD 21218, USA
email: kgilbert@stsci.edu

[2]The William H. Miller III Department of Physics & Astronomy, Bloomberg Center for Physics and Astronomy, Johns Hopkins University, 3400 N. Charles Street, Baltimore, MD 21218

[3]Department of Astronomy, Columbia University, New York, NY, USA

[4]Department of Astronomy, University of Washington, Box 351580, U.W., Seattle, WA 98195-1580, USA

[5]Department of Astronomy & Astrophysics, University of California, Santa Cruz, 1156 High Street, Santa Cruz, CA 95064, USA

Abstract. The Triangulum Extended (TREX) Survey is a spectroscopic survey† that targets resolved stars of all ages throughout the disk of Triangulum (M33). We summarize the first results from the TREX Survey, including the discovery of a kinematically hot, halo-like population throughout M33's inner disk in both the old and intermediate-age populations and evidence that the youngest stars have been dynamically heated. We also discuss the implications for our understanding of M33's dynamical history in the context of recent results from the Panchromatic Hubble Andromeda Treasury: Triangulum Extended Region (PHATTER) Survey.

Keywords. Galaxy stellar halos, Triangulum Galaxy, Local Group, Galaxy evolution, Disk galaxies

---

1. Introduction

As the closest low-mass spiral galaxy, M33 provides a unique laboratory in which to study the secular evolution and stability of low mass stellar disks. M33 is slightly more massive than the Large Magellanic Cloud (LMC), and intermediate in metallicity between the LMC and the Local Group's two massive spiral galaxies, M31 and the Milky Way. M33 presents a relatively undisturbed central disk, with distortions at large radius observed in both the H I gas and the stellar surface density. While a close previous encounter with M31 has been posited to explain the distortions in the H I and stars seen at large radius (McConnachie et al. 2009; Putman et al. 2009; Semczuk et al. 2018), analyses which take into account the proper motion of M31 argue that M33 is more likely on its first infall into the M31 system (Patel et al. 2017a,b; van der Marel et al. 2019).

Significant observational campaigns have provided contiguous mapping of the spatial distribution and line-of-sight velocity of the gas in M33 to large projected distances

† The data presented herein were obtained at the W.M. Keck Observatory, which is operated as a scientific partnership among the California Institute of Technology, the University of California and the National Aeronautics and Space Administration. The Observatory was made possible by the generous financial support of the W.M. Keck Foundation.

(including H I, Hα, and CO, e.g., Gratier et al. 2010; Druard et al. 2014; Kam et al. 2015), as well as large scale, contiguous imaging of the stellar populations in the inner disk of M33 (PHATTER; Williams et al. 2021), and to large distances from M33's center (Pan-Andromeda Archaeological Survey, or PAndAS; McConnachie et al. 2018). TREX aims to increase our understanding of M33's stellar populations and disk dynamics with resolved stellar spectroscopy throughout M33's inner disk and in its extended stellar structures. Via the measurement of the stellar disk dynamics as a function of stellar age, and chemo-dynamical analysis of the older stellar populations, TREX aims to shed light on the forces that have driven the large-scale disturbances in M33's stellar and gaseous disks, determine if M33 has a stellar halo, and if so, constrain it's origin.

This contribution summarizes the recent results by Gilbert et al. (2022) and Quirk et al. (2022), and places those results in context with recent work by Smercina et al. (submitted) and Cullinane et al. (in prep.).

2. First Results From The TREX Survey

2.1. *The TREX Dataset*

The TREX Survey has obtained stellar spectra for more than 5000 stars in M33's disk and extended stellar structures to date. Spectra were obtained with the DEIMOS multi-object spectrograph on the Keck II telescope. Spectroscopic masks that targeted primarily red populations (e.g., red giant branch (RGB) stars or asymptotic giant branch (AGB) stars) were generally observed with the 1200 line mm^{-1} grating, resulting in a resolution of $R \sim 6000$ and wavelength range of $\lambda\lambda \sim 6300-9800$ Å. Masks that targeted the full range of ages, or primarily blue populations (e.g., massive main sequence stars or blue helium burning stars) were observed with the 600 line mm^{-1} grating, resulting in a resolution of $R \sim 2000$ and larger wavelength range of $\lambda\lambda \sim 4600-9800$ Å. The work discussed below analyzes stars in the disk of M33, from the innermost regions to projected distances of ~ 11 kpc from M33's center (~ 18 kpc, or ~ 10 scale lengths, in the plane of the disk).

2.2. *A Kinematically Hot Component Present Throughout M33's Disk*

The presence or absence of a significant stellar halo component in M33 has been a matter of debate within the literature, with observational evidence on both sides (see Gilbert et al. (2022) for a detailed discussion, and references therein). Gilbert et al. (2022) analyzed the velocity distribution of 1667 RGB stars (Figure 1). A two-component model for describing the velocity distribution is strongly preferred. One component is consistent with rotating in the plane of the H I disk, at rotation speeds similar to that of the H I gas, and with a velocity dispersion of ~ 19 km s^{-1}. The second component, comprising 22% of the full RGB sample, is consistent within 1.5σ of having no rotation in the plane of the H I disk, and has a velocity dispersion of 59 km s^{-1}. The lack of rotation in the plane of the disk, combined with the high dispersion, favors interpreting this component as a stellar halo rather than a thick disk. The fraction of RGB stars in the kinematically hot population decreases with increasing distance from M33's center. In the outermost radial bin, where the majority of the stars lie beyond the break in the surface brightness profile of M33's disk (Ferguson et al. 2007), 90% of the RGB stars belong to the disk component, providing strong evidence that M33's disk break does *not* reflect the transition to a halo-dominated region, as had been previously hypothesized (Barker et al. 2011; Cockcroft et al. 2013).

Recent work by Cullinane et al. (in prep.), modelling 671 intermediate-age AGB and spectroscopically identified Carbon stars, finds that the kinematically hot component seen in the older RGB population is also present in M33's intermediate-age population,

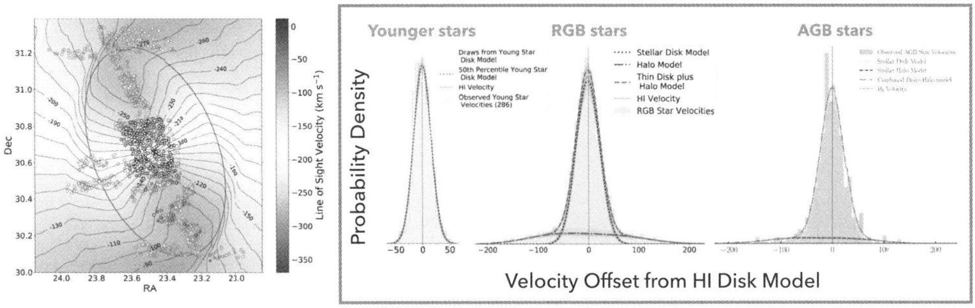

Figure 1. To model the stellar kinematics of M33's disk, each star's line-of-sight velocity (points, left panel) is compared to the expectation from a tilted ring model of M33's disk based on H I observations (Kam et al. 2017, contours, left panel). At each star's position, we calculate the velocity offset of the star's line-of-sight velocity from that of the H I Disk Model. If the stars follow the gas exactly, expect no mean velocity offset, and an approximately Gaussian distribution. In Gilbert et al. (2022), we find this is true for a young, red star comparison sample, while the RGB sample requires a second, higher dispersion component to match the observed distribution. In Cullinane et al. (in prep.), we find a second, higher dispersion component is also required to fit the intermediate-age AGB star population. Figure adapted from Gilbert et al. (2022) and Cullinane et al. (in prep.).

with a two-component model decisively preferred over a single-component disk model (Figure 1). Approximately 12% of the intermediate-age population is in a high velocity dispersion component ($\sigma_v \sim 50\,\mathrm{km\ s^{-1}}$) that is statistically consistent with having no rotation in the plane of the disk, and which appears to have a systematic offset in its mean velocity.

2.3. *Evidence of Recent Dynamical Heating of M33's Stellar Disk*

Quirk et al. (2022) measured M33's stellar disk kinematics as a function of stellar age, utilizing a sample of ~ 4500 disk stars to make local, empirical measurements of the stellar mean velocity and velocity dispersion. Quirk et al. (2022) also paired stellar and gas line-of-sight velocity measurements, comparing the inferred stellar rotation velocities to the inferred gas rotation velocity in order to make a local, empirical measure of the asymmetric drift of the stars relative to the gas. No trend with age is observed in either velocity dispersion or asymmetric drift ((Figure 2). Compared to the vast majority of M33 analogs in the IllustrisTNG simulations, the youngest age bin ($\sim 20-180$ Myr) in M33 has an unusually high asymmetric drift, suggesting some phenomenon is dynamically heating the youngest stars in M33's disk.

3. Implications for M33's Dynamical History and Future Prospects

The TREX results demonstrate that a kinematically hot component is present throughout M33's disk, and that there is evidence the young stars in M33's disk have been dynamically heated.

Recent results from the PHATTER survey provide critical additional context for interpreting the TREX results. Smercina et al. (submitted) finds that the observed morphology of M33's disk is age-dependent, with the distribution of RGB, AGB, and HeB stars all showing clear evidence of a bar, as well as two asymmetric spiral arms, providing potential evidence that a tidal disturbance has affected M33's inner disk. Smercina et al. use the PHATTER stellar catalog in conjunction with archival Spitzer IRAC 3.6 μm imaging to model the stellar surface brightness profile of M33's RGB stars from M33's center to a semi-major axis distance of 4.7 kpc. Smercina et al. find the RGB

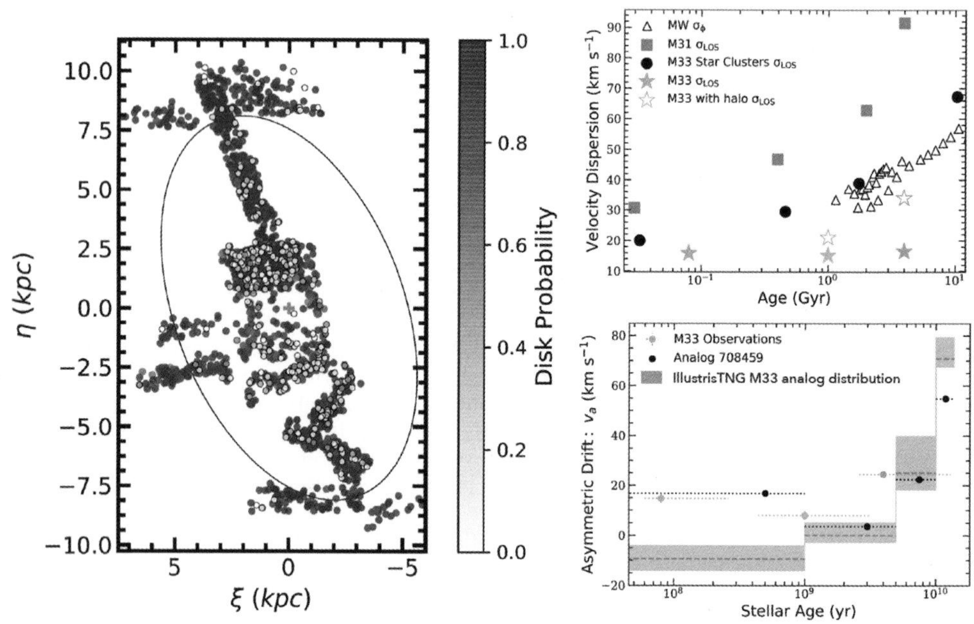

Figure 2. Quirk et al. (2022) measured the velocity dispersion and asymmetric drift of stars with respect to the gas as a function of stellar age, restricting the primary analysis to stars with a high probability of belonging to M33's disk (rather than M33's halo; left panel). In contrast to measurements made in the disks of M31 and the MW, there is no trend of increasing velocity dispersion with age for M33's disk stars (top right panel). Furthermore, the asymmetric drift of the youngest stars is higher than expected, relative to both the asymmetric drift of the intermediate-age and old stellar populations, and expectations based on cosmological simulations of M33 analogs (bottom right panel). This likely indicates that the young stars in M33 have been dynamically heated. Figures from (Quirk et al. 2022).

surface brightness profile is well described by the combination of a bar, an exponential disk, and a broken-power-law halo. The best-fit radial profile produces average halo fractions as a function of radius that are fully consistent with the halo fractions measured from the kinematical modelling described above (Gilbert et al. 2022). A power-law profile is consistent with expectations for the surface brightness profile of accreted stellar halos based on the profiles of observed and simulated galaxies at larger mass scales.

Together, the TREX and PHATTER observations strongly indicate the likelihood of significant dynamical disturbances acting on M33's disk, and conclusively demonstrate that M33 does contain a stellar halo. The surface brightness profile of the halo drops off rapidly with projected radius, consistent with the lack of an observed stellar halo component at large radii in the PAndAS survey (McMonigal et al. 2016). However, the origins of M33's stellar halo remain unclear: is it the product of dynamical heating mechanisms acting on M33's stellar disk, or was it formed primarily from stars accreted during one or more mergers? The lack of rotation in the plane of the disk and the power-law profile are consistent with expectations for accreted stellar halos based on the stellar halos of larger mass galaxies. However, the presence of a kinematically hot component in M33's AGB stars with properties consistent with the RGB halo component presents a challenge to an accretion scenario, since the accreted satellite(s) would be of low stellar mass, and thus would not be expected to have formed a significant amount of stars at intermediate ages. While the photometry-based metallicity distribution of RGB stars in M33's halo and disk components are very similar (estimated by comparing the stars' colors and magnitudes with stellar isochrones; Gilbert et al. 2022), the mean metallicity of M33's halo component is also consistent with an accretion scenario (Smercina et al., submitted). Readers are referred to Gilbert et al. (2022) and Smercina et al. for more detailed discussion. Future measurements of the [α/Fe] and [Fe/H] distribution of stars in M33's halo and disk have the potential to discriminate between these two scenarios.

The authors recognize and acknowledge the very significant cultural role and reverence that the summit of Maunakea has always had within the indigenous Hawaiian community. We are most fortunate to have the opportunity to conduct observations from this mountain. Support for this work was provided by NSF grants AST-1909066 (K.M.G. and L.R.C.) and AST-1909759 (P.G.), DGE-1842400 (A.C.N.Q.), and NASA GO-14610 (A.S.).

References

Barker, M. K., Ferguson, A. M. N., Cole, A. A., et al. 2011, *MNRAS*, 410, 504
Cockcroft, R., McConnachie, A. W., Harris, W. E., et al. 2013, *MNRAS*, 428, 1248
Druard, C., Braine, J., Schuster, K. F., et al. 2014, *A&A*, 567, A118
Gilbert, K. M., Quirk, A. C. N., Guhathakurta, P., et al. 2022, *ApJ*, 924, 116
Gratier, P., Braine, J., Rodriguez-Fernandez, N. J., et al. 2010, *A&A*, 522, A3
Ferguson, A., Irwin2, M., Chapman, S., et al. 2007, Island Universes, 3, 239
Kam, Z. S., Carignan, C., Chemin, L., et al. 2015, *MNRAS*, 449, 4048
Kam, S. Z., Carignan, C., Chemin, L., et al. 2017, *AJ*, 154, 41
McConnachie, A. W., Irwin, M. J., Ibata, R. A., et al. 2009, *Nature*, 461, 66
McConnachie, A. W., Ibata, R., Martin, N., et al. 2018, *ApJ*, 868, 55
McMonigal, B., Lewis, G. F., Brewer, B. J., et al. 2016, *MNRAS*, 461, 4374
Patel, E., Besla, G., & Mandel, K. 2017, *MNRAS*, 468, 3428
Patel, E., Besla, G., & Sohn, S. T. 2017, *MNRAS*, 464, 3825
Putman, M. E., Peek, J. E. G., Muratov, A., et al. 2009, *ApJ*, 703, 1486
Quirk, A. C. N., Guhathakurta, P., Gilbert, K. M., et al. 2022, *AJ*, 163, 166
Semczuk, M., Łokas, E. L., Salomon, J.-B., et al. 2018, *ApJ*, 864, 34
van der Marel, R. P., Fardal, M. A., Sohn, S. T., et al. 2019, *ApJ*, 872, 24
Williams, B. F., Durbin, M. J., Dalcanton, J. J., et al. 2021, *ApJS*, 253, 53

Discussion

SARAH BIRD: What is the comparison between the extent of your sample and that of M33's gas disk?

KAROLINE GILBERT: M33's H I disk is large compared to its stellar disk. The observations presented here lie well within the extent of M33's H I disk. As can be seen from Figure 1, while a small fraction of the stars lie beyond the warp in the H I disk, the majority of the stellar sample analyzed to date lies interior to the warp.

ANA BONACA: Do you have enough spatial coverage to measure the density profile of the halo component as a function of radius and height above the disk? Different formation scenarios might be differentiated this way.

KAROLINE GILBERT: M33's inclination is 54°, so constraining the height above the disk, or thickness, for different components would be very challenging. It also has significantly less dust than M31, likely limiting our ability to statistically determine the scale height of the disk. Still, perhaps some clever forward modelling of the line-of-sight velocities combined with constraints from the PHATTER data could shed some light on this.

KYLE OMAN: Thinking about an accreted halo component, usually the halo is dominated (in mass or star counts) by a single accretion event, or at least that's the conventional wisdom I'd carry over from L_* galaxies. The extent of the halo - the part donated by that main progenitor - is fixed by the time of accretion (with some dependence on the detailed orbit). So I wonder if M33 might fit into a picture where it has had a fairly quiet accretion history with a last major accretion event a long time ago.

KAROLINE GILBERT: It might fit into such a picture. If a low mass halo progenitor deposited most of its stars near the central regions of M33 a long time ago, a small fraction of stars from that low mass progenitor at larger radius would be extremely difficult to detect. However, the presence of a kinematically hot component in the AGB population is hard to reconcile with this scenario. On the other hand, a non-rotating halo component that formed from heating of the disk yet contains intermediate-age stars also seems hard to reconcile.

KYLE OMAN: What is intermediate-age, in Gyr?

KAROLINE GILBERT: The AGB and Carbon stars in the Cullinane et al. intermediate-age sample span a range of roughly $1-3$ Gyr, while the RGB stars are older (~ 3 Gyr on average), and are expected to span a range of roughly $1.4-9$ Gyr. So there is some overlap in these age bins.

Session 3:
Dwarf galaxies: dwarf spheroidals, ultra-faint dwarfs, dwarf irregulars

Dynamical Masses of Local Group Galaxies
Proceedings IAU Symposium No. 379, 2025
P. Bonifacio, M.-R. Cioni, F. Hammer, M. S. Pawlowski & S. Taibi, eds.
doi:10.1017/S1743921323001655

On the observational challenges to estimating the masses of dwarf galaxies

A. W. McConnachie

National Research Council Herzberg Astronomy and Astrophysics, 5071 West Saanich Road, Victoria, BC, Canada, V9E2E7
email: alan.mcconnachie@nrc-cnrc.gc.ca

Abstract. This presentation is an attempt to lay out, in the context of observational studies of the masses of Milky Way dwarf galaxies, what it is I think we want to do, why we want to do it, and how we go about doing it. They are my own personal opinions, and are presented here in the hope that they will stimulate others to step back and consider what are the most important scientific investigations that we should be doing, given the plethora of facilities and data available that are now available to us.

Keywords. dwarf galaxies, photometry, spectroscopy, astrometry, resolved stellar populations

---

1. Introduction

Over the last decade there has been a qualitative shift in the nature of the scientific studies that are being undertaken that involve the population of dwarf galaxies that orbit the Milky Way. It is now considered entirely normal that we are able to give reasonable estimates of the orbit of a dwarf galaxy around the Milky Way soon after it has been discovered (e.g., Smith et al. 2022). For these newly discovered system, it is considerably easier - or at least quicker! - to estimate their net proper motion than it is to measure their radial velocity. And we can now even group satellites into families based on implied accretion histories (e.g., Battaglia et al. 2022, and correlate morphological parameters with pericenters (e.g., Pace et al. 2022). This qualitiative change in the science we are doing is in large part because of a qualitative change in the data that we are using.

And so this prompts the question: observationally, what will be the driver of the next qualitative shift in our field?

2. Driving Science

In order to have a chance at providing an answer to the question posed above, it is first necessary to restrict the domain we are considering. In what follows, I consider only scientific pursuits involving the Milky Way dwarf galaxy population. This is not to imply anything inadequate about non-Milky-Way dwarfs, but it is just a recognition that the quality of the data available for the Milky Way population will generally always be better than for any other population of dwarfs.

It is also necessary to go out on a limb and answer a related question: what is it we actually want to do? What is the driving science that makes us care what the masses of dwarf galaxies actually are?

I submit that, as of the time of writing, there are two main science cases:

(1) *Driving Science I: What is the intrinsic dynamical mass profiles of dwarf galaxies?*

Unlike larger system, dwarfs on the mass of Fornax and below are thought to have minimal contributions from baryons, both in terms of their total mass and in terms of the changes to their dark matter profile caused by the baryons. This science case goes towards the nature of the dark matter particle. It necessitates measuring the mass as a function of radius, not just the total mass.

(2) *Driving Science II: What is the lowest mass galaxy that can exist?*
Once we know about these low mass galaxies, it is natural to wonder if they formed at low mass, or evolved to become low mass (through tidal stripping). At the root of this question are the related questions about how low a mass dark matter halo...
- ...can exist? This probes the nature of dark matter again
- ...can host a galaxy? This probes the efficiency and energetics of feedback during the galaxy formation process
- ...can survive? This probes the cuspiness of halos (see discussion in Errani et al. 2023)

3. Observables and their limitations

The Jeans Equation provides a convenient way to quickly see the observational quantities on which dynamical mass estimates depend:

$$\frac{-GM(r)}{r^2} = \frac{d(n_\star \sigma_r^2)}{dr} = 2\frac{\beta n_\star \sigma_r^2}{r} \tag{1}$$

where $\beta(r) = 1 - \sigma_t^2/\sigma_r^2$ and $\sigma_t = \sigma_\theta = \sigma_\phi$.

On converting to projected quantities we get

$$\Sigma_\star \sigma_{los}^2(R) = \int_{R^2}^{\infty} n_\star \sigma_r^2 \left[1 - \frac{R^2}{r^2}\beta(r)\right] \frac{dr^2}{\sqrt{r^2 - R^2}} \tag{2}$$

Here, $\Sigma_\star$ is the projected density profile of the tracer population with a 3D density profile of $n_\star$; R is the projected radius and r is the 3D radius; σ is the velocity dispersion in the direction given by the subscript. β is the velocity anisotropy. Massari et al. (2018) obtain an estimate of this via:

$$\hat{\beta} = \beta(\hat{r}) = 1 - \frac{\sigma_T^2}{\sigma_{los}^2 + \sigma_R^2 + \sigma_T^2} \tag{3}$$

The above quantities are all observables that we associate with a given galaxy G. These observables are themselves derived from measureable quantities associated with the individual stars, i, in the galaxy G. Specifically, the key measurables are stellar positions $(\alpha, \delta)_i$ (for the density profile and measurements as a function of radius), stellar radial velocities ($v_{los,i}$) and stellar proper motions $(\mu_\alpha \cos\delta, \mu_\delta)_i$, both necessary for derivation of three-dimensional velocity components and related parameters, such as β. Also necessary to convert angular measurements (differential positions, proper motions) to physical measurements (kiloparsecs) is a distance to the galaxy.

To measure stellar positions, radial velocities and proper motions of stars that belong to galaxy G, requires imaging and spectroscopy. For each, there are different types of challenges that emerge in deriving the observables for galaxy G from the individual stellar measurements. In what follows, I will focus on challenges relating to *identification*, *systematics*, *precision* and *statistics*.

4. Target galaxies and their members stars

Figure 1 shows all the known Milky Way dwarf galaxies (and candidates) fainter than the SMC (data from McConnachie 2012). Figure 2 shows these plotted in the absolute magnitude - distance plane. In terms of any study requiring high SNR measurements,

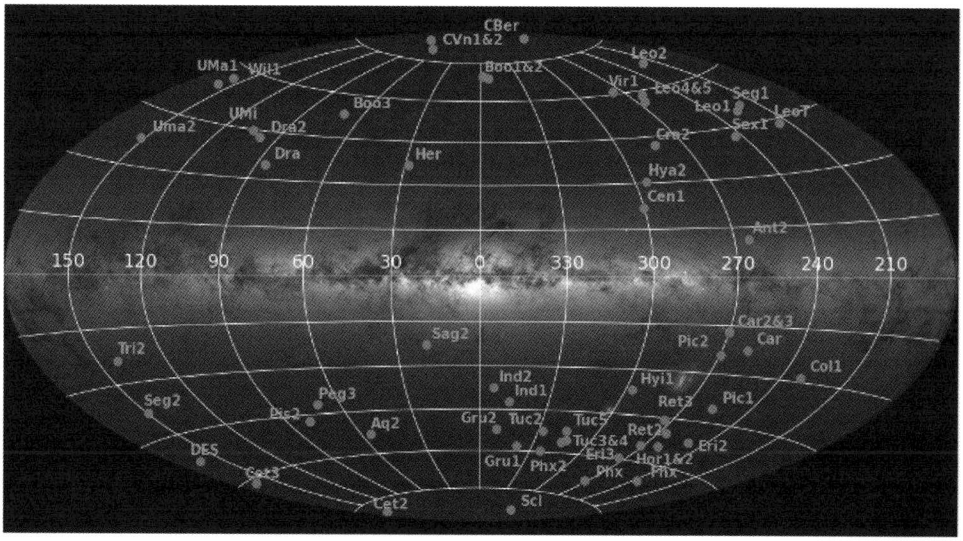

Figure 1. An Aitoff projection in Galactic coordinates of the currently known Milky Way satellites. Figure from Jensen et al. 2023, in prep.

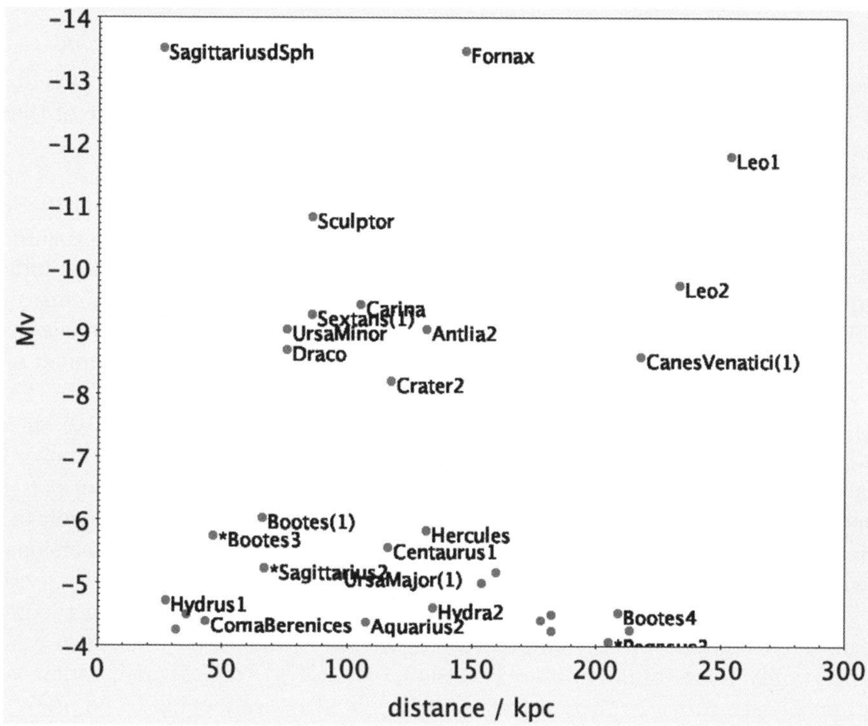

Figure 2. Absolute magnitude as a function of heliocentric distance for the brighter MW satellites. Brighter galaxies and closer galaxies have more stars that are accessible at a given SNR.

closer galaxies and brighter galaxies will both have more bright stars for which more precise measurements will be possible. Note also that while Sagittarius is both bright and close, as a system undergoing active and significant tidal disruption (Ibata et al. 2001), its suitability for many science cases may be questionable.

Identification of member stars within a dwarf galaxy has traditionally relied on a combination of spatial, photometric (and/or metallicity), and radial velocity cuts. A good example is the VLT/FLAMES DART study of Sextans, by Battaglia et al. (2011). For all these parameters, member stars are expected to cluster together in some way and so distinguish themselves from the foreground/background.

More recently, Gaia has significantly improved our abilities to select member stars, and members can now be identified with relatively little contamination, Different variants on the technique exist (e.g., Pace and Li 2019; McConnachie and Venn 2020; Battaglia et al. 2022), but essentially it relies on simultaneous analysis of spatial, color magnitude and proper motion distributions, where again member stars cluster in a way that is distinct from the contamination. Follow-up radial velocities usually confirm that a very high proportion of the Gaia-identified members remain as members even when radial velocities are considered (e.g., Sestito et al. 2023a,b). This means that, above the magnitude limit of Gaia, very good estimates of the stellar membership of the dwarf is possible using the proper motions and photometric data, without the need for radial velocities; inclusion of the latter will decrease the contamination to generally negligible levels.

5. Stellar radial velocities

5.1. *When the measurement is right but the interpretation is wrong*

Core to observational studies of dwarf galaxies is the premise that the motion observed for an individual star is due to its motion under the gravitational influence of the dwarf galaxy. This is not always correct, however.

Specifically, it has been noted that potentially a very significant fraction of stars in dwarf galaxies belong to binary star systems (e.g., Minor 2013; Minor et al. 2019; Spencer et al. 2017, 2018). These can be identified via multi-epoch radial velocity measurements, where the change in the velocity is larger than comfortably explained via individual measurement errors. However, such a technique is only sensitive to those binary stars on an orbit such that their radial velocity has changed sufficiently relative to measurement errors on the relevant time-baseline. For most observations, this implies changes of order few - several km/s on timescales of days - few years.

To put some of these numbers in context, McConnachie and Côté (2010) show that the expected radial velocity dispersion of many of the ultra-faint dwarfs known at the time are, in the absence of dark matter, of order a few hundreds of meters per second. Further, while for the classical dwarfs, binary stars are not going to cause you to imply the presence of dark matter when none is there, it will cause velocity dispersions to be inflated by, potentially, hundreds of meters per second or even a km/s or so, and especially so if one is looking at dispersion as a function of radius. Considering the goal is a precision determination of the temperature of the dark matter particle, such a systematic uncertainty seems unfortunate. Precision (sub-km/s), multi-epoch radial velocity measurements spanning a range of time baselines is ultimately going to be preferred.

5.2. *Radial velocity uncertainties as a function of magnitude*

Extensive radial velocity datasets for dwarf galaxies exist: the advent of good 8-m class MOS facilities such as Keck/DEIMOS, VLT/FLAMES, Magellan/IMACS and M2FS, Gemini/GMOS and others have had significant impacts. Large datasets exist for the classical dwarfs; here, datasets such as the DART study (e.g., Tolstoy et al. 2006 and

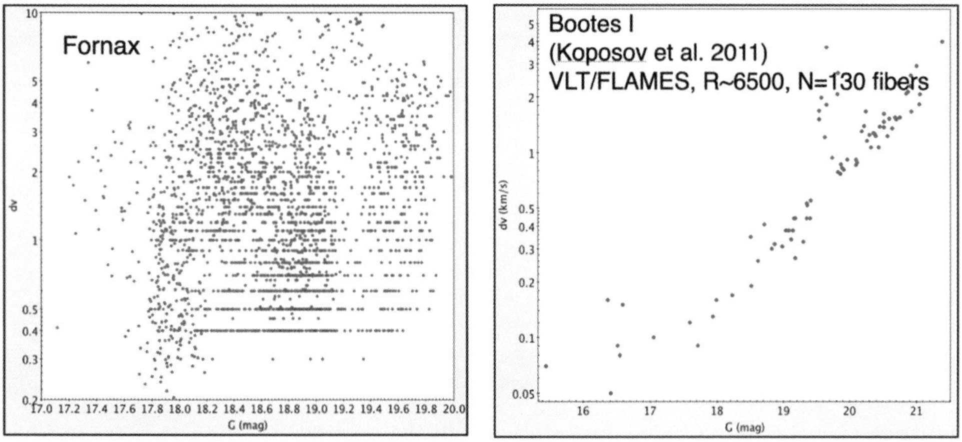

Figure 3. Left panel: The radial velocity dataset from M2FS for Fornax (Walker et al. 2009, with velocity uncertainties plotted as a function of Gaia G magnitude. Right panel: same, but for the study of Bootes I with VLT/FLAMES by Koposov et al. (2011).

the M2FS studies (e.g., Walker et al. 2009) are particularly valuable. The left panel of Figure 3 shows M2FS data from Walker et al. (2009) for Sculptor, which is indicative of the precision of this and other comparable datasets.

The right panel of Figure 3 shows the velocity uncertainties as a function of magnitude for the study of Bootes I by Koposov et al. (2011). This was a particularly notable study at the time and was able to acheive particularly low radial velocity uncertainties that have generally not been surpasses since. This study demonstates, however, that for all stars brighter than around 20th magnitude in Gaia, sub-km/s accuracy is possible using mature spectrographs on ground-based 8-m class facilties; uncertainties of order 100m/s are achievable for brighter (17th magnitude stars). As discussed in the previous subsection, such accuracy is necessary for precision dynamical studies, and is an approrpriate observational benchmark.

5.3. *Science and sample size*

Figure 4 and 5 shows the number of possible target stars within the classical (i.e., bright) and faint dwarfs, respectively. These are listed as a function of Gaia magnitude, using members identified by the algorithm of Jensen et al. 2023, in prep. Also shown is the number of those stars that have radial velocity measurements, taking the radial velocities from the list of studies described in Table 4 of McConnachie and Venn (2020) (these numbers may not be fully complete, therefore, but they are indicative). In all these systems, while many stars have been observed,more remain to be observed; for the brighter systems, a large majority of stars have not been observed. Its also important to note that, while at bright magnitudes many stars have been observed, all stars brighter than $G = 20$ could be measured with sub km/s precision, but most are not.

But what is the point of getting more radial velocities? Figure 6 shows two different studies, conducted as part of MSE (McConnachie et al. 2016; The MSE Science Team et al. 2019 and TMT (Skidmore et al. 2015) science development activiites. The left two panels show the recovery of intrinsic profiles (based on a cuspy NFW) for a dwarf galaxy as a function of spectroscopic sample size. Shaded regions show 95% credible intervals from a standard Jeans analysis and it assumes median velocity error of 2km/s. The contrast in our ability to recover the profile if we have 1000 or 10000 stars is striking. Along similar lines, the right panel of Figure 6 shows the accuracy of recovery of logarithmic

Galaxy	Mv	d (kpc)	G < 18 total	G < 18 w/ velocity	18 < G < 19 total	18 < G < 19 w/ velocity	19 < G < 20 total	19 < G < 20 w/ velocity	20 < G <21 total	20 < G <21 w/ velocity	G < infinity total	G < infinity w/ velocity
			Number of stars in Gaia within 3 x rh									
Fornax	-13.5	147	397	236	2553	1518	5761	623	13797	7	22519	2384
Sculptor	-10.8	86	378	296	651	404	1640	537	3611	103	6280	1340
UrsaMinor	-9	76	114	102	225	165	910	141	662	1	1911	409
Draco	-8.7	76	104	85	183	131	539	195	851	71	1677	482
Sextans1	-9.3	86	89	74	171	98	478	195	1018	119	1756	486
Carina	-9.4	105	78	74	149	99	485	237	1537	363	2249	773
Hydrus1	-4.7	27	26	17	18	8	21	10	62	4	127	39
Crater2	-8.2	117	18	12	35	18	115	28	336	4	504	62
Bootes1	-6	66	16	15	17	9	90	28	83	20	206	72

Figure 4. Number of possible target stars within the classical (i.e., bright) as a function of Gaia magnitude, using members identified by the algorithm of Jensen et al. 2023, in prep. Also shown is the number of those stars that have radial velocity measurements, taking the radial velocities from the list of studies described in Table 4 of McConnachie and Venn (2020) (these numbers may not be fully complete, therefore, but they are indicative). In all these systems, while many stars have been observed, many more remain to be observed. This is especially so when considering that all stars brighter than $G = 20$ could be measured with sub km/s precision, but most are not.

Galaxy	Mv	d (kpc)	G < 18 total	G < 18 w/	18 < G < 19 total	18 < G < 19 w/	19 < G < 20 total	19 < G < 20 w/	20 < G <21 total	20 < G <21 w/	G < infinity total	G < infinity w/ velocity
			Number of stars in Gaia within 3 x rh									
Boo1	-6	66	16	15	17	9	90	28	83	20	206	72
Hercules	-5.8	132	0	0	3	3	20	16	19	9	42	28
Centaur	-5.6	116	1	0	4	0	5	0	17	0	27	0
Sag2	-5.2	67	5	3	10	5	34	11	21	7	70	26
CVen2	-5.2	160	0	0	1	1	5	5	9	8	15	14
UMajor1	-5.1	97	1	1	4	3	10	6	37	18	52	28
Leo4	-5	154	0	0	2	1	2	1	5	1	9	3
Hydrus1	-4.7	28	26	17	18	8	21	10	62	4	127	39
Hydra2	-4.6	134	0	0	1	0	5	2	13	4	19	6
Carina2	-4.5	36	8	2	16	11	15	3	29	0	68	16
...	...		...								...	
Carina3	-2.4	28	3	0	2	0	0	0	6	0	11	0
Segue2	-1.9	35	1	1	7	7	7	6	9	7	24	21
Triangul	-1.8	30	2	1	6	2	2	1	5	2	15	6
Hor2	-1.6	78	0	0	0	0	1	0	1	0	2	0
Tucana5	-1.6	55	0	0	1	0	0	0	3	0	4	0
Segue1	-1.3	23	2	2	3	3	3	2	18	12	26	19
Virgo1	-0.3	91	0	0	0	0	0	0	1	0	1	0
Cetus2	0	30	1	0	0	0	3	0	4	0	8	0

Figure 5. Same as Figure 4, but for the faint-end of the dwarf galaxy population. Unlike for the bright end, most of these galaxies just do not have very many stars that can be observed.

slope of density profile as a function of radial velocity sample size, where uncertainities continue to decrease as a function of sample size and do not level out until 20000 stars or more have been observed.

How do we observe 10000 stars in a single dwarf? No dwarf has that many stars in Gaia, and it requires going to $V \sim 23$ or deeper. Getting to 25000 or more stars with good uncertainties moves this science solidly into the regime of the Very Large Optical Telescopes (VLOTs). Indeed, getting to even many hundreds of stars in the ultra-faint dwarfs is also likely a VLOT science case. Figure 7 shows deep HST/ACS CMDs of 6 of the brightest ultra-faint dwarfs from Brown et al. (2014). Spectroscopic samples of many hundreds of stars will only be obtained if this deep photometry is followed up with very deep spectroscopy.

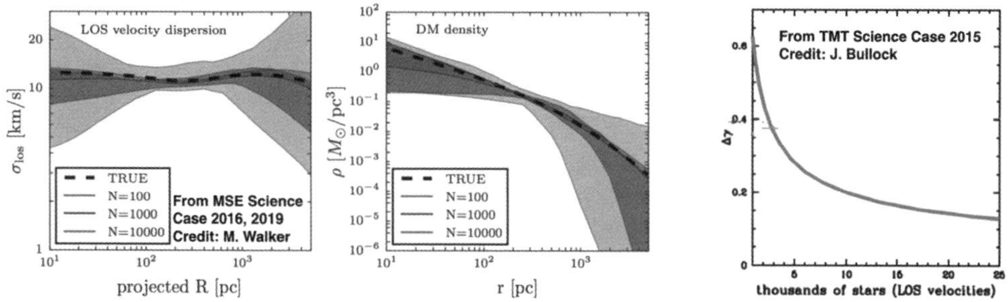

Figure 6. Left panels: Recovery of intrinsic profiles (based on cusp NFW) for a dwarf galaxy as a function of spectroscopic sample size. Shaded regions show 95% credible intervals from a standard Jeans analysis. Assumes median velocity error of 2km/s. Right panel: Accuracy of recovery of logarithmic slope of density profile as a function of radial velocity sample size.

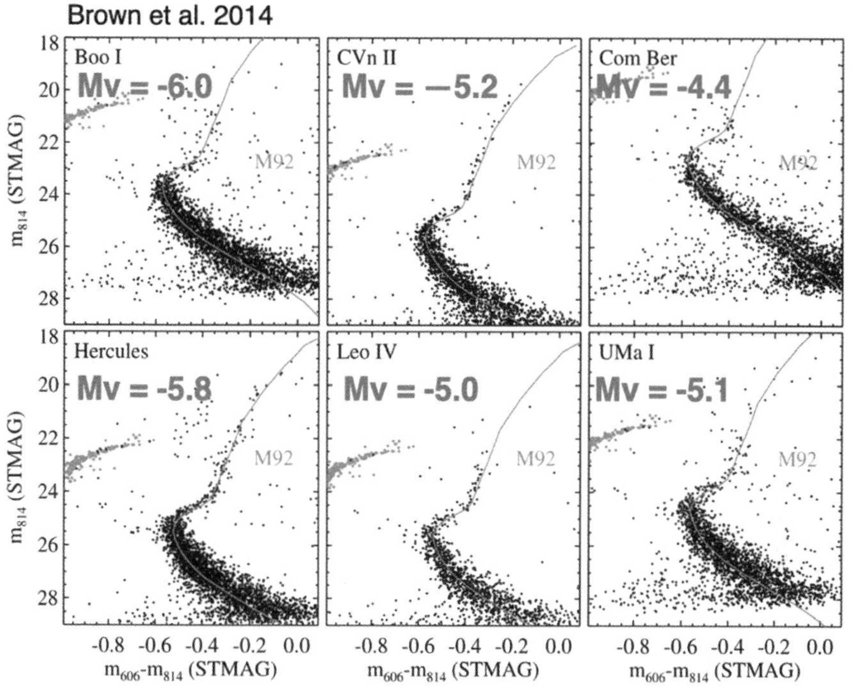

Figure 7. HST/ACS CMDs of 6 ultra-faint dwarf galaxies from Brown et al. (2014). Significant spectroscopic samples of stars in these systems will only occur if very deep spectroscopy can be obtained.

6. Proper motions, now and in the future

Proper motion science has undergone a revolution with the advent of Gaia, but even prior to that HST was delivering impressive results, though for a less extensive area of sky. As a baseline, recall that a typical dwarf galaxy tangential velocity dispersion might be expected to be in the region of $5\mathrm{km\,s}^{-1}$. At a distance of 100kpc, this means accuracies of around $10\,\mu\mathrm{as\,yr}^{-1}$.

What can we expect from existing and upcoming instruments in terms of relevant proper motion accuracies for this field of research. Figure 8 shows the predicted Gaia † and Rubin (LSST Science Collaboration et al. 2009) proper motion accuracies. The Gaia

† https://www.cosmos.esa.int/web/gaia/science-performance

Magnitude	DR3*	DR4**	DR5**
G = 15 mag	42 µas/yr	12 µas/yr	4 µas/yr
G = 16 mag	65	19	7
G = 17 mag	102	32	11
G = 18 mag	167	58	20
G = 19 mag	291	114	40
G = 20 mag	565	250	88
G = 21 mag	1469	450	159

Magnitude	Gaia DR5	Rubin 10yr *
G = 21 mag	159 µas/yr	–
r = 21 mag	–	200 µas/yr
r = 22 mag	–	300 µas/yr
r = 23 mag	–	500 µas/yr
r = 24 mag	–	1000 µas/yr

Figure 8. Left panel: Expected proper motion accuracy for Gaia DR3, 4 and 5 as described on the Gaia Science Performance web pages (see footnote). Right panel: Rubin observatory 10 year predicted proper motion accuracies, from the LSST Science Book v2, Table 3.3.

data is particularly impressive; however, there are literally only a dozen or so stars in the known dwarf galaxies under consideration (i.e. excluding the Magellanic Clouds and Sagittarius) that are 16th magnitude or brighter (see Figures 4 and 5). Rubin continues Gaia's excellent performance to fainter magnitudes, but will not be suitable for measuing individual proper motions of stars within dwarfs.

VLOTs, especially ELT and TMT which have first light "wide field" AO systems are very interesting for this science. Both ELT/MICADO† and TMT/IRIS‡ are being designed to achieve proper motion accuracies of order 10 μas yr^{-1}, in large part because of consideration of science cases like these. Their very small fields of view will lead to some challenges, however, when dealing with physically extended systems such as dwarf galaxies. Current multi-conjugate adaptive optics systems - especially Gemini/GEMS - provide a very interesting forum for the development of the necessary toolkits to be able to do precision proper motions in the context of wide field adaptive optics (e.g., Taheri et al. 2022).

Space will remain one of the premier places to do proper motion science, however, and HST/ACS has been highly impactful in the regard. Long baselines now exist which, when combined with the excellent intrument characteristics of HST/ACS, allow "route" measurments with an accuracy of around 12 or 13 μas yr^{-1} (e.g., Sohn et al. 2012; Libralato et al. 2023 and references therein). JWST, with a larger mirror but longer wavelengths should be expected to have somewhat comparable precisions, as could Euclid and Roman. All might ecxpect their initial measurements to be combined with HST to provide the time baselines. Euclid and Roma will be especially notable, however, because of their vast fields of view, making global studies of dwarf galaxies a possibility

7. Summary for Driving Science I: What is the intrinsic dynamical mass profiles of dwarf galaxies?

And so we return to the first of our driving science cases. Consideration of the diverse issues that have been touched up here cause me to conclude that, to meet the challenge of this science, we need:

† "Top Level Requirements for ELT-CAM", Doc Number: ESO-193104, V2; Released On: 2015-03-30

‡ TMT Science Requirements document, Doc Number: TMT.PSC.DRD.05.001.CCR25, Released on April 21 2021

• Complete, multi-epoch (including years-cadence) spectroscopic survey of all member stars identified in Gaia at $G < 20$ with $100 - 1000$m/s precision

– Achievable in part through new wide field MOS on WEAVE, 4MOST and DESI and new campaigns on existing 8m-class high resolution spectrograph. It is worth noting that the throughput of new 4m-class spectrographs is humbling compared to previous generations of 8m-class spectrographs (4m spectroscopy of M31 with DESI; Dey et al. 2023); multiplexing is awesome

• 8m-class MOS campaigns for targets to $V = 23 - 24$ to create necessary sample sizes, again with high velocity precision.

– Aim for 25K+ stars, not 2K stars.

– Start with Subaru/PFS, VLT/MOONS. Ultimately, we need MSE and WST i.e. new facilities

– Rubin will be able to provide proper motion cleaning when available to reduce contamination, since Gaia in this regime is no longer useful

• To ensure "first epoch" fields at the diffraction limit exist for future astrometry (i.e., HST, JWST, possibly 8m-AO)

– Field placement needs to be driven by dynamical considerations of measurement of internal orbital anisotropy, not just whatever data exists in the archive. So much of the archival data is not in the centers of the dwarfs, meaning that we have reduced sensitivity to the dynamics in the central regions where core/cusp related issues can be best probed

– We need Roman and Euclid to survey the dwarfs! These are our only likely chance of "complete" proper motion studies including in the outskirts

• ELT/MICADO imaging of Fornax, Carina, Sculptor and Sextans as soon as it sees first light; we need TMT/IRIS imaging of Ursa Minor and Draco (and Sextans)

– Their fields of view are small, so we'll need to tile the centres of these galaxies

8. Summary for Driving Science II: What is the lowest mass galaxy that can exist?

As for our second driving science case, consideration of the issues touched on here lead me to conclude that we need:

• Complete, multi-epoch (including years-cadence) spectroscopic survey of all member stars identified in Gaia at $G < 20$ with $100 - 1000$m/s precision

– Easiest to do with high-resolution spectrographs; need for multiplexing not strong for majority of systems

• Deep, wide field imaging of every UFD to 26 magnitude. For this, Subaru/HSC is especially good, also other 8m imaging

• 8m-class MOS campaigns for targets to $V = 24 - 25$ (selected from the deep wide field imaging)

– many stars will be contamination - LSST proper motions will help clean southern target samples

– multi-epochs especially necessary – binary influence larger because velocity dispersions are lower

– can start attacking this now using Subaru/PFS, VLT/MOONS especially; possibly may require MSE, WST

• To ensure that every UFD has HST/ACS or JWST/NIRCAM imaging in its central regions at least (for 1st epoch proper motions)

– Gaia can be useful, but is ultimately not deep enough

– As before, we need Roman and Euclid to survey the dwarfs

• ELT/MICADO and TMT/IRIS imaging (astrometric)

• ELT/HARMONI and TMT/IRIS spectroscopy to 26 magnitude

References

Battaglia, G., Taibi, S., Thomas, G. F., & Fritz, T. K. 2022, *A&A*, 657, A54.

Battaglia, G., Tolstoy, E., Helmi, A., Irwin, M., Parisi, P., Hill, V., & Jablonka, P. 2011, *MNRAS*, 411, 1013–1034.

Brown, T. M., Tumlinson, J., Geha, M., *et al.* 2014, *ApJ*, 796(2), 91.

Dey, A., Najita, J. R., Koposov, S. E., *et al.* 2023, *ApJ*, 944(1), 1.

Errani, R., Navarro, J. F., Peñarrubia, J., Famaey, B., & Ibata, R. 2023, *MNRAS*, 519(1), 384–396.

Ibata, R., Irwin, M., Lewis, G. F., & Stolte, A. 2001, *ApJ*, 547, L133–L136.

Koposov, S. E., Gilmore, G., Walker, M. G., *et al.* 2011, *ApJ*, 736, 146.

Libralato, M., Bellini, A., van der Marel, R. P., *et al.* 2023, *arXiv e-prints*, arXiv:2303.00009.

LSST Science Collaboration, Abell, P. A., Allison, J., Anderson, S. F., Andrew, J. R., *et al.* 2009, *arXiv e-prints*, arXiv:0912.0201.

Massari, D., Breddels, M. A., Helmi, A., Posti, L., Brown, A. G. A., & Tolstoy, E. 2018, *Nature Astronomy*, 2, 156–161.

McConnachie, A., Babusiaux, C., Balogh, M., *et al.* 2016, *ArXiv e-prints*,.

McConnachie, A. W. 2012, *AJ*, 144, 4.

McConnachie, A. W. & Côté, P. 2010, *ApJ*, 722, L209–L214.

McConnachie, A. W. & Venn, K. A. 2020, *AJ*, 160(3), 124.

Minor, Q. E. 2013, *ApJ*, 779(2), 116.

Minor, Q. E., Pace, A. B., Marshall, J. L., & Strigari, L. E. 2019, *MNRAS*, 487(2), 2961–2968.

Pace, A. B., Erkal, D., & Li, T. S. 2022, *ApJ*, 940(2), 136.

Pace, A. B. & Li, T. S. 2019, *ApJ*, 875(1), 77.

Sestito, F., Roediger, J., Navarro, J. F., Jensen, J., Venn, K. A., Smith, S. E. T., Hayes, C., & McConnachie, A. W. 2023,a *MNRAS*.

Sestito, F., Zaremba, D., Venn, K. A., D'Aoust, L., Hayes, C., Jensen, J., Navarro, J. F., Jablonka, P., Fernández-Alvar, E., Glover, J., McConnachie, A. W., & Chené, A.-N. 2023,b *arXiv e-prints*, arXiv:2301.13214.

Skidmore, W., TMT International Science Development Teams, & Science Advisory Committee, T. 2015, *Research in Astronomy and Astrophysics*, 15, 1945.

Smith, S. E. T., Jensen, J., Roediger, J., Sestito, F., Hayes, C. R., McConnachie, A. W., Cuillandre, J.-C., Gwyn, S., Magnier, E., Chambers, K., Hammer, F., Hudson, M., Martin, N., Navarro, J., & Scott, D. 2022, *arXiv e-prints*, arXiv:2209.08242.

Sohn, S. T., Anderson, J., & van der Marel, R. P. 2012, *ApJ*, 753, 7.

Spencer, M. E., Mateo, M., Olszewski, E. W., Walker, M. G., McConnachie, A. W., & Kirby, E. N. 2018, *AJ*, 156(6), 257.

Spencer, M. E., Mateo, M., Walker, M. G., & Olszewski, E. W. 2017, *ApJ*, 836(2), 202.

Taheri, M., McConnachie, A. W., Turri, P., Massari, D., Andersen, D., Bono, G., Fiorentino, G., Venn, K., Véran, J.-P., & Stetson, P. B. 2022, *AJ*, 163(4), 187.

The MSE Science Team, Babusiaux, C., Bergemann, M., *et al.* 2019, *arXiv e-prints*, arXiv:1904.04907.

Tolstoy, E., Hill, V., Irwin, M., Helmi, A., Battaglia, G., Letarte, B., Venn, K., Jablonka, P., Shetrone, M., Arimoto, N., Abel, T., Primas, F., Kaufer, A., Szeifert, T., Francois, P., & Sadakane, K. 2006, *The Messenger*, 123, 33.

Walker, M. G., Mateo, M., & Olszewski, E. W. 2009, *AJ*, 137, 3100–3108.

Dynamical Masses of Local Group Galaxies
Proceedings IAU Symposium No. 379, 2025
P. Bonifacio, M.-R. Cioni, F. Hammer, M. S. Pawlowski & S. Taibi, eds.
doi:10.1017/S1743921323003526

Kinematics of The Tidally Disrupting Dwarf Galaxy Antlia II

Alexander Ji and the S^5 Collaboration

University of Chicago, 5640 S. Ellis Avenue, Chicago, IL 60637, USA
email: alexji@uchicago.edu

Abstract. The Milky Way satellite dwarf galaxy Antlia II is one of the lowest surface brightness galaxies known. It has a size comparable to the Large Magellanic Cloud, but only 10^6 solar masses of stars. We present kinematic and chemical measurements from the Southern Stellar Stream Spectroscopic Survey using the AAT/2dF that clearly demonstrate that Antlia II is tidally disrupting. The orbit and velocity gradient also clearly shows that the Milky Way has moved in response to the Large Magellanic Cloud. However, Antlia II currently lies on the galaxy mass-metallicity relation, suggesting that it has not lost too much stellar mass. These measurements illustrate the importance of full dynamic models when interpreting the masses of local group galaxies.

Keywords. Dwarf galaxies, stellar streams, Local Group

1. Introduction

The Antlia II (Ant II) dwarf galaxy was discovered in 2019 in a search for Milky Way satellite dwarf galaxies in Gaia DR2 (Torrealba et al. 2019). It has the same physical size as the Large Magellanic Cloud but a stellar mass of only $1.6 \times 10^6 M_\odot$. As a result, Ant II is one of the lowest surface brightness galaxies known, with a surface brightness of 30.7 mag/arcsec^2 (Collins et al. 2020; Ji et al. 2021).

Galaxies of this size and surface brightness are difficult to form in standard ΛCDM cosmology, and thus provide an interesting laboratory to test the nature of dark matter or gravity. One crucial factor to consider is tidal disruption, which has been invoked as an important way to explain the sizes and velocity dispersions of galaxies like Ant II (e.g., Sanders et al. 2018). Early simulations showed that if Ant II would be difficult to explain if it originally resided in a cuspy dark matter halo as expected in standard cold dark matter, but it would be possible to explain the kinematic properties if its central cusp was cored through baryonic feedback (Torrealba et al. 2019). Another possibility is that non-standard dark matter models naturally produce cored dark matter halos that would make Ant II easier to explain (Sameie et al. 2020; Broadhurst et al. 2020). A third possibility is modified gravity, which is able to explain the velocity dispersion and size of a similar galaxy Crater II (McGaugh 2016; Caldwell et al. 2017) and should be tested in Ant II. In all cases, it is important to observationally establish whether or not Ant II is tidally disrupting, which affects the assumptions of dynamical equilibrium that are often used.

Here we describe recent spectroscopic observations of Antlia II from the Southern Stellar Stream Spectroscopic Survey extending out to much larger radii than previously obtained (Ji et al. 2021). We show with a detailed kinematic analysis that Ant II is clearly tidally disrupting. Our observations of Ant II also show the clear impact of the Large Magellanic Cloud (LMC) on our Milky Way's barycenter.

2. Observations and Analysis

The Southern Stellar Stream Spectroscopic Survey (Li et al. 2019) used the Anglo-Australian Telescope to observe red giant stars in four new fields around Antlia II, going out to two half light radii, which covers four times more area than previous spectroscopic observations (Ji et al. 2021). Radial velocities and metallicities were determined with `rvspecfit` (Koposov 2019). Member stars of Ant II were determined by fitting a three-component Gaussian mixture model using the positions, proper motions, radial velocities, and metallicities. Two background components were used to model contamination from the disk and halo. A linear velocity gradient was allowed, and the model included geometric effects like perspective rotation and differential solar reflex motion since the spatial coverage is much larger than one degree on the sky. Structural properties were also updated using Gaia DR3. Once members were confirmed, we redetermined metallicities using a calcium triplet equivalent width calibration (Carrera et al. 2013), which is both more accurate and precise than the full synthetic spectrum fit (see Li et al. 2022). The bulk properties of Antlia II are available in references Ji et al. (2021) and Vivas et al. (2022).

3. Results and Discussion

Fig 1 shows the luminosity-metallicity relation and the position of Ant II. We find Ant II lies on this relation, in contrast to the original discovery paper, which found that Ant II had a substantially higher metallicity and placed it to the left of the luminosity-metallicity relation (Torrealba et al. 2019). This was previously used as evidence that Ant II is tidally disrupting, but this is no longer a valid argument. We note that our data sample includes all stars from the previous work, and the previous results were analyzed with an older version of the same code. In addition, the lower metallicity has also been confirmed with high-resolution spectroscopy of a few stars. Thus there is no doubt our new metallicity is accurate. Assuming galaxies are born on this luminosity-metallicity relation, we conservatively estimate that Ant II can have lost at most 90% of its stars to tidal disruption.

Though the chemistry does not favor substantial tidal disruption, we now find clear kinematic evidence that Ant II is tidally disrupting. Fig 2 shows the spatial distribution and velocity gradient of Ant II. The proper motion of Ant II is indicated with a purple arrow. There is an obvious velocity gradient from one side of Ant II to the other that is well described by a linear trend. We allowed the orientation of this velocity gradient to be a free parameter in our fit, and it clearly prefers to be aligned with the proper motion orientation. This is exactly what is expected if Ant II's large size is due to tidal disruption, as recently tidally disrupted stars in a static potential should lie on the orbital path of the tidally disrupting system. The blue contours in the top row show the results of a tidal disruption model for Ant II. This is an excellent qualitative match in the spatial distribution and velocity gradient. Though velocity gradients are often interpreted as evidence for rotation (e.g., Collins et al. 2020), we reject this interpretation for Ant II because it is unlikely that the rotation axis would happen to be aligned in the direction that matches expectations from tidal disruption.

A small but significant observation is that the spatial orientation of Ant II is slightly misaligned with the direction of proper motion by 20° (detected at $\gtrsim 2\sigma$). This difference is actually well explained by including the impact of the LMC on the Milky Way barycenter. Since the LMC is quite massive and just past its pericenter, it causes the Milky Way's disk to move substantially (currently about 20 kpc and 30 km/s relative to the barycenter prior to the LMC's infall) (Gómez et al. 2015; Garavito-Camargo et al. 2019; Petersen and Peñarrubia 2021). This breaks the assumption of orbiting within a

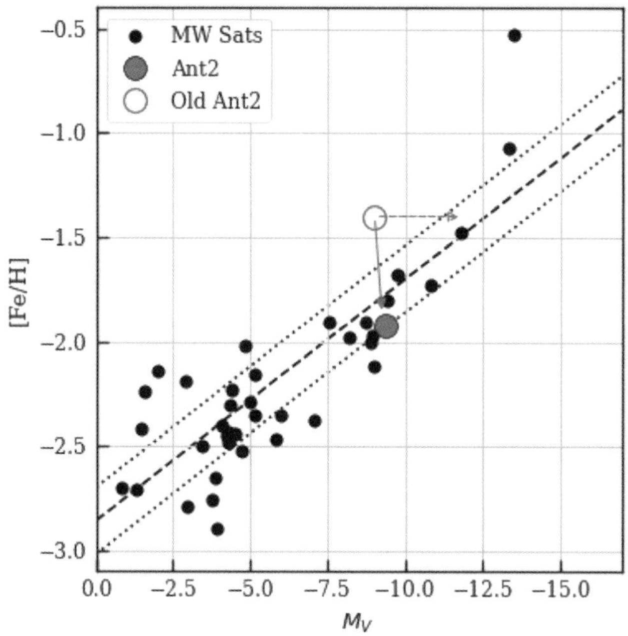

Figure 1. Luminosity-metallicity relation for Milky Way satellites, with Antlia II highlighted. Previous measurements obtained an incorrect metallicity (open red circle) which suggested it had lost a substantial amount of stellar mass through tidal disruption. Our new metallicity shows that Antlia II lies on the luminosity-metallicity relation, and thus should be considered

static potential, and it effectively causes an extra reflex motion correction that manifests as a difference between the spatial orientation and proper motion of a tidally disrupting system. The red contours in the bottom row of Fig 2 show the result of a numerical model, which indeed matches both the observed spatial orientation and velocity gradient better than the blue model in the top row. It is important to note that none of these models were tuned to match Ant II. All these models were based on fits to other kinematic data, most notably the Orphan-Chenab stream (Erkal et al. 2019), but they work incredibly well in explaining the kinematics of Ant II. This is a testament to the broad accuracy of our current mass models of the Milky Way-LMC system.

We also consider the scenario that modified gravity is responsible for Ant II's kinematic properties. The motivation for considering this is that McGaugh (2016) used the external field effect in Modified Newtonian Dynamics (MOND) to accurately predict the velocity dispersion of another low density galaxy around the Milky Way Crater II (Caldwell et al. 2017). In the case of Crater II, the expected velocity dispersion for an isolated galaxy would be 4 km s^{-1}, but the external field effect reduces the prediction to 2 km s^{-1}, very close to the measured value of $2.3^{+0.4}_{-0.3}$ km s^{-1}. The very low velocity dispersion in Crater II is very challenging, though not impossible, to explain in ΛCDM (Sanders et al. 2018). We thus apply the same calculation to Ant II (using the most recent RR Lyrae based distance, (Vivas et al. 2022)). The isolated MOND prediction for Ant II is 6 km s^{-1}, while the external field effect reduces the prediction to 3 km s^{-1}. However, we measure a velocity dispersion of $6.0^{+0.4}_{-0.4}$, clearly inconsistent with the external field effect. Antlia II and Crater II have very similar sizes, stellar masses, distances, metallicites, and orbits; so it seems unlikely that one galaxy should experience the external field effect and the other would not, which naively challenges the MOND interpretation for large

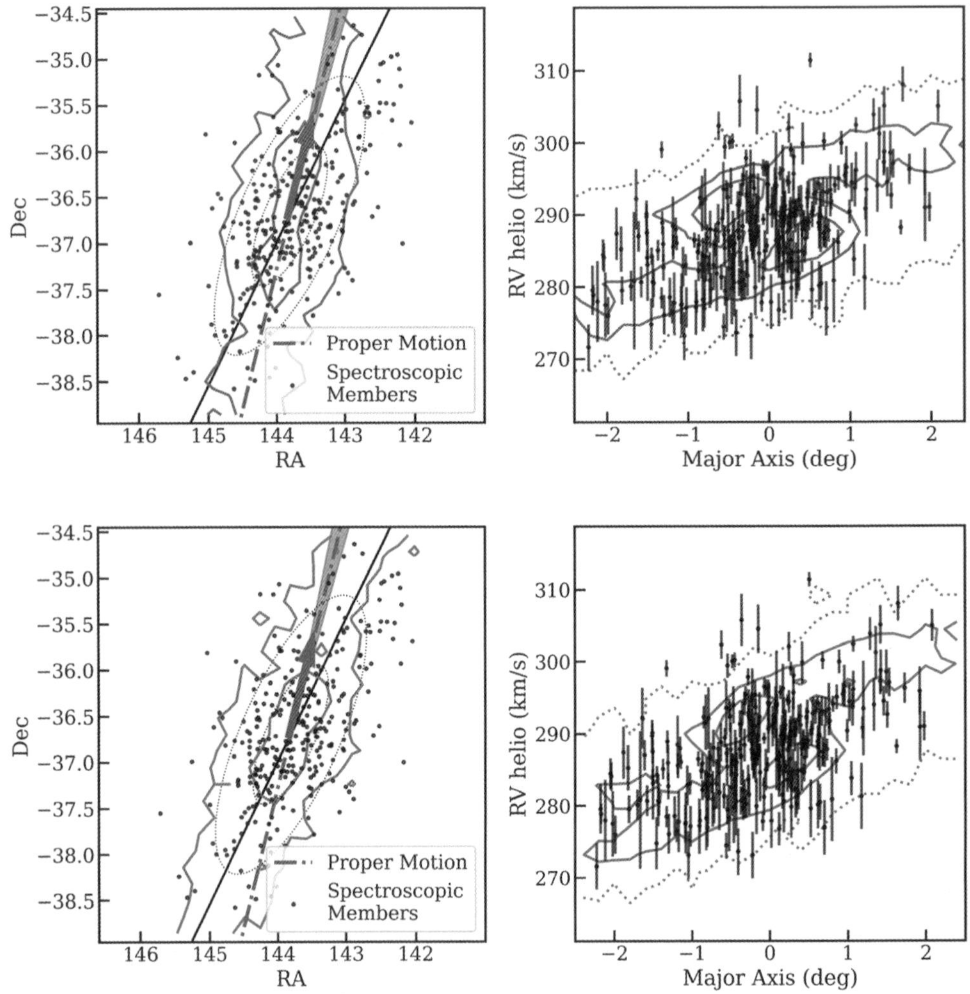

Figure 2. Spatial distribution and velocity gradient in Antlia II. The same data are plotted as black points in the top and bottom rows. The black points in the left column are the positions of confident spectroscopic members of Ant II, and the black dotted ellipses indicate the spatial orientation and 1 and 2 half light radii derived using Gaia DR3. The thick black line indicates the ellipse major axis. The purple arrow and dashed line shows the direction of observed proper motion and uncertainty. There is a clear misalignment between the direction of proper motion and the spatial orientation of Ant II. The right column shows the observed heliocentric radial velocity as a function of position along the spatial major axis. There is a clear velocity gradient across Ant II. In the top row, the blue contours are 68% and 95% contours for stars from a model that does not include the LMC. In the bottom row, the red contours are from a model that includes the LMC. The LMC does not directly impact Ant II in this model, and all effects are due to the response of the Milky Way's barycenter to the LMC.

dwarf galaxy kinemamtics. However, since these systems are not in equilibrium and the external field effect calculations assume equilibrium, numerical simulations are needed to settle this issue.

In summary, our spectroscopic followup of Antlia II has shown that it is clearly tidally disrupting. The evidence is a nearly-colinear proper motion, spatial orientation, and velocity gradient. The mass and metallicity suggest that Ant II has not lost more than 90% of its original stellar mass. The detailed kinematics show the clear effect of the LMC

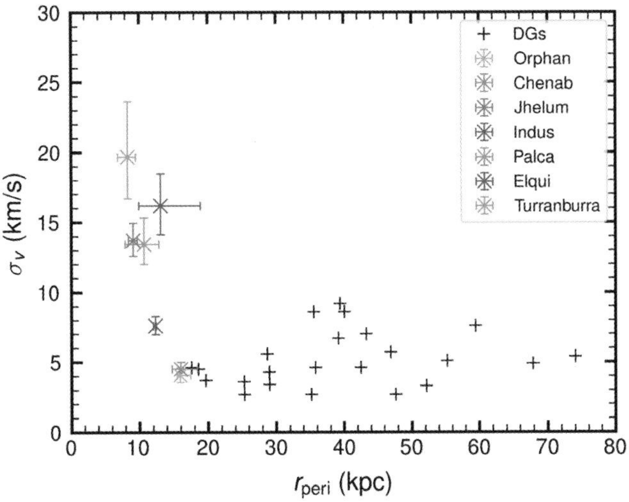

Figure 3. Line of sight velocity dispersion as a function of orbital pericenter for intact dwarf galaxies and dwarf galaxy stellar streams (Li et al. 2022). Below 15-20 kpc, there are no known intact dwarf galaxies, and streams with closer pericenters have higher velocity dispersions.

on the Milky Way's barycenter. These new observations provide important constraints on different dark matter or modified gravity models, but the data are now good enough that models need to take into account previously minor effects like the LMC.

4. Extra Note on Dwarf Galaxy Pericenters

There was extensive discussion at this conference about pericenter distributions and whether dwarf galaxies are tidally affected in the conference. To represent the discussion, Figure 3 shows the velocity dispersion of dwarf galaxies as a function of their pericenter (Li et al. 2022), including both intact dwarf galaxies and tidal streams with chemistry consistent with being dwarf galaxies (Ji et al. 2020). The data suggest that dwarf galaxies with pericenters below 20 kpc become tidally disrupted, with higher velocity dispersions for galaxies with closer pericentric passages. Antlia II has a pericenter of 38.6 kpc (Ji et al. 2021), so while it has tidally disrupted, its velocity dispersion might not greatly inflated by tidal effects (after accounting for the velocity gradient).

References

Broadhurst, T., De Martino, I., Luu, H. N., Smoot, G. F., & Tye, S. H. H. 2020, *Phys. Rev. D*, 101(8), 083012.

Caldwell, N., Walker, M. G., Mateo, M., Olszewski, E. W., Koposov, S., Belokurov, V., Torrealba, G., Geringer-Sameth, A., & Johnson, C. I. 2017, *ApJ*, 839(1), 20.

Carrera, R., Pancino, E., Gallart, C., & del Pino, A. 2013, *MNRAS*, 434(2), 1681–1691.

Collins, M. L. M., Tollerud, E. J., Rich, R. M., Ibata, R. A., Martin, N. F., Chapman, S. C., Gilbert, K. M., & Preston, J. 2020, *MNRAS*, 491(3), 3496–3514.

Erkal, D., Belokurov, V., Laporte, C. F. P., Koposov, S. E., Li, T. S., Grillmair, C. J., Kallivayalil, N., Price-Whelan, A. M., Evans, N. W., Hawkins, K., Hendel, D., Mateu, C., Navarro, J. F., del Pino, A., Slater, C. T., Sohn, S. T., & Orphan Aspen Treasury Collaboration 2019, *MNRAS*, 487(2), 2685–2700.

Garavito-Camargo, N., Besla, G., Laporte, C. F. P., Johnston, K. V., Gómez, F. A., & Watkins, L. L. 2019, *ApJ*, 884(1), 51.

Gómez, F. A., Besla, G., Carpintero, D. D., Villalobos, Á., O'Shea, B. W., & Bell, E. F. 2015, *ApJ*, 802(2), 128.

Ji, A. P., Koposov, S. E., Li, T. S., Erkal, D., Pace, A. B., Simon, J. D., Belokurov, V., Cullinane, L. R., Da Costa, G. S., Kuehn, K., Lewis, G. F., Mackey, D., Shipp, N., Simpson, J. D., Zucker, D. B., Hansen, T. T., Bland-Hawthorn, J., & S5 Collaboration 2021, *ApJ*, 921(1), 32.

Ji, A. P., Li, T. S., Hansen, T. T., Casey, A. R., Koposov, S. E., Pace, A. B., Mackey, D., Lewis, G. F., Simpson, J. D., Bland-Hawthorn, J., Cullinane, L. R., Da Costa, G. S., Hattori, K., Martell, S. L., Kuehn, K., Erkal, D., Shipp, N., Wan, Z., & Zucker, D. B. 2020, *AJ*, 160(4), 181.

Koposov, S. E. 2019,. Astrophysics Source Code Library, record ascl:1907.013.

Li, T. S., Ji, A. P., Pace, A. B., Erkal, D., Koposov, S. E., Shipp, N., Da Costa, G. S., Cullinane, L. R., Kuehn, K., Lewis, G. F., Mackey, D., Simpson, J. D., Zucker, D. B., Ferguson, P. S., Martell, S. L., Bland-Hawthorn, J., Balbinot, E., Tavangar, K., Drlica-Wagner, A., De Silva, G. M., & Simon, J. D. 2022, *ApJ*, 928(1), 30.

Li, T. S., Koposov, S. E., Zucker, D. B., Lewis, G. F., Kuehn, K., Simpson, J. D., Ji, A. P., Shipp, N., Mao, Y. Y., Geha, M., Pace, A. B., Mackey, A. D., Allam, S., Tucker, D. L., Da Costa, G. S., Erkal, D., Simon, J. D., Mould, J. R., Martell, S. L., Wan, Z., De Silva, G. M., Bechtol, K., Balbinot, E., Belokurov, V., Bland-Hawthorn, J., Casey, A. R., Cullinane, L., Drlica-Wagner, A., Sharma, S., Vivas, A. K., Wechsler, R. H., Yanny, B., & S5 Collaboration 2019, *MNRAS*, 490(3), 3508–3531.

McGaugh, S. S. 2016, *ApJ*, 832(1), L8.

Petersen, M. S. & Peñarrubia, J. 2021, *Nature Astronomy*, 5, 251–255.

Sameie, O., Chakrabarti, S., Yu, H.-B., Boylan-Kolchin, M., Vogelsberger, M., Zavala, J., & Hernquist, L. 2020, *arXiv e-prints*, arXiv:2006.06681.

Sanders, J. L., Evans, N. W., & Dehnen, W. 2018, *MNRAS*, 478(3), 3879–3889.

Torrealba, G., Belokurov, V., Koposov, S. E., Li, T. S., Walker, M. G., Sanders, J. L., Geringer-Sameth, A., Zucker, D. B., Kuehn, K., Evans, N. W., & Dehnen, W. 2019, *MNRAS*, 488(2), 2743–2766.

Vivas, A. K., Martínez-Vázquez, C. E., Walker, A. R., Belokurov, V., Li, T. S., & Erkal, D. 2022, *ApJ*, 926(1), 78.

Dynamical Masses of Local Group Galaxies
Proceedings IAU Symposium No. 379, 2025
P. Bonifacio, M.-R. Cioni, F. Hammer, M. S. Pawlowski & S. Taibi, eds.
doi:10.1017/S1743921323001473

The Dark Matter Density Profiles of Antlia B

Mariana P. Júlio[1,2,3], Jarle Brinchmann[1,4], Justin I. Read[5] and The MUSE Collaboration

[1]Instituto de Astrofísica e Ciências do Espaço, Universidade do Porto, CAUP, Rua das Estrelas, PT4150-762 Porto, Portugal
email: mpouseirojulio@aip.de

[2]Departamento de Física e Astronomia, Faculdade de Ciências, Universidade do Porto, Rua do Campo Alegre 687, PT4169-007 Porto, Portugal

[3]Leibniz-Institut für Astrophysik Potsdam (AIP), An der Sternwarte 16, D-14482 Potsdam, Germany

[4]Leiden Observatory, Leiden University, P.O. Box 9513, 2300 RA Leiden, The Netherlands
[5] University of Surrey, Physics Department, Guildford, GU2 7XH, UK

Abstract. We use spectroscopic observations of Antlia B, a distant ($d \sim 1.35$ Mpc) faint dwarf galaxy ($M_V = -9.7$, $M_\star = 10^5 M_\odot$), from MUSE-Faint, to explore alternative dark matter (DM) models to ΛCDM, which are one possible explanation for the small-scale problems in the standard model.

We measure line-of-sight velocities of 127 stars, and combine these with GravSphere, a Jeans modelling code, to place constraints on DM models and derive the first DM density profiles for this object.

We present constraints on the nature of self-interacting DM, in which DM particles can interact with one another in the form of annihilation (we find a core radius of $r_c \lesssim 69$ pc) and on scalar field DM, comprised of ultralight bosons that form a Bose-Einstein Condensate (we find a characteristic length scale of the self-interaction of $R_{\rm TF} \lesssim 180$ pc). These results suggest that we can rule out these models as an explanation for the cores in the larger dwarf galaxies in the Local Group.

Keywords. Dark Matter, Galaxies, Antlia B, Kinematics, Dynamics, Spectroscopy

1. Introduction

Since dark matter was proposed as an additional component to explain the flat observed rotation curves of galaxies (Rubin et al. 1980), many more indications for its existence have been found. The presence of this non-baryonic matter is required to explain the behaviour of most structures existing in the Universe and the vast majority of the cosmological observations favour a cold dark matter plus dark energy scenario (ΛCDM model). ΛCDM has been very successful at predicting and explaining the large-scale structure of the Universe and its evolution with time (e.g. Vogelsberger et al. 2014), which has led to it now being considered the standard model of cosmology.

However, when we approach smaller scales, pure dark matter structure formation simulations in the standard model start to face numerous challenges (Bullock and Boylan-Kolchin 2017). Namely, for example, the missing satellites problem (e.g. Klypin et al. 1999), the too-big-to-fail problem (e.g. Boylan-Kolchin et al. 2011), and the core-cusp problem (e.g. Moore 1994). Several solutions have been proposed to address the small-scale problems in ΛCDM and here we focus on the possibility of the nature of dark matter itself being different from the current paradigm. The proposed alternatives to the CDM cover a wide range of particle masses and different interactions between the dark matter particles. In this work, we explore models of self-interacting dark matter (SIDM) through annihilation (Kaplinghat et al. 2000) and of scalar field dark matter (SFDM) comprised of ultralight bosons (Li et al. 2014) to address the core-cusp problem.

To understand and solve small-scale challenges to the ΛCDM paradigm, we need to study the very smallest structures in the Universe. The most encouraging class of objects for this purpose are possibly ultra-faint dwarfs (UFDs) since they are the most dark matter-dominated systems known (Simon 2019). Additionally, baryonic feedback should not significantly influence the dark matter density profiles of UFDs, since they do not have enough stars to have significant supernovae feedback (Orkney et al. 2021).

However, the faintness of UFDs makes it challenging to obtain large numbers of stellar line-of-sight velocities and thus high-precision density profiles. Antlia B (Ant B), a slightly brighter dwarf galaxy ($M_V = -9.7$, $M_\star \sim 8 \times 10^5 M_\odot$, $r_h = 273$ pc) was discovered by Sand et al. (2015) using the DECam at Cerro Tololo International Observatory and was spotted as a companion to NGC 3109. More recent studies of this galaxy were carried out by Hargis et al. (2020), which allowed a more precise measurement of its distance ($d = 1.35$ Mpc).

To further explore Ant B and reach our goal, we use spectroscopic observations from the Multi-Unit Spectroscopic Explore (MUSE, Bacon et al. 2010). The data reduction used in this work follows the one described in Zoutendijk et al. (2020). We used PAMPELMUSE (Kamann et al. 2013) to extract the spectra of the stellar sources in the MUSE cube, followed by SPEXXY (Husser 2012), which allowed us to determine the line-of-sight velocities from the previously collected spectra. The kinematic data were carefully analysed to ensure the membership of Antlia B stars (e.g. removing the faintest sources and foreground/background galaxies and stars). Further details will be provided in a subsequent work (Júlio et al. in prep.). The final selection and correspondent line-of-sight velocity dispersion can be seen in Figure 1.

2. Methods

2.1. *Dark matter models*

Self-interacting dark matter

In order to solve the so called core-cusp problem, several models of interacting dark matter have been proposed. It is believed that if the dark matter particles can interact with one another they can alter the halo structure and, possibly, produce a constant-density core in the heart of the halo (Elbert et al. 2015). The interacting DM models proposed mainly include scattering and annihilation. Here, we focus on the work of Lin and Loeb (2016) that includes signatures of annihilation and self-interaction. The density profile of this model is given by

$$\rho_{\mathrm{SIDM}}(r) = \frac{\rho_0}{r/r_c + (r/r_s)(1 + r/r_s)^2}, \tag{1}$$

with characteristic density $\rho_0 = \rho_c(r_c/r_s)$, where ρ_c is the core density, r_c is the core radius and r_s is the scale radius. When the core radius $r_c = 0$, the SIDM profile is equal to the CDM profile. When we have $r_c > 0$, the profile becomes cored instead of exhibiting a cusp, as predicted by the standard model. There is a relation that links the self-interaction coefficient Γ to the parameters of the profile and is given by

$$f\Gamma = \frac{r_c/r_s}{t\rho_0}, \tag{2}$$

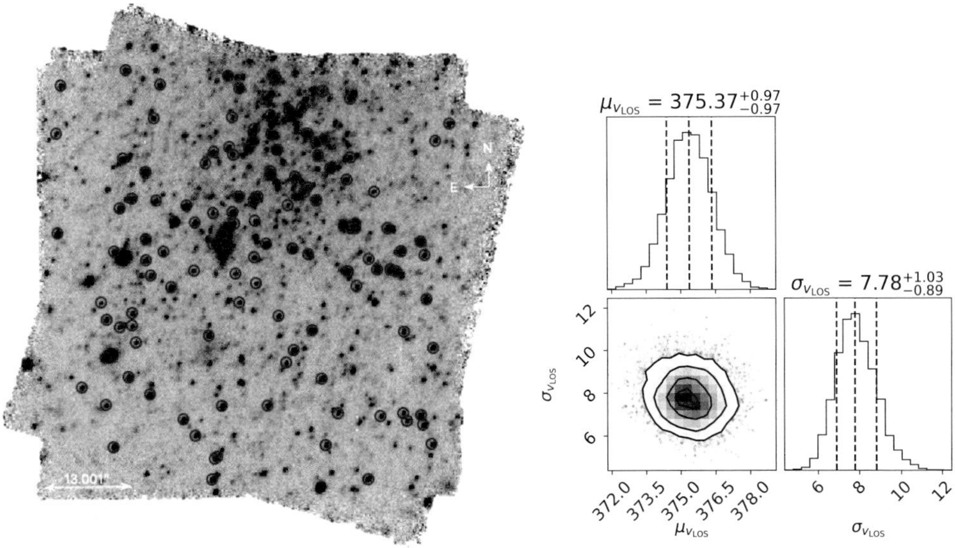

Figure 1. *Left:* Image of the final cube with the 127 stars identified as Ant B members represented by the red circles. *Right:* Corner plot of the velocity fit, using the 127 member stars of Ant B, obtained by following the method of Hargreaves et al. (1994). The histograms along the diagonal represent the posterior distribution for each parameter: the mean value $\mu_{v_{\rm LOS}}$ in the top panel and the dispersion $\sigma_{v_{\rm LOS}}$ in the bottom-right, both in km/s. The vertical dashed lines indicate the median and 68% confidence interval. The bottom-left panel represents the 2D posterior distributions of both of these parameters, with the contours corresponding to 0.5σ, 1σ, 1.5σ and 2σ confidence levels, where σ is the standard deviation of the 2D distribution.

where t is the time elapsed since the start of the self-interaction. However, it is possible to notice the degeneracy that exists between Γ and the fudge factor f. For dwarf galaxies, this factor is predicted to be ~ 10 (Kaplinghat et al. 2000) but is not precisely known (Zoutendijk et al. 2021a). Thus, we will try to constrain the product $f\Gamma$ - the effective self-interaction coefficient - to understand how the annihilation processes affect the system in study.

Scalar field dark matter

The scalar field dark matter is comprised of ultralight bosons and is described by the coupled Schrödinger-Poisson equations. This model, like the SIDM, is an alternative to the ΛCDM and in this case, it leads to a Bose-Einstein condensate and quantum superfluid. Just like fuzzy dark matter (FDM; e.g. Dine and Fischler 1983) this model's behaviour approaches the CDM model on large scales and differs the most on small scales. The main difference between SFDM and FDM is that SFDM also includes a repulsive self-interaction (Dawoodbhoy et al. 2021).

While CDM is described by the collisionless Boltzmann equation, SFDM obeys the nonlinear Schrödinger equation rewritten in terms of quantum hydrodynamics equations for the conservation of mass and momentum. Its density profile is given by

$$\rho_{\rm SFDM} = \begin{cases} \rho_c \mathrm{sinc}(\pi r / R_{\rm TF}) & r \leq \alpha R_{\rm TF} \\ \rho_{\rm CDM} & r \geq \alpha R_{\rm TF} \end{cases}, \tag{3}$$

where ρ_c is the central density and α is a given fraction of $R_{\rm TF}$ determined by requiring mass conservation and density continuity at $r = \alpha R_{\rm TF}$. $R_{\rm TF}$ is defined by

$$R_{\rm TF} = \pi\sqrt{\frac{g}{4\pi G m^2}}. \tag{4}$$

This quantity is a physical constant of the SFDM since it is fixed by the particle parameters (m, g) in their combination g/m^2, called the strength parameter. $R_{\rm TF}$ is the relevant quantity below which SFDM differs from CDM. To avoid the small-scale structure problems for the standard model discussed before, $R_{\rm TF} \gtrsim 1$ kpc (Dawoodbhoy et al. 2021).

2.2. *GravSphere*

To measure the dark matter density profiles of Ant B, we use the updated version of GravSphere Jeans modelling code (Read and Steger 2017). A detailed explanation of its implementation can be found in Read and Steger (2017) and Read et al. (2018).

GravSphere solves the Jeans equation (Jeans 1922) for our member stars after binning the data radially with Binulator (Collins et al. 2021). GravSphere assumes that the stellar systems are spherical, non-rotating, and in a steady state; allowing us to recover the density profile and the velocity anisotropy profile of the studied stellar systems (Read and Steger 2017). GravSphere relies on fourth-order Virial Shape Parameters (Merrifield and Kent 1990) to partially break the degeneracy that exists between the radial velocity dispersion (and therefore the cumulative mass distribution) and the velocity anisotropy.

3. Results and Discussion

Using GravSphere we sample the parameter spaces of our implemented dark matter density models. In order to do this, we use the kinematical measurements of Ant B, mentioned before. Here we present the constraints on the dark matter models and the recovered density profiles, as well as their comparison.

The constraints of the physical parameters of the SIDM and SFDM models are displayed in Figure 2.

In the SIDM profile, we can only present the constraint of the core radius as an upper limit. The posterior distribution that we get for the core radius obtained for this profile contains values that are smaller ($r_c \approx 0.01$ pc) than our smallest projected radius (1.84 pc), leading to a lack of constraining power at the lower end of the range of r_c. Therefore, the self-interaction coefficient is also presented as an upper limit. To derive this quantity, using equation 2, we had to assume a value for t since this quantity is unknown. For that purpose, we consider that the time elapsed since the start of self-interaction is equal to the age of the older stellar population, so we use $t = 13.5$ Gyr.

The upper limit found for the core radius is $r_c < 10^{-1.16} \lesssim 0.069$ kpc at the 68% confidence level and $r_c < 10^{-0.74} \lesssim 0.18$ kpc at the 95% confidence level. We then have an associated effective self-interaction coefficient $f\Gamma < 10^{-26.83} \lesssim 1.48 \times 10^{-27}\mathrm{cm^3 s^{-1} eV^{-1}} c^2$ at the 68% confidence level and $f\Gamma < 10^{-26.34} \lesssim 4.57 \times 10^{-27}\mathrm{cm^3 s^{-1} eV^{-1}} c^2$ at the 95% confidence level. We can translate $f\Gamma$ in the specific annihilation cross sections $\sigma/m \lesssim 7.4 \times 10^{-35}(f/10)^{-1}(v/10\mathrm{kms^{-1}})^{-1}\mathrm{cm^2 eV^{-1}} c^2$ (68% confidence level) and $\sigma/m \lesssim 2.3 \times 10^{-34}(f/10)^{-1}(v/10\mathrm{kms^{-1}})^{-1}\mathrm{cm^2 eV^{-1}} c^2$ (95% confidence level).

Although our constraints on this model are less strong than the ones found by gamma-ray searches, they apply to all annihilation products and to all dark matter particle masses (Zoutendijk et al. 2021a). The upper limit that we placed on the core radius $r_c \lesssim 69$ pc (68% confidence level) and $r_c \lesssim 180$ pc (95% confidence level) rules out cores of the sizes observed in several more massive dwarf galaxies in the Local Group.

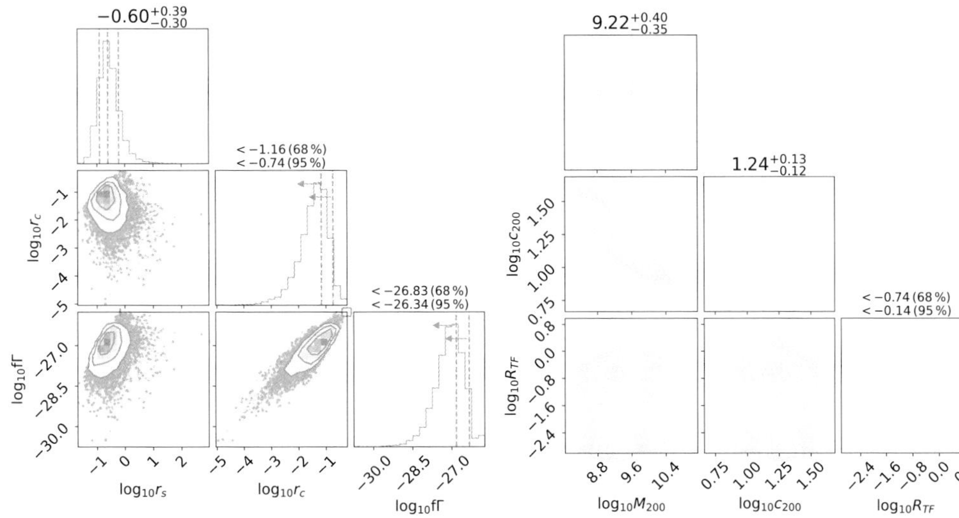

Figure 2. *Left:* Constraints on the SIDM profile for Ant B. The histograms along the diagonal represent the posterior distribution for each parameter: the scale radius r_s and the core radius r_c in kpc, and the self-interaction coefficient $f\Gamma$ in $\mathrm{cm}^3\mathrm{s}^{-1}\mathrm{eV}^{-1}c^2$. *Right:* Constraints on the SFDM profile for Ant B. The histograms along the diagonal represent the posterior distribution for each parameter: the virial mass M_{200} in $M_\odot$, the concentration parameter c_{200} and the characteristic lenght scale of the self-interaction R_{TF} in kpc. For both plots, the units were omitted for clarity. The vertical dashed lines indicate the median and 68% confidence interval (without arrows) or the 68% and 95% confidence limits (upper and lower arrows, respectively). The other panels have represented the 2D posterior distributions of these parameters, with the contours corresponding to 0.5σ, 1σ, 1.5σ and 2σ confidence levels, where σ is the standard deviation of the 2D distribution.

For the SFDM profile, we find a virial mass of $M_{200} = 10^{9.22^{+0.40}_{-0.35}} \approx 1.66^{+2.51}_{-0.92} \times 10^9\ \mathrm{M}_\odot$ and a concentration parameter of $c_{200} = 10^{1.24^{+0.13}_{-0.12}} \approx 17.38^{+6.06}_{-4.20}$. These results are in agreement with those expected for a CDM profile (further details will be given in Júlio et al. in prep.; see also Zoutendijk et al. 2021b). For the characteristic length scale of the repulsive self-interaction, it is also appropriate to present it as an upper limit for the same reason as before - the posterior distribution allows $R_{\mathrm{TF}} \approx 0.63$ pc while the projected radii of our innermost tracer is 1.84 pc. We find $R_{\mathrm{TF}} < 10^{-0.74} \lesssim 0.18$ kpc at the 68% confidence level and $R_{\mathrm{TF}} < 10^{-0.14} \lesssim 0.72$ kpc at the 95% confidence level. This translates to $\frac{g}{m^2c^4} \lesssim 5.2 \times 10^{-20}\ \mathrm{eV}^{-1}\mathrm{cm}^3$ (68% confidence level) and $\frac{g}{m^2c^4} \lesssim 8.3 \times 10^{-20}\ \mathrm{eV}^{-1}\mathrm{cm}^3$ (95% confidence level). This gives us a minimum particle mass of $m \gg 4.8 \times 10^{-21}\ \mathrm{eV}/c^2$. These results aren't enough to solve the core-cusp problem, since this model requires $R_{\mathrm{TF}} \gtrsim 1$ kpc (Dawoodbhoy et al. 2021), and therefore this model cannot explain cores in dwarfs in the Local Group.

In Figure 3 we show the recovered density profiles for the three models previously discussed as a function of the radius. These density profiles are represented by median density of the random samples and their 68% confidence interval at every radius.

The SFDM model agrees with the other models within the uncertainties, while the SIDM only agrees with the CDM within the uncertainties in the inner projected radii of Ant B. The agreement of the CDM model with the SFDM model is almost perfect at large radii since this second model takes the parametrization of the first one for $r \geq \alpha R_{\mathrm{TF}}$. The uncertainties of the SFDM allow for the CDM model even at the smallest radii, when $R_{\mathrm{TF}} \to 0$. The agreement of the three discussed models is largest at more populated radii, i.e., where there is a greater concentration of stars. As we approach the center of

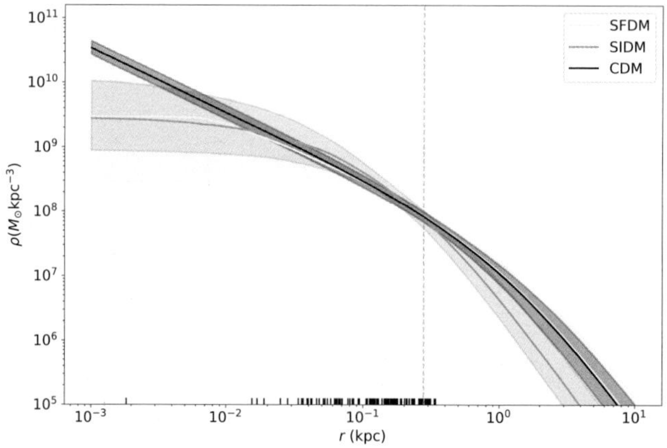

Figure 3. Recovered DM density profiles of Ant B for the approached models. The solid lines represent the median density profile: in red the CDM model is represented, in yellow the SFDM model and in orange the SIDM model. The filled areas represent the 68% confidence interval. The vertical dashed line indicates the half-light radius and the bottom marks represent the projected radii of the members of Ant B.

the galaxy, at smaller radii, the CDM model and the SIDM model start to diverge but the medians of the SFDM and the SIDM are close to each other.

The model that represents CDM has the highest density and smallest uncertainties, while the other profiles seem to prefer lower densities. Although the SFDM model allows for higher densities when $R_{\rm TF} \rightarrow 0$, this model allows for the lowest density when its characteristic length scale approaches the values discussed in the literature. The SIDM allows for the presence of a core when $r_c > 0$, and becomes the CDM model when $r_c = 0$. However, to ensure physical solutions in our numerical parametrization, we had to implement the condition $r_c > 0$. For this reason, our model never reaches the CDM solution and our results are always biased for a cored profile, even though this might not be the real density profile. To further improve the constraints placed on the models, we would need line-of-sight velocities of stars closer to the centre. This remains challenging since we are dealing with faint galaxies.

The dark matter density profiles of ultra-faint dwarfs have only started to be estimated very recently. The measurements of the velocity of stars of these galaxies with the spectroscopic data from MUSE-Faint allow us to improve the constraints on their inner dark matter density profiles. With a larger number of ultra-faint dwarfs analysed, we will be able to place stronger constraints on the physical properties of dark matter.

References

Bacon, R., Accardo, M., Adjali, L., et al. 2010, SPIE AT+I, 773508–773508
Boylan-Kolchin, M. Bullock, J. S., Kaplinghat, M., 2011, MNRAS, 415, L40-L44
Bullock, J. S. and Boylan-Kolchin, M., 2017, Annual Review of A&A, 55, 343-387
Collins, M. L. M., Read. J. I., Ibata, R. A., et al. 2021, MNRAS, 505, 5686-5701
Dawoodbhoy, T., Shapiro, P. R., Rindler-Daller, T., 2021, MNRAS, 506, 2118-2444
Dine, M., Fischler, W., 1983, Phys. Lett. B, 120, 137-141
Elbert, O. D., Bullock, J. S., Garrison-Kimmel S. et al. 2015, 453, 29-37
Hargis, J. R., Albers, S., Crnojević, D. et al. 2020, ApJ, 888, 31
Hargreaves, J. C, Gilmore, G., Irwin, M. J et al. 1994, MNRAS, 269, 957-974
Husser, T.-O., 2012, Universitätsverlag Göttingen
Kamann, S., Wisotzki, L., Roth, M. M., 2013, A&A 549, A71
Kaplinghat, M., Knox, L., Turner, M. S., 2000, Phys. Rev. Lett. 85, 16, 3335-3338

Klypin, A., Kravtsov, A. V., Valenzuela, O. et al. 1999, ApJ, 522, 82-92
Jeans, J. H., 1922, MNRAS, 82, 122-132
Li, B. Rindler-Daller, T., Shapiro, P. R., 2014, Phys. Rev. D, 89, 8, 083536
Lin, H. W., Loeb, A., 2016, JCAP, 2016, 3, 9
Merrifield, M. R. and Kent, S. M., 1990, AJ, 99, 1548
Moore, B., 1994, Nature, 370, 620-631
Okrney, M. D. A., Read, J. I., Rey, M. P., et al. 2021, MNRAS, 504, 3509-3522
Read, J. I., Walker, M. G., Steger, P., 2021, MNRAS, 481, 860-877
Read, J. I., Steger, P., 2017, MNRAS, 471, 4541-4558
Rubin, V. C., Ford, W. K., Thonnard, N., ApJ, 238, 471-487
Sand, D. J., Spekkens, K., Crnojević, D. et al. 2015, ApJ Letters, 812, L13
Simon, J. D. 2019, Annual Review of A&A, 57, 375–415
Vogelsberger, M., Genel, S. Springel, V., et al. 2014, MNRAS, 444, 1518-1547
Zoutendijk, S. L., Brinchmann, J., Boogaard, L. A., et al. 2020, A&A, 635, A107
Zoutendijk, S. L. and Brinchmann, J. and Bouché, N. F. et al. 2021 A&A, 651, A80
Zoutendijk, S. L., Júlio, M. P., Brinchmann, J., et al. 2021b, arXiv e-prints, arXiv:2112.09374

Dynamical Masses of Local Group Galaxies
Proceedings IAU Symposium No. 379, 2025
P. Bonifacio, M.-R. Cioni, F. Hammer, M. S. Pawlowski & S. Taibi, eds.
doi:10.1017/S174392132300145X

High resolution spectroscopy of stars in the outskirts of the Tucana II ultra-faint dwarf galaxy

A. Chiti

Department of Astronomy & Astrophysics, University of Chicago, 5640 S Ellis Avenue, Chicago, IL 60637, USA
email: achiti@uchicago.edu

Kavli Institute for Cosmological Physics, University of Chicago, Chicago, IL 60637, USA

Abstract. We discuss recent results from high-resolution Magellan/MIKE spectroscopy of five stars in the outskirts (up to $\sim$1 kpc) of the Tucana II ultra-faint dwarf galaxy (UFD), complemented by prior observations of seven stars closer to the galaxy's center. These outer stars were identified via their low SkyMapper photometric metallicities and consistent Gaia DR2 proper motions, and their membership was confirmed through follow-up medium-resolution spectroscopy. The high-resolution spectroscopy presented here provides detailed chemical abundances and more precise velocities, facilitating a revised dynamical analysis for signs of tidal disruption and a collective analysis of the detailed chemistry to evaluate astrophysical scenarios for the origin of the spatially extended feature. We discuss these signatures here and assess the evidence for several formation scenarios for the extended feature of Tucana II, highlighting how such studies of the outskirts of the UFD population as a whole can inform which scenarios may be preferred.

Keywords. Galaxy: halo; Galaxy: stars; Stars: abundances

1. Introduction

The ultra-faint dwarf galaxies (UFDs) of the Milky Way are local relics produced by galaxy evolution in the high-redshift ($z \gtrsim 6$) universe, as indicated by their ancient ($\sim$ 13 Gyr old; Brown et al. 2014), very metal-poor ($\langle$[Fe/H]$\rangle < -2.0$; Simon 2019) stellar populations. This local population of dwarf galaxies can provide insights into the fundamental properties of the earliest generations of galaxies, including their dynamical masses, dark matter halos, early chemical evolution, and the general processes (e.g., feedback, mergers) that regulate their early evolution. Of particular relevance to the Symposium, UFD dynamical mass estimates suggest that these galaxies are the most dark-matter-dominated galaxies known, with typical mass-to-light ratios $\gtrsim 100\, M_\odot/L_\odot$ (e.g., Simon & Geha 2007). Additionally, searches for signatures of tidal disruption of UFDs have spanned over a decade (e.g., Sand et al. 2012, Roderick et al. 2015) to assess the dynamical equilibrium of these systems and understand their evolution in the Milky Way potential. The advent of *Gaia* DR2 and DR3 proper motion data (Gaia Collaboration et al. 2018, 2022) has allowed for detailed orbit modeling of these systems, thereby increasing interest in their dynamical state and context in the Milky Way ecosystem (e.g., Simon 2018, Li et al. 2021, Pace et al. 2022).

This proceedings entry focuses on a sequence of recent results that uncovered the presence of stars extending out to $\sim$ 1 kpc (out to $\sim$6 to $\sim$9 half-light radii, depending on the adopted structural parameters; Bechtol et al. 2015; Moskowitz & Walker 2020) in the Tucana II UFD (Chiti et al. 2020, 2021, 2023). The detection of these stars was somewhat unexpected; before confirming that these stars were members (Chiti et al. 2021), only one UFD (Tucana III) had confirmed members beyond ~ 4 half-light radii (Li et al. 2018) and Tucana III had obvious tidal tails in photometry (Drlica-Wagner et al. 2015). As discussed in Chiti et al. (2020, 2021), stars in the outskirts of Tucana II were identified by removing contamination by Milky Way foreground stars through photometric metallicities from the intermediate-band SkyMapper v filter in conjunction with *Gaia* DR2 proper motions. This allowed for relatively efficient spectroscopic follow-up of stars to confirm their membership and demonstrate that the system harbored an "extended halo" component out to large r_h. Motivated by investigating the origin of this feature, Chiti et al. (2023) presented detailed chemical abundances from high-resolution spectroscopy of stars in this feature to evaluate various scenarios of its formation, which was the primary focus of the talk at this Symposium. The identification of stars in the outskirts of dwarf galaxies is now an active area of investigation, with recent studies targeting Bootes I (Filion & Wyse 2021; Longeard et al. 2022), Hercules (Longeard et al. 2023), Fornax (Yang et al. 2022), and several other systems (Waller et al. 2023; Sestito et al. 2023a,b).

We highlight the key results and discussions during this Symposium relevant to the spectroscopy of stars in the outskirts of Tucana II, and to evaluate the formation scenarios of this extended feature using those stars. For a more detailed discussion of the analyses of the high-resolution spectra, we direct the reader to Section 3 of Chiti et al. (2023); the full chemical abundances and velocities of these stars are shown in Tables 3 and 6 of the same paper. We will begin by briefly summarizing the method by which these stars were identified, then evaluate the formation scenarios of the Tucana II outskirts in light of these results and questions raised during the Symposium, and finally conclude.

2. Confirming stars in the outskirts of Tucana II & high-resolution spectroscopy

A significant challenge in UFD studies, prior to the *Gaia* mission providing extensive proper motion data of stars, was the identification of member stars in UFDs for subsequent spectroscopic observations. These observations are critical because the confirmation of star membership via low metallicities or coherent radial velocities is required before (1) dynamical mass estimates can be derived from velocity dispersions, and (2) more detailed high-resolution spectroscopy of these stars can be obtained for chemical evolution studies. Before the advent of *Gaia* DR2, the success rates in identifying members in the core of these systems were typically around $\sim 10-20\,\%$ due to contamination by Milky Way stars along the line of sight. *Gaia* DR2 proper motions significantly increased this success rate, but identifying stars in the outskirts of these systems remained a challenge due to the extremely low surface density of members in this region. In Chiti et al. (2020), photometry from the SkyMapper u, v, g, i filter-set was employed to derive photometric metallicities and surface gravities for stars up to $\sim 1.5°$ from the center of Tucana II, facilitated by the wide-field imaging capability of the SkyMapper Telescope (Keller et al. 2007). Given that UFD stars are significantly more metal-poor than the Milky Way foreground stars in the field-of-view, foreground stars could be accurately flagged and removed up to the outskirts of Tucana II, enabling members to be identified with reasonable confidence (e.g., see Figure 1 in Chiti et al. 2021). This process, for the first time, allowed the efficient identification of member stars well into the outskirts of a UFD. This pre-selection technique based on photometric metallicities has been successfully applied by the Pristine survey (Starkenburg et al. 2017) recently in Bootes I and the

Hercules UFDs (Longeard et al. 2022, 2023), demonstrating that stars in the outskirts of UFDs are now observationally accessible with this technique.

Initial medium-resolution spectroscopic follow-up of several outer Tucana II stars with Magellan/MagE confirmed them as members through their low metallicities ([Fe/H] < -2.0) and radial velocities (Chiti et al. 2021). These stars displayed no evidence of a velocity gradient, which is a signature of tidally disrupting dwarf galaxies (e.g., Li et al. 2018), and appeared to be misaligned with the orbital track of the system. This led the authors to argue that a simple tidal disruption scenario could not accurately replicate the spatial distribution of stars, and instead, these stars likely trace the underlying extended dark matter halo of the system. The mass estimate, derived from the requirement that the farthest star be bound to Tucana II, results in a lower limit of $10^7\,M_\odot$ up to 1 kpc. This is consistent with Jeans's modeling of the system using an NFW profile (Chiti et al. 2021) and aligns with the idea that distant stars may be bound assuming UFDs inhabit NFW-like halos.

Subsequent high-resolution spectroscopy of five stars in the outskirts of Tucana II was presented in Chiti et al. (2023). These spectra, the results of which were discussed in the Symposium, enabled a higher velocity precision (down to ~ 0.9 km/s) and detailed chemical abundances for these stars to be derived. The goal of these observations was to determine whether the detailed chemical abundances of these stars could shed light on the possible formation scenarios of the extended feature, whether it was due to galactic mergers at the smallest scales (e.g., Tarumi et al. 2021) or early supernovae feedback. Additionally, the higher velocity precision would allow for a more sensitive analysis for a velocity gradient in searches for tidal disturbances by the Milky Way. Either result would be meaningful in understanding the nature of this UFD (and UFDs more generally), in possibly revealing the extent to which mergers and/or feedback may have shaped its evolution, providing better estimates of the extent of its dark matter halo, or alternatively, revealing the extent of tidal disturbances by the Milky Way.

3. Evaluating formation scenarios for the Tucana II outskirts

Three scenarios were discussed as potential explanations for the extended feature of Tucana II. The first was tidal disruption by the Milky Way, a process known to elongate dwarf galaxies and ultimately lead to the formation of tidal tails (e.g., Johnston et al. 1995). The second was a merger between two smaller UFDs, possibly first-generation galaxies, as was recently modeled by Tarumi et al. (2021). The third was feedback by early supernovae (SNe), which may inject energy into the system or core the central dark matter potential, either of which could lead to a "puffier" final UFD. We refer the reader to Section 5 in Chiti et al. (2023) for a more detailed discussion of each of these scenarios in light of the new high-resolution spectroscopic observations; here we provide a summary with updates based on questions and discussions in the Symposium.

3.1. *Tidal Disruption*

Tidal disruption by the Milky Way is the most obvious explanation for the extended nature of Tucana II, but some observational signatures of this process are not readily reproduced in the data. First, the predicted direction of tidal debris lies nearly perpendicular to the distribution of stars in the outskirts of Tucana II (e.g., Figure 7 in Chiti et al. 2023). Second, no radial velocity gradient is detected, even with the more precise velocities from high-resolution spectroscopy. Such a gradient is a signature of tidal disruption (e.g., Li et al. 2018), and the measurement from Chiti et al. (2023) excludes a velocity gradient at the level of the spectroscopically confirmed tidally disrupting UFD, Tucana III (Li et al. 2018). And third, naively, one does not need to appeal to tidal disruption to detect members at large distances from UFDs– an underlying NFW-like dark matter halo, in which these systems are assumed to form, would extend the tidal radius of Tucana II just beyond the distance of the farthest detected member (Chiti et al. 2021). Accordingly, it should not necessarily be surprising to find stars at these distances in UFDs even in the absence of tidal disruption.

However, several facets complicate the above arguments. The process of tidal disruption is inherently complex, and e.g., tidal shocks from the Milky Way potential may disperse stars in a direction not necessarily along the predicted axis from orbit modeling. Additionally, it is not obvious to what extent a velocity gradient must be excluded to claim the system is not disrupting; a requirement for this prediction would be to first generate an existing disruption model that matches the spatial orientation of stars in Tucana II. The ideal avenue to test the tidal disruption hypothesis would be an N-body simulation of UFD disruption, which can test the extent to which tidal shocking and perturbations would be expected given the orbital parameters and an underlying dark matter distribution for Tucana II. It was noted in a question by F. Hammer following this talk that Tucana II has the highest orbital energy of UFDs that are bound to the Milky Way in Figure 5 of the recent work in Hammer et al. (2023). This may explain the extended feature as a result of tidal influences from being on an energetic orbit. Further work on identifying these features around other UFDs, coupled with the N-body simulations to predict the dynamics and location of these features as UFDs evolve in the Milky Way's potential, will comprehensively test the tidal disruption scenario.

3.2. *Dwarf mergers*

Another mechanism for forming the Tucana II extended feature is a dwarf-dwarf merger that leads to the formation of a small-scale stellar halo. Such scenarios have been discussed in the literature– Rey et al. (2019) posited that late-time dry mergers can lead to the formation of a highly diffuse, very metal-poor ultra-faint dwarf galaxy. Notably, Tarumi et al. (2021) modeled the merger of two first-generation galaxies and did a direct comparison to Tucana II, finding that the resulting system reproduced several properties including the elongated distribution and weak metallicity gradient in Chiti et al. (2021). The origin of this weak metallicity gradient in the simulation by Tarumi et al. (2021) was that the merger triggered star formation in the center of the galaxy, leading to a central, slightly more metal-rich population.

Accordingly, one prediction of this early merger scenario is that Tucana II ought to show evidence of some delayed/extended star formation. An observational test of this would be if the more metal-rich stars in Tucana II show evidence of enrichment by Type Ia SNe, since those events occur after some delay time (e.g., Maoz et al. 2012). A clear signature of this is an α-"knee", or downturn in [α/Fe] vs. [Fe/H] since Type Ia SNe produce trace amounts of α-elements whereas core-collapse SNe produce [α/Fe]~ 0.4 (e.g., Kirby et al. 2011). This α-"knee" signature is seen in the chemical abundances of stars in Tucana II at a metallicity of [Fe/H] ~ -2.6. Additionally, the most metal-rich star in Tucana II shows a strong deficiency in [Mn/Fe], implying enrichment by sub-Chandrasekhar mass type Ia SNe (Kirby et al. 2019). We note that many factors can reignite star formation in a dwarf galaxy, so a merger is not a unique explanation for delayed/extended star formation history of Tucana II. If early dwarf mergers are the primary mechanism for forming dwarf outskirts, then one would expect that UFDs with extended features to preferentially show α-"knee" features. Notably, the Segue I UFD does not show an α-"knee" (Frebel et al. 2014). Consequently, this early merger scenario would predict that Segue 1 ought not to host an extended feature; a future detection around Segue 1 may disfavor this scenario.

Two questions that were relevant to this scenario were asked following the talk. The first question asked whether a metallicity gradient existed in Tucana II since this was not explicitly mentioned in the talk. There is some evidence for a weak metallicity gradient in Tucana II in Chiti et al. (2021)– in particular, the most metal-poor stars in the system appear to be preferentially farther from the center of the galaxy than the more metal-rich stars. The second question raised the possibility of whether Tucana II may have accreted stars from the Milky Way halo. This is unlikely, as the stars in the outskirts of Tucana II almost certainly formed in a UFD environment due to their abnormally low neutron-capture abundances. This is a key characteristic that separates UFD stars from those in the Milky Way halo (see Chiti et al. 2023, Ji et al. 2019).

3.3. *Supernova feedback*

The final possibility discussed for the formation of the Tucana II outskirts was supernovae feedback. The notable insight from the detailed chemical abundances of Tucana II stars is that their abundances generally track what is seen in other UFDs (e.g., Figures 3 and 4 in Chiti et al. 2023). Accordingly, the chemical abundance data does not suggest that the SNe in Tucana II were particularly energetic compared to the SNe in other UFDs. If SNe feedback drove the formation of the extended feature in Tucana II, then the expectation from the Tucana II chemical abundances is that other UFDs ought to harbor these features as well, regardless of other factors such as e.g., orbital parameters and α-knees.

4. Conclusions & Future work

As noted before, the discovery of stars in the outskirts of dwarf galaxies is an active area of research to better understand the dynamical state and early evolution of these systems (e.g., Filion & Wyse 2021, Longeard et al. 2022, Yang et al. 2022, Sestito et al. 2023a). Such stars in the periphery of ultra-faint dwarf galaxies (UFDs) are of particular interest, since UFDs are powerful, local windows on the properties of the earliest generations of galaxies. Therefore, fundamental insights into these systems, such as their mass, dark matter content, and dark matter halos, can meaningfully inform predictions from models of early galaxy formation. The outskirts of these systems are powerful probes of the above– if these systems were to be tidally disrupting, then assumptions of dynamical equilibrium in estimating their masses may not be valid. In the absence of tidal disruption, the stars in the outskirts of these systems would serve as the only visible tracers of their extended dark matter potentials. Several engaging talks and posters were presented in this conference on identifying dwarf outskirts and evaluating the formation of these features (e.g., by J. Jensen, F. Waller, A. Goater). Individual stars in these extended "halos" around dwarf galaxies are now observationally accessible, and observations of their chemistry and kinematics will meaningfully fill-in our observational understanding of these ancient galaxies.

References

Bechtol, K., Drlica-Wagner, A., Balbinot, E., et al. 2015, ApJ, 807, 50.
Brown, T. M., Tumlinson, J., Geha, M., et al. 2014, ApJ, 796, 91.
Chiti, A., Frebel, A., Jerjen, H., et al. 2020, ApJ, 891, 8.
Chiti, A., Frebel, A., Simon, J. D., et al. 2021, Nature Astronomy, 5, 392.
Chiti, A., Frebel, A., Ji, A. P., et al. 2023, AJ, 165, 55.
Drlica-Wagner, A., Bechtol, K., Rykoff, E. S., et al. 2015, ApJ, 813, 109.
Filion, C. & Wyse, R. F. G. 2021, ApJ, 923, 218.
Frebel, A., Simon, J. D., & Kirby, E. N. 2014, ApJ, 786, 74.

Gaia Collaboration, Brown, A. G. A., Vallenari, A., et al. 2018, A&Ap, 616, A1.
Gaia Collaboration, Vallenari, A., Brown, A. G. A., et al. 2022, arXiv:2208.00211.
Hammer, F., Li, H., Mamon, G. A., et al. 2023, MNRAS, 519, 5059.
Johnston, K. V., Spergel, D. N., & Hernquist, L. 1995, ApJ, 451, 598.
Ji, A. P., Simon, J. D., Frebel, A., et al. 2019, ApJ, 870, 83.
Keller, S. C., Schmidt, B. P., Bessell, M. S., et al. 2007, PASA, 24, 1.
Kirby, E. N., Cohen, J. G., Smith, G. H., et al. 2011, ApJ, 727, 79.
Kirby, E. N., Xie, J. L., Guo, R., et al. 2019, ApJ, 881, 45.
Li, T. S., Simon, J. D., Kuehn, K., et al. 2018, ApJ, 866, 22.
Li, H., Hammer, F., Babusiaux, C., et al. 2021, ApJ, 916, 8.
Longeard, N., Jablonka, P., Arentsen, A., et al. 2022, MNRAS, 516, 2348.
Longeard, N., Jablonka, P., Battaglia, G., et al. 2023, arXiv:2304.13046.
Maoz, D., Mannucci, F., & Brandt, T. D. 2012, MNRAS, 426, 3282.
Moskowitz, A. G. & Walker, M. G. 2020, ApJ, 892, 27.
Pace, A. B., Erkal, D., & Li, T. S. 2022, ApJ, 940, 136.
Rey, M. P., Pontzen, A., Agertz, O., et al. 2019, ApJL, 886, L3.
Roderick, T. A., Jerjen, H., Mackey, A. D., et al. 2015, ApJ, 804, 134.
Sand, D. J., Strader, J., Willman, B., et al. 2012, ApJ, 756, 79.
Sestito, F., Roediger, J., Navarro, J. F., et al. 2023, MNRAS.
Sestito, F., Zaremba, D., Venn, K. A., et al. 2023, arXiv:2301.13214.
Simon, J. D. & Geha, M. 2007, ApJ, 670, 313.
Simon, J. D. 2018, ApJ, 863, 89.
Simon, J. D. 2019, ARA&A, 57, 375.
Starkenburg, E., Martin, N., Youakim, K., et al. 2017, MNRAS, 471, 2587.
Tarumi, Y., Yoshida, N., & Frebel, A. 2021, ApJL, 914, L10.
Waller, F., Venn, K. A., Sestito, F., et al. 2023, MNRAS, 519, 1349.
Yang, Y., Hammer, F., Jiao, Y., et al. 2022, MNRAS, 512, 4171.

Dynamical Masses of Local Group Galaxies
Proceedings IAU Symposium No. 379, 2025
P. Bonifacio, M.-R. Cioni, F. Hammer, M. S. Pawlowski & S. Taibi, eds.
doi:10.1017/S1743921323001564

Leo T Dissected with the MUSE-Faint Survey

Daniel Vaz[1,2], Jarle Brinchmann[1,3] and The MUSE Collaboration

[1]Instituto de Astrofísica e Ciências do Espaço, Universidade do Porto, CAUP, Rua das Estrelas, 4150-762 Porto, Portugal
email: Daniel.Vaz@astro.up.pt

[2]Departamento de Física e Astronomia, Faculdade de Ciências, Universidade do Porto, Rua do Campo Alegre 687, PT4169-007 Porto, Portugal

[3]Leiden Observatory, Leiden University, PO Box 9513, 2300 RA Leiden, The Netherlands

Abstract. Leo T is the lowest mass galaxy known to contain neutral gas and to show signs of recent star formation, which makes it a valuable laboratory for studying the nature of gas and star formation at the limits of where galaxies are found to have rejuvenating episodes of star formation.

Here we discuss a novel study of Leo T that uses data from the MUSE integral field spectrograph and photometric data from HST. The high sensitivity of MUSE allowed us to increase the number of Leo T stars observed spectroscopically from 19 to 75. We studied the age and metallicity of these stars and identified two populations, all consistent with similar metallicity of [Fe/H] ~ -1.5 dex, suggesting that a large fraction of metals were ejected. Within the young population, we discovered three emission line Be stars, supporting the conclusion that rapidly rotating massive stars are common in metal-poor environments. We find differences in the dynamics of young and old stars, with the young population having a velocity dispersion consistent with the kinematics of the cold component of the neutral gas. This finding directly links the recent star formation in Leo T with the cold component of the neutral gas.

Keywords. Spectroscopy, Galaxies, Leo T, Stars, Kinematics, Star Formation, Be Stars

1. Introduction

Ultra-Faint Dwarf galaxies (UFDs) represent a fascinating enigma in the study of the Universe. These elusive objects are characterised by their extremely low metallicities, simple assembly histories, and dominant dark matter content (Simon 2019), making them an essential piece of the puzzle in understanding galaxy formation and evolution.

Among the faint and ultra-faint dwarf sample, Leo T stands out as a particularly intriguing object that has received significant attention from astronomers. Leo T is the faintest and least massive dwarf galaxy known to contain neutral gas and exhibit signs of recent star formation. This unique set of characteristics makes Leo T a valuable testing ground for galaxy formation models, as they present a challenge to current theories that have struggled to reproduce similar galaxies. Further observations of Leo T will enable astronomers to refine their models and determine whether they are on the right track towards a comprehensive and predictive theory of galaxy formation.

Leo T was discovered using SDSS imaging by Irwin *et al.* (2007). The stellar mass of Leo T is estimated to be $M = 1.3 \times 10^5$ $M_\odot$ (McConnachie 2012). The star formation history (SFH) of Leo T has been extensively studied (Irwin *et al.* 2007; de Jong *et al.* 2008; Weisz *et al.* 2012; Clementini *et al.* 2012). These studies show that 50% of the total stellar mass was formed prior to 7.6 Gyr ago, with star formation beginning over 10 Gyr ago and continuing until recent times. They also show evidence of a quenching of star formation in Leo T about 25 Myr ago. None of the studies found evidence of an evolution in isochronal metallicity such that, over the course of its lifetime, it is consistent with a constant value of [M/H] ~ -1.6.

The only previous spectroscopic observations of Leo T are those of Simon & Geha (2007). They derive a mean radial velocity of $v_{rad} = 38.1 \pm 2$ km s^{-1}, and velocity dispersion of $\sigma_{v_{rad}} = 7.5 \pm 1.6$ km s^{-1}, which corresponds to a total dynamical mass of 8.2×10^6 M$_\odot$.

Ryan-Weber *et al.* (2008) and Adams & Oosterloo (2018) concluded that Leo T contains neutral gas. The HI mass is almost four times the stellar mass, with $M_{HI} = 4.1 \times 10^5$ M$_\odot$ (Adams & Oosterloo 2018). They show that the gas is present in a Cold Neutral Medium (CNM) (T $\sim$ 800K) and a Warm Neutral Medium (WNM) (T $\sim$ 6000K) HI, with the CNM corresponding to almost 10% of the total HI mass. Relevantly, the CNM was found to have a lower velocity dispersion ($\sigma_{\rm CNM} = 2.5 \pm 0.1$ km s^{-1}) than the WNM ($\sigma_{\rm WNM} = 7.1 \pm 0.4$ km s^{-1}) (Adams & Oosterloo 2018). The presence of this large cold component raises the question of whether this component is related to the recent star formation observed in Leo T.

To further investigate Leo T, we used spectroscopic observations using the Multi-Unit Spectroscopic Explorer (MUSE, Bacon *et al.* 2010) integral field spectrograph (IFS). Succinctly, we densely map the stellar content and use the stellar spectra to measure the stellar metallicity and stellar kinematics. We find identifiers of a young population, namely Be stars. The data and results presented here are discussed further in Zoutendijk *et al.* (2021b) and Vaz *et al.* (submitted to *A&A*).

2. Results and Discussion

The central region of Leo T was observed as part of MUSE-Faint (Zoutendijk *et al.* 2020), a MUSE GTO survey of UFDs (PI Brinchmann). MUSE (Bacon *et al.* 2010) is a large-field medium-spectral-resolution integrated field spectrograph installed at Unit Telescope 4 of the Very Large Telescope (VLT). We extracted spectra from the final data cube using PampleMuse (Kamann *et al.* 2013). As a general rule, the extracted spectra have a modest signal-to-noise ratio (S/N) and spectral resolution (R $\sim$ 3000). We used the `spexxy` full-spectrum fitting code (Husser *et al.* 2016) together with the PHOENIX (Husser *et al.* 2013) model spectra to estimate the physical parameters, namely line-of-sight velocities and [Fe/H].

2.1. *Different Populations in Leo T*

We were able to identify 58 stars as members of Leo T based on their kinematics. For 55 of these stars we were also able to obtain an estimate of [Fe/H]. The three stars without [Fe/H] estimates are emission line Be stars, which are discussed further below. By combining with the data from Simon & Geha (2007) we have measurements of the kinematics for 75 stars.

We complemented this data with HST ACS F606W and F814W photometry†. We fit PARSEC stellar isochrones (Bressan *et al.* 2012) to the resulting colour magnitude diagram for the 58 stars (Figure 1). The best-fit [M/H] for the isochrone is [M/H]$= -1.6$, which is consistent with the value found in Weisz *et al.* (2012), and therefore we use this value to draw the representative isochrones shown in Figure 1. The first clear conclusion is that the sample covers a wide range of ages, with some stars consistent with ages as

† HST Proposals 12914, Principal Investigator Tuan Do and 14224, Principal Investigator Carmen Gallart

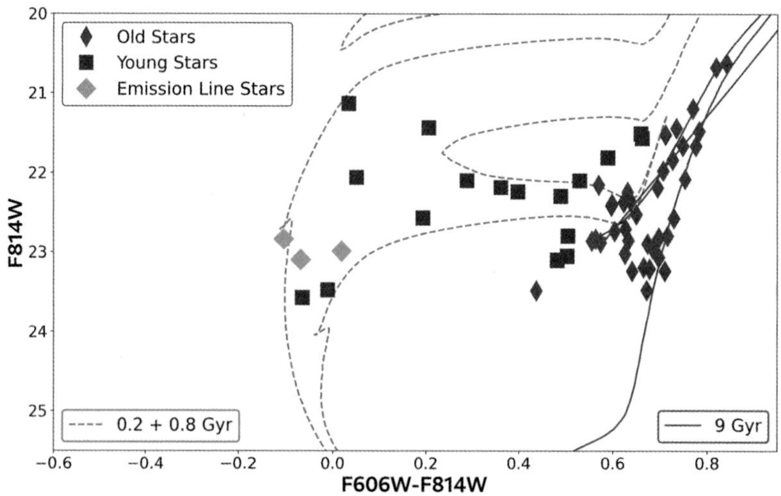

Figure 1. Color-magnitude diagram of the 58 Leo T stars detected with MUSE, plotted against PARSEC isochrones drawn for constant [Fe/H] $= -1.6$. Three representative isochrones are plotted, two in blue, with ages of 0.2 and 0.8 Gyr, and one in gray for age of 9 Gyr. The stars that were found consistent with the younger isochrones are shown as dark blue squares, with the emission line stars shown as blue squares. The stars consistent with the older isochrones are shown as red diamonds.

high as > 10 Gyr and others as low as 200 Myr. As such, we divided the stars into two populations: a young population, of 20 stars consistent with ages < 1 Gyr, and an old population, of 38 stars consistent with ages > 5 Gyr. Both populations are displayed using different colors in Figure 1. To assign each star to a population, we covered the color magnitude space with isochrones of different ages and with [M/H]$= -1.6$. We assign each star the age of the nearest isochrone. Because there is a degeneracy between the different isochrones in certain parts of the color-color space, we repeated all analyses discussed below by reassign the stars that fall in a degenerated region. We find that this does not affect our results.

Within the young population we identified three emission line stars. We tentatively identified these as Be stars. Due to their peculiar spectra, `spexxy` failed to fit for metallicity and, therefore, these stars were not included in our metallicity analysis. Of relevance is the fact that they make up 15% of the young sample, which is comparable to Milky Way studies that reported rates at a level of $10 - 20\%$ in stellar clusters (Mathew *et al.* 2007). More recent studies, such as Schootemeijerr *et al.* (2022), show that OBe stars and, therefore, rapidly rotating massive stars, are common in metal-poor environments, and the detection of Be stars in Leo T supports this conclusion and extend them to even lower metallicity.

2.2. *Chemical Evolution of Leo T*

The histogram of the metallicity estimates for the 55 stars is shown in Figure 2. To quantify this distribution, we implemented an MCMC model, assuming that the underlying distribution is a Gaussian. Therefore, we fit the mean metallicity and the metallicity dispersion of the distribution, which are also shown in Figure 2. We obtained a metallicity of [Fe/H] $= -1.53 \pm 0.05$, which is in good agreement with our photometric analysis. We repeated the analysis by resampling and removing outliers, without consequences on the results (the outliers are usually low S/N). We find a metallicity dispersion of $\sigma_{[\mathrm{Fe/H}]} = 0.21 \pm 0.06$, which is low, implying that all stars have similar metallicity and

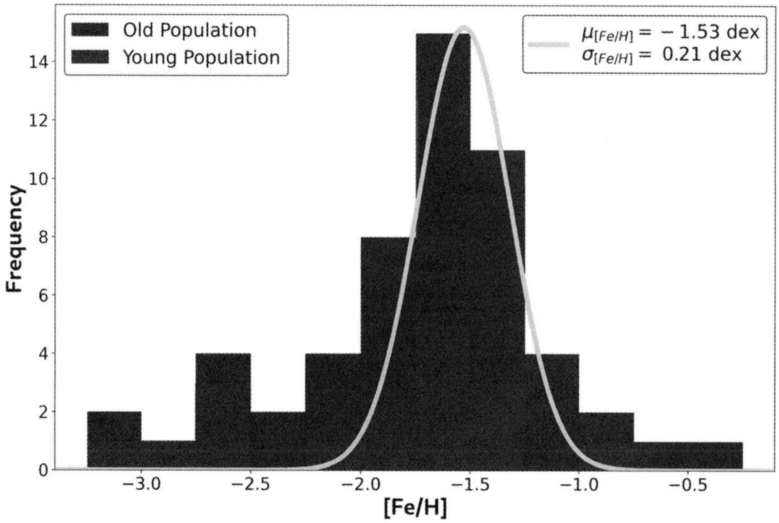

Figure 2. The distribution of the metallicity ([Fe/H]) of 55 Leo T stars, estimated using `spexxy`. The younger population, consisting of 17 stars, is represented with a lighter color. Also plotted is the result of a MCMC fit for the mean and standard deviation of the distribution.

that Leo T underwent almost no metallicity evolution throughout its history. This, in conjunction with the extended history of star formation of Leo T, suggests that a large fraction of metals have been ejected, keeping metallicity constant. In fact, this is consistent with theoretical expectations for low-mass dwarf galaxies (Emerick *et al.* 2018).

We repeated this analysis separately for each population. Even though the results are consistent with each other, the results are not conclusive because the sample of young stars is too small (consisting only of 17 stars). In addition, the distribution is asymmetric, especially for the young stars, with the low S/N spectra preferring a lower metallicity, meaning that our constraints prefer a somewhat lower metallicity for the younger population. However, the uncertainties here are too high to draw any conclusions.

2.3. *Stellar Kinematics vs Neutral Gas Kinematics*

The histogram of the radial velocity estimates for 75 stars is shown in Figure 3. To fit the distribution, we applied the same MCMC model as before to obtain the mean barycentric velocity and the velocity dispersion of the sample. The fit is also shown in Figure 3. We find a mean velocity $v_{\rm los} = 39.4^{+1.3}_{-1.3}$ km s^{-1}, and an intrinsic velocity dispersion of $\sigma_v = 7.1^{+1.3}_{-1.1}$ km s^{-1}, which is consistent with what was found by Simon & Geha (2007). We repeated this analysis for each population. In this case, we used the same young population as before, but now the old population consists of 55 stars, including 17 stars from Simon & Geha (2007). These distributions and the respective fits are shown in Figures 4 and 5 for the young and old population, respectively. It is worth noting that the best fit plotted does not include uncertainties. For the younger population we obtain a mean velocity of $v_{\rm los} = 39.3^{+2.1}_{-2.1}$ km s^{-1}, and a velocity dispersion of $\sigma_v = 2.3^{+2.7}_{-1.7}$ kms^{-1}. For the older population we find a mean velocity of $v_{los} = 39.7^{+1.6}_{-1.6}$ km s^{-1}, and a velocity dispersion of $\sigma_v = 8.2^{+1.7}_{-1.4}$ km s^{-1}. Notably, we find that both populations have different kinematics, with the younger population having a significantly smaller velocity dispersion than the older stars. This is comparable to what was found for the two components of the gas in Leo T, where the cold component was found to have a velocity dispersion smaller

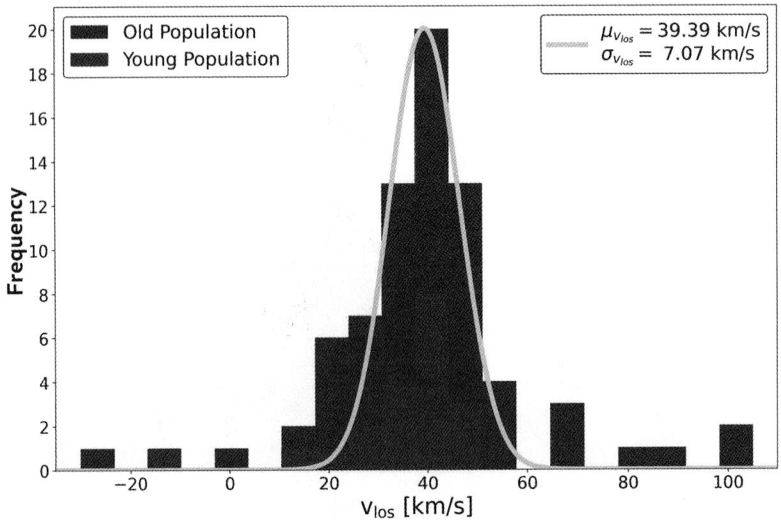

Figure 3. The distribution of the line-of-sight velocity of 75 Leo T stars, estimated using `spexxy`. The younger population, consisting of 20 stars, is represented with a lighter color. Also plotted is the result of a MCMC fit for the mean and standard deviation of the distribution.

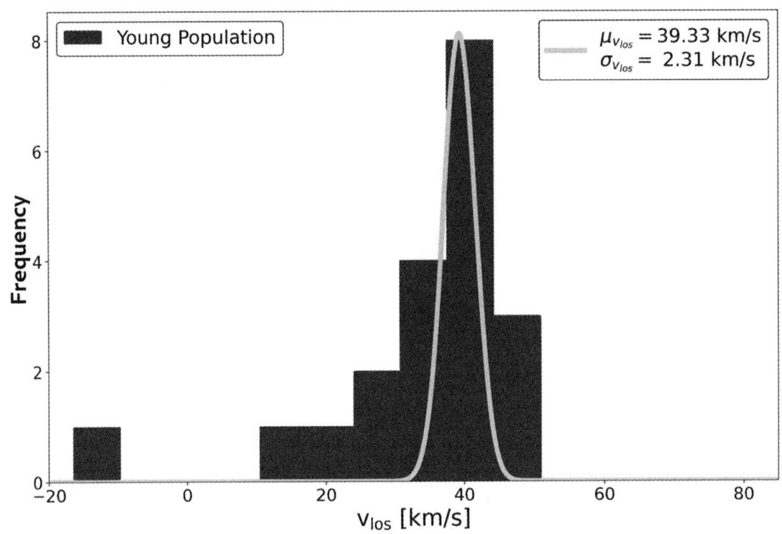

Figure 4. The distribution of the line-of-sight velocity of 20 young Leo T stars, estimated using `spexxy`. Also plotted is the result of a MCMC fit for the mean and standard deviation of the distribution.

than that of the warm component. We compare the differences in kinematics between young and old stars with what was found for the HI kinematics of warm and cold neutral gas in Figure 6. We find a good match when comparing the velocity dispersion of the young population with the cold component of the HI gas, and between the kinematics of old Leo T stars and the warm component of the HI gas. The natural inference from these results is that the most recent star formation in Leo T is linked to the CNM. The results presented here combined with the results from Weisz *et al.* (2012) of no star formation in Leo T in the last $\sim$25 Myr are consistent with recent models that suggest that star

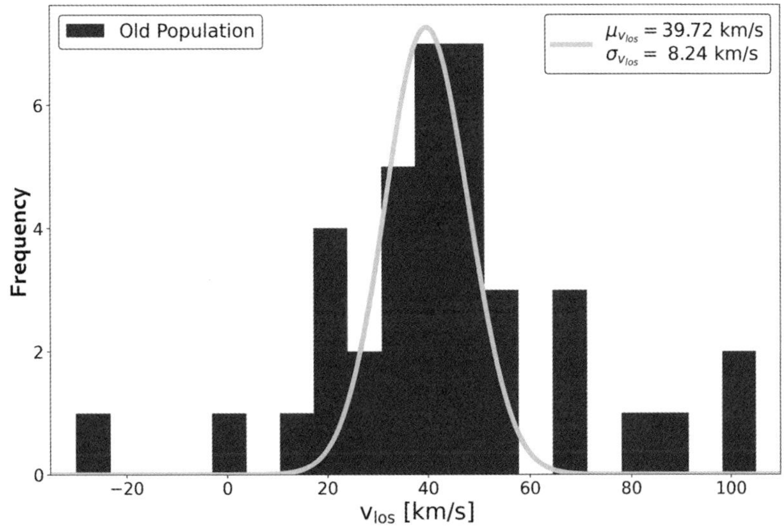

Figure 5. The distribution of the line-of-sight velocity of 55 old Leo T stars, estimated using `spexxy`. Also plotted is the result of a MCMC fit for the mean and standard deviation of the distribution.

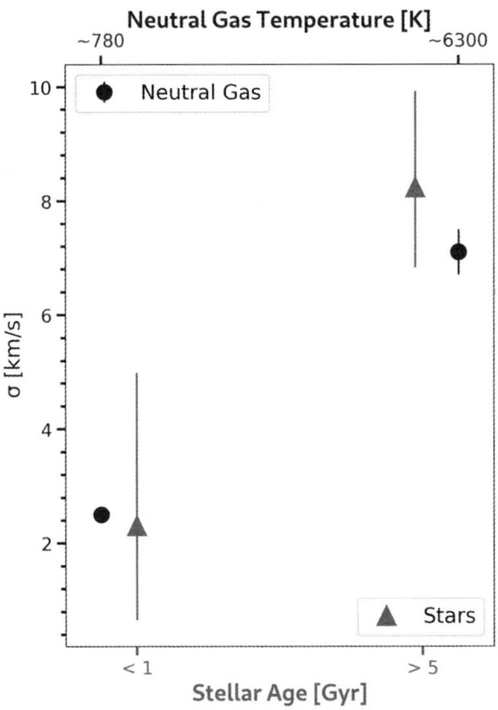

Figure 6. Comparison between stellar velocity dispersion of this work and neutral gas velocity dispersion from Adams & Oosterloo (2018).

formation in low mass galaxies should be bursty with short quiescent periods (Collins & Read 2022): due to stellar feedback, the star formation is momentarily quenched and metals are ejected from the environment and, after a short quiescent period, the gas is allowed to cool down and re-ignite star formation.

References

Adams, E. A. K. & Oosterloo, T. A. 2018, Astronomy & Astrophysics, 612, A26
Bacon, R., Accardo, M., Adjali, L., *et al.* 2010, in SPIE Astronomical Telescopes+ Instrumentation (International Society for Optics and Photonics), 773508–773508
Bressan, A., Marigo, P., Girardi, L., *et al.* 2012, MNRAS, 427, 127–145
Clementini, G., Cignoni, M., Ramos, R. C., *et al.* 2012, The Astrophysical Journal, 756, 108
Collins, M. L. M. & Read, J. I. 2022, Nature Astronomy, 6, 647
de Jong, J. T. A., Harris, J., Coleman, M. G., *et al.* 2008, The Astrophysical Journal, 680, 1112–1119
Emerick, A., Bryan, G. L., Low, M.-M. M., *et al.* 2018, The Astrophysical Journal, 869, 94
Husser, T.-O., Kamann, S., Dreizler, S., *et al.* 2016, Astronomy & Astrophysics, 588, A148
Husser, T.-O., Wende-von Berg, S., Dreizler, S., *et al.* 2013, Astronomy & Astrophysics, 553, A6
Irwin, M. J., Belokurov, V., Evans, N. W., *et al.* 2007, The Astrophysical Journal, 656, break L13–L16
Kamann, S., Wisotzki, L., & Roth, M. M. 2013, Astronomy & Astrophysics, 549, A71
Mathew, B., Subramaniam, A., & Bhatt, B. C. 2008, MNRAS, 388, 1879
McConnachie, A.W., 2012. The Observed Properties of Dwarf Galaxies in and around the Local Group. The Astronomical Journal 144, 4
Ryan-Weber, E. V., Begum, A., Oosterloo, T., *et al.* 2008, MNRAS, 384, 535–540
Schootemeijer, A., Lennon, D. J., Garcia, M., *et al.* 2022, Astronomy & Astrophysics, 667, A100
Simon, J. D. 2019, Annual Review of Astronomy and Astrophysics, 57, 375–415
Simon, J. D. & Geha, M. 2007, The Astrophysical Journal, 670, 313–331
Vaz, D., Brinchmann, J., Zoutendijk, S. L., *et al.*, submitted to Astronomy & Astrophysics
Weisz, D. R., Zucker, D. B., Dolphin, A. E., *et al.* 2012, The Astrophysical Journal, 748, 88
Zoutendijk, S. L., Brinchmann, J., Boogaard, L. A., *et al.* 2020, Astronomy & Astrophysics, 635, A107
Zoutendijk, S. L., Júlio, M. P., Brinchmann, J., *et al.* 2021b, arXiv e-prints, arXiv:2112.09374

Dynamical Masses of Local Group Galaxies
Proceedings IAU Symposium No. 379, 2025
P. Bonifacio, M.-R. Cioni, F. Hammer, M. S. Pawlowski & S. Taibi, eds.
doi:10.1017/S1743921323001357

Kinematic lopsidedness in the dwarf irregular galaxy WLM

M. Khademi[1,2,3], Y. Yang[2], F. Hammer[2] and S. Nasiri[3]

[1]School of Astronomy, Institute for Research in Fundamental Sciences (IPM), 19395-5531 Tehran, Iran
email: maryam.khademi@ipm.ir

[2]GEPI, Observatoire de Paris, Universite PSL, CNRS, Place Jules Janssen 92195, Meudon, France

[3]Department of Physics, Shahid Beheshti University, 19839 Tehran, Iran

Abstract. To investigate whether an $m = 1$ perturbation in the halo potential could be a mechanism creating observed kinematical asymmetry in the rotation curve of the dwarf irregular galaxy WLM, we fit a theoretical rotational velocity associated with such a perturbation in the halo potential model to the observed data. We show that a lopsided halo potential model can explain the asymmetry in the kinematic data reasonably well. To investigate how well such a small perturbation in the halo potential represents the observed asymmetries in the rotation curve and in the velocity field of WLM, we measure the value of the perturbation parameter ϵ_{per} from the kinematic data on the rotation curve of both sides in the outer part. We obtain the value of the perturbation parameter denoting the perturbed halo potential of about 0.2.

Keywords. Galaxies: kinematics and dynamics - galaxies: dark matter halo - galaxies: structure - galaxies: dwarf - galaxies: Local Group

1. Introduction

WLM is a near edge-on gas-rich dwarf irregular galaxy located near the edge of the Local Group. This galaxy is isolated from massive galaxies and rotationally supported and clearly rotating; with an ordered velocity field, described by a spider-like diagram (Kepley *et al.* 2007). However its HI rotation curve is asymmetric; the rotation curves for the approaching and the receding sides of WLM are not symmetric and they are distinctly different, especially in the outer parts (Kepley *et al.* 2007). Furthermore, its velocity dispersion map does not show any peak at the center and shows an unexpected minimum in the central region (Ianjamasimanana *et al.* 2020), which led us to investigate whether this dwarf irregular galaxy has a perturbed kinematic. We conclude that a lopsided halo potential model (an m = 1 perturbation in the halo potential) can explain the asymmetry in the kinematic data reasonably well. Such kinematic lopsidedness in the dwarf irregular galaxy WLM can be caused by the HI gas disc responding to a halo potential perturbed by a tidal encounter. Based on a kinemetry analysis, we found that it is possible for WLM to lie in the transition region, where the disk and merger coexist. So, a merger may be one of the possible origins of the dark matter halo lopsidedness for this isolated galaxy (Khademi *et al.* 2021). Yang *et al.* (2022) discovered the trailing extended HI gas clouds, evidence for WLM to undergo a ram-pressure stripping process, despite being 930 kpc and 830 kpc from the Milky Way and M31, respectively. They also verified it by performing N-body hydrodynamical simulations. Ram pressure can serve to further strip the galaxy of cool gas and prevent new gas to accrete, and lead to the reduction and shutting off of star formation activity and quenching of the dwarf galaxy (Pasha *et al.* 2023).

To investigate one of the possible origins of the perturbation in the halo potential for this isolated dwarf irregular galaxy WLM, we considered that the dark matter halo and the baryonic disk do not have the same center of mass, i.e., the disk lies off-center in the potential of the extended dark matter halo. We showed that displacing the disk center by 0.1 kpc from the halo center is sufficient to improve modeling the rotation curve (Khademi and Nasiri 2022).

This paper is organized as follows. In Sect. 2, we examine whether an $m = 1$ perturbation in a halo potential could be a mechanism creating observed kinematical asymmetry in the dwarf irregular galaxy WLM. In Sect. 3, based on kinematic asymmetries, for the purpose of investigating whether a merger could have created such a lopsidedness in the halo potential, we determine the strength of deviation of the velocity field of a perturbed rotation model (an m = 1 perturbation in the potential) from that of an ideal rotating disk case by obtaining the relative level of deviation of the velocity field from that of an ideal rotating disk case. In Sect. 4, we investigate how well an $m = 1$ perturbation in the halo potential represents the observed asymmetries in the rotation curve of the dwarf irregular galaxy WLM. We end the paper with a brief summary in Sect. 5.

2. Asymmetric rotation curves and a perturbed halo potential

To investigate the origin of the strong asymmetry in the rotation curve of the dwarf irregular galaxy WLM, we examined whether an $m = 1$ perturbation in a halo potential could be considered the mechanism creating such kinematical asymmetry. To do so, we fit the rotation curve of a perturbed halo potential model to that of the dwarf irregular galaxy WLM.

Figure 1 shows the extracted asymmetric rotation curve of WLM from the model associated with an m = 1 perturbation in the halo potential and also asymmetric rotation curve from the observational data by Kepley *et al.* (2007). There is a good agreement between the corresponding gas kinematics of a perturbed halo potential and the observed gas kinematics of WLM. In contrast to Kepley *et al.* (2007), our model provides a good fit to the rotation curve of WLM, simultaneously for both approaching and receding sides, with a perturbed halo potential model with the best-fitting parameters of $\epsilon_{lop} = 0.11$, $\phi_1 = 0$, and $v_e = 44(km/s)$ (Khademi *et al.* 2021). For this best-fit model, we mapped the velocity field in Figure 2, which shows asymmetric characteristics that is comparable to the observed one, as shown, for instance, in Figure 5 in (Kepley *et al.* 2007). We conclude that a lopsided halo potential model (a global m = 1 perturbation mode in the halo potential) can create kinematical asymmetry between two halves of a galaxy.

3. Merger-disk classifications based on kinemetry analysis

The kinematic analysis provides the only way to infer whether a galaxy is dominated by a relaxed rotation or by gravitational perturbation by a tidal encounter that is often linked with galaxy interactions or merger events (Krajnović *et al.* 2006; Shapiro *et al.* 2008). Based on kinematic asymmetries, we determine the strength of deviation of the velocity field of a perturbed rotation model (an m = 1 perturbation in the potential) from that of an ideal rotating disk case by obtaining the relative level of deviation of the velocity field from that of an ideal rotating disk case (Khademi *et al.* 2021). The level of deviation from an ideal rotating disk or the asymmetry in the velocity field has been plotted in Figure 3 in terms of radius. According to the classification by Bellocchi, Arribas, and Colina (2016) that is based on the total kinematic asymmetry, it is possible

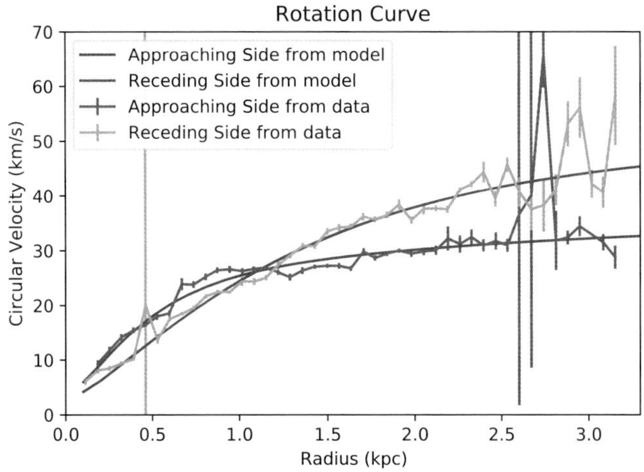

Figure 1. Asymmetric rotation curve extracted from the perturbed halo potential model and from the observational data (Kepley *et al.* 2007) for the approaching and the receding sides of WLM.

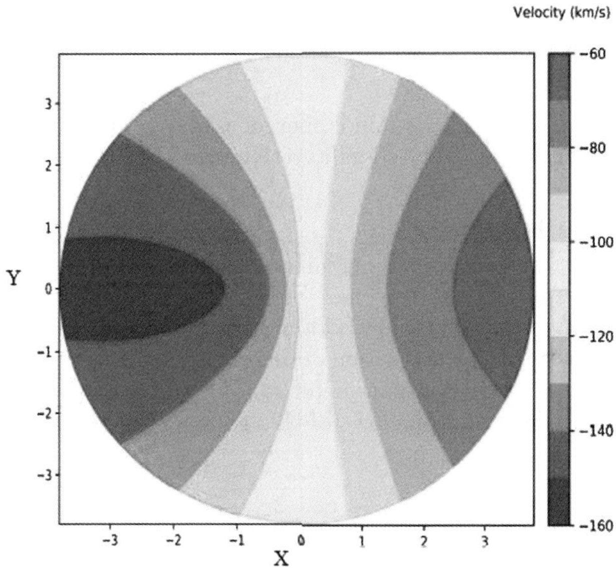

Figure 2. Perturbed velocity field map caused by a lopsided halo potential model cube (a global m = 1 mode in the potential), projected in the X-Y plane. The perturbed velocity field map is clearly asymmetric; with strong curvature of the iso-velocity contours on one side and weaker curvature on the other side.

for WLM to lie in the transition region, in which disk and merger coexist and such kinematic lopsidedness in the disk may have been excited by a minor merger. In a minor merger, where a smaller galaxy merges with a larger disk galaxy, the impact on the disk can be relatively minor. The smaller galaxy is typically absorbed by the larger galaxy, and the disk can continue to rotate in a similar manner as before the merger. In such a merger, the disk is not destroyed and the kinematics of the remnant does not appear too complex. The galaxy is observed to still be rotating along its main optical axis and the velocity dispersion map does not show any peak at the center (Puech *et al.* 2006). These are properties of the perturbed galaxies (Hammer *et al.* 2009, 2017), such as

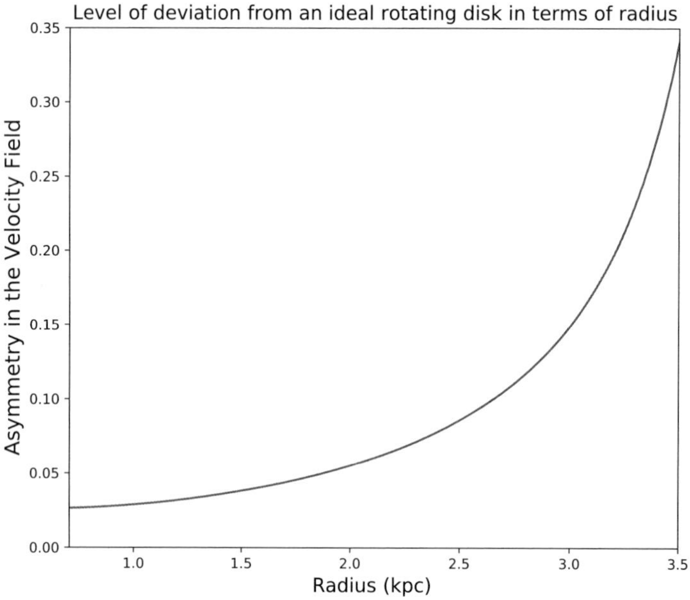

Figure 3. Level of deviation from an ideal rotating disk in terms of radius.

WLM(Khademi *et al.* 2021). So, a minor merger may be one of the possible origins of the halo perturbation for this isolated galaxy (Khademi *et al.* 2021).

4. Kinematic lopsidedness

To investigate how well an $m=1$ perturbation in the halo potential represents the observed asymmetries in the rotation curve of the dwarf irregular galaxy WLM, one can measure the value of the perturbation parameter ϵ_{per}. Assuming that a distorted halo potential produced by a galaxy interaction or merger events caused kinematic asymmetry, one can estimate the perturbation parameter ϵ_{per} for the perturbed halo potential from the kinematic data on the rotation curve of this galaxy (van Eymeren *et al.* 2011).

$$\epsilon_{per} = \frac{v_{rec} - v_{app}}{2v_c} \tag{1}$$

where v_{rec} and v_{app} are the maximum rotation velocities, measured from the rotation curves of the receding and approaching sides, respectively, and v_c is the maximum rotation velocity of the combined rotation curve. For the dwarf irregular galaxy WLM, we obtained the value of the perturbation parameter ϵ_{per} to be about 0.2.

5. Summary

We investigated whether an $m=1$ perturbation (lopsidedness) in the halo potential could be a mechanism creating observed kinematical asymmetry in the rotation curve of the dwarf irregular galaxy WLM. We showed that a lopsided halo potential model can explain the asymmetry in the kinematic data reasonably well. To investigate how well such a small perturbation in the halo potential represents the observed asymmetries in the rotation curve and in the velocity field of the dwarf irregular galaxy WLM, we measured the value of the perturbation parameter ϵ_{per} from the kinematic data on the rotation curve of the receding and approaching sides in the outer part of the dwarf irregular galaxy WLM. For this dwarf irregular galaxy, we obtained the value of the perturbation parameter denoting the perturbed halo potential of about 0.2. The perturbation parameter (ϵ_{lop}) obtained from a perturbed halo potential model with a global m = 1 mode, is smaller than that obtained from the kinematic data on the rotation curve ($\epsilon_{per} \approx 0.2$). The reason is that for our perturbed halo potential model, we made a simplifying assumption that the m = 1 potential perturbation is the dominant perturbation term. Therefore, we considered just an m = 1 perturbation mode in the halo potential and ignored the higher-order modes, while in addition to an m=1 perturbation mode, the perturbation in the halo potential includes higher modes of a Fourier analysis.

References

Bellocchi, E., Arribas, S., and Colina, L.: 2016, *Astronomy and Astrophysics* 591, A85. doi:10.1051/0004-6361/201526974.

Hammer, F., Flores, H., Puech, M., Yang, Y.B., Athanassoula, E., Rodrigues, M., and, ...: 2009, *Astronomy and Astrophysics* 507, 1313. doi:10.1051/0004-6361/200912115.

Hammer, F., Puech, M., Flores, H., & Rodrigues, M. 2017, Studying Distant Galaxies: a Handbook of Methods and Analyses, first edn. (New Jersey: World Scientific)

Ianjamasimanana, R., Namumba, B., Ramaila, A.J.T., Saburova, A.S., Józsa, G.I.G., Myburgh, T., and, ...: 2020, *Monthly Notices of the Royal Astronomical Society* 497, 4795. doi:10.1093/mnras/staa1967.

Kepley, A.A., Wilcots, E.M., Hunter, D.A., and Nordgren, T.: 2007, *The Astronomical Journal* 133, 2242. doi:10.1086/513716.

Khademi, M., Yang, Y., Hammer, F., and Nasiri, S.: 2021, *Astronomy and Astrophysics* 654, A7. doi:10.1051/0004-6361/202140336.

Khademi, M. and Nasiri, S.: 2022, *Canadian Journal of Physics* 100, 201. doi:10.1139/cjp-2021-0332.

Krajnović, D., Cappellari, M., de Zeeuw, P.T., and Copin, Y.: 2006, *Monthly Notices of the Royal Astronomical Society* 366, 787. doi:10.1111/j.1365-2966.2005.09902.x.

Pasha, I., Mandelker, N., van den Bosch, F.C., Springel, V., and van de Voort, F.: 2023, *Monthly Notices of the Royal Astronomical Society* 520, 2692. doi:10.1093/mnras/stac3776.

Puech, M., Hammer, F., Flores, H., Östlin, G., and Marquart, T.: 2006, *Astronomy and Astrophysics* 455, 119. doi:10.1051/0004-6361:20054218.

Shapiro, K.L., Genzel, R., Förster Schreiber, N.M., Tacconi, L.J., Bouché, N., Cresci, G., and, ...: 2008, *The Astrophysical Journal* 682, 231. doi:10.1086/587133.

van Eymeren, J., Jütte, E., Jog, C.J., Stein, Y., and Dettmar, R.-J.: 2011, *Astronomy and Astrophysics* 530, A29. doi:10.1051/0004-6361/201016177.

van Eymeren, J., Jütte, E., Jog, C.J., Stein, Y., and Dettmar, R.-J.: 2011, *Astronomy and Astrophysics* 530, A30. doi:10.1051/0004-6361/201016178.

Yang, Y., Ianjamasimanana, R., Hammer, F., Higgs, C., Namumba, B., Carignan, C., and, ...: 2022, *Astronomy and Astrophysics* 660, L11. doi:10.1051/0004-6361/202243307.

Dynamical Masses of Local Group Galaxies
Proceedings IAU Symposium No. 379, 2025
P. Bonifacio, M.-R. Cioni, F. Hammer, M. S. Pawlowski & S. Taibi, eds.
doi:10.1017/S1743921323001576

c-M relation of the sub-galactic dark matter halos and the effect of the cusp-to-core transition

Yuka Kaneda[1], **Koki Otaki**[1] **and Masao Mori**[2]

[1]Graduate School of Science and Technology, University of Tsukuba,
1-1-1 Tennodai, Tsukuba, Ibaraki, 305-8577, Japan
email: kaneda@ccs.tsukuba.ac.jp

[2]Center for Computational Science, University of Tsukuba,
1-1-1 Tennodai, Tsukuba, Ibaraki, 305-8577, Japan

Abstract. Almost 30 years have passed since not only the universal density distribution of dark matter (DM) halos in cosmological N-body simulations, but also the scalings between properties of DM halos represented by concentration-mass (c-M) relations are proposed. We derive the c-M relation for sub galactic halos (subhalos) of Milky Way (MW)-sized host halo using the result of the ultra-high resolution cosmological N-body simulation, Phi-4096, with a particle mass of $5.13 \times 10^3\ h^{-1} M_\odot$. This c-M relation is confirmed to be consistent with a c-M relation from near the free streaming scale to the galactic scale proposed in a literature. One of our main findings is that the c-M relation can reproduce observational properties of DM halos from dwarf galaxies to clusters of galaxies. In addition, we provide a testable prediction of the density distributions of MW subhalos for future observations.

Keywords. Galaxy Formation, dark matter, Cosmological N-body simulation

1. Introduction

The Λ-dominated cold dark matter (CDM) theory is most successful in explaining observational phenomena on large scales. On the other hand, it has serious issues on galactic and sub-galactic mass scales. One of the oldest is the cusp-core problem. Cosmological N-body simulations produce a universal density distribution of DM halo, which diverges near the centre, called a cusp profile (Navarro et al. 1997). As a representative, the Navarro-Frenk-White (NFW) profile (Navarro et al. 1997) well fits the density profiles of DM halos formed in CDM simulations. In addition, many past studies claim that these CDM halos have a tight correlation between their concentrations of mass density distributions, virial masses, and redshifts, called the c-M relation. This relation is a essential diagnostic to characterise the mass density distribution of CDM halos. On the other hand, observations of low-mass galaxies indicate that most of DM halos have constant density distribution in the centre, called the core profile (Moore 1994). The Burkert profile (Burkert 1995) well fits the density structures of DM halos with cores. This controversy over the central DM halo density distribution has been a subject of intensive analysis for a long time. There are several explanations for the constant density core observed in dwarf galaxies. One of them is to consider the new candidates for DM particles, which intrinsically produce cores, such as the self-interactive DM and the fuzzy DM. The other is to take baryonic feedback effects into account to alter a central cusp into a core. Several mechanisms have been proposed and one of them is effect

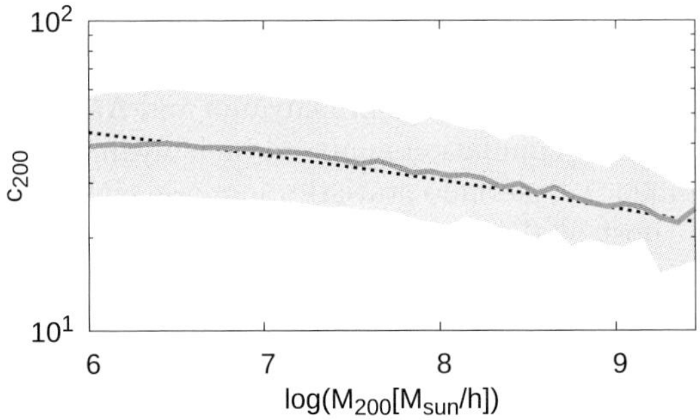

Figure 1. The *c-M* relation for subhalos of MW-sized host from Phi-4096 is depicted in the orange solid line with 25-75 percentile region. The black dotted line describes the approximate function proposed by Ishiyama and Ando (2020) for different simulation.

of stellar feedback that changes a gravitational potential of gas (Navarro et al. 1996). In this scenario, cusp-to-core transitions are thought to occur around dwarf galaxy mass scale, $< 10^{11}\, M_{\odot}$ in a virial mass, where global gas loss is caused by stellar feedback (Dekel and Silk 1986). The motivation of this work is to investigate DM halo mass dependence on the occurrence of the cusp-to-core transitions. As a first step, we compare the *c-M* relation with estimated properties of DM halo by observations for a wide range of object from dwarf galaxies to clusters of galaxies to examine how it holds in the real universe. Parameters that are integrated to the outer part of the halo such as the virial mass M_{200} and the maximum circular velocity $V_{\rm max}$ is used since they are not sensitive to the form of density profiles.

2. Simulation Data

An ultra-high resolution cosmological N-body simulation, Phi-4096, by Ishiyama et al. (2021) is utilised to derive the *c-M* relation of the subhalos of MW-sized host halos. Phi-4096 has the highest resolution among the Uchuu suite simulations to evaluate resolution effect at high redshifts in other simulations. It has 4096^3 DM particles covering a comoving $16\, h^{-1}$Mpc side length box, with resulting mass resolution of $5.13 \times 10^3\, h^{-1} M_{\odot}$. We take advantage of this unprecedented resolution and a relatively large volume of Phi-4096 to investigate properties of galactic substructures statistically. We have 27 MW-sized host halos which fulfil the mass criteria $3.40 \times 10^{11} < M_{\rm vir} h^{-1} M_{\odot} < 2.04 \times 10^{12}$ at redshift $z = 0$. Subhalos are identified as the halos that reside inside twice the virial radius of the host halo at $z = 0$. The mass cut $\geq 10^6\, M_{\odot}$ is applied to the virial masses of subhalos to avoid halos consisting of $\lesssim 1000$ particles.

3. Results

In Figure 1, we present the *c-M* relation of the subhalos of the MW-sized host. The concentration is defined by $c_{200} \equiv r_{200}/r_{\rm N}$, where r_{200} is the virial radius and $r_{\rm N}$, the scale radius of the NFW profile. M_{200} is connected with r_{200} by $M_{200} = 4/3\pi 200 \rho_{\rm crit} r_{200}^3$. The orange solid line depicts the median of c_{200} in each mass bin with the 25-75 percentile described in the orange transparent band. The parametrization of the *c-M* relation proposed by Ishiyama and Ando (2020),

$$c_{200} = \sum_{i=0}^{3} c_i \times \left[\ln \frac{M_{200}}{M_{\odot}} \right]^i , \tag{1}$$

where $c_i = [94.6609, -4.1160, 3.3747 \times 10^{-2}, 2.0932 \times 10^{-4}]$, shown in the black dotted line coincides the resultant c-M relation of our work (orange solid line), although there are several differences in the data set used in Ishiyama and Ando (2020) and in this work. The first difference is that the simulation analysed by Ishiyama and Ando (2020) was half the box length of Phi-4096, i.e. subhalo statistics increase eightfold in this work. Second, the mass range of host halos to which subhalos belong is different. While Ishiyama and Ando (2020) did not limit the mass range of host halos, we only adopted the subhalos which reside in the MW-sized hosts. Moreover, Ishiyama and Ando (2020) have fitted the concentration of halos from free streaming scale to the galactic scale, namely, with mass range $10^{-6} - 10^{-4}\,\mathrm{M}_{\odot}$ and $10^{7} - 10^{12}\,\mathrm{M}_{\odot}$, interpolating between $10^{-4} - 10^{7}\,\mathrm{M}_{\odot}$. Nevertheless, the result shows that the fitting function proposed by Ishiyama and Ando (2020) agrees with the c-M relation of MW subhalos $10^{6} - 10^{9.5}\,M_{\odot}$ nicely.

Then the c-M relation is tested by observationally estimated properties of DM halo. Here, we compare the parameters of the NFW profiles obtained from the c-M relation with those obtained from observations of dwarf galaxies, normal galaxies, groups and clusters of galaxies. Data from Oh et al. (2015) for dwarf galaxies, de Blok et al. (2008), Spano et al. (2008), and Sofue (2016) for galaxies and Gastaldello et al. (2007) for groups and Merten et al. (2015) and Umetsu et al. (2016) for clusters of galaxies are used. In Figure 2, the black solid curve describes the fitting function proposed by Ishiyama and Ando (2020), and the symbols represent estimated parameter values by the observations. The top panel shows c_{200} versus M_{200}, the bottom-left panel, r_{N} versus ρ_{N}, the scale density of NFW profile, the bottom-right panel, V_{max}, versus r_{max}, the radius at which the circular velocity becomes maximum. In the top panel, theoretical prediction for c_{200} decreases as the mass increases. Observed c_{200} values also decline as the objects become more massive. The theoretical c-M relation is consistent with the observations for a wide mass range. In the bottom-left panel, the theoretical curve also declines as the r_{N} becomes larger. The c-M relation is in good accordance with the observations from dwarf galaxies to galaxy clusters. In the bottom-right panel, the theoretical curve displays that the r_{max} increases as the V_{max} increases and r_{max} is almost proportional to V_{max} on logarithmic scale. The c-M relation nicely predicts the observed V_{max} and r_{max} values. However, on the massive end ($> 10^{13}\,M_{\odot}$), there is an offset between the theoretical prediction and the observations. The reasons for this are discussed in Section 4.

4. Summary and Discussion

In this work, we investigate the c-M relation for the subhalos of the MW-sized host based on the high resolution cosmological N-body simulation and compare it with the observations. As a result, our main findings are summarised as follows. (i) The c-M relation for subhalos of MW-sized hosts is in good agreement with the parametrization for a wider mass range proposed by Ishiyama and Ando (2020). (ii) Ishiyama-Ando relation can reproduce the observational properties of DM halos from dwarf galaxy to clusters of galaxy all on the M_{200}-c_{200} plane, the r_{N}-ρ_{N} plane, and the V_{max}-r_{max} plane.

In the top panel of Figure 2, c_{200} shows a large scatter from 1 to 100 in the mass range of dwarf galaxies. Since the rotation velocities and the velocity dispersion of such systems are comparable, it is difficult to determine the density profile. Therefore, we cannot limit DM models or cusp-to-core transition models in this mass range. Upcoming telescope programs with high dispersion spectroscopy may solve this problem. For example, Subaru PFS is promising to estimate the DM distribution of nearby dwarf galaxies.

In the mass range of groups and clusters of galaxies, there is a systematic offset between the theoretical prediction and the observations. For one reason, the fitting function proposed by Ishiyama and Ando (2020) has fitted the simulation data up to $10^{12}\,M_{\odot}$.

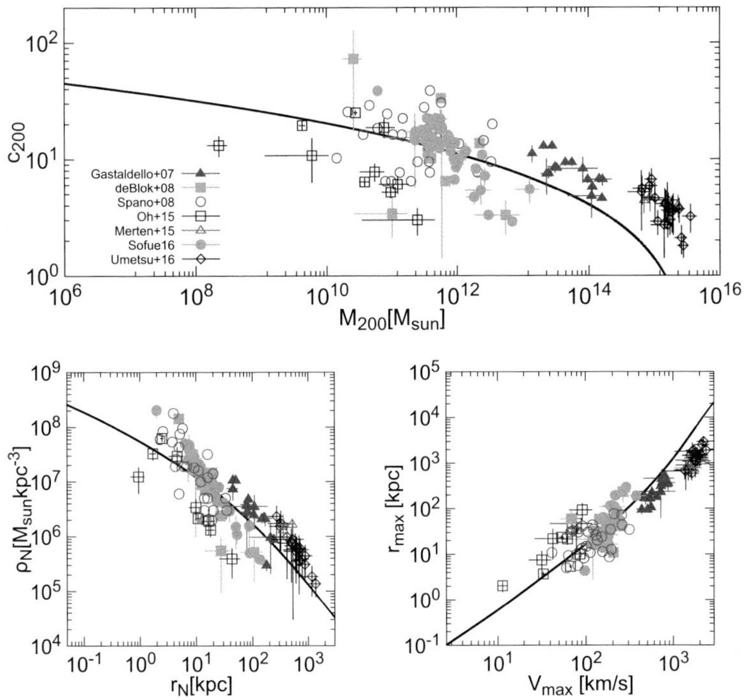

Figure 2. Comparison between the c-M relation (Ishiyama and Ando (2020), black solid line) and observation from dwarf galaxies to clusters of galaxies (symbols). Top panel shows c_{200}-M_{200} plane, bottom-left, $r_{\rm N}$-$\rho_{\rm N}$, bottom-right, $V_{\rm max}$-$r_{\rm max}$.

This means that the c-M relation used in our analysis is extrapolation over $10^{12}\ M_{\odot}$. This fact may cause a systematic offset. A larger-scale cosmological simulation with sufficient statistical properties of cluster-sized halos is needed to evaluate this effect. Another reason should be the difference in the redshifts between theoretical prediction and the observed galaxy clusters. While we use the c-M relation at $z = 0$, our cluster sample contains objects at slightly higher redshifts; the farthest one is at $z = 0.86$. Previous works on cosmological N-body simulations showed that the concentration of $\sim 10^{15}\ M_{\odot}$ halo increases $\sim 30\%$ from $z = 0$ to 1. If we take this evolution effect into account, the theoretical prediction should come closer to the observations. These problems are investigated in our forthcoming study.

Since the parameters used to compare the c-M relation and the observations in our study such as M_{200} and $V_{\rm max}$ do not make an apparent difference between a cusp and a core profile, we need to see the parameters that represent the inner density distribution to evaluate the changes in properties of a halo caused by the cusp-to-core transition. To achieve this, Kaneda et al. in prep. evaluate the necessary observational resolution and discuss the capability by future instruments in detail.

This work was supported in part by JSPS KAKENHI Grant Number JP20K04022, and by Multidisciplinary Cooperative Research Program in CCS, University of Tsukuba.

References

Burkert, A. 1995, *ApJ*, 447, L25–L28.
de Blok, W. J. G., Walter, F., Brinks, E., *et al.* 2008, *AJ*, 136(6), 2648–2719.
Dekel, A. & Silk, J. 1986, *ApJ*, 303, 39.

Gastaldello, F., Buote, D. A., Humphrey, P. J., *et al.* 2007, *ApJ*, 669(1), 158–183.
Ishiyama, T. & Ando, S. 2020, *MNRAS*, 492, 3662–3671.
Ishiyama, T., Prada, F., Klypin, A. A., *et al.* 2021, *MNRAS*, 506, 4210–4231.
Merten, J., Meneghetti, M., Postman, M., *et al.* 2015, *ApJ*, 806(1), 4.
Moore, B. 1994, *Nature*, 370, 629–631.
Navarro, J. F., Eke, V. R., & Frenk, C. S. 1996, *MNRAS*, 283, L72–L78.
Navarro, J. F., Frenk, C. S., & White, S. D. M. 1997, *ApJ*, 490, 493–508.
Oh, S.-H., Hunter, D. A., Brinks, E., *et al.* 2015, *AJ*, 149(6), 180.
Sofue, Y. 2016, *PASJ*, 68, 2.
Spano, M., Marcelin, M., Amram, P., *et al.* 2008, *MNRAS*, 383, 297–316.
Umetsu, K., Zitrin, A., Gruen, D., *et al.* 2016, *ApJ*, 821, 116.

Dynamical Masses of Local Group Galaxies
Proceedings IAU Symposium No. 379, 2025
P. Bonifacio, M.-R. Cioni, F. Hammer, M. S. Pawlowski & S. Taibi, eds.
doi:10.1017/S1743921323001801

Census of Northern Local Group Dwarf Galaxies using UNIONS

S. E. T. Smith[1], A. W. McConnachie[2,1] and the UNIONS Collaboration

[1]Department of Physics and Astronomy, University of Victoria, Victoria, BC, V8P 1A1, Canada
email: simonsmith@uvic.ca

[2]NRC Herzberg Astronomy and Astrophysics, 5071 West Saanich Road, Victoria, BC, V9E 2E7, Canada

Abstract.
In this work we present preliminary results from a systematic search for new Local Group dwarf galaxies in the deep, wide-field Ultraviolet Near-Infrared Optical Northern Survey (UNIONS). The first major result from this work is the discovery of a candidate ultra faint dwarf galaxy, Boötes V. This satellite of the Milky Way has a physical half-light radius of $26.9^{+7.5}_{-5.4}$ pc, an absolute V-band magnitude of -4.5 ± 0.4 mag, and resides at a heliocentric distance of approximately 100 kpc. We also announce the discovery of a faint, compact star cluster, UNIONS 1. The end goal of this work will be to put firm constraints on the detectability and completeness of Local Group dwarf galaxies given the broad sky coverage and excellent photometric depth of UNIONS.

Keywords. galaxies: dwarf, Local Group, galaxies: stellar content, techniques: photometric

1. Introduction

The known dwarf galaxy population of the Local Group has grown rapidly since the inception of wide-field photometric surveys using digital telescopes (e.g. Sloan Digital Sky Survey, Pan-Andromeda Archaeological Survey, Dark Energy Survey: York *et al.* 2000; McConnachie *et al.* 2009; Abbott *et al.* 2018). The faintest of these systems have been called "ultra-faint dwarf galaxies" (UFDs; e.g. see review by Simon *et al.* 2019), defined as galaxies with total $M_V > -7.7$ mag. They are the smallest, least massive, and most metal-poor galaxies yet observed and their very existence tests current theories of galaxy formation. In particular, UFDs probe the physics of stellar feedback and gas retention, and they trace the Local Group dark matter distribution at parsec length-scales (Bullock *et al.* 2017).

The work we present here is a search for new Local Group dwarf galaxies in the deep, wide-field photometric Ultraviolet Near-Infrared Optical Northern Survey (UNIONS), a consortium of northern imaging surveys all based in Hawaii. The 5000 deg^2 *ugriz* survey consists of the Canada-France Imaging Survey (CFIS) collaboration that uses the Canada-France-Hawaii Telescope (CFHT), team members from the Panoramic Survey Telescope and Rapid Response System (Pan-STARRS), and the Wide Imaging with Subaru Hyper Suprime-Cam (HSC) of the Euclid Sky (WISHES). We have conducted this work on the current data releaseof UNIONS, specifically with the r- and i-bands being provided by CFIS and Pan-STARRS respectively, jointly covering $3600\,\mathrm{deg}^2$ at median 5σ point source depths of 24.7 and 24 mag in r and i.

We have performed a systematic search for dwarf galaxies using a matched-filter methodology that scans the UNIONS footprint for stellar overdensities consistent with an old (12 Gyr), metal-poor ([Fe/H] $= -2$ dex) stellar population. The matched-filter methodology has proven to be effective (e.g. Belokurov *et al.* 2007; Bechtol *et al.* 2015) and the full details of our search algorithm can be found in Smith *et al.* (2022).

2. Discoveries in UNIONS

Our full sky search has recovered all known dwarf galaxies that had been previously discovered using matched-filter techniques in shallower surveys within 1.9 Mpc of the Milky Way, along with many globular clusters and nearby galaxies in the Local Universe. Therefore, we have turned to our list of detected stellar overdensities with no prior association and discovered two prominent stellar overdensities, an ultra-faint dwarf galaxy candidate, and a faint star cluster.

Boötes V (Smith *et al.* 2022) was discovered in the combined CFIS-r and Pan-STARRS-i dataset as a resolved overdensity of stars with the characteristics of a faint Milky Way satellite dwarf galaxy. We derive structural properties for this system, finding it to have an angular half-light radius of $0.93^{+0.26}_{-0.19}\,'$ and an ellipticity of $0.35^{+0.17}_{-0.21}$. Using several Blue Horizontal Branch stars, we constrain the distance to the dwarf galaxy to be 100 ± 20 kpc and calculate an absolute V-band magnitude of -4.5 ± 0.4 mag. In turn, we derive a physical half-light radius of $26.9^{+7.5}_{-5.4}$ pc and an effective surface brightness of $25.7 \pm 0.7\,\mathrm{mag\,arcsec^{-2}}$. We further refine the parameters of the stellar population by eye, finding that a 13 Gyr, [Fe/H]$= -2.2$ dex isochrone, the most metal-poor isochrone available in our set, is a good fit to the resolved stars on a Colour-Magnitude Diagram (CMD).

Following McConnachie & Venn (2020), we identify several stars in the Third Data Release from *Gaia* as likely members of Boötes V based on their spatial coherence, position on a CMD, and proper motion. From these members we derive the systemic proper motion of the dwarf to be $(\mu_\alpha \cos\delta, \mu_\delta) = (-0.23 \pm 0.04$ (stat) $\pm$ 0.033 (sys), -0.28 ± 0.07 (stat) $\pm$ 0.033 (sys)) mas year^{-1}. Additionally, we obtained follow-up spectroscopy of the brightest member star, a Red Giant Branch star near the tip of the RGB, using the Gemini Multi-Object Spectrograph at Gemini North. From our long-slit spectroscopic observations centered on the Calcium II infrared triplet absorption feature, we measure the radial velocity of the star to be $+5.1 \pm 13.4\,\mathrm{km\,s^{-1}}$ and the metallicity to be [Fe/H]$= -2.85 \pm 0.10$ dex. See Fig. 1 for the spatial position and CMD of Boötes V in UNIONS with *Gaia*-identified member stars overplotted.

We determine that Boötes V is likely a dwarf galaxy given its large scale radius and large heliocentric distance compared to known globular clusters. Additionally, while we have only measured a single star, the very metal-poor nature of this star is a good indication that the system as a whole is metal-poor. This provides a good starting point for future, focused spectroscopic studies of Boötes V, which will ultimately be necessary to definitively classify this distant, resolved stellar system. We note that this dwarf galaxy candidate was independently co-discovered by Cerny *et al.* (2022).

We have also identified a very compact group of co-moving stars with a physical half-light radius of ~4 pc which we name UNIONS 1, under the assumption that it is a star cluster, not a dwarf galaxy. However, this system will be presented and it's classification will be further assessed in an upcoming study. We note that these two stellar systems are the only two stellar overdensities whose reality could be confirmed with a combination

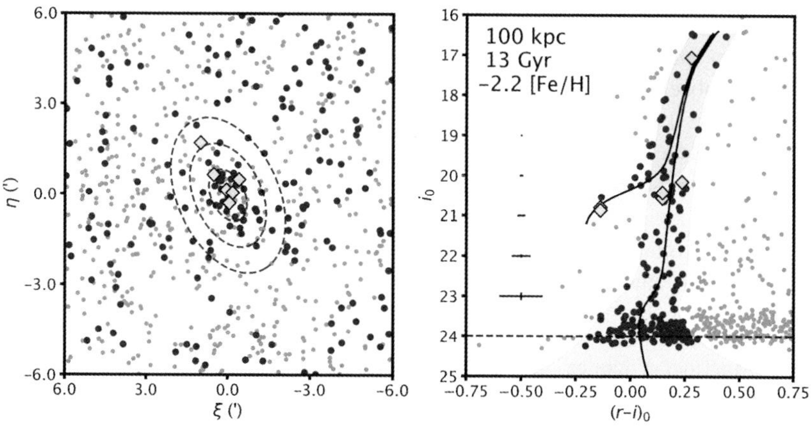

Figure 1. *Left*: Sky positions of all stars in a $12' \times 12'$ region about Boötes V, projected onto the tangent plane centered at RA, Dec = (14h 15m 38.6s, +32° 54′ 42″). Dark blue (dark grey) sources are those that meet the isochrone selection criterion (Right). Yellow (light grey) diamonds are stars identified in *Gaia* DR3 as sources with complete astrometric information, and are selected as high confidence members. The concentric black ellipses indicate 1×, 3×, and 5× the half-light radius. *Right*: Colour-magnitude diagram of extinction-corrected CFIS-r and Pan-STARRS-i for all stellar sources in a $12' \times 12'$ region about Boötes V. We overlay an old (13 Gyr), metal-poor ([Fe/H] = −2.2) isochrone, shifted to a distance of 100 kpc. We use a broad color selection criterion, selecting all stars with $(r-i)_0 \leq 0.1$ from the isochrone and we add, in quadrature, the empirical photometric errors from each band. Stars consistent with this selection criterion are highlighted in dark blue (dark grey). Photometric uncertainties are shown for integer magnitudes in black. Again, yellow (light grey) diamonds are high confidence members observed by *Gaia*. If printed in greyscale, refer to shades denoted in brackets.

of UNIONS and *Gaia* data, but there are a plethora of other candidates that we are investigating further with follow-up observations.

3. Outlook on Local Group Census

There are exciting prospects for the future of this work, namely the completion of UNIONS and synergy with the *Euclid* space telescope. UNIONS is not yet at full depth nor coverage. With final 5σ point source depths of 24.3, 25.2, 24.9, 24.3, 24.1 across *ugriz* bands, UNIONS is an ideal dataset for probing resolved stellar populations with multi-band photometry out to 1 Mpc within the Local Group. Additionally, UNIONS was in part conceptualized to support the *Euclid* space mission. This work will be able to take advantage of the space-based seeing capabilities of *Euclid* to improve upon our current star-galaxy separation, allowing us to distinguish faint stellar populations from background contamination.

Each newly discovered dwarf galaxy provides a unique opportunity to study physical processes and predictions such as star formation, the effects of feedback, tidal perturbations, and dark matter structure on the smallest known scales, but a concerted effort to build up the dwarf galaxy population of the Local Group works towards a bigger goal. A statistically significant sample of dwarf galaxies allows for the study of broad trends and properties of the Local Group, down to the lowest observed limits of the galaxy luminosity function.

Some studies have performed a census of dwarf galaxies around the Milky Way (e.g. Koposov *et al.* 2008; Drlica-Wagner *et al.* 2020) and around M31 (Doliva-Dolinsky *et al.* 2022, 2023). The end goal of this work will be to constrain the detectability and completeness of satellite and isolated dwarf galaxies out to 1 Mpc within the Local Group, predicting the number of discoveries that may lie in wait beyond the edge of current observational limits. This Local Group census will also allow for meaningful comparisons between the observed population of dwarf galaxies and predictions from both cosmological and zoom-in hydrodynamical simulations.

References

Abbott, T. M. C., Abdalla, F. B., Allam, S., *et al.* 2018, *ApJS*, 239, 18
Bechtol, K., Drlica-Wagner, A., Balbinot, E., *et al.* 2015, *ApJ*, 807, 50
Belokurov, V., Zucker, D. B., Evans, N. W., *et al.* 2007, *ApJ*, 654, 897
Bullock, J. S., & Boylan-Kolchin, M. 2017, *ARA&A*, 55, 343
Cerny, W., Martínez-Vázquez, C. E., Drlica-Wagner, A., *et al.* 2022, arXiv e-prints, arXiv:2209.12422
Doliva-Dolinsky, A., Martin, N. F., Thomas, G. F., *et al.* 2022, *ApJ*, 933, 135
Doliva-Dolinsky, A., Martin, N. F., Yuan, Z., *et al.* 2023, arXiv e-prints, arXiv:2303.01528
Drlica-Wagner, A., Bechtol, K., Mau, S., *et al.* 2020, *ApJ*, 893, 47
Koposov, S. E., Belokurov, V., Evans, N. W., *et al.* 2008, *ApJ*, 686, 279
McConnachie, A. W., & Venn, K. A. 2020, *AJ*, 160, 124
McConnachie, A. W., Irwin, M. J., Ibata, R. A., *et al.* 2009, *Nature*, 461, 66
Simon, J. D. 2019, *ARA&A*, 57, 375
Smith, S. E. T., Jensen, J., Roediger, J., *et al.* 2022, *in press*, arXiv e-prints, arXiv:2209.08242
York, D. G., Adelman, J., Anderson, John E., J., *et al.* 2000, *AJ*, 120, 1579

Dynamical Masses of Local Group Galaxies
Proceedings IAU Symposium No. 379, 2025
P. Bonifacio, M.-R. Cioni, F. Hammer, M. S. Pawlowski & S. Taibi, eds.
doi:10.1017/S1743921323001837

Revealing mass distributions of dwarf spheroidal galaxies in the Subaru-PFS era

K. Hayashi[1,2,5], L. Dobos[3], C. Filion[3], E. Kirby[4], M. Chiba[5], R. Wyse[3] and PFS Galactic Archaeology Science Working Group

[1]National Institute of Technology, Sendai College, Natori, Miyagi 981-1239, Japan
email: khayashi@sendai-nct.ac.jp

[2]National Institute of Technology, Ichinoseki College, Hagisho, Ichinoseki, Iwate 021-8511, Japan

[3]Department of Physics and Astronomy, The Johns Hopkins University, Baltimore, MD 21218, USA

[4]Department of Physics and Astronomy, University of Notre Dame, 225 Nieuwland Science Hall, Notre Dame, IN 46556, USA

[5]Astronomical Institute, Tohoku University, Sendai, Miyagi 980-8578, Japan

Abstract. The Galactic dwarf spheroidal galaxies (dSphs) provide valuable insight into dark matter (DM) properties and its role in galaxy formation. Their close proximity enables the measurement of line-of-sight velocities for resolved stars, which allows us to study DM halo structure. However, uncertainties in DM mass profile determination persist due to the degeneracy between DM mass density and velocity dispersion tensor anisotropy. Overcoming this requires large kinematic samples and identification of foreground contamination. With 1.25 deg^2 and 2394 fibers, PFS plus pre-imaging with Hyper Suprime Cam will make significant progress in this undertaking.

Keywords. Dark matter – Dwarf spheroidal galaxies – Galaxy dynamics – Local Group

1. Introduction

The Galactic dwarf spheroidal (dSph) galaxies are ideal sites for studying the basic properties of dark matter and its role in galaxy formation. This is because these galaxies have high dynamical mass-to-light ratio ($M/L \sim 10 - 1000$), which means that these are the dark matter rich systems. Owing to their proximity of the Sun, the dSphs have the advantage that individual member stars can be resolved. Therefore, it is possible to measure accurate line-of-sight velocities for their member stars, so that we are able to set constraints on their internal structures of dark matter halo using the high-quality data (Battaglia & Nipoti 2022; Battaglia et al. 2013, for reviews).

It is well documented that the Λ cold dark matter (ΛCDM) theory has well-reproduced the cosmological and astrophysical observations on large spacial scales such as cosmic microwave background and large-scale structure of galaxies. At galactic and sub-galactic scales ($\lesssim 1$ Mpc), however, this concordant theory has several outstanding discrepancies between the predictions from pure dark matter simulations based on ΛCDM models and some observational facts (Bullock & Boylan-Kolchin 2017). The oldest controversial issues in ΛCDM models is the so-called "core-cusp" problem: dark matter halos predicted by ΛCDM simulations have strongly cusped central density profiles, whereas the dark matter halos in the observed less massive galaxies (dSphs and low surface brightness galaxies) are suggested to have cored dark matter density profiles. Recently, the dynamical studies for the luminous dSphs, so-called classical dSphs, have suggested that although there are still large uncertainties, these galaxies show a diversity of the inner dark matter densities (e.g., Read et al. 2019; Hayashi et al. 2020). To interpret this diversity, there are various possible mechanisms such as baryonic feedback and star formation burst (Lazar et al. 2020) and alternative dark matter models (e.g., self-interacting dark matter: Nishikawa et al. 2020).

On the other hand, current dynamical studies of dSphs still face challenges in accurately measuring their central dark matter density profiles due to the well-known degeneracy between dark matter mass density and the anisotropy of the stellar velocity dispersion tensor as well as limited availability of data. Motivated by the aforementioned problem, we, the Prime Focus Spectrograph (PFS) Galactic Archaeology (PFS-GA) science working group, have addressed the development of dynamical mass models that take into account non-trivial effects such as Milky Way contamination stars, binary stars, and non-equilibrium systems. This is because the future PFS spectroscopic surveys will enable us to improve the quantity and quality of kinematic data. Such data from PFS will open the door to placing more stringent constraints on inner dark matter density profiles of the dSphs.

2. Subaru Prime Focus Spectrograph

The Subaru Prime Focus Spectrograph attached on Subaru 8-m class telescope is a massively-multiplexed fiber-fed optical and near-infrared three-armed spectrograph (Tamura et al. 2016). The wave length coverage is from the optical to the near infrared ($380 - 1260$ nm), and thus PFS can observe the blue and infrared ends simultaneously. PFS is designed to allow simultaneous low and intermediate-resolution spectroscopy with blue ($380 - 650$ nm, $R \sim 2,300$), medium-resolution red ($710 - 885$ nm, $R \sim 5,000$) arm, and infrared ($940 - 1260$ nm, $R \sim 4,300$) arm, respectively. Thanks to the Subaru 8-m class telescope and its Prime Focus, PFS has a large field of view (1.25 degrees in diameter in a hexagonal field) and 2400 optical fibers arranged in the hexagon. This new spectroscopy capability will allow us to obtain statistically significant samples of stellar and galactic spectra over wide areas. Figure 1 shows a schematic comparison of the FoV of Subaru PFS with that of VLT/FLAMES, Gemini/GMOS, and Keck/DEIMOS. It is clear from this figure that PFS has much larger FoV than the other current spectroscopy instruments, and thus PFS will offer unique opportunities in large astronomical survey.

Taking advantage of Subaru PFS in synergy with Subaru Hyper Suprime Cam (HSC) pre-imaging, PFS-GA science working group is planning a large observation for the GA science cases in the context of a Subaru Strategic Program (Takada et al. 2014), which will start from 2024. The main targets of our provisional plans are the Galactic dwarf galaxies, M31 disk and halo regions as well as M33 companion, and the Milky Way outer disk/halo region including stellar streams. Here, we provide an update on the progress of the PFS-GA science project, with a particular focus on the survey of Galactic dSphs.

3. Motivation for PFS-dSph survey

As stated in the introduction, there are significant uncertainties in estimating the dark matter density profiles of dSphs due to the limited availability of kinematic data. The small field of view of current spectroscopy instruments has resulted in the outer regions of most dSphs being unobserved, which is one of the main reasons for this challenge. Before presenting the current progress of the PFS-dSph survey, we demonstrate why kinematic data in the outer regions of dSphs are crucial for estimating their dark matter distributions.

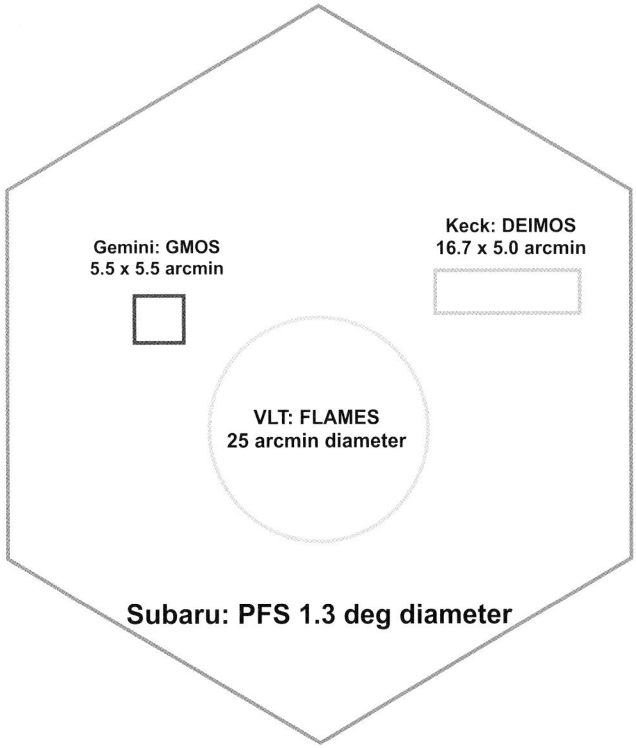

Figure 1. Schematic comparison of the field of view of Subaru PFS with that of current spectroscopy instruments such as VLT/FLAMES, Gemini/GMOS, and Keck/DEIMOS.

Recently, new member stars of Ursa Minor (UMi), which is classified as classical dSph, are discovered by Sestito et al. (2023). Using Gaia proper motion and GRACES stellar spectra, they identified five bright member stars. Surprisingly, these stars are far away from the center of Ursa Minor. In particular, one of them is located at 11 times half-light radius of this galaxy, which corresponds to 4 kpc as physical scale. Adding these stars to the current available kinematic data ($N_{\rm star} = 313$), we re-performed the axisymmetric Jeans analysis constructed by Hayashi et al. (2020).

Here, we describe briefly our mass models as follow. For surface stellar density distribution of a dSph, we assumed oblate Plummer profile generalized to an axisymmetric shape. For dark matter density profile, we assume a generalized Hernquist profile:

$$\rho_{\rm DM}(R,z) = \rho_0\left(\frac{r}{b_{\rm halo}}\right)^{-\gamma}\left[1+\left(\frac{r}{b_{\rm halo}}\right)^{\alpha}\right]^{-\frac{\beta-\gamma}{\alpha}}, \tag{1}$$

$$r^2 = R^2 + z^2/Q^2, \tag{2}$$

where ρ_0 and $b_{\rm halo}$ are the scale density and radius, respectively; α is the sharpness parameter of the transition from the inner slope γ to the outer slope β; and Q is a constant axial ratio of a dark matter halo. These $(Q, \rho_0, b_{\rm halo}, \alpha, \beta, \gamma)$ are free parameters in our models. Utilizing these density profiles, we solve numerically the axisymmetric Jeans equations (Binney & Tremaine 2008) and calculate the line-of-sight velocity dispersions from these equations, where a stellar velocity anisotropy $\beta_z = 1 - \overline{v_z^2}/\overline{v_R^2}$ (= constant) is taken into account. We employ the axisymmetric mass models and Markov chain Monte

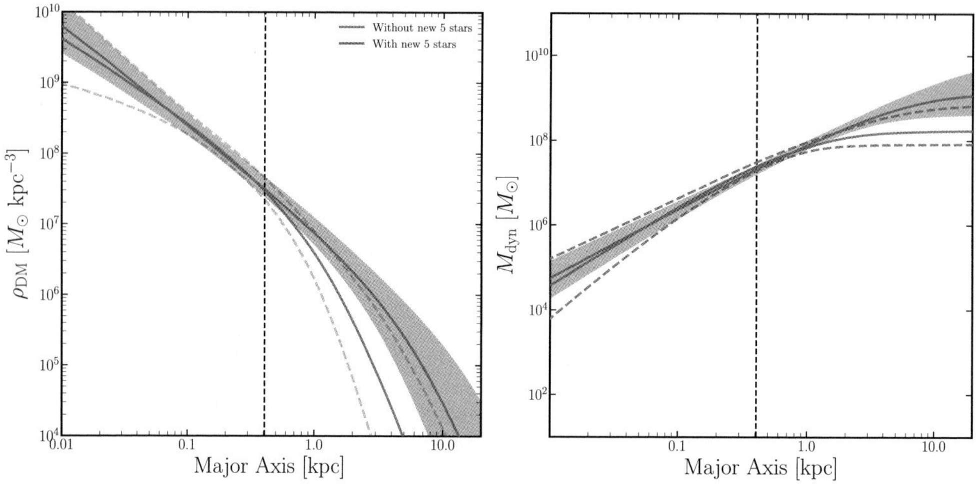

Figure 2. The estimated dark matter density profile (left) and mass profile (right) of Ursa Minor dwarf spheroidal galaxy. The gray solid and dashed lines are median and 68 per cent confidence level from the kinematic sample *without* five new member stars (i.e, $N_{\rm star} = 313$), while the green solid line and shaded region are the same as gray ones but calculated from the kinematic sample *with* five new member stars ($N_{\rm star} = 313 + 5$).

Carlo techniques based on Bayesian statistics to analyze the line-of-sight velocity data of UMi dSph and obtain limits on its dark matter halo parameters.

Figure 2 shows the estimated dark matter density profile (left) and mass profile (right) of Ursa Minor dwarf spheroidal galaxy. The gray lines calculated from the sample without the five new member stars (i.e., these results are the same as the ones from Hayashi et al. (2020) completely), while green ones are estimated from the sample including new member stars. This result suggests that the addition of outermost stars can alter the dark matter density profile, although we should consider the potential influence of tidal forces on these stars. Therefore, to place robust constraints on the dark matter density profiles, it is crucial to collect a significant number of kinematic samples from stars across wide areas of the Galactic dSphs.

4. PFS forecast of dark matter density profile in a mock dSph

To take advantage of PFS's exceptional field-of-view, depth, and spectroscopic multiplexing capabilities, we are performing mock observations and dynamical analyses for the Galactic dSphs. We here focus on Draco dSph as an example. Mock data for dSph's member stars are generated from `AGAMA` public code (Vasiliev 2019), which can treat non-spherical stellar and dark matter components. Using the mock data, we conducted PFS survey simulation that was developed by the PFS science collaboration. Thanks to the wide and deep PFS survey, we obtained approximately 5,000 stellar spectra with only four PFS pointings (see the inset in Figure 3). In comparison to the current available kinematic data of Draco dSph, which consists of 500 stars, PFS will significantly improve the statistical quality of the kinematic sample.

Applying our axisymmetric mass models to the mock PFS data for Draco, we can demonstrate a significant improvement in the estimated dark matter density profile compared to the current available kinematic data. From Figure 3, the increased number of stars observed by PFS allows us to probe the kinematics of the outermost regions of

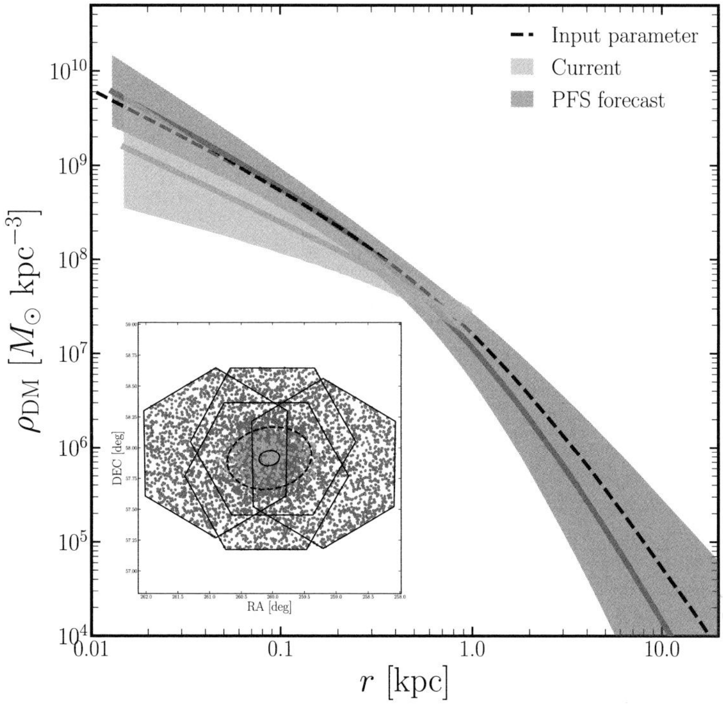

Figure 3. The dark matter density profiles derived by non-spherical mass models, using "Current" ($N_{\rm star} = 500$) and "PFS forecast" ($N_{\rm star} = 5000$ stars) samples shown in the inset (Draco-like mock PFS data, where solid and dashed oval lines are the half-light and nominal tidal radii, respectively). The black dashed line shows the input dark-matter density profile, whereas the colored solid lines with the shaded regions are the median values with the 68 per cent confidence intervals obtained from each sample within the respective radial range of stars.

Draco and thus better constrain the dark matter distribution. This will provide important insights into the nature of dark matter and the formation history of the Milky Way's satellite galaxies.

5. Future plans

It is important to keep in mind that the current mock data shown in Figure 3 do not take into account non-negligible effects such as contamination stars, binary member stars, and dynamical non-equilibrium. These effects may affect the accuracy of the mass models and the resulting dark matter density profiles. Therefore, it is important to generate more realistic mock data and to develop more sophisticated models that can take these effects into account when analyzing the actual PFS data.

Furthermore, in order to more precisely and accurately estimate the dark matter density profiles of the Galactic dwarf satellites, it is essential to use all of the information available in the full line-of-sight velocity distribution (LOSVD). This is because the LOSVD, characterized by higher-order velocity moments, is sensitive to the shape of the velocity ellipsoid, and thus should be a powerful technique for mitigating the degeneracy between dark matter density and velocity anisotropy. Considering the shape of LOSVD can, therefore, place further constraints on the inner density slope of a dark matter halo. Thus, for the PFS survey operation, we are developing dynamical mass models that include the LOSVD information based on Jeans analysis as well as Schwarzschild modeling, and considering the non-negligible effects using statistical approaches such as Bayesian statistics.

Acknowledgements

This work was supported in part by the MEXT Grant-in-Aid for Scientific Research (No. 20H01895, 21K13909, 21H05447 and 23H04009 for K.H.)

References

Battaglia, G. & Nipoti, C. 2022, *Nature Astronomy*, 6, 659.
Battaglia, G., Helmi, A., & Breddels, M. 2013, *NewAR*, 57, 52.
Bullock, J. S. & Boylan-Kolchin, M. 2017, *ARA&A*, 55, 343.
Read, J. I., Walker, M. G., & Steger, P. 2019, *MNRAS*, 484, 1401.
Hayashi, K., Chiba, M., & Ishiyama, T. 2020, *ApJ*, 904, 45.
Lazar, A., Bullock, J. S., Boylan-Kolchin, M., et al. 2020, *MNRAS*, 497, 2393.
Nishikawa, H., Boddy, K. K., & Kaplinghat, M. 2020, *PRD*, 101, 063009.
Tamura, N., Takato, N., Shimono, A., et al. 2016, procspie, 9908, 99081M.
Takada, M., Ellis, R. S., Chiba, M., et al. 2014, *PASJ*, 66, R1.
Sestito, F., Zaremba, D., Venn, K. A., et al. 2023, arXiv:2301.13214.
Binney, J. & Tremaine, S. 2008, Galactic Dynamics: Second Edition (Princeton, NJ: Princeton Univ. Press)
Vasiliev, E. 2019, *MNRAS*, 482, 1525.

Dynamical Masses of Local Group Galaxies
Proceedings IAU Symposium No. 379, 2025
P. Bonifacio, M.-R. Cioni, F. Hammer, M. S. Pawlowski & S. Taibi, eds.
doi:10.1017/S1743921323002454

Gaia EDR3 proper motions, energies, angular momenta of Milky Way dwarf galaxies: a recent infall to the Milky Way halo

Y. Yang[1], F. Hammer[1], H. Li[2], M. S. Pawlowski[3], J. L. Wang[4], C. Babusiaux[5], G. A. Mamon[6], P. Bonifacio[1], Y. Jiao[1] and H. Wang[7]

[1]Observatoire de Paris, Paris Sciences et Lettres, CNRS France
email: yanbin.yang@obspm.fr

[2]School of Physical Sciences, University of Chinese Academy of Sciences, Beijing, China

[3]Leibniz-Institut fuer Astrophysik Potsdam, Germany

[4]CAS Key Laboratory of Optical Astronomy, National Astronomical Observatories, Beijing 100101, China

[5]Université de Grenoble-Alpes, CNRS, IPAG, F-38000 Grenoble, France

[6]Institut d'Astrophysique de Paris, CNRS, France

[7]Centro Ricerche Enrico Fermi, Roma, Italy

Abstract. Gaia EDR3 has provided proper motions of Milky Way (MW) dwarf galaxies with an unprecedented accuracy, which allows us to investigate their orbital properties. We found that the total energy and angular momentum of MW dwarf galaxies are much larger than that of MW K-giant stars, Sagittarius stream stars and globular clusters. It suggests that many MW dwarf galaxies have had a recent infall into the MW halo. We confirmed that MW dwarf galaxies lie near their pericenters, which suggests that they do not behave like satellite systems derived from Lambda-Cold-Dark-Matter cosmological simulations. These new results require revisiting the origin of MW dwarf galaxies, e.g., if they came recently, they were likely to have experienced gas removal due to the ram pressure induced by MW's hot gas, and to be affected by MW tides. We will discuss the consequences of these processes on their mass estimation.

Keywords. Milky Way Galaxy, Globular clusters, Dwarf spheroidal galaxies, Dwarf irregular galaxies, Circumgalactic medium

1. Introduction

Within a distance of 260 kpc around the Milky Way (MW), there are 46 dwarf spheroidal galaxies (dSphs), including the 11 so-called classical dSphs: Leo I, Leo II, Fornax, Sculptor, Sextans, Carina, Ursa Minor (UMi), Sagittarius (Sgr), Draco, and ultra-faint dwarf galaxies (UFDs), those with extremely low values of central surface brightness. UFDs possess the least stellar mass, which could be only a few hundred solar masses, such as Segue1 (McConnachie 2012). It is worth noting, however, that three UFDs are eventually as massive as the classical dSphs, which are Canes Venatici (Martin et al. 2008), Antlia 2 and Crater 2 (Torrealba et al. 2016; Ji et al. 2021) despite their ultra-faint central surface brightness.

A large effort has been made to search for the signatures of tides from the MW in the classical dSphs, however, apart from Sgr, no firm evidence was found (Walker 2013). Therefore, the dSphs were thought to be long-lived satellites of the MW, with their first infall around 8-10 Gyr ago, as speculated based on their relatively regular morphologies and predominantly old stellar populations. These characteristics are primarily observed in classical dSphs, and largely shared by UFDs (Simon 2019).

The MW dwarf galaxy system shows several peculiar properties that contradict to the prediction of Lambda-Cold-Dark-Matter (LCDM) simulations (Bullock & Boylan-Kolchin 2017). For example, the missing satellite problem, i.e., the total number of observed dwarf galaxies in the MW is so small, by a factor of at least 200, compared to that predicted by LCDM simulations. To resolve this issue, one has to assume that most of low mass sub-halos are unable to activate their star formation, and thus remain fully dark today.

One of the well-known peculiar properties of the MW dwarf galaxies is the Vast POlar Structure (VPOS, Pawlowski et al. 2012) which includes not only the classical dwarf galaxies (previously known as the Disk of Satellites, see Kroupa et al. 2005; Lynden-Bell 1976), but also UFDs. This is a thin structure, almost perpendicular to the MW disk, where galaxies show coherent motions and share similar orbital angular momentum directions.

Based on Gaia Data Release 2 (DR2), Fritz et al. (2018) found that not all, but only 60% of the dwarf galaxies could be in VPOS. With much more accurate proper motions from Gaia Early DR3 (EDR3) and an enlarged sample of dSphs, we confirmed their result in Li et al. (2021). Surprisingly, we further discovered a new large-scale structure perpendicular to both the VPOS and the MW disk (Hammer et al. 2021). It consists of 20% (i,e., 10) of dwarf galaxies orbiting or counter-orbiting with the Sgr. This result further complicates our understanding of the MW's dwarf galaxy system, as up to 80% of the galaxies are preferentially distributed in planes.

VPOS challenges the LCDM model of cosmology because its kinematics-space distribution makes the MW an outlier in LCDM cosmology. Many efforts have been made to explain the special characteristics of the MW's dwarf galaxies within the LCDM frame. Despite various proposed remedies, the MW dwarf galaxy system stands out as atypical when compared to LCDM simulations.

Cosmologists have acknowledged the exceptional situation of the MW dwarf galaxy system when compared to LCDM simulations. Based on Gaia DR2 proper motions, Cautun & Frenk (2017) found that MW dwarf galaxies show excesses of tangential velocity on their orbits, making the MW exceptional (only 1.5%) among LCDM satellites systems. Gaia EDR3 proper motions confirmed this conclusion (Hammer et al. 2021).

Why do MW dwarf galaxies exhibit such peculiar spatial and velocity distributions? Apart from challenging the LCDM cosmology, could we trace their orbital properties and learn about their origins? Ultimately, the dynamics, particularly the orbital history of dwarf galaxies, is the primary reason that drives the assembly of the dwarf galaxies into the MW, as well as their formation and transformation.

2. Orbital properties of MW dwarf galaxies revealed by Gaia EDR3

Gaia EDR3 (Gaia Collaboration et al. 2016; Lindegren et al. 2021; Riello et al. 2021; Gaia Collaboration et al. 2021) provides a significant improvement in the precision of astrometry measurements by a factor ≈ 2 for individual sources. This helps to reduce the errors in proper motion determination of the MW dwarf galaxies, especially for UFDs because their proper motions are mostly dominated by statistical errors. In Li et al. (2021), we measured the proper motions using Gaia EDR3 for a larger sample of MW dwarf galaxies (46 vs. 39 in Fritz et al. 2018, based on Gaia DR2). After taking into account distance and radial velocity measurementsfrom the literature, we established a

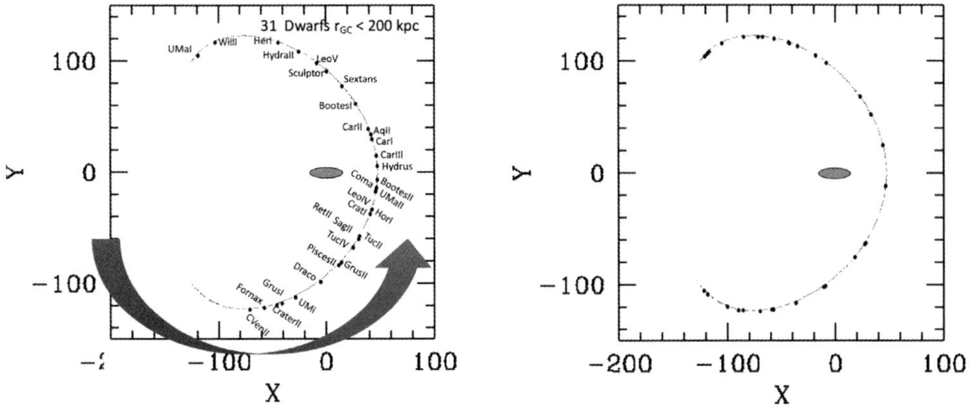

Figure 1. *Left*: Orbital phase distribution of 31 dwarf galaxies currently within 200 kpc around MW. The orbital phase of each dwarf galaxy has been normalized to the median orbit of this subsample, i.e., the orbit with an eccentricity of 0.55, assuming a MW total mass of $8.1 \times 10^{11} M_{\odot}$ (see text for more details). *Right*: Orbital phase distribution assuming a random distribution in time, on the same orbit.

complete six-dimensional (6D) dynamical data (3D in space and 3D in velocity) of each dwarf galaxy in the reference of the MW. This data allows us to study all dynamical aspects of the dSphs' orbits.

2.1. *A strong tendency to be near pericenter*

Based on Gaia EDR3 proper motion measurements, it was noticed that MW dwarf galaxies show excesses of tangential velocity on their orbits, which suggests that MW dwarf galaxies may be preferentially approaching their pericenters or apocenters. To investigate further, we calculated their orbital properties, such as orbital angular momentum, eccentricities, pericenters or apocenters (Li et al. 2021). Given the large uncertainties in determining the MW total mass (Jiao et al. 2021), we performed the calculations by assuming different MW mass profiles, ranging from 2.8 to 15 $\times 10^{11} M_{\odot}$, that fit the MW's rotation curve from Gaia DR2 (Eilers et al. 2019). It is not surprising that orbital parameters change with the MW mass, because a higher MW mass gives more circularized solutions, while a lower MW mass gives more eccentric solutions.

Although we cannot draw conclusions on the exact values of orbital properties, we noticed that MW dwarf galaxies are, overall, on quite eccentric orbits. For instance, for a MW mass of $8.1 \times 10^{11} M_{\odot}$ (Li et al. 2021), the median eccentricity is 0.55. What is more puzzling is that the galaxies tend to be near to their pericenters, which holds true regardless of the choice of MW masses. With careful analysis, the tendency was found to be significant with a probability of 0.001 for a MW mass of $8.1 \times 10^{11} M_{\odot}$, when compared to a null hypothesis, i.e., MW dwarf galaxies are randomly distributed in time on their orbits (Li et al. 2021). This contradicts what is expected for a virialized satellite system, in which satellite galaxies have to be distributed closer to their apocenters, in accordance with Kepler's law (or its generalization for extended mass profile). On an elliptical orbit, an object spends a relatively short amount of time near the pericenter. This is illustrated in Figure 1. Why do MW dwarf galaxies show such a tendency? Could they be coordinated in time to arrive at the MW?

2.2. *High angular momenta and total energies of MW dwarf galaxies*

If MW dwarf galaxies are ancient inhabitants as it has been often assumed, we should be able to find such evidence by comparing them to the otherMW inhabitants, such as

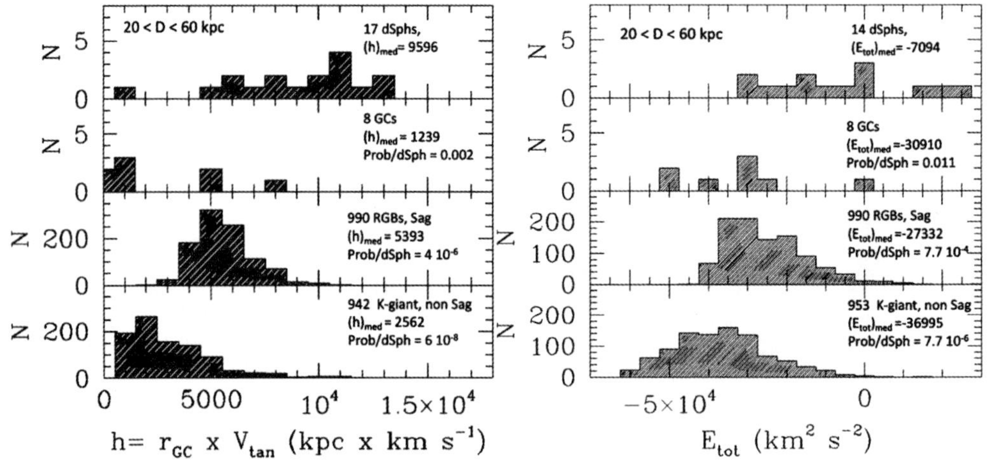

Figure 2. *Left*: Comparison of orbital angular momentum (h) distribution between (from top to bottom) MW dwarf galaxies, globular clusters, Sgr Stream stars, and K-giant stars. In each panel, we have indicated the number of objects for each species, their median h and the probability that the distribution is consistent with that of the dwarf galaxies. All samples are limited to a range of 20 to 60 kpc to ensure completeness. *Right*: The same comparison is made as in *Left*:, but for total orbital energy, $E_{\rm tot}$. Two dwarf galaxies are excluded from the $E_{\rm tot}$ comparison due to their large error bar.

globular clusters or halo stars. With fast growing observations, especially from Gaia, we compiled catalogues of complete 6D dynamical data for large samples of different tracers located at least 20 kpc from the MW center, such as K-gaint halo stars, globular clusters, and stars in the Sgr stream (see Hammer et al. 2021, for more details). Then, we compared the angular momenta and total energies of MW dwarf galaxies with those of the different tracers. In Figure 2 we present one of such comparisons, taking into account the completeness in volume among the different samples. It shows that the MW dwarf galaxies possess systematically high values in both angular momentum and total orbital energy, compared to other MW halo inhabitants. Both angular momentum and total energy are fundamental dynamical quantities, and are expected to be conserved over a long period of time if the system evolved adiabatically. Even with possible dissipation of energy in the system, one may expect to see ordered evolutions in both angular momentum and total energy. Hence, the high angular momenta and total energies of MW dwarf galaxies indeed indicate that they have different dynamical properties than other MW halo populations.

2.3. *Relation between Infall time and total energy*

To further investigate the dynamical properties of MW dwarf galaxies, Figure 3 presents the relation between the total angular momentum (h) versus the total orbital energy ($E_{\rm tot}$), assuming a MW mass of $8.1 \times 10^{11} M_{\odot}$. For comparison, we plot 156 globular clusters, stellar streams, and other known substructures of the MW (see Hammer et al. 2023, for more details). For a better visualisation, we separate the sample into those with eccentricities smaller (upper panel) and larger (lower panel) than 0.6. Figure 3 suggests an impressively tight relation between angular momentum and total energy, particularly for cases with eccentricities below 0.6.

Kruijssen et al. (2020) and Malhan et al. (2022) identified several substructures in the Galactic halo according to their dynamical properties, and derived their accretion

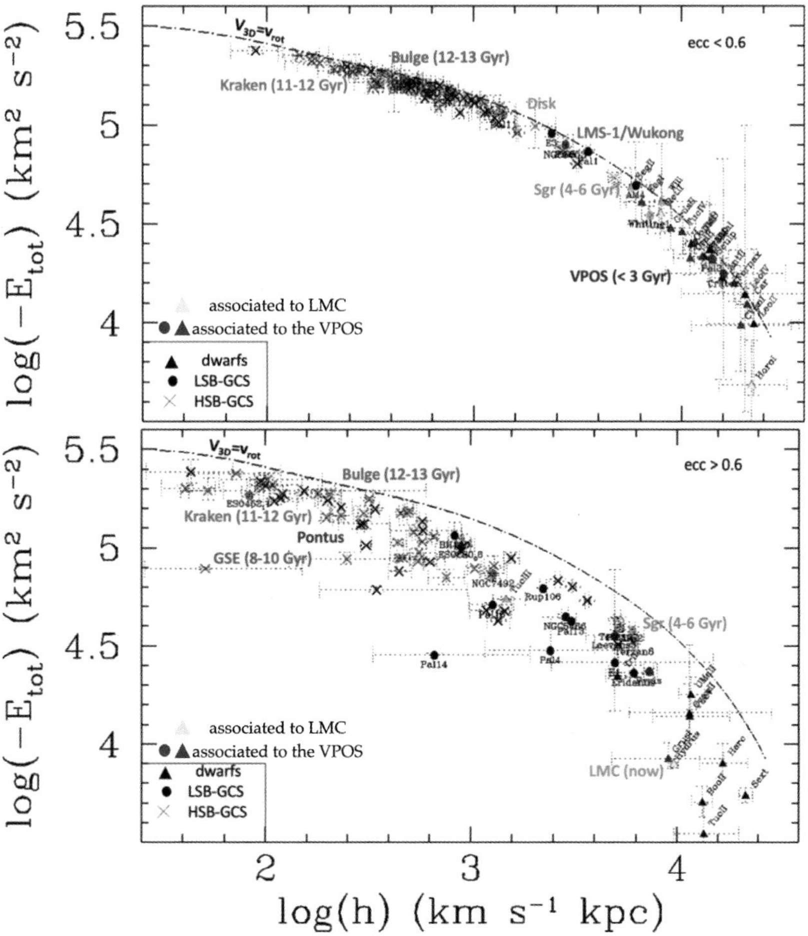

Figure 3. The diagram of angular momentum (h) versus total energy for various samples, including high-surface brightness globular clusters (HSB-GCs) represented by crosses, low-surface brightness globular clusters (LSB-GCs) represented by filled circles, and dwarf galaxies represented by triangles. The top and bottom panels show the objects with orbital eccentricities smaller than 0.6 (including 79 globular clusters and 20 dwarf galaxies) and larger than 0.6 (including 77 globular clusters and 15 dwarf galaxies), respectively. The dotted-dashed line represents the limit of the lowest energy (corresponding to a circular orbit) as a function of angular momentum. The identified structures, as well as their first infall epochs, as determined by (Malhan et al. 2022) and (Kruijssen et al. 2020), are indicated by different colors in the figure. VPOS dwarf galaxies and globular clusters are also shown in blue, while the few dwarf galaxies associated with the LMC are shown with cyan. Note that few dwarf galaxies with positive energy are not shown, including Leo I in the VPOS, Carina II and III and Phoenix II, all related to the LMC, and Hydra II and Leo V.

epochs, such as Kraken, Pontus, Gaia Sausage Enceladus (GSE), LMS1-Wukong, and Sgr. Using their results, we labeled each of these identified substructures associated with globular clusters in Figure 3, along with their names and infall epochs. Interestingly, there appears to be a good correlation between the infall epoch and the total energy of substructures. The Large Magellanic Cloud (LMC) is widely accepted at the pericenter of its first infall to the MW (Besla et al. 2007), so it is labeled "now". From these results, we may infer that MW dwarf galaxies are likely latecomers, appearing later than the Sgr

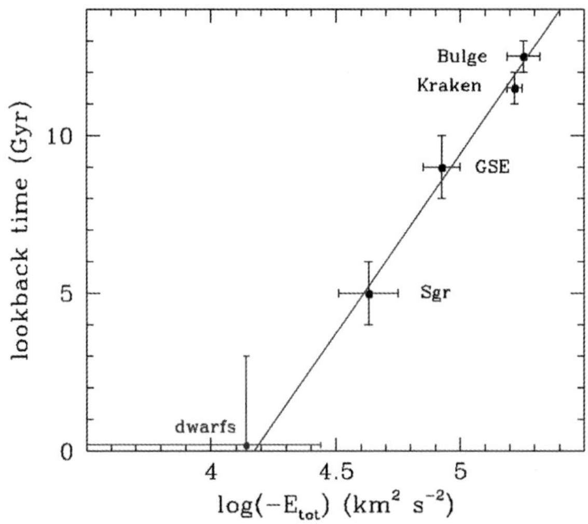

Figure 4. A relationship established from the analysis of Figure 3 shows that the in-falling history of the MW's components can be accurately traced by their total orbital energy.

dSph whose first infall may have occurred about 4 Gyr ago (e.g., Wang et al. 2022). For a detailed analysis, we refer readers to Hammer et al. (2023).

The relationship between infall epoch and total energy can be represented as in Figure 4. The point labeled "dwarfs" is derived from 25 dwarf galaxies that do not belong to the low-energetic Sgr system (excluding also Sgr, Segue II, Tucana III and Willman I) or to the high energetic LMC system (Carina II, Carina III, Horologium I, Hydrus I, Phoenix II, and Reticulum II). The blue line in the figure is a linear fit to the data. A simple interpretation is that dwarf satellites with $\log(-E_{\rm tot}/[{\rm km^2 s^{-2}}]) < 4.17$ are in their initial approach phase, with a value very close to the logarithm of the average energy (4.14) of 25 dwarf galaxies, whose scatter provides an upper limit of $-E_{\rm tot} = 4.34$. The latter combined with the linear fit suggests a lookback time of halo entry smaller than 3 Gyr. This possibly explains why MW dwarf galaxies show a strong tendency to be near to their pericenters.

The relation in Figure 4 describes that the MW was likely to be formed by accretion and structured similarly to the onion skin model initiated by Gott (1975) for explaining the outer density profile of elliptical galaxies. Together with the growth of the MW and possible energy dissipation mechanisms, such as dynamical frictions and tidal destruction of substructures, early infall components stay at lower energy orbits while later infall objects stay at higher energy orbits. The dynamical formation history in Figure 4 is consistent with the fact that the MW has not experienced major mergers since 10 Gyr (Hammer et al. 2007; Haywood et al. 2018; Belokurov et al. 2018), and also in excellent agreement with cosmological simulation results Rocha et al. (2012) and Boylan-Kolchin et al. (2013).

3. Possible consequences of the recent infall of MW dwarf galaxies

Based on Gaia EDR3 data, we possess complete 6D dynamical data for MW dwarf galaxies. Our comprehensive analysis of their orbital properties suggests that most of them are likely to be the latecomers arrived in the MW within the last 3 Gyr. This finding completely changes the framework for understanding their origins because they were formerly assumed to be long-live inhabitants of the MW since 8–10 Gyr ago.

Dwarf galaxies in the Local Group show the well known dichotomy that dwarf galaxies within about 250 kpc from the MW or M31 are typically devoid of gas, with only a few exceptional cases (which are relatively very massive objects) like the LMC, SMC, and IC10 (e.g., Putman et al. 2021). This phenomenon has been taken as one of the strong evidences that the diffuse gas in the Galactic halo has played a crucial role in the formation of dSphs, which are thought to have been transformed from dwarf irregulars (dIrrs) through ram-pressure stripping induced by the Galactic halo diffuse gas. This mechanism has been studied intensively and proved to be effective by Mayer et al. (2006). However, according to our analysis of orbits, the time required to transform a dIrr into a dSph is within about 3 Gyr rather than the 10 Gyr as assumed by Mayer et al. (2006). This suggests that a rapid and voilent ram-pressure stripping during the transformations from dIrrs to dSphs may have occurred, which may have had a substantial impact on their star formation, chemical evolution, and dynamical evolution.

One possible imprint of such violent transformations from dIrrs $\rightarrow$ dSphs may be breaks in the density profiles of dSphs that have been noticed since a long time (e.g., Westfall et al. 2006). At this epoch these breaks are detected only in classical dSphs. In Fornax, an extended stellar component was discovered recently by Yang et al. (2022), which confirms the fact that all classical dSphs, except Leo II, show a break in their density profiles. This indicates that a second and extended stellar component is common in the classical dSphs. Sgr is not taken into account here, as it has a different infall history compared to the other dSphs, see our analysis in Sect. 2.3. In fact, an extremely large and faint stellar component has been identified recently using spectroscopy not only in the classcial dSph UMi (Sestito et al. 2023), but also in some of the UFDs: Ursa Major I, Coma Berenices, Bootes I (Waller et al. 2023), Tucana II (Chiti et al. 2023), and Grus I Cantu et al. (2021). These extended stellar components are found to be symmetrically surrounding the dSphs, indicative of their non-tidal origins. Could they possibly be the relics of the violent transformation from dIrrs $\rightarrow$ dSphs? For example, the removal of the last remaining gas may have caused a sudden lack of gravity, resulting in an expansion of the progenitor's core, and finally the formation of an extended stellar component. Such a possibility of destabilisation has been proposed by Yang et al. (2014), see also (Hammer et al. 2023). In this context, the internal dynamics of the dSphs could also be affected, prompting us to reassess the estimates of their dynamical masses.

In brief, it is likely that most of the MW dwarf galaxies are latecomers that were accreted by the MW during the past 3 Gyr, as evidenced by their orbital properties revealed from the Gaia EDR3 data. This provides us with a new frame to understand how dIrrs have been transformed into dSphs in such a short period of 3 Gyr. For most of the MW dSphs, such a transformation just happened during their first infall. We expect that future efforts will be made towards a precise determination of the star formation history, chemical evolution, structure and internal dynamics of dSphs, accounting in particular for a recent ram pressure stripping event associated to their revised orbital histories.

References

Belokurov, V., Erkal, D., Evans, N. W., et al. 2018, MNRAS, 478, 611
Besla, G., Kallivayalil, N., Hernquist, L., et al. 2007, ApJ, 668, 949
Boylan-Kolchin, M., Bullock, J. S., Sohn, S. T., et al. 2013, ApJ, 768, 140
Bullock, J. S. & Boylan-Kolchin, M. 2017, ARA&A, 55, 343
Cantu, S. A., Pace, A. B., Marshall, J., et al. 2021, ApJ, 916, 81
Cautun, M. & Frenk, C. S. 2017, MNRAS, 468, L41
Chiti, A., Frebel, A., Ji, A. P., et al. 2023, AJ, 165, 55
Eilers, A.-C., Hogg, D. W., Rix, H.-W., et al. 2019, ApJ, 871, 120

Fritz, T. K., Battaglia, G., Pawlowski, M. S., et al. 2018, A&A, 619, A103
Gaia Collaboration, Prusti, T., de Bruijne, J. H. J., et al. 2016, A&A, 595, A1
Gaia Collaboration, Brown, A. G. A., Vallenari, A., et al. 2021, A&A, 649, A1
Gott, J. R. 1975, ApJ, 201, 296
Ji, A. P., Koposov, S. E., Li, T. S., et al. 2021, ApJ, 921, 32
Jiao, Y., Hammer, F., Wang, J. L., et al. 2021, A&A, 654, A25
Hammer, F., Puech, M., Chemin, L., et al. 2007, ApJ, 662, 322
Hammer, F., Wang, J., Pawlowski, M. S., et al. 2021, ApJ, 922, 93
Hammer, F., Li, H., Mamon, G. A., et al. 2023, MNRAS, 519, 5059
Haywood M., Di Matteo P., Lehnert M. D., Snaith O., Khoperskov S., Gómez A., 2018, ApJ, 863, 113
Li, H., Hammer, F., Babusiaux, C., et al. 2021, ApJ, 916, 8
Lindegren, L., Klioner, S. A., Hernández, J., et al. 2021, A&A, 649, A2
Lynden-Bell, D. 1976, MNRAS, 174, 695
Mayer, L., Mastropietro, C., Wadsley, J., et al. 2006, MNRAS, 369, 1021
McConnachie, A. W. 2012, AJ, 144, 4
Malhan, K., Ibata, R. A., Sharma, S., et al. 2022, ApJ, 926, 107
Martin, N. F., de Jong, J. T. A., & Rix, H.-W. 2008, ApJ, 684, 1075
Kruijssen, J. M. D., Pfeffer, J. L., Chevance, M., et al. 2020, MNRAS, 498, 2472
Kroupa, P., Theis, C., & Boily, C. M. 2005, A&A, 431, 517
Pawlowski, M. S., Pflamm-Altenburg, J., & Kroupa, P. 2012, MNRAS, 423, 1109
Putman, M. E., Zheng, Y., Price-Whelan, A. M., et al. 2021, ApJ, 913, 53
Riello, M., De Angeli, F., Evans, D. W., et al. 2021, A&A, 649, A3
Riley, A. H., Fattahi, A., Pace, A. B., et al. 2019, MNRAS, 486, 2679
Rocha, M., Peter, A. H. G., & Bullock, J. 2012, MNRAS, 425, 231
Sestito, F., Zaremba, D., Venn, K. A., et al. 2023, arXiv:2301.13214
Simon, J. D. 2019, ARA&A, 57, 375
Torrealba, G., Koposov, S. E., Belokurov, V., et al. 2016, MNRAS, 459, 2370
Waller, F., Venn, K. A., Sestito, F., et al. 2023, MNRAS, 519, 1349
Walker, M. 2013, Planets, Stars and Stellar Systems. Volume 5: Galactic Structure and Stellar Populations, 1039
Wang, H.-F., Hammer, F., Yang, Y.-B., et al. 2022, ApJ, 940, L3
Westfall, K. B., Majewski, S. R., Ostheimer, J. C., et al. 2006, AJ, 131, 375
Yang, Y., Hammer, F., Fouquet, S., et al. 2014, MNRAS, 442, 2419
Yang, Y., Hammer, F., Jiao, Y., et al. 2022, MNRAS, 512, 4171

Discussion

Ekta Patel: Have you computed the number of satellites expected to be at apocenter ? We have cosmological expectations for the total expected MW satellite population but these are certainly spread between pericenter and apocenter with a bias towards apocenter (at least from basic Keplerian arguments). One could do this for a range of MW masses.

Yanbin Yang: Thanks! We did such calculations and comparisons. As shown in Figure 1, we tested the significance of the observed tendency to the results from a null hypothesis based on the Keplerian arguments, just as you have described for cosmological expectations (see text Sect. 2.1).

Dynamical Masses of Local Group Galaxies
Proceedings IAU Symposium No. 379, 2025
P. Bonifacio, M.-R. Cioni, F. Hammer, M. S. Pawlowski & S. Taibi, eds.
doi:10.1017/S1743921323002843

Do old globular clusters in low-mass galaxies disprove modified gravity?

M. Bílek[1,2,3], **H. Zhao**[4], **B. Famaey**[3], **S. T. Nagesh**[3], **F. Combes**[1,2], **O. Müller**[5], **M. Hilker**[6], **P. Kroupa**[7,8] and **R. Ibata**[3]

[1]LERMA, Observatoire de Paris, CNRS, PSL Univ., Sorbonne Univ., 75014 Paris, France
email: michal.bilek@asu.cas.cz

[2]Collège de France, 11 place Marcelin Berthelot, 75005 Paris, France

[3]Université de Strasbourg, CNRS, Observatoire astronomique de Strasbourg, UMR 7550, F-67000 Strasbourg, France

[4]Scottish Universities Physics Alliance, University of St Andrews, North Haugh, St Andrews, Fife KY16 9SS, UK

[5]Institute of Physics, Laboratory of Astrophysics, Ecole Polytechnique Fèdèrale de Lausanne (EPFL), 1290 Sauverny, Switzerland

[6]European Southern Observatory, Karl-Schwarzschild-Strasse 2, 85748 Garching bei München, Germany

[7]Helmholtz-Institut für Strahlen- und Kernphysik (HISKP), Universität Bonn, Nußallee 14-16, D-53115 Bonn, Germany

[8]Astronomical Institute, Faculty of Mathematics and Physics, Charles University in Prague, V Holešovičkách 2, CZ-18000 Praha, Czech Republic

Abstract. The controversy "dark matter vs. modified gravity" constitutes a major topic of discussion. It was proposed that dynamical friction could be used to discriminate between the two alternatives. Analytic calculations indicate that, with modified gravity, globular clusters (GCs) of low-mass galaxies experience much stronger dynamical friction than in the equivalent system with Newtonian gravity and dark matter. As a result, in modified gravity the old GCs of low mass galaxies should have already settled in the centers of the galaxies. This is not observed. Here we report on our efforts to verify the analytic results by self-consistent simulations with the MOND-type (modified Newtonian dynamics) gravity. The core stalling mechanism, that was not considered in the analytic calculations, prevents GCs to settle in centers of ultra-diffuse galaxies. For isolated dwarf galaxies, which are gas-rich objects, supernova explosions prevent the GCs from settling.

Keywords. Dwarf galaxies, globular clusters, dynamical friction, modified gravity

1. Introduction

The nature of gravity and inertia is not clear yet. A generally accepted theory of quantum gravity is missing. There are indications that our imperfect knowledge of the laws or gravity or inertia is the reason why we encounter the missing mass problem. The strongest indication for this is the fact that dynamical properties of most galaxies can be predicted from the distribution of their visible material (McGaugh et al. 2016; Lelli et al. 2017; Bílek et al. 2019; Milgrom 2019) using the prescriptions of modified Newtonian dynamics (MOND, Milgrom 1983), aparadigm of modified gravity and/or inertia. This indicates that if indeed gravity or inertia need to be updated, then the correct theory has to follow MOND in certain regimes. Nevertheless, alternatives to such theories as the explanation of the missing mass problem exist, such as the hypothesis of dark matter.

One can test these hypotheses using dynamical friction (Kroupa 2015). It is a force acting against the direction of motion of two interacting N-body systems. It happens because the relative orbital energy of the systems is transferred into the internal energy of their constituents, that is stars, gas, or dark matter particles.

We are still rather at the beginning of the investigation of dynamical friction in MOND. Its exact functioning depends to an unknown degree on the particular MOND theory. We thus consider hereafter only the MOND modified gravity theories of Bekenstein & Milgrom (1984) and Milgrom (2010). Simulations have shown that once we deal with interactions of two objects with similar masses (such as a major merger of galaxies), dynamical friction turns out to be weaker in MOND than in an equivalent Newtonian system with collisionless dark matter (Ciotti & Binney 2004). On the other hand, analytic calculations indicate that once the two interacting objects have a large mass ratio, dynamical friction is stronger in the MOND system than in the equivalent Newtonian system. The enhancement of dynamical friction is generally larger for objects with a lower surface brightness. According to analytic calculations, the strong dynamical friction makes globular clusters (GCs) of low-mass galaxies spiral down the center of their host on the timescale of a few Gyr. This opposes the observed ages of the GCs in such galaxies of many Gyr. We decided to verify the analytic results by high-resolution simulations.

2. Case of isolated ultra-diffuse galaxies

Ultra-diffuse galaxies (UDGs) are characterized by masses of dwarf galaxies and sizes of giant galaxies. It was found that not only they contain old GCs, but these GCs can be exceptionally massive, exceeding $10^6\,M_\odot$. This should make the settling of the GCs in the center of their host galaxy (i.e. the process known as “sinking”) even faster. We simulated the evolution of such objects in MOND in Bílek et al. (2021).

We used the MOND version of the adaptive-mesh-refinement code RAMSES (Teyssier 2002; Lüghausen et al. 2015). The galaxy was simulated having an effective radius of 2 kpc and stellar mass of $2 \times 10^8\,M_\odot$, i.e. resembling the UDG NGC 1052-DF2 that hosts very massive GCs. The galaxy was modeled as a spherical N-body isotropic system, the GCs as single massive particles. The maximum spatial resolution was 50 pc and the particles of the galaxy had $20\,M_\odot$ each. The galaxy was simulated as isolated.

We first let a GC having $10^6\,M_\odot$ fall on the UDG from 5 kpc with a zero relative velocity. The GC first experienced strong damping of its oscillation by dynamical friction. Nevertheless, the orbital decay of the GC suddenly slowed down once the GC moved within the inner 1 kpc of the galaxy. It is unlikely that the change was caused by an insufficient resolution: the friction became less efficient when the size of the orbit was about ten times larger than the resolution and the behavior did not change after increasing the resolution. The phenomenon is known from simulations with Newtonian gravity too under the name of core stalling. It happens in galaxies having nearly harmonic gravitational potentials ($\phi(r) \propto r^2$) near their centers. Dynamical friction becomes ineffective because stars have equal orbital periods in such a potential. This behavior is not captured by the analytic calculations that predicted the fast sinking of GCs of low-mass galaxies in MOND. This is because of the simplifying assumptions done during their derivation.

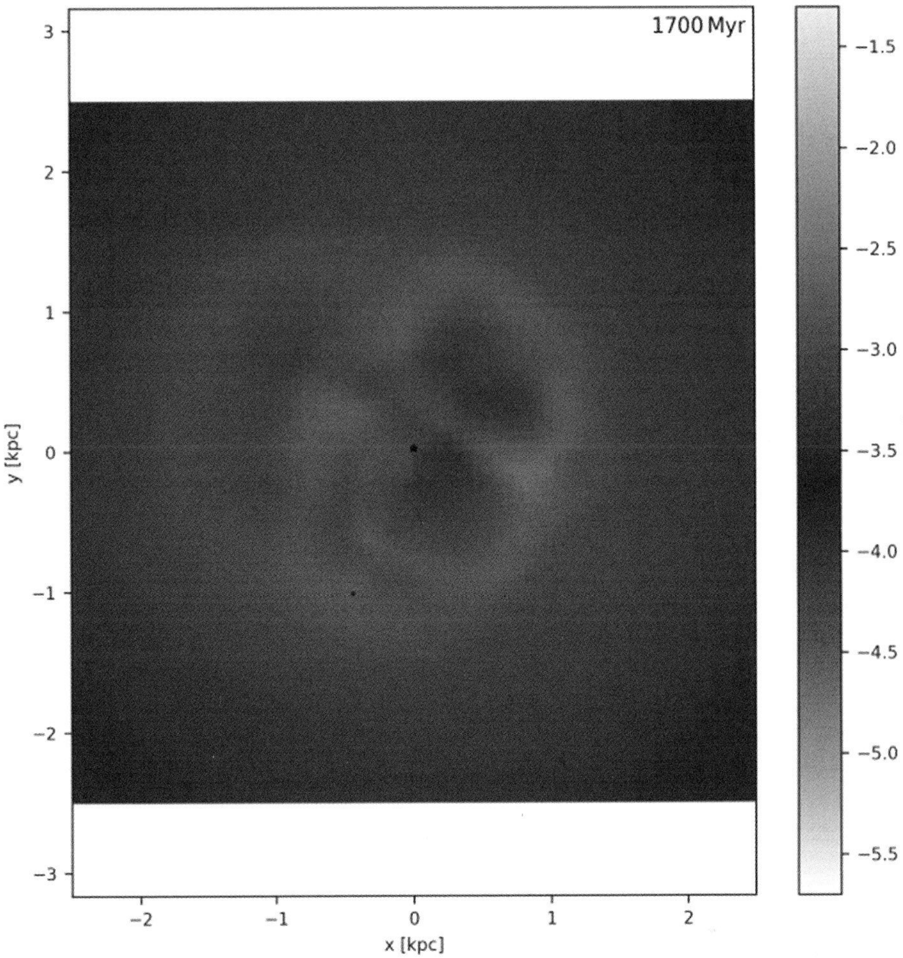

Figure 1. Example snapshot from a simulation with supernovae. The star indicates the barycenter of the galaxy, the dot indicates the GC.

With Newtonian gravity, the Chandrasekhar formula is usually used for evaluating the strength of dynamical friction. Its analog for MOND has been proposed by Sánchez-Salcedo et al. (2006). This formula indicated the problem of fast sinking of GCs of low-mass galaxies in MOND. We were tested the validity of this formula. We repeated the simulation of the free-fall of a GC varying the mass of the GC. Then we explored the motion of a $10^6\,M_\odot$ GC for different orbital shapes. It turned out that the Sánchez-Salcedo formula works well, unless the GC moves inside the inner 1 kpc of the galaxy, where the core stalling occurs. This is the same result as for the Chandrasekhar formula in Newtonian gravity. We found that the value of Coulomb logarithm in the Sánchez-Salcedo formula varies qualitatively with the orbital shape of the GCs as in the Chandrasekhar formula (Chan et al. 1997).

We then simulated an evolution of a system of ten GCs. The GCs were distributed in the simulated galaxy in a similar fashion as observed in the DF2 galaxy. We considered two types of GC mass functions – the standard one, observe for most galaxies, and the mass function seen in the galaxy DF2 that is biased toward high GC masses. The system was evolved for a Hubble time. It turned out that the outer GCs are little affected by

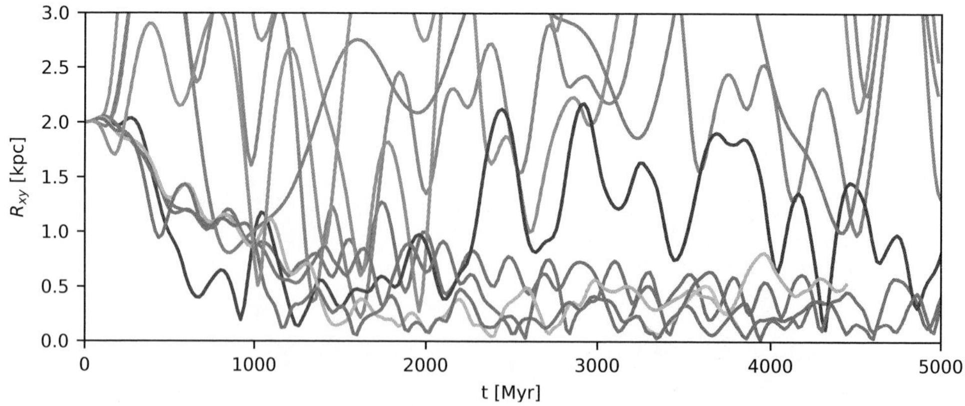

Figure 2. Evolution of the distance between the GC and galaxy barycenter. The different lines indicate the results for the different prescriptions for star formation and supernova feedback, such as the star formation timescale, efficiency or supernova kinetic energy fraction.

dynamical friction. Only the inner ones approached toward the galaxy center, but a full sinking was disabled by core stalling. In the result, the GC system became more centrally concentrated. This evolution was stronger for the DF2-like GC mass function. We predicted the profiles of surface number-density that should be observed in isolated UDGs. It should be noted however, that the DF2-like GC mass function has been reported so far only for the galaxy DF2 itself, for which our simulations do not apply because it is not isolated – it is exposed to a strong external field effect from a neighboring massive elliptical (Famaey, McGaugh, & Milgrom 2018; Haghi et al. 2019). For such a galaxy, the Newtonian simulations of Dutta Chowdhury, van den Bosch, & van Dokkum (2020) are more relevant.

We then utilized symmetries of the equations governing the motion of the GCs to scale our simulation for UDGs of different masses and sizes. It helped us to verify for nearly all observed UDGs that the presence of GCs in them does not exclude MOND. Finally, we found that GCs can evaporate through tidal heating during their encounters.

3. Case of isolated dwarf galaxies

In an ongoing project, we study the survival of GCs of isolated dwarf galaxies. The galaxies are gas-rich rotating objects. Equations indicate the GCs of the least massive dwarfs sink the fastest. We reviewed the literature on observations of GCs of isolated dwarfs and found that the least massive isolated dwarfs that are known to host GC candidates have baryonic masses of about $2 \times 10^7\,M_{\odot}$, and that GCs are common in galaxies over about $10^8\,M_{\odot}$.

We simulated the motion of GCs in a galaxy of the mass of $10^8\,M_{\odot}$. The dwarf was initiated with 90% of gas, representing the likely state 10 Gyr ago, when old GCs are formed. The scale length of both the gas and stellar disks were 2 kpc. The GCs were again simulated as a point masses. The grid was refined if the mass in a cell exceeded $4 \times 10^4\,M_{\odot}$ with the limit of 0.5 pc.

We first explored the motion of a GC of a relatively typical mass of $10^5\,M_{\odot}$ in simulations without star formation and supernova feedback. If the GC initially moves inside the plane of the disk, either on a circular co-rotating obit or a radial orbit, the GC approaches the center of the galaxy in about 1 Gyr. For a counter-rotating, polar or axial orbit, the GC does not sink within 10 Gyr. A GC of $10^4\,M_{\odot}$ does not show substantial sinking on any orbit within 10 Gyr.

We then included star formation and supernova feedback. We explored a wide range of star formation efficiencies and of parameters of the supernova models included in RAMSES. All simulations showed the same: supernovae prevent the GC from sinking regardless of the initial direction of motion of the GC. In MOND, where most of the mass of the dwarf is in gas, supernova explosions lead to large fluctuations of gravitational potential (Fig. 1). The fluctuations of the gravitational field give random kicks to the GC, which prevent it from settling at the galaxy center (Fig. 2). Supernova explosions can even change the sense of rotation of the GC around the center of the galaxy. With Newtonian gravity, this mechanism would be much reduced since dark matter, the dominant component, does not directly react to the supernovae.

References

Bekenstein J., Milgrom M., 1984, ApJ, 286, 7.
Bílek M., Samurović S., Renaud F., 2019, A&A, 625, A32.
Bílek M., Zhao H., Famaey B., Müller O., Kroupa P., Ibata R., 2021, A&A, 653, A170.
Chan R., Mamon G. A., Gerbal D., 1997, ApL&C, 36, 47
Ciotti L., Binney J., 2004, MNRAS, 351, 285.
Dutta Chowdhury D., van den Bosch F. C., van Dokkum P., 2020, ApJ, 903, 149. doi:10.3847/1538-4357/abb947
Famaey B., McGaugh S., Milgrom M., 2018, MNRAS, 480, 473.
Haghi H., Kroupa P., Banik I., Wu X., Zonoozi A. H., Javanmardi B., Ghari A., et al., 2019, MNRAS, 487, 2441.
Kroupa P., 2015, CaJPh, 93, 169.
Lelli F., McGaugh S. S., Schombert J. M., Pawlowski M. S., 2017, ApJ, 836, 152.
Lüghausen F., Famaey B., Kroupa P., 2015, CaJPh, 93, 232.
McGaugh S. S., Lelli F., Schombert J. M., 2016, PhRvL, 117, 201101.
Milgrom, M., 1983, ApJ 270, 365.
Milgrom, M., 2010, MNRAS, 403, 886.
Milgrom, M., 2019, PRD 99, 044041.
Sánchez-Salcedo F. J., Reyes-Iturbide J., Hernandez X., 2006, MNRAS, 370, 1829.
Teyssier R., 2002, A&A, 385, 337.

Dynamical Masses of Local Group Galaxies
Proceedings IAU Symposium No. 379, 2025
P. Bonifacio, M.-R. Cioni, F. Hammer, M. S. Pawlowski & S. Taibi, eds.
doi:10.1017/S1743921323001461

Testing Jeans dynamical models with prolate rotation on a cosmologically simulated dwarf galaxy

Amrit Sedain[1,2] and **Nikolay Kacharov**[1]

[1]Leibniz Institute for Astrophysics, An der Sternwarte 16, 14482 Potsdam, Germany
email: sedain@uni-potsdam.de

[2]Institute of Physics and Astronomy, Karl-Liebknecht Str.24/25 14476 Potsdam, Germany

Abstract. Prolate rotation is characterized by a significant stellar rotation around a galaxy's major axis, which contrasts with the more common oblate rotation. Prolate rotation is thought to be due to major mergers and thus studies of prolate-rotating systems can help us better understand the hierarchical process of galaxy evolution. Dynamical studies of such galaxies are important to find their gravitational potential profile, total mass, and dark matter fraction. Recently, it has been shown in a cosmological simulation that it is possible to form a prolate-rotating dwarf galaxy following a dwarf-dwarf merger event. The simulation also shows that the unusual prolate rotation can be time enduring. In this particular example, the galaxy continued to rotate around its major axis for at least 7.4 Gyr (from the merger event until the end of the simulation). In this project, we use mock observations of the hydro-dynamically simulated prolate-rotating dwarf galaxy to fit various stages of its evolution with Jeans dynamical models. The Jeans models successfully fit the early oblate state before the major merger event, and also the late prolate stages of the simulated galaxy, recovering its mass distribution, velocity dispersion, and rotation profile. We also ran a prolate-rotating N-body simulation with similar properties to the cosmologically simulated galaxy, which gradually loses its angular momentum on a short time scale ~ 100 Myr. More tests are needed to understand why prolate rotation is time enduring in the cosmological simulation, but not in a simple N-body simulation.

Keywords. Galaxy dynamics, Prolate rotation, Cosmological simulations

---

1. Introduction

Dwarf galaxies are important building blocks in galaxy hierarchical formation and evolution theories, so it is essential to understand their properties. There are distinct types of dwarf galaxies, depending on mass and gas content - mainly gas-rich dwarf irregulars and gas-poor dwarf spheroidals, as well as transition types. The majority of them are dark matter (DM) dominated. However, their low stellar densities and shallow potential wells make them very sensitive to perturbations from stellar feedback, external interactions, etc.

Dynamical studies of dwarf galaxies can reveal their total mass and the mass of their sub-components (stars, gas, DM content), as well as their internal structure and kinematical properties, which is not possible from photometry alone.

On the other hand, cosmological simulations allow us to study the formation and evolution of galaxies at all scales and testreal-time observations. By comparing observations and the simulated results we can better understand the physical processes in our Universe. Zoom-in simulations are used to study smaller-scale processes in individual galaxies and galaxy clusters, such as interactions, gas cooling, star formation, and feedback from supernovae and black holes.

While most galaxies have a certain amount of angular momentum, which flattens them in the direction of the rotation axis, there are rare exceptions of galaxies that rotate around their major axis, known as prolate rotation. In the Local Group, we know of only two such cases, the And II dwarf (Amorisco et al. 2014) and the Phoenix dwarf (Kacharov et al. 2017). In both studies, they strongly suggest that a major merger at some point in the galaxy's evolution causes a change in the rotation axis. Similarly, Cardona-Barrero et al. (2021) discovered the presence of prolate rotation in a cosmologically simulated dwarf galaxy, using the GEAR code, based on GADGET-2 (Springel 2005; Revaz & Jablonka 2012). The culprit, causing the flip of the rotation axis is also a major dwarf-dwarf merger with a DM mass ratio of about $1:5$.

In this study, we aim to test how well Jeans dynamical models can reproduce the mass distribution and kinematic properties of the simulated prolate-rotating dwarf. If successful, such dynamical models can then be applied to real observational data.

2. Jeans Anisotropic Multi-Gaussian Expansion (JAM) on the cosmological data

Jeans equations are stellar hydrodynamic equations, derived from the collisionless Boltzmann equation, that are used to study the dynamics of equilibrium systems. One of the advantages of the Jeans equations is that instead of dealing with the phase-space distribution function, we use velocity moments of the line-of-sight velocity distribution (LOSVD), which is directly linked to the gravitational potential and density of the system. This method has been extensively used to study spherical and axisymmetric stellar systems under the assumption of quasi-equilibrium (Cappellari 2016).

In this work, we utilise the JAM code by Cappellari (2020) to solve the Jeans equations for the cosmologically simulated prolate rotating dwarf galaxy, discovered by Cardona-Barrero et al. (2021). We obtain best-fit Jeans dynamical models at different stages during its evolution in the cosmological simulation, which predicts the dwarf's mass distribution and compares the model results with the known properties of the simulated dwarf. The primary objective is to investigate whether the Jeans models can accurately describe the dynamical properties of the prolate system in a quasi-equilibrium approximation since in the cosmological simulation the prolate rotation survives for more than 7.56 Gyr.

In the early stages of the galaxy's evolution, its rotation axis was the minor axis ($q = \frac{b}{a} < 1$). After a massive merger with its satellite halo, the galaxy's axis of rotation changes to the major axis. We assumed here axisymmetry around its major axis ($q = \frac{b}{a} > 1$). We created JAM models for each of the three main stages of the galaxy's evolution: pre-merger ($z = 1.58$), shortly after the merger ($z = 0.58$), and at the present day ($z = 0.00$).

We used the initial oblate case as a benchmark to evaluate the performance of our prolate models.

3. Methods

The Jeans dynamical models link the generally derived from observations stellar density and kinematics of the modelled system to its gravitational potential.

From the simulation, we know the surface brightness profile of the galaxy and perform a simple Sérsic profile fit to obtain its central surface brightness, half-light radius,

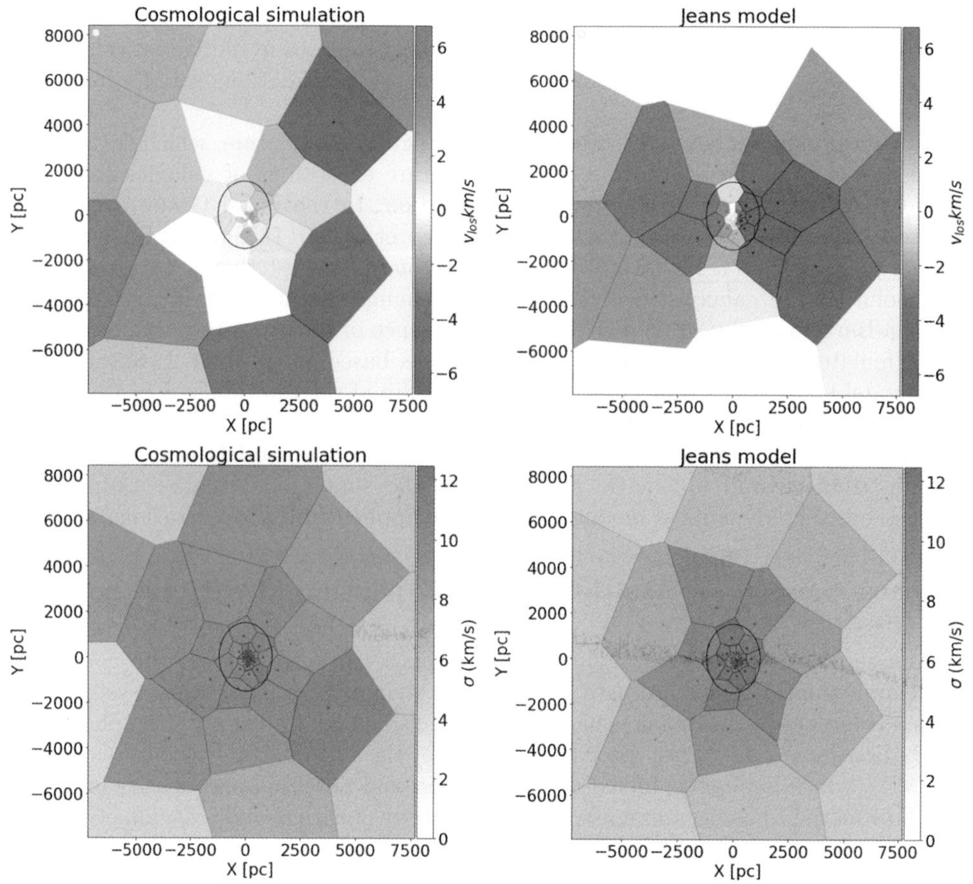

Figure 1. The top-left figure displays the line-of-sight velocity from the cosmological simulation of the galaxy at redshift $z = 0.00$ in Voronoi bins, while the top-right figure shows the corresponding JAM model. The black ellipse indicates the half-light radius. The bottom left figure displays the velocity dispersion from the cosmological simulation, while the bottom right shows the dispersion obtained from the fitted JAM model.

and Sérsic index. We represent the resulting Sérsic profile with a Multi Gaussian Expansion (MGE), which makes it easy to deproject the density profile.

We have also created Voronoi maps of the line of sight velocity (v_{los}) and velocity dispersion (σ) at different snapshots from the simulation. We assigned uncertainties to these values based on the local surface brightness (uncertainties increase with decreasing surface brightness).

In our dynamical model, the gravitational potential is driven by DM only. We used a generalised Navarro-Frank-White (NFW) density profile as representative of the gravitational potential to solve the Jeans equations. In this particular JAM model, we used a spherical, cored NFW profile (central density slope $\gamma = 0$), which we later found to give the most promising mass profile of the model, as compared with the cosmological simulation.

We used Markov Chain Monte Carlo (MCMC) method to get the maximum likelihood of the fitted parameters. We fit for the central density (ρ_0) and scale radius (r_0) of the NFW profile, as well as velocity anisotropy ($\beta_z = 1 - \frac{\sigma_z^2}{\sigma_r^2}$) and rotation amplitude (κ) of the galaxy.

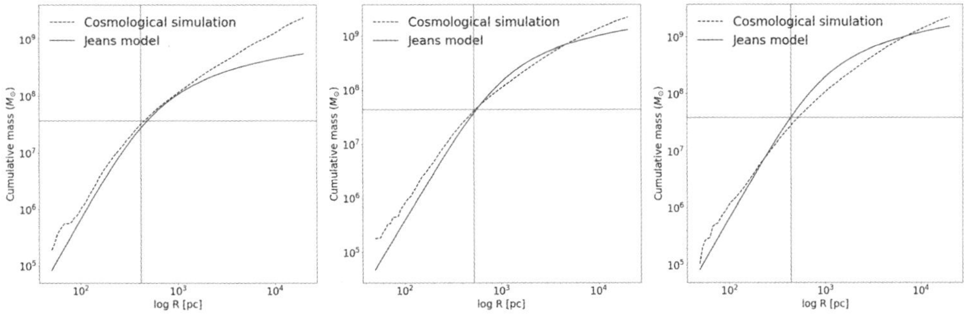

Figure 2. Best-fit mass profiles of the simulated galaxy at pre-merger ($z = 1.58$, left), just after the merger ($z = 0.58$, middle), and at the present time ($z = 0$, right), compared to the true mass distribution from the cosmological simulation. The cross shows the virial mass estimate enclosed within the half-light radius.

4. Results

We first modelled the simulated galaxy during its pre-merger stage ($z = 1.58$), when it still had normal oblate rotation. After 2 Gyr, due to a massive collision, the galaxy began rotating in a prolate manner. This is the stage in the simulation, at which we performed our second JAM model ($z = 0.58$). Interestingly, the galaxy also has prolate rotation at $z = 0$ (the present day of the simulation), indicating that it is in a state of quasi-equilibrium.

In all three cases, we fitted the line-of-sight velocity (v_{los}) and velocity dispersion, and we estimated the mass distribution of the simulated galaxy using the modelled parameters listed above. An example of the fit for the last case ($z = 0$) is shown in Figure 1. The Voronoi maps of the mock observations of the galaxy rotation and velocity dispersion are compared to the best-fit results from the Jeans model.

The Jeans model provided a well matched line-of-sight velocity and velocity dispersion for all the stages within the half-light radius. Despite the fewer data points outside the half-light radius, we also got a relatively good fit with the data.

The capability to determine the galaxy's mass profile is one of the main advantages of dynamical modelling. We calculated the mass for each of the three models stated above. The masses for all three models are displayed in Figure 2, along with the mass derived from the cosmological simulation.

5. N-body Simulation of the galaxy

We also checked the survivability of the galaxy's rotation using an N-body simulation.

We used the inverse transform sampling method to draw positions from the best-fit de-projected 3D sérsic profile and initial 3D velocities from the best-fit Jeans model.

We ran the N-body code assuming star particles move within an embedded spherical DM halo. The simulation was done using SWIFT (Schaller et al. 2016). Interestingly, the galaxy loses its prolate rotation after only 100 Myr, in contrast to its longevity in the cosmological simulation. More research is necessary to determine the underlying reasons. For example, we could try changing the shape of the DM halo to prolate as well.

6. Conclusions

We successfully applied the JAM code on a cosmologically simulated, prolate rotating dwarf galaxy and fitted its velocity moments. We usedthe early oblate, pre-merger stage as a benchmark for our project. We then modelled the prolate rotating stages shortly after the merger ($z = 0.58$) and at the present time ($z = 0.0$). In all cases, we fitted a spherical NFW mass profile, as well as the amplitude of rotation and velocity anisotropy. We then calculated the mass profile of the galaxy at different evolutionary stages and compared them to the known mass profiles from the simulation, which agree quite well. This project confirms the viability of the Jeans method for estimating the masses of prolate rotating galaxies in the approximation of quasi-dynamical equilibrium. Furthermore, we also did an N-body simulation but in this simple setup, the galaxy's prolate rotation dies after 100 Myr. Further work is needed to better understand the survivability of its prolate nature.

Acknowledgements

We want to express our gratitude to Dr. Salvador Cardona-Barrero for providing the simulation data. We acknowledge Dr. Yves Revaz for his help to run the N-body simulation. We thank Prof. Dr. Maria-Rosa Cioni for the opportunity to conduct this research at the AIP.

References

Amorisco, N. C., Evans, N. W., & van de Ven, G. 2014, Nature, 507, 335. doi:10.1038/nature12995

Cappellari, M. 2016, Annu. Rev. Astron. Astrophys., 54, 597. doi:10.1146/annurev-astro-082214-122432

Cappellari, M. 2020, MNRAS, 494, 4819. doi:10.1093/mnras/staa959

Cardona-Barrero, S., Battaglia, G., Di Cintio, A., et al. 2021, MNRAS, 505, L100. doi:10.1093/mnrasl/slab059

Kacharov, N., Battaglia, G., Rejkuba, M., et al. 2017, MNRAS, 466, 2006. doi:10.1093/mnras/stw3188

Revaz, Y. & Jablonka, P. 2012, A&A, 538, A82. doi:10.1051/0004-6361/201117402

Schaller, M., Gonnet, P., Chalk, A. B. G., et al. 2016, Proceedings of the Platform for Advanced Scientific Computing Conference, 2. doi:10.1145/2929908.2929916

Springel, V. 2005, MNRAS, 364, 1105. doi:10.1111/j.1365-2966.2005.09655.x

Session 4:
The Magellanic System: the Clouds, the Stream, and the Leading Arm

Dynamical Masses of Local Group Galaxies
Proceedings IAU Symposium No. 379, 2025
P. Bonifacio, M.-R. Cioni, F. Hammer, M. S. Pawlowski & S. Taibi, eds.
doi:10.1017/S1743921323001722

Exploring the Magellanic Stellar Periphery

L. R. Cullinane[1], A. D. Mackey, G. S. Da Costa[2], D. Erkal[3] and S. E. Koposov[4,5]

[1]The William H. Miller III Department of Physics & Astronomy, Bloomberg Center for Physics and Astronomy, Johns Hopkins University, 3400 N. Charles Street, Baltimore, MD 21218
email: lcullin4@jhu.edu

[2]Research School of Astronomy and Astrophysics, Australian National University, Canberra, ACT 2611, Australia

[3]Department of Physics, University of Surrey, Guildford GU2 7XH, UK

[4]Institute for Astronomy, University of Edinburgh, Royal Observatory, Blackford Hill, Edinburgh EH9 3HJ, UK

[5]Institute of Astronomy, University of Cambridge, Madingley Road, Cambridge CB3 0HA, UK

Abstract. Recent panoramic maps of the Magellanic system have revealed a wealth of low-surface-brightness stellar substructures surrounding both the Large and Small Magellanic Clouds (LMC/SMC); clear evidence of tidal interactions between the two Clouds, as well as with the Milky Way. The Magellanic Edges Survey (MagES), a spectroscopic survey that targets red clump and red giant branch stars across the outskirts of both Clouds, aims to characterise these features and shed light on their formation. We summarise recent results from MagES, which suggest multiple previous LMC-SMC interactions are required to fully explain the observed dynamical properties of the Clouds.

Keywords. Magellanic Clouds, Local Group, Galaxies: kinematics and dynamics

1. Introduction

The Large and Small Magellanic Clouds (LMC/SMC) are ideally situated for detailed study of the effects of tidal interactions on galaxy evolution. The LMC is the most massive Milky Way (MW) satellite; and, despite only being on its first infall into the Milky Way's gravitational potential (Besla et al. 2007), has already significantly impacted the MW (and been affected in return). It has generated overdensities in the MW dark matter halo (e.g. Garavito-Camargo et al. 2019) and reflex motions in the MW stellar halo (e.g. Petersen and Peñarrubia 2021),and perturbed the orbits of smaller satellites and stellar streams (e.g. Erkal et al. 2019; Patel et al. 2020). In addition, dynamical models suggest the Clouds have likely repeatedly interacted with each other over several Gyr (e.g. Besla et al. 2012; Pardy et al. 2018), with signatures of these interactions long-apparent in the Clouds. The SMC is particularly disturbed, with an asymmetric, irregular morphology; line-of-sight (LOS) depths up to 30 kpc (e.g. Hatzidimitriou and Hawkins 1989; Nidever et al. 2013); and kinematic evidence for tidal expansion (De Leo et al. 2020; Zivick et al. 2021). While more ordered than the SMC, the LMC also displays substantial disturbances, including multiple warps (Olsen and Salyk 2002; Choi et al. 2018), and a tilted, off-centre stellar bar (van der Marel and Cioni 2001).

Recently, deep photometric studies of the Magellanic periphery (e.g. Mackey et al. 2016, 2018; Pieres et al. 2017), in combination with multidimensional phase-space information from Gaia (e.g. Belokurov and Erkal 2019; Gaia Collaboration et al. 2021b),

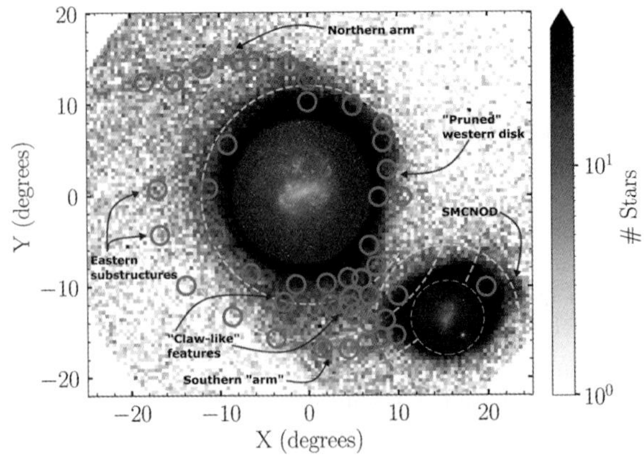

Figure 1. Density map of Magellanic red clump/RGB stars selected from Gaia EDR3, using the criteria of (Cullinane et al. 2022), with major features labelled. Red circles indicate the position of MagES 2-degree diameter fields. Dashed yellow circles indicate distances of 12°/16°/20° from the LMC and 4°/8° from the SMC centres respectively.

have also revealed a wealth of low-density stellar substructure surrounding both Clouds (Fig. 1). This includes a $\sim 23°$ long arm-like feature north of the LMC (Mackey et al. 2016; Belokurov and Erkal 2019) and diffuse structures to the east of the LMC (El Youssoufi et al. 2021), claw-like structures in the southern LMC outskirts and an apparent truncation in the western edge of the LMC disk (Mackey et al. 2018), diffuse overdensities north of the SMC (Pieres et al. 2017; El Youssoufi et al. 2021), and a long thin feature which appears to wrap around the southern LMC, stretching between the eastern outskirts of the SMC and the eastern LMC disk (Belokurov and Erkal 2019). These extended features are of particular interest in constraining the interaction history of the Clouds: stars at large radii are more weakly gravitationally bound, and thus relatively easily perturbed by interactions compared to stars in the central regions of the galaxies; and, because dynamical timescales in the outskirts are long, the resulting structural and kinematic signatures are more persistent compared to the central regions, potentially remaining for several Gyr.

However, in order to place precise constraints on the complex interactions which produce these features, full three-dimensional (3D) kinematic data are critical. With the advent of the Gaia satellite, providing proper motions for stars $G \lesssim 21$ mag across the entire Magellanic system (Gaia Collaboration et al. 2018), the limiting factor are now line-of-sight (LOS) velocities. Existing spectroscopic surveys of the Clouds have predominantly targeted stars in their interior; and the vast majority of stars in the low-density outskirts are too faint for Gaia to provide LOS velocities (available only for $G \lesssim 14$ mag: Gaia Collaboration et al. 2022). The Magellanic Edges Survey (MagES), a spectroscopic survey which specifically targets the low-density periphery of the Clouds, is designed to fill this gap. Here, we summarize recent results from MagES, and how those inform our understanding of the masses, orbits, and interaction history of the Clouds.

2. The Magellanic Edges Survey

MagES utilizes the 2dF multi-object fibre positioner (Lewis et al. 2002), and the dual-beam AAOmega spectrograph (Sharp and Birchall 2010) on the 3.9 m Anglo-Australian Telescope (AAT). The 2dF positioner allows for the observation of ~ 350 science targets per 2-degree-diameter field. We configure the blue arm on AAOmega with the 1500 V grating (wavelength coverage ~4910–5615 Å and resolution R$\sim$ 3700), designed to provide

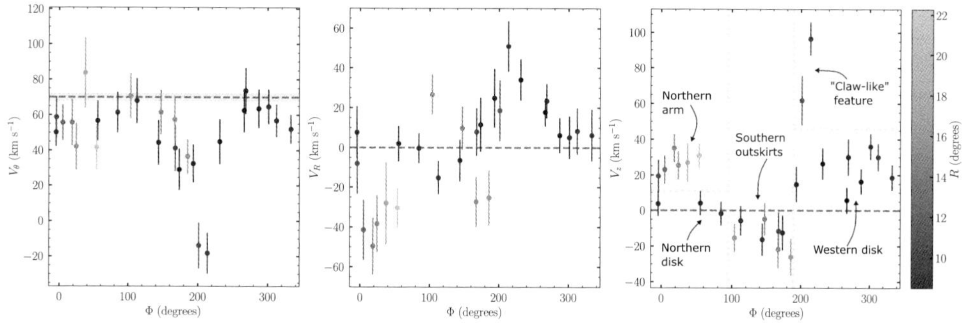

Figure 2. Aggregate kinematics relative to the plane of the LMC disk for MagES fields analysed to date (compiled from Cullinane et al. 2021, 2022), calculated assuming a disk inclination of $i = 36.5^\circ$ and line-of-nodes position angle $\Omega = 145^\circ$ (Cullinane et al. 2022). Fields associated with distinct regions are labelled as in Fig. 1 and separated by dashed yellow lines in the right panel. From left to right, the azimuthal velocity V_θ, the in-plane radial velocity V_R, and the out-of-plane (vertical) velocity perpendicular to the disk plane V_z are shown, each plotted as a function of position angle east of north. Points are coloured by LMC on-sky galactocentric radius. Grey dashed lines indicate predicted kinematics for an equilibrium disk.

precise LOS velocity estimates using the MgIb triplet. The red arm is configured with the 1700D grating (wavelength coverage ~8370–8790 Å and resolution of R~ 10000), designed to provide sufficiently high-resolution coverage of the CaII triplet to allow for an estimation of metallicity (as in e.g. Da Costa 2016) and a complementary LOS velocity estimate.

Fourty-four MagES fields have been observed to date; their locations are indicated in Fig. 1. Within each field, MagES targets red clump and red-giant branch (RGB) stars selected from a combination of Dark Energy Camera (DECam) photometry (Koposov et al. 2015; Mackey et al. 2018) if observed prior to the release of Gaia DR2, and Gaia DR2/EDR3 photometry (Gaia Collaboration et al. 2018, 2021a) for fields observed subsequent to these data releases. In total, LOS velocities have been obtained for > 12000 such stars, with metallicity estimates for > 500 RGB stars with $G < 18$ mag. For the more numerous and fainter red clump stars within each field, we stack spectra for likely ($P \geq 50\%$) Magellanic stars to create a single "representative" spectrum used for these measurements; the resulting [Fe/H] estimates tend towards the mean field metallicity. In addition, as the red clump is a standardizable candle, we utilize this to obtain relative mean distance estimates for fields which contain similar stellar populations, as indicated by having consistent mean metallicities and $G_{BP} - G_{RP}$ colours.

As it is generally informative to consider the kinematics and distances of MagES fields relative to the frame of the LMC disk, we utilize the framework presented in van der Marel et al. (2002) to transform the observed kinematics of each field into a cylindrical coordinate system aligned with the LMC disk, with its origin at the LMC centre of mass (COM). We subsequently obtain V_θ, the azimuthal rotation velocity (positive values indicate motion in the same direction as the LMC's clockwise on-sky rotation); V_R, the in-plane radial velocity (positive values indicate motion radially outward from the COM in the plane of the disk); V_z, the vertical velocity perpendicular to the disk plane (positive values indicate motion in a direction predominantly towards the viewer: i.e. "in front of" the disk plane); as well as dispersions ($\sigma_\theta, \sigma_R, \sigma_z$) in each of these components. We additionally obtain the out-of-plane distance z, describing how far "in front of" or "behind" the expected disk plane stars in each field are located. Fig. 2 presents these measurements for MagES fields analysed to date (i.e. observed prior to 2022) which are associated with the LMC.

2.1. *Magellanic Models*

In order to assist in our interpretation of the disturbed kinematics in the Magellanic outskirts, we compare our observations to a suite of simple dynamical models of the LMC+SMC+MW system. These models are only a first exploration of the large and complex parameter space that describes the allowable orbits of the Clouds, and there are consequently associated limitations: most significantly a lack of self-gravity. This can affect both the orbits of the Clouds, and the response of stars within them to close interactions; accordingly, we only perform qualitative comparisons to our observations.

The LMC is modelled as a collection of test particles within a two-component potential: an exponential disk with mass 2×10^9 $M_\odot$, scale radius 1.5 kpc, and scale height 0.4 kpc; and a Hernquist dark matter halo (Hernquist 1990) of mass 1.5×10^{11} $M_\odot$ (Erkal et al. 2019) and scale radius 20 kpc, such that the circular velocity is ~ 90 km s^{-1} at 10 kpc. The SMC is modelled as a Hernquist profile with mass 2.5×10^9 $M_\odot$ and scale radius 0.043 kpc, such that it has a circular velocity of 60 km s^{-1} at 2.9 kpc (Stanimirović et al. 2004). The Milky Way is modelled as a three-component system with a bulge, disk, and dark matter halo similar to Bovy (2015). We treat each of these three galaxies as a particle sourcing a potential, allowing us to account for the motion of the Milky Way in response to the LMC (Gómez et al. 2015) and the dynamical friction of the Milky Way on the LMC (Jethwa et al. 2016). The LMC and SMC are initialized at their present day locations, then rewound for one Gyr in the presence of each other and the Milky Way. The LMC disk is then initialized with $\sim 2.5 \times 10^6$ tracer particles† using AGAMA (Vasiliev 2019) to account for the LMC's velocity dispersion, with a geometry matching that from Cullinane et al. (2022). The system is then evolved to the present.

We run multiple model realizations, sampling from literature uncertainties on the current-day systemic properties of both the LMC and SMC, to explore the allowable parameter space, and hence differing orbits (and interaction histories) for the Clouds. In all model realizations, the SMC has had a recent close pericentric passage around the LMC ~ 150 Myr ago (consistent with Zivick et al. 2018), with a total pericentric distance $r_{\rm peri} \sim 8.0$ kpc (consistent with Choi et al. 2022). The projected galactocentric radius of the pericentric passage is ~ 4 kpc, in a direction toward the south-west of the LMC. At earlier times, the orbit of the SMC varies significantly depending on how the systemic motions of both Clouds are sampled. Approximately half our realizations have a second SMC crossing of the LMC disk ~ 400 Myr ago, which can occur across a broad range ($\sim 29 \pm 10$ kpc) of in-plane radial distances. A handful of model realizations ($\sim 4\%$) also have a roughly simultaneous third disk crossing and SMC pericentric passage ~ 1 Gyr ago, though a significantly larger fraction would experience one or both of the disk crossing and pericentric passage if our models were rewound for a greater length of time than precisely one Gyr. The particulars of this interaction are much less robustly constrained than more recent interactions, with pericentric distances $r_{\rm peri} \sim 6 \pm 3$ and crossing distances of $\sim 54^{+13}_{-46}$ kpc permitted due to the increasing uncertainty in the SMC's orbit at earlier times. Fig. 3 shows the current particle distribution for a single model realisation which has experienced each of these interactions, coloured by the particle distance from the SMC at the time of each SMC crossing of the LMC disk plane, indicating which regions of the LMC are most strongly affected by each interaction.

3. Properties of the Magellanic outskirts

Below, we summarize recent results from MagES, as published in Cullinane et al. (2020, 2021, 2022, 2023), and their implications for the interaction history of the Clouds.

† We do not initialize the SMC potential with tracer particles.

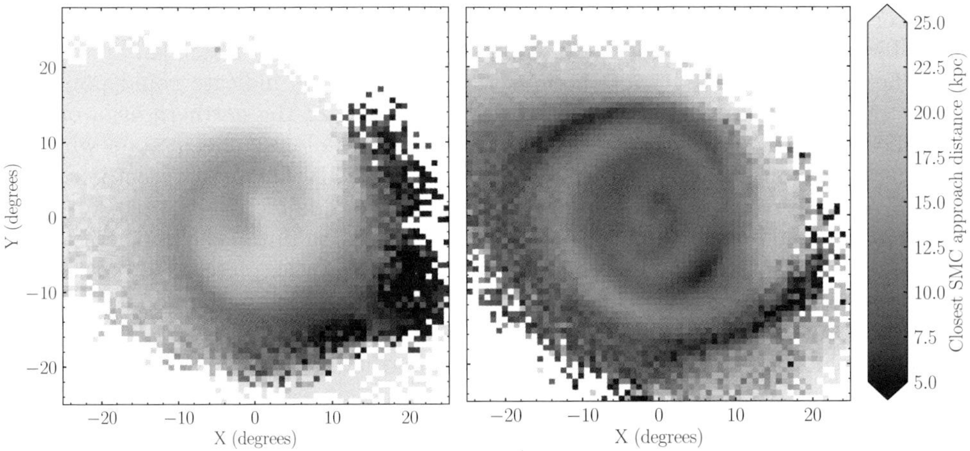

Figure 3. Current particle distribution for a single Magellanic model realisation, colour coded by the particle distance from the SMC at the time of the SMC disk crossings ~ 400 Myr ago (left) and ~ 980 Myr ago (right). Adapted from Cullinane et al. (2022).

3.1. *The northern LMC and its tidal arm*

MagES fields across the northeastern disk of the LMC display very ordered kinematics, with aggregate out-of-plane distances, vertical and radial velocities consistent with zero, and azimuthal velocities of ~ 70 km s^{-1}. This is consistent with previous measurements of the LMC's rotation derived from data at smaller radii (e.g. van der Marel and Kallivayalil 2014). Under the assumption that this indicates equilibrium kinematics within these fields, we thus utilize the derived circular velocity to calculate the enclosed mass of the LMC within 10.5 kpc as $M_{\rm enc} = (1.8 \pm 0.3) \times 10^{10}$ M$_\odot$. This is consistent with enclosed mass measurements derived using similar techniques at smaller radii when projected to the same distance, but low compared to mass measurements derived using more indirect methods ($\sim 1.5 - 3 \times 10^{11}$ M$_\odot$), such as perturbations to stellar streams (e.g. Erkal et al. 2019), the timing argument (Peñarrubia et al. 2016), or cosmological simulations of similar systems (e.g. Shao et al. 2018). This difference is to be expected, as each of the above methods provides the total infall mass of the LMC, including its dark halo; in contrast, the MagES fields here are still well within the LMC's dark halo. If, however, the assumption is made that the LMC rotation curve remains flat out to ~ 30 kpc (where LMC-associated stars have been found: Navarrete et al. 2019), and that the LMC is embedded in a typical dark matter halo, the inferred enclosed mass is $\sim 1.1 \times 10^{11}$ M$_\odot$: more in line with total infall mass estimates.

Fields along the LMC's northern tidal arm have similar metallicities to those of the northern LMC disk, with a mean overall metallicity of [Fe/H]~ -1, and only weak ($< 2\sigma$) evidence for a negative metallicity gradient along the feature. The azimuthal velocity along the arm is broadly consistent with the LMC's rotation, and distance estimates also suggest the arm largely follows the inclined LMC disk plane. In combination, this suggests the arm is comprised of LMC disk material. There are, however, strong non-equilibrium motions in the radial and vertical velocity components: the radial velocity is consistently strongly negative, moving at ≥ 30 km s^{-1} inward along the arm, and the vertical velocity gradually increases along the feature from near zero to a maximum of nearly 30 km s^{-1} in the most distant field. The increasing vertical velocity is a result of the LMC's infall to the MW potential: the orbit and inclination of the LMC relative to the Milky Way is such that the MW's gravitational force acts in a direction predominantly aligned with the positive z direction, and models including only the LMC+MW are able to replicate the observed trend. However, the strongly negative radial velocity cannot be replicated in our LMC+MW models, and likely requires perturbation from the SMC.

Of the several LMC-SMC interactions which our models indicate could potentially impact the LMC and hence contribute to the formation of the northern arm, only the SMC disk crossing and pericentric passage ~ 1 Gyr ago closely affects stars which are today in the vicinity of the northern arm, as seen in Fig. 3. The most recent pericentric passage, although a close interaction, occurs effectively diametrically opposite to the northern arm, and – as dynamical timescales at the large radius of the northern arm are $\gtrsim 1$ Gyr – insufficient time has passed for debris disturbed by this interaction to reach the position of the northern arm. The ~ 400 Myr disk crossing most closely affects stars in the western outskirts of the LMC. We therefore suggest the interaction ~ 1 Gyr ago (or potentially older interactions not captured in our current models) have likely contributed to the formation of the northern arm.

3.2. *The disturbed southern LMC*

MagES fields throughout the southern outskirts of the LMC have metallicities consistent with the northern LMC disk, indicating these are also dominated by LMC material. However, unlike the ordered kinematics of the northeastern disk, the kinematics in this region are quite disturbed, with all fields displaying clear kinematic trends as a function of position angle. Moving toward the southwest, the azimuthal velocity substantially decreases, to ~ 30 km s^{-1}, and the out-of-plane distance increases significantly, with the outermost fields located up ~ 12 kpc closer to us than the nominal disk plane. However, the vertical velocity in these fields is generally mildly negative ($|V_z| \lesssim 20$ km s^{-1}), indicating material in these fields is currently slowly moving back towards the disk plane.

Fields at smaller LMC galactocentric radii ($R \sim 10°$) show a trend of increasing in-plane radial velocity as a function of increasing position angle, indicating an increasingly strong radially outward motion as one approaches the present position of the SMC. Our models suggest any of the potential LMC-SMC interactions in the past Gyr can produce this trend. In contrast, fields at larger LMC galactocentric radii ($R \sim 16°$, including fields nominally along the "southern arm") have increasingly negative in-plane radial velocities as a function of position angle. This is most closely replicated by models which have experienced multiple LMC-SMC interactions in the past Gyr, including disk crossings both ~ 1 Gyr and ~ 400 Myr ago.

Unlike the relatively well-defined kinematic trends observed in the southern outskirts, fields located in the western "claw-like" feature are highly disturbed. With a negative V_θ observed, this structure appears to be counter-rotating relative to the LMC disk; and it posses enormous out-of-plane velocities ($V_z > 60$ km s^{-1}), indicating it is rapidly moving towards us and further away from the LMC disk plane. Our current Magellanic models cannot replicate these distinctive kinematics, requiring further investigation into the origin of this feature.

3.3. *The pruned western LMC disk*

Fields along the western outskirts of the LMC also display largely consistent kinematic trends. Fields in the southwest, nearest the present-day position of the SMC, have comparable in-plane radial velocities to the nearby southern disk fields at similar radii, suggesting these fields may be similarly perturbed by LMC-SMC interactions in the past Gyr. More northern fields have in-plane radial velocities consistent with zero. Fields throughout the western outskirts are located on average ~ 4 kpc in front of the LMC disk plane, and have moderately positive vertical velocities ($V_z \sim 15 - 35$ km s^{-1}) indicating this region of the disk continues to move towards us and away from the LMC disk plane (though significantly more slowly than the claw-like features). While our models suggest stars in the western LMC outskirts may be closely affected by a SMC disk crossing ~ 400 Myr ago, with resulting in-plane radial velocities similar to those observed, our models also predict kinematic trends in the vertical direction which are opposite to those that we observe: i.e. increasingly negative out-of-plane distances and vertical velocities in the northwestern LMC. More detailed modelling of this disk crossing is required in order to determine if there are associated parameters which can reproduce our observed kinematic trends, or if earlier interactions with the SMC are responsible for the largely positive vertical perturbations we observe.

3.4. *The extreme eastern SMC*

MagES fields in the outskirts of the SMC ($R_{\rm SMC} \sim 6°$) have distinctly lower metallicities ([Fe/H]~ -1.6) than fields in the greater LMC outskirts, suggesting they are indeed dominated by perturbed SMC material, and have kinematics reflective of the SMC's tidal expansion observed at smaller radii (e.g. De Leo et al. 2020). However, at greater distances, the situation is more complex. In MagES field 3, located $\sim 9°$ east of the SMC along the "southern arm" feature, two chemodynamically distinct populations are observed. Approximately two-thirds of the stars in this field have kinematics and metallicities consistent with those expected for extended SMC debris. In contrast, the remaining ~one-third of the material in this field is substantially more metal-rich ([Fe/H]~ -1.2), located $\gtrsim 7$ kpc closer to us than the bulk population, and is moving away from the SMC significantly more rapidly ($V_{\rm rel} \sim 230$ km s^{-1}, compared to ~ 70 km s^{-1} for the bulk population) in a direction largely toward the LMC. While an association with the LMC cannot be ruled out given the higher metallicity of this population, this population does have substantially different kinematics from nearby MagES fields dominated by LMC debris. This material may instead by debris from the inner (i.e. more metal-rich) SMC, which has been recently perturbed by, e.g., the SMC's close pericentric passage around the LMC ~ 150 Myr ago.

4. Summary

The Magellanic Clouds are ideal testbeds in which to study the effects of interactions on galaxy structure, kinematics, and star formation. Each distinct feature in the wealth of substructure spanning the outskirts of the Clouds encodes unique constraints on their past interactions, with multidimensional phase-space information required in order to unravel these complex and often overlapping signatures. MagES has already made great strides in characterising the properties of multiple Magellanic substructures, and has demonstrated the efficacy of such data as a benchmark for assessing dynamical models to disentangle their origins, finding multiple LMC-SMC interactions are likely required to produce the disturbed features observed. Additional observations from MagES and forthcoming surveys – such as the 4-metre Multi-Object Spectroscopic Telescope (4MOST) consortium survey 1001 Magellanic Fields (1001MC: Cioni et al. 2019) – in conjunction with more detailed dynamical modelling, will allow for the interaction history of the Clouds to be precisely constrained, and provide new understanding on how these processes have shaped their evolution.

References

Belokurov, V. & Erkal, D. 2019, *MNRAS*, 482(1), L9–L13.
Besla, G., Kallivayalil, N., Hernquist, L., *et al.* 2007, *ApJ*, 668(2), 949–967.
Besla, G., Kallivayalil, N., Hernquist, L., *et al.* 2012, *MNRAS*, 421(3), 2109–2138.
Bovy, J. 2015, *ApJS*, 216(2), 29.

Choi, Y., Nidever, D. L., Olsen, K., *et al.* 2018, *ApJ*, 866(2), 90.
Choi, Y., Olsen, K. A. G., Besla, G., *et al.* 2022, *ApJ*, 927(2), 153.
Cioni, M.-R., Storm, J., Bell, C. P. M., *et al.* 2019, *The Messenger*, 175, 54–57.
Cullinane, L. R., Mackey, A. D., Da Costa, G. S., *et al.* 2021, *MNRAS*, 510(1), 445–468.
Cullinane, L. R., Mackey, A. D., Da Costa, G. S., *et al.* 2022, *MNRAS*, 512(4), 4798–4818.
Cullinane, L. R., Mackey, A. D., Da Costa, G. S., *et al.* 2020, *MNRAS*, 497(3), 3055–3075.
Cullinane, L. R., Mackey, A. D., Da Costa, G. S., *et al.* 2023, *MNRAS*, 518, L25–L30.
Da Costa, G. S. 2016, *MNRAS*, 455(1), 199–206.
De Leo, M., Carrera, R., Noël, N. E. D., *et al.* 2020, *MNRAS*, 495(1), 98–113.
El Youssoufi, D., Cioni, M.-R. L., Bell, C. P. M., *et al.* 2021, *MNRAS*, 505(2), 2020–2038.
Erkal, D., Belokurov, V., Laporte, C. F. P., *et al.* 2019, *MNRAS*, 487(2), 2685–2700.
Gaia Collaboration, Brown, A. G. A., Vallenari, A., *et al.* 2018, *A&A*, 616, A1.
Gaia Collaboration, Brown, A. G. A., Vallenari, A., *et al.* 2021,a *A&A*, 649a, A1.
Gaia Collaboration, Luri, X., Chemin, L., *et al.* 2021,b *A&A*, 649b, A7.
Gaia Collaboration, Vallenari, A., Brown, A., *et al.* 2022, *A&A*,.
Garavito-Camargo, N., Besla, G., Laporte, C. F. P., *et al.* 2019, *ApJ*, 884(1), 51.
Gómez, F. A., Besla, G., Carpintero, D. D., *et al.* 2015, *ApJ*, 802(2), 128.
Hatzidimitriou, D. & Hawkins, M. R. S. 1989, *MNRAS*, 241(4), 667–690.
Helmi, A., Irwin, M., Deason, A., *et al.* 2019, *The Messenger*, 175, pp. 23–25.
Hernquist, L. 1990, *ApJ*, 356(2), 359.
Jethwa, P., Erkal, D., & Belokurov, V. 2016, *MNRAS*, 461(2), 2212–2233.
Koposov, S. E., Belokurov, V., Torrealba, G., & Evans, N. W. 2015, *ApJ*, 805(2), 130.
Lewis, I. J., Cannon, R. D., Taylor, K., *et al.* 2002, *MNRAS*, 333(2), 279–298.
Mackey, A. D., Koposov, S. E., Erkal, D., *et al.* 2016, *MNRAS*, 459(1), 239–255.
Mackey, D., Koposov, S. E., Da Costa, G., *et al.* 2018, *ApJ*, 858(2), L21.
Navarrete, C., Belokurov, V., Catelan, M., *et al.* 2019, *MNRAS*, 483(3), 4160–4174.
Nidever, D. L., Monachesi, A., Bell, E. F., *et al.* 2013, *ApJ*, 779(2), 145.
Olsen, K. A. G. & Salyk, C. 2002, *AJ*, 124(4), 2045–2053.
Pardy, S. A., D'Onghia, E., & Fox, A. J. 2018, *ApJ*, 857(2), 101.
Patel, E., Kallivayalil, N., Garavito-Camargo, N., *et al.* 2020, *ApJ*, 893(2), 121.
Peñarrubia, J., Gómez, F. A., Besla, G., *et al.* 2016, *MNRAS*, 456(1), L54–L58.
Petersen, M. S. & Peñarrubia, J. 2021, *Nature Astronomy*, 5(3), 251–255.
Pieres, A., Santiago, B. X., Drlica-Wagner, A., *et al.* 2017, *MNRAS*, 468(2), 1349–1360.
Shao, S., Cautun, M., Deason, A. J., *et al.* 2018, *MNRAS*, 479(1), 284–296.
Sharp, R. & Birchall, M. N. 2010, *PASA*, 27(1), 91–103.
Stanimirović, S., Staveley-Smith, L., & Jones, P. A. 2004, *ApJ*, 604(1), 176–186.
van der Marel, R. P., Alves, D. R., Hardy, E., & Suntzeff, N. B. 2002, *AJ*, 124(5), 2639–2663.
van der Marel, R. P. & Cioni, M.-R. L. 2001, *AJ*, 122(4), 1807–1826.
van der Marel, R. P. & Kallivayalil, N. 2014, *ApJ*, 781(2), 121.
Vasiliev, E. 2019, *MNRAS*, 482(2), 1525–1544.
Zivick, P., Kallivayalil, N., & van der Marel, R. P. 2021, *ApJ*, 910(1), 36.
Zivick, P., Kallivayalil, N., van der Marel, R. P., *et al.* 2018, *ApJ*, 864(1), 55.

Discussion

Simon Rozier: Do you account for the reflex motion of the LMC to the SMC in your simulations?

Lara Cullinane: Yes, though we do not account for dynamical friction between the Clouds.

Joanna Sakowska: Can you tell me more about how you classified the chemodynamically distinct stars near the SMC to have come from the inner SMC? Have you done any age dating?

LARA CULLINANE: We found these stars were more noticably more metal-rich, with a mean [Fe/H] of ~ -1.2, than the bulk population in this field. Comparing that to the SMC's current metallicity gradient suggests these stars could have originated in the inner SMC and very recently been pushed outwards. The relatively high velocity of this population suggests that scenario is not implausible, but more detailed models are needed to confirm this. We have not yet done any age-dating for this population, but this is something we are interested in pursuing.

JIANLING WANG: The Northern tidal arm is so thin from your results. It looks like it is crossing the Galactic disk, do you have any plan to identify and make spectral observations of these stars above the Galactic disk? These stars will be important for constraining the modelling.

LARA CULLINANE: We did previously plan to follow up those stars, but were unfortunately weathered out during the observing run; we may revisit this in the future. Stars along the northern arm will also be followed up in the 4MOST consortium surveys 1001MC (Cioni et al. 2019) and the Milky Way Halo Low-Resolution Survey (Helmi et al. 2019).

YANBIN YANG: I am very interested in the thin tail north of the LMC. Do you have any information like stellar ages or metallicities for it?

LARA CULLINANE: We do not have ages for these stars, but we are interested in measuring these in the future. We do, however, have metallicities derived from the equivalent width of the Calcium triplet, and find mean [Fe/H] values of ~ -1 along the entire arm.

Dynamical Masses of Local Group Galaxies
Proceedings IAU Symposium No. 379, 2025
P. Bonifacio, M.-R. Cioni, F. Hammer, M. S. Pawlowski & S. Taibi, eds.
doi:10.1017/S1743921323002053

Ages, metallicities and structure of stellar clusters in the Magellanic Bridge

Raphael A. P. Oliveira[1], **Francisco F. S. Maia**[2], **Beatriz Barbuy**[1], **Bruno Dias**[3] **and the VISCACHA collaboration**

[1]Universidade de São Paulo, IAG, Rua do Matão 1226, São Paulo 05508-090, Brazil
email: rap.oliveira@usp.br

[2]Universidade Federal do Rio de Janeiro, Av. Athos da Silveira, 149, Rio de Janeiro 21941-909, Brazil

[3]Instituto de Alta Investigación, Sede Esmeralda, Universidad de Tarapacá, Av. Luis Emilio Recabarren 2477, Iquique, Chile

Abstract. The formation of the Magellanic Bridge during an encounter between the Magellanic Clouds ~ 200 Myr ago would be imprinted in the chemical evolution and kinematics of its stellar population, with sites of active star formation. Since it contains hundreds of stellar clusters and associations, we combined deep photometry from VISCACHA and SMASH surveys to explore this topic, by deriving structural parameters, age, metallicity, distance and mass for 33 Bridge clusters with robust statistical tools. We identified a group of 13 clusters probably stripped from the Small Magellanic Cloud ($0.5 - 4.7$ Gyr, [Fe/H] < -0.6 dex) and another 15 probably formed in-situ (< 200 Myr, [Fe/H] ~ -0.4 dex). Two metallicity dips were detected in the age-metallicity relation, coeval to the Stream and Bridge formation epochs. Cluster masses range from 500 to $\sim 10^4 M_\odot$, and a new estimate of $3 - 5 \times 10^5 M_\odot$ is obtained for the Bridge stellar mass.

Keywords. Magellanic Clouds, galaxies: evolution, galaxies: star clusters: general

1. Introduction

The Magellanic System contains the Large and Small Magellanic Clouds (LMC and SMC), Magellanic Bridge, Stream and Leading Arm. Besides containing the pair of satellites closest to the Milky Way, the Bridge is the nearest tidally-stripped structure. It is also the only one of the three gaseous structures that contain a significant stellar mass ($1.5 \times 10^4\ M_\odot$; Harris 2007), with hundreds of star clusters and associations (Bica *et al.* 2020). The Bridge was first detected as an HI overdensity by Hindman *et al.* (1963), but a blue, young stellar population was found decades later, more concentrated close to the SMC Wing and halfway the Bridge, and strongly correlated with the distribution and kinematics of the HI gas (Skowron *et al.* 2014). An old population was also found later, more spread across the Bridge.

Based on different epochs of *HST* proper motions, two N-body models attempt to reproduce the formation of the LMC-SMC pair and determine if they were independent satellites of the Milky Way until the LMC captured the SMC ~ 1.2 Gyr ago (Diaz & Bekki 2011), or an old interacting system in the first perigalactic passage, entering the Galactic potential ~ 2 Gyr ago (Besla *et al.* 2012). Despite being highly dependent on the total masses of the Milky Way and LMC, both models reproduce the Bridge formation

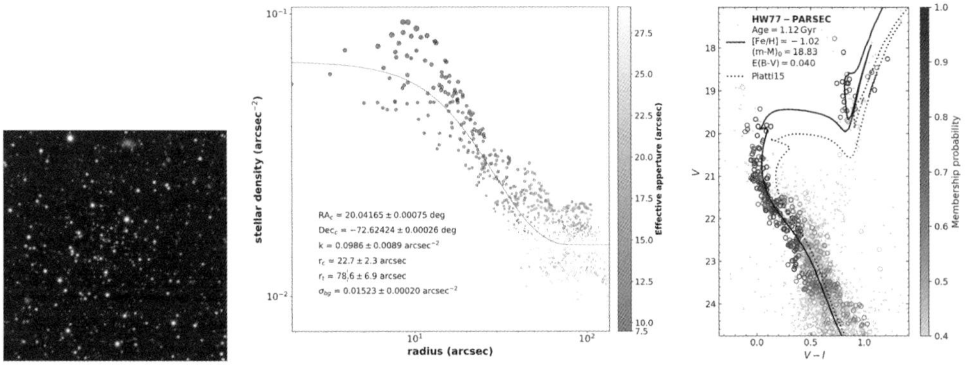

Figure 1. *(Left:)* *BVI* composite image of the SMC Wing cluster HW77, obtained with the SAM@SOAR imager ($3 \times 3\,\mathrm{arcmin}^2$). *(Middle:)* Result obtained in the fitting of King functions to the RDP of HW77. *(Right:)* Isochrone fit obtained for the decontaminated CMD of HW77, with 1.1 ± 0.1 Gyr and $[\mathrm{Fe/H}] = -1.02 \pm 0.11$ dex, and a final mass of $(2.2 \pm 0.6) \times 10^3\,M_\odot$.

during the most recent encounter between the LMC-SMC (150 − 300 Myr ago; Besla *et al.* 2012; Zivick *et al.* 2018; Choi *et al.* 2022), with gas and stars tidally stripped mostly from the SMC and possibly dragged from the LMC. This scenario would imply a gradient of increasing metallicity toward the LMC due to a minor contribution of its more metal-rich gas, and the presence of an old stellar population amidst a predominant young population formed in situ.

In this work (based on Oliveira *et al.* 2023, MNRAS in press) we analyse deep photometry for 33 Bridge clusters in terms of structural and fundamental parameters in order to investigate their assembly history, spatial distribution and the existence of such gradients.

2. Photometric data and methodology

The VISCACHA survey (Maia *et al.* 2019) uses the adaptive optics system of the SOAR 4-m telescope (SAM) to observe clusters in the outskirts of the Magellanic Clouds and Bridge. This very deep ($V \sim 24$ mag) photometry, with good spatial resolution ($\sim 0.6''$), allowed us to derive precise age, metallicity, distance, mass and structural parameters for 33 Bridge clusters at RA $< 3^{\mathrm{h}}$ (Oliveira *et al.* 2023). Observations with Goodman@SOAR are being carried out to cover the objects in RA $> 3^{\mathrm{h}}$. A spectroscopic follow-up in the CaT region was also conducted to derive radial velocities and metallicity for clusters older than 1 Gyr (Dias et al., in preparation). Photometry from the SMASH survey (Nidever *et al.* 2017) is also used as a complement, with a similar depth but lower spatial resolution in crowded regions.

In order to ensure a homogeneous and self-consistent analysis, we implemented the Markov chain Monte Carlo (MCMC) sampling (Foreman-Mackey *et al.* 2013) both to the fitting of analytical functions to radial density profiles (RDPs) and statistical isochrone fitting to colour-magnitude diagrams (CMDs). The RDPs are obtained by computing the local stellar density around each star, using a variable aperture size. A likelihood function comparing the observed local density with the model (King 1962) is evaluated and coupled to the MCMC to obtain a new centre, core and tidal radii, and central and background densities. Before the isochrone fitting, a membership analysis to exclude probable field stars are performed as described in Maia *et al.* (2010). The method compares the stars within a fraction of the tidal radius (cluster + field stars) with

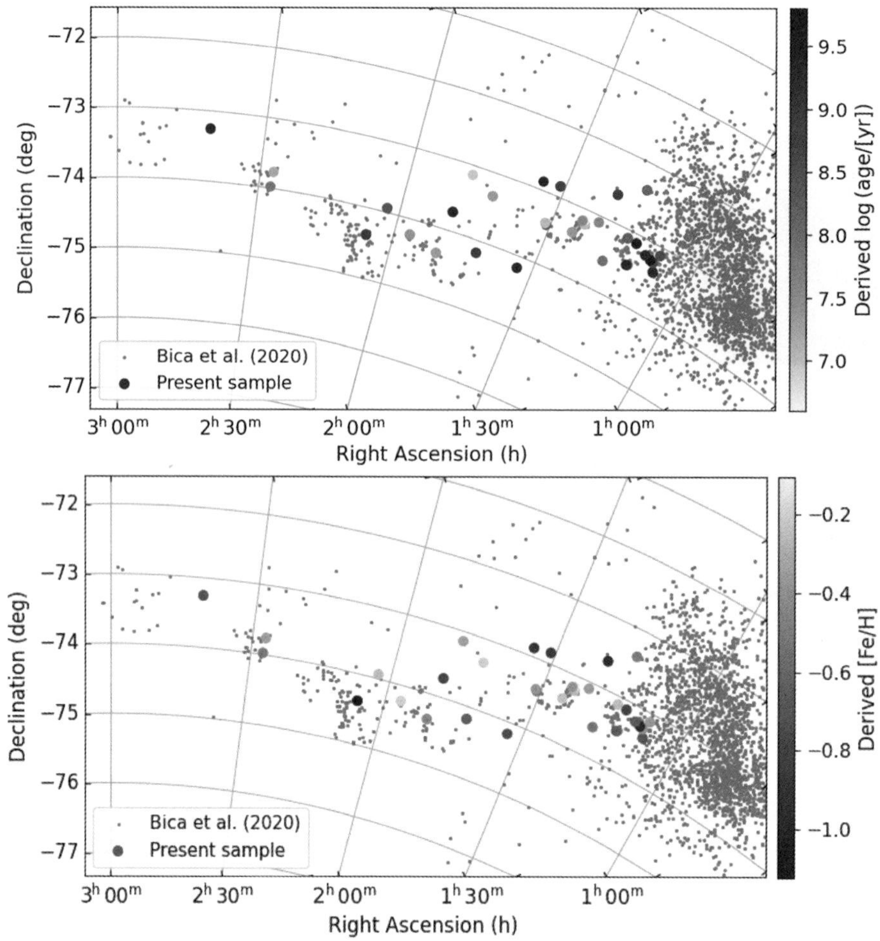

Figure 2. Projected distribution of the derived ages and metallicities for the 33 sample clusters.

a nearby field and, based on their relative location in the CMD and radius to the cluster centre, computes a median membership value. For the isochrone fitting, we employ the `SIRIUS` code (Souza *et al.* 2020) to fit PARSEC isochrones (Bressan *et al.* 2012) to the decontaminated V vs. $V-I$ CMDs. In this case, the likelihood compares the position of each star in the CMD to the closest point of the tentative isochrone, so that the MCMC retrieves posterior distributions in age, metallicity, distance and reddening. The cluster mass is obtained by integrating the flux of all member stars, converting it to absolute magnitude and applying a calibration with age and metallicity provided in Maia *et al.* (2014).

3. Results: cluster parameters, gradients and age-metallicity relation

Figure 1 gives the structural and fundamental parameters derived for the Wing/Bridge cluster HW77 with VISCACHA data. The observation was taken in optimal conditions for the use of adaptive optics, providing the deepest CMD and best seeing ($\sim 0.5''$) of the sample. A large core radius of $22.7''$ was derived with a small concentration parameter $c = \log(r_t/r_c) = 0.54$, whereas the age of 1.12 Gyr and [Fe/H] $= -1.02$ dex resulted in

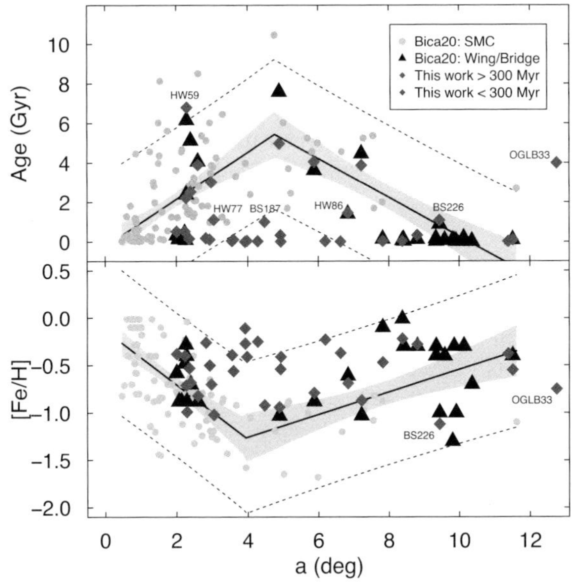

Figure 3. Age and metallicity as a function of the projected distance to the SMC centre. Grey and black points, as well as the lines and shaded area (gradients), are from Bica *et al.* (2020), and the coloured points are the present sample. The clusters older than 300 Myr that deviate from the gradients are identified.

a mass of $2.2 \pm 0.6 \times 10^3\,M_\odot$. The SMASH data for this cluster are nearly identical to VISCACHA, with the CMD g vs. $g - i$ with very similar results and uncertainties (Oliveira et al., in preparation).

The sample of 33 Wing/Bridge clusters resulted in concentration parameters between 0.46 and 1.25 (smaller than Galactic open and globular clusters, but consistent with the Magellanic Clouds clusters), ages from 4 Myr (HW81) to 6.8 Gyr (HW59), distances from 50 to 69 kpc, and masses from ~ 500 (L101, ICA45) to $1.4 \times 10^4\,M_\odot$ (L113, L110). The mass of the 33 clusters add up to $10^5\,M_\odot$ which, extrapolating to the ~ 100 clusters and ~ 300 associations located at the Bridge, provide a new estimate of $3 - 5 \times 10^5\,M_\odot$ for the Bridge stellar mass, more than one order of magnitude higher than Harris (2007).

Figure 2 presents the projected distribution of all the objects from Bica *et al.* (2020) together with the 33 sample clusters colour-coded by the derived age and metallicity. The older clusters are mostly isolated, located close to the SMC, whereas the young ones appear to be grouped along the Bridge. Two groups become evident: old clusters (> 500 Myr) more metal-poor than -0.6 dex, vs. clusters younger than the Bridge, with [Fe/H] ~ -0.4 dex, probably formed in situ. Figure 3 reproduces a figure from Bica *et al.* (2020, gray and black dots) with the sample clusters older or younger than 300 Myr as red or blue symbols, in order to check if our results follow the age and metallicity gradients of the SMC vicinity. As expected, the old clusters (probably stripped from the SMC) follow both gradients, with increasing age and decreasing metallicity until $a \sim 4^\circ$ and an inversion after that. The young clusters does not show a pattern in age and have a nearly constant metallicity along the Bridge.

The age-metallicity relation (AMR) is a valuable tool to analyse the chemical evolution of a galaxy, with hints of chemical enrichment or decrease in metallicity. The present results follow the chemical evolutionary models in most of the cases, but some clusters imprint two metallicity dips in the AMR: a larger one around 1.5 Gyr with the metallicity decreasing 0.4 dex, and a smaller one around 200 Myr ago, with a 0.3 dex decrease in metallicity. These epochs coincide with those of the formation of the Magellanic Stream and Bridge, respectively. According to the models (e.g. Tsujimoto & Bekki 2009), such metallicity decrease is usually explained by an infall of more metal-poor gas, followed by a rapid chemical enrichment (and possibly an increase in the star formation rate).

A complete interpretation of the results is given in Oliveira *et al.* (2023).

References

Besla, G., Kallivayalil, N., Hernquist, L., *et al.* 2012, *MNRAS*, 421, 2109
Bica, E., Westera, P., Kerber, L. O., *et al.* 2020, *AJ*, 159, 82
Bressan, A., Marigo, P., Girardi, L., *et al.* 2012, *MNRAS*, 427, 127
Choi, Y., Olsen K., Besla, G., *et al.* 2022, *ApJ*, 927, 153
Diaz, J. & Bekki, K. 2011, MNRAS, 413, 2015
Foreman-Mackey, D., Hogg, D. W., Lang, D., *et al.* 2013, *PASP*, 125, 306
Harris, J. 2007, *ApJ*, 658, 345
Hindman, J. V., Kerr, F. J. & McGee, R. X. 1963, *AuJPh*, 16, 570
King, I. 1962, *AJ*, 67, 471
Maia, F. F. S., Corradi, W. J. B. & Santos, J. F. C. 2010, *MNRAS*, 407, 1875
Maia, F. F. S., Piatti, A. E., Santos, J. F. C. 2014, *MNRAS*, 437, 2005
Maia, F. F. S., Dias, B., Santos, J. F. C., *et al.* 2019, *MNRAS*, 484, 5702
Nidever, D. L., Olsen, K., Walker, A. R., *et al.* 2017, *AJ*, 154, 199
Oliveira, R. A. P., Maia, F. F. S., Barbuy, B., *et al.* 2023, arXiv e-prints, p. arXiv:2306.05503
Santos, J. F. C., Maia, F. F. S., Dias, B., *et al.* 2020, *MNRAS*, 498, 205
Skowron, D. M., Jacyszyn, A. M., Udalski, A., *et al.* 2014, *ApJ*, 795, 108
Souza, S. O., Kerber, L. O., Barbuy, B., *et al.* 2020, *ApJ*, 890, 38
Tsujimoto, T. & Bekki, K. 2009, *ApJ*, 700, 69
Zivick, P., Kallivayalil, N., van der Marel, R. P., *et al.* 2018, *ApJ*, 864, 55

Dynamical Masses of Local Group Galaxies
Proceedings IAU Symposium No. 379, 2025
P. Bonifacio, M.-R. Cioni, F. Hammer, M. S. Pawlowski & S. Taibi, eds.
doi:10.1017/S1743921323001400

Strömgren photometric metallicity of the Small Magellanic Cloud stars using Gaia DR3-XP spectra

Abinaya O. Omkumar[1,2,3], **Smitha Subramanian**[1], **Maria-Rosa L. Cioni**[2] **and Jos de Bruijne**[4]

[1]Indian Institute of Astrophysics, Koramangala II Block, Bangalore-560034, India
email: aomkumar@aip.de

[2]Leibniz-Institut für Astrophysik Potsdam (AIP), An der Sternwarte 16. D-14482 Potsdam, Germany

[3]Institut für Physik und Astronomie, Universität Potsdam, Haus 28, Karl-Liebknecht-Str. 24/25, D-14476 Golm (Potsdam), Germany

[4]European Space Agency, ESTEC, Noordwijk, Netherlands

Abstract. Observational studies have identified several sub-structures in different regions of the Magellanic Clouds, the nearest pair of interacting dwarf satellites of the Milky Way. By studying the metallicity of the sources in these sub-structures, we aim to shed light on the possible origin of these sub-structures. Spectroscopic metallicities exist only for a few thousand sources, mostly giant stars located in specific regions of the galaxies. These metallicities come from different instruments at various spectral resolutions, and systematic uncertainties hamper comparisons and draw firm conclusions about their origin. The third data release of *Gaia* has provided us with $\sim$ 0.17 million XP spectra of the different stellar populations in the SMC alone as faint as $\sim$ 18 mags in the G band, which are spread across $\sim 10^\circ$ from the SMC centre. We aim to determine the metallicities of these sources based on synthetic Strömgren photometry derived from XP spectra and produce a high-resolution metallicity map of the SMC. Our metallicity gradient estimate of the SMC turns out to be -0.062 ± 0.009 dex/deg. This is comparable with the previous estimates, which also validates our method of metallicity estimation. We aim to apply this method to other stellar populations and to the LMC to create a high-resolution metallicity map of the Magellanic Clouds.

Keywords. galaxies: Magellanic Clouds, galaxies: abundances, galaxies: evolution

1. Introduction

The Large Magellanic Cloud (LMC) and the Small Magellanic Cloud (SMC) are both gas-rich interacting dwarf irregulars. Evidence shows that the LMC–SMC pair, also known as the Magellanic Clouds (MCs), interacts with the Milky Way. And their proximity ($\sim$ 55 kpc) makes them an ideal laboratory in the Local Group to study galaxy interaction and evolution processes in detail. Previous studies (Mackey et al. 2016; Pieres et al. 2017; Mackey et al. 2018; Belokurov & Erkal 2019; El Youssoufi et al. 2021) have identified multiple signatures of interactions in the form of stellar sub-structures, over-densities and gaseous structures in and around the MCs. Apart from these above-mentioned sub-structures, some dual populations of intermediate/old stars (red clump; Omkumar et al. 2021 and red giant branch stars; Dobbie et al. 2014)have also been identified and studied. The results showed that they have different kinematics and are located at different distances. This could suggest that the dual populations formed during the interaction between the MCs and the stripped population's kinematics have been altered. If so, we expect similar metallicity ([Fe/H]) among the sources in these populations. Investigating the nature and origin of these sub-structures is essential for a comprehensive understanding of the consequences of dynamical interactions. But to do so, we need a metallicity estimate of a homogenous sample spread across the entire SMC, including its outskirts, where a plethora of sub-structures have been identified. Until now, we have spectroscopic metallicities of about a few thousand sources in the SMC from various instruments with different spectral resolutions. Comparing these estimates, which have different systematic uncertainties, is not trivial, and the results obtained will not be statistically significant. Hence, we need to use another standard indirect method to obtain the [Fe/H] of a larger sample.

Grady et al. (2021) presented a machine-learning method to obtain photometric metallicity estimates for the Magellanic Cloud red giants from *Gaia* Data Release 2 (DR2). Their predicted metallicity estimates of the MCs were comparable with the previous studies, but their method depends on their training sample. Also, the DR2 data was incomplete in the central regions of the MCs, and after applying all their quality filters, their final sample consisted of $\sim$ 36,000 SMC sources. In the latest *Gaia* (DR3), $\sim$ 0.17 million XP spectra consisting of different stellar populations spread across a $\sim 10^\circ$ of the SMC is provided as ancillary data. This is the largest spectroscopic dataset of the SMC obtained so far. In this work, we used a subset of this dataset ($\sim$ 80,000 giant stars) to estimate metallicity using the synthetic Strömgren photometry from the XP spectra. Strömgren photometry has proved to be one of the reliable methods to determine the metallicity of giant and sub-giant stars (Hilker 2000). Their calibration is based on a homogeneous sample of globular clusters (ω Centauri, M22 and M55), which are mostly metal-poor. When we apply this calibration relation to a sample consisting of both metal-poor giants and relatively metal-rich younger giants, the calibration would work poorly on the metal-rich giants, since such calibrators were absent (Lombardo et al. 2021). This is due to age-metallicity degeneracy. Although photometric metallicity estimates were proven to be a great tool for obtaining metallicity estimates, one must be careful to consider the age-metallicity degeneracy. In the case of our sample, it does not include many younger giants, so the effect of age-metallicity degeneracy is less likely to affect our estimates.

2. Gaia sample selection

We selected *Gaia* DR3 sources within $\sim 10^\circ$ of the SMC centre (RA $= 12^\circ\!.80$ and Dec $= -73^\circ\!.15$) with magnitudes G < 20.5. This resulted in 4,710,809 sources, which further reduced to 171,060 sources having Gaia DR3 XP spectra. We also applied additional criteria on parallax and proper motions (Gaia Collaboration et al. 2021; Arenou et al. 2018; Omkumar et al. 2021) only to select probable SMC sources. Our final sample contained 151,530 sources. We used this source list to query the XP spectra from the Gaia archive (Datalink-products). Currently, we have estimated the [Fe/H] for only a subset ($\sim$ 80,000 sources) of the final sample in the dereddened colour range $0.5 < (b - y)_0 < 1.1$ mag for which the [Fe/H] calibrations were available in the literature. The spatial distribution of this subset of data is shown in Figure 1.

3. Synthetic Photometry using GaiaXPy

We derived Strömgren magnitudes (v, b and y) using the GaiaXPy tool provided by the *Gaia* consortium. This tool allows the generation of synthetic photometry in

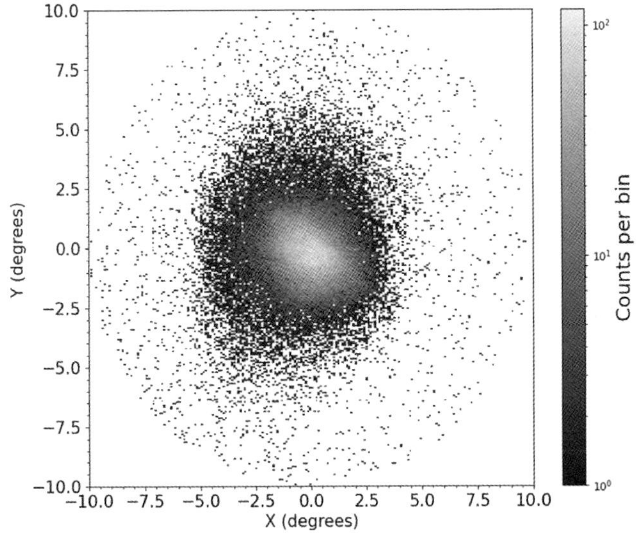

Figure 1. Spatial distribution of the selected sources in the dereddened colour range $0.5 < (b - y)_0 < 1.1$ mag in XY plane centred on the SMC. The increasing number density is shown with the colour scale from blue to yellow.

a set of desired systems from the input internally-calibrated continuously-represented mean spectra (see Montegriffo et al. (2022) for more details). We then corrected for the interstellar extinction by translating the extinction coefficient in the visual band A_V into A_v, A_b and A_y, where $A_V = A_0/1.003$ (Sale et al. 2014) and A_0 is the extinction at 547.7 nm. Using these extinction-corrected magnitudes, we proceeded to estimate the [Fe/H].

4. Estimation of [Fe/H]

Hilker (2000) provided the calibration equation to calculate the iron abundance [Fe/H] that is applicable for sources in the dereddened colour range $0.5 < (b - y)_0 < 1.1$ mag. This is reproduced here in equation 1. We used our extinction-corrected colours in this equation and estimated the [Fe/H] of each of the sources. We plotted a 2D Hess diagram of the estimated [Fe/H] (dex) as a function of radius (degrees) from the SMC centre in the left panel of Figure 2. The increasing stellar density is indicated by the colour from blue to yellow. From the figure, it is clear that we do see a larger spread in the [Fe/H] values, but the stellar density indicates that very few sources have very low and high [Fe/H] values. To better understand these values, we binned the dataset in concentric shells of $0.^{\circ}5$ from the SMC centre. We then estimated the median abundance and the standard errors in each of the bins. We plotted the median [Fe/H] and median radius of each concentric shell in Figure 2). To estimate the metallicity gradient, we made a linear fit using Python and the best fit is also shown in Figure 2). We obtained a negative metallicity gradient from the centre to the SMC outskirts from our sample, which is -0.062 ± 0.009 dex/deg, consistent with the previous studies suggesting that the central SMC is metal-rich. We also note that in the outer bin at $> 8^{\circ}$ that the metallicity increases slightly and reaches e median value similar to that of the inner regions. This could be due to the lesser number of stars in the outskirts but further investigation is required to confirm it.

$$[Fe/H] = (m_{1,0} + a_1 * (b - y)_0 + a_2)/a_3 * (b - y)_0 + a_4 \quad (1)$$

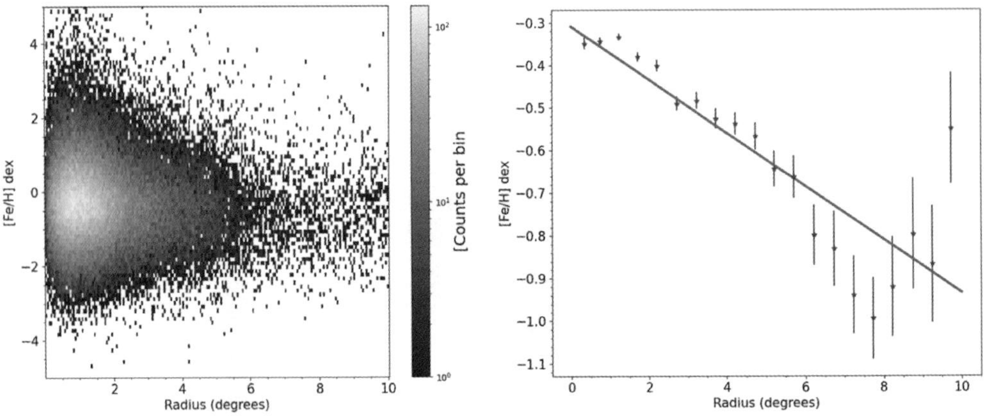

Figure 2. *Left Panel*: Estimated [Fe/H] of the subset of selected (red in Figure 1) Gaia DR3 SMC sources as a function of radius. *Right Panel*: Median metallicities for the sources in the left panel as a function of radius. The linear fit corresponds to a gradient of –0.062 ± 0.009 dex/deg.

where

$$m_{1,0} = (v - b)_0 - (b - y)_0 \tag{2}$$

and
a_1 = -1.277 ± 0.050; a_2 = 0.331 ± 0.035; a_3 = 0.324 ± 0.035; a_4 = -0.032 ± 0.025

We are also estimating the uncertainties for our [Fe/H] estimations using error propagation and it will further help as a constraint to have more significant results.

5. Summary and Ongoing work

In this contribution, we presented the method we used and the preliminary results from our investigation of the metallicity distribution across the SMC where we estimate [Fe/H] from *Gaia* XP spectra using the synthetic Strömgren photometry. The resulting [Fe/H] gradient of –0.062 ± 0.009 dex/deg from our study is comparable with the results of previous studies like –0.075 ± 0.011 dex/deg from Dobbie et al. (2014) and –0.031 ± 0.005 dex/deg from Choudhury et al. (2020). Our study also validates the potential usage of *Gaia* XP spectra to estimate the individual [Fe/H] values which are needed for different science cases. As we have mentioned in the previous sections, we have only explored a subset of the data available for the SMC. We also aim to apply this method to different types of stars, not only giants, within the galaxy to increase the size of our metallicity sample and reduce the uncertainties in our measurements. One of our next steps is to produce a high spatial resolution photometric map using the estimated values. Then identify the sources in different substructures and compare them with those of the main body of the SMC to shed light on their possible origins.

We are also applying this general method we have developed in our study to stellar populations of the Large Magellanic Cloud and this can also be extended to other systems. This will provide us with the largest and most homogeneous abundance estimates of the MCs, which can be used to analyse different stellar sub-structures, metallicities of young and old stellar populations, metallicities in different regions of the MCs and so on.

6. Acknowledgement

AOO is grateful to ESA for support via the Archival Research Visitor Programme during her stay at ESTEC.

References

Arenou, F., Luri, X., Babusiaux, C., *et al.* 2018, *A&A*, 616, A17
Belokurov, V. A. & Erkal, D. 2019, *MNRAS*, 482, L9
Choudhury, S., de Grijs, R., Rubele, S., *et al.* 2020, *MNRAS*, 497, 3746
Dobbie, P. D., Cole, A. A., Subramaniam, A., *et al.* 2014, *MNRAS*, 442, 1680
El Youssoufi, D., Cioni, M.-R. L., Bell, C. P. M., *et al.* 2021, *MNRAS*, 505, 2020
Grady, J., Belokurov, V., & Evans, N. W. 2021, *APJ*, 909, 150
Lombardo, L., François, P., Bonifacio, P., *et al.* 2021, *A&A*, 656, A155
Mackey, A. D., Koposov, S. E., Erkal, D., *et al.* 2016, *MNRAS*, 459, 239
Mackey, D., Koposov, S., Da Costa, G., *et al.* 2018, *ApJL*, 858, L21
Montegriffo, P., De Angeli, F., Andrae, R., *et al.* 2022, arXiv:2206.0620
Hilker, M. 2000, *A&A*, 355, 994
Gaia Collaboration, Luri, X., Chemin, L., *et al.* 2021, *A&A*, 649, A7
Omkumar, A. O., Subramanian, S., Niederhofer, F., *et al.* 2021, *MNRAS*, 500, 2757
Pieres, A., Santiago, B. X., Drlica-Wagner, A., *et al.* 2017, *MNRAS*, 468, 1349
Sale, S. E., Drew, J. E., Barentsen, G., *et al.* 2014, *MNRAS*, 443, 2907

Dynamical Masses of Local Group Galaxies
Proceedings IAU Symposium No. 379, 2025
P. Bonifacio, M.-R. Cioni, F. Hammer, M. S. Pawlowski & S. Taibi, eds.
doi:10.1017/S174392132300234X

Parameterization of star clusters in the Magellanic Clouds using Gaia DR3

S. R. Dhanush[1,2], Annapurni Subramaniam[1], Prasanta Nayak[3] and Smitha Subramanian[1]

[1]Indian Institute of Astrophysics, Koramangala, Bangalore, India
email: dhanush.sr@iiap.res.in

[2]Pondicherry University, R.V.Nagar. kalapet, Puducherry, India

[3]Institute of Astrophysics- PUC, Santiago, Chile

Abstract. We have introduced a quantitative and automated method to parameterize star clusters in the Magellanic Clouds (MCs) using the Gaia DR3 data. We used the existing cluster catalogs and extracted their Gaia DR3 data and nearby field regions. We automated the Field Star Decontamination (FSD) algorithm with multiple annular field regions for isolated clusters. We estimated the LMC and SMC clusters' age, extinction, distance modulus, and metallicity using a Bayesian approach. We expect to parameterize many clusters in the outer LMC with the help of the wide coverage of the Gaia data. We aim to identify correlated cluster formation episodes between the MCs, thereby throwing light on their interaction history. Here we present the preliminary results of this study.

Keywords. galaxies: star clusters, Magellanic Clouds, Field Star Decontamination, cluster parameters.

1. Introduction

The Magellanic Clouds (MCs) are the two nearest interacting dwarf irregular galaxies, comprising the Large Magellanic Cloud (LMC) and the Small Magellanic Cloud (SMC) located at $\sim$ 50 kpc (Pietrzyński et al. 2019) and $\sim$ 60 kpc (Graczyk et al. 2020), respectively. Like in our Milky Way, the MCs host many star clusters of different ages and richness. Over the past years, several studies (Glatt, Grebel, & Koch 2010; Nayak et al. 2018), estimated the properties of the star clusters, mainly the age, and reddening. These studies mostly covered the clusters in the inner regions of the MCs. The clusters in the outskirts of the MCs, especially in LMC are not well studied yet, due to the lack of data covering the entire MCs.

In this study, we attempt to parameterize star clusters of the MCs over a large spatial extent, using the Gaia DR3 data (Gaia Collaboration et al. 2022) and using the recent extensive cluster catalogs in MCs (Bica et al. 2008; Sitek et al. 2016, 2017). Further, we aim to understand the cluster formation history in the MCs and look for evidence of interaction among the MCs and with the Milky Way (MW).

2. Methods and Analyses

To deal with the large data set containing a few thousand clusters, the entire process of accessing data from the Gaia archive, Field Star Decontamination (FSD), isochrone fitting to the cleaned CMDs, and estimation of four parameters (age, metallicity, distance modulus, extinction) for the clusters were automated.

2.1. *cluster classification and Field Star Decontamination (FSD).*

The initial stages of automation involve classifying the clusters based on the spatial location, which requires identifying clusters that do not overlap with a neighboring cluster (3019 clusters) and those that overlap (1022 clusters) in the line of sight (LOS). Further, the FSD is done based on workspace availability around each cluster. The FSD algorithm iterates over stars in the cluster region and searches for the presence of similar stars in magnitude and color bins of ranges $[\Delta G, \Delta(G_{bp} - G_{rp})] = [0.02, 0.01]$ up to a maximum of [0.5, 0.25] within the field region, then subtract them from the cluster region if found any. For isolated clusters, up to 5 annular field regions around the cluster region are used to perform the FSD. In the case of 5-FSD, the algorithm iterates star by star from the cluster region. It simultaneously checks for the presence of a similar star in all the five field regions, and if not found, the current iterating star is considered a cluster member. The same logic applies till 1-FSD. Thus, five stages of FSD are done for each cluster. We initially characterize the well-isolated clusters in the MCs (clusters with five annular field regions available for the FSD.). The cluster workspace thus involves the cluster region and the field regions. We found 2294 well-isolated clusters to begin our study. The chosen cluster workspaces are downloaded from the Gaia archive with some initial selection criteria before performing FSD, 1) filtering stars from workspaces with RUWE (Re-normalized Unit Weight Error) < 1.4 for better astrometric accuracy, 2) retaining stars with $< 2\sigma$ of proper motion (pm) and parallax from the mean pm and parallax of stars in each cluster workspace for the foreground decontamination from the MW, 3) retaining stars brighter than $G = 20.5$ mag to maintain photometric accuracy in cluster parametrization. After this, the five stages of FSD are done on each star cluster. The reason for choosing multiple annular field regions will be given in the subsequent section.

2.2. *Cleaned CMD and prior selection criterion*

The different stages of FSD have different numbers of cluster stars retained after field star subtraction. The 5-FSD stage retains the least number, whereas 1-FSD retains the maximum number of members in cleaned CMD, and the final choice is made based on χ^2 of observed CMD data points to the grid of the model isochrones. The likelihood function is the average χ^2 associated with each CMD data point (magnitude, color, magnitude error, color error: $m_{i,o}$, $c_{i,o}$, em_i, ec_i) and nearest isochrone model data points (model magnitude, model color: $m_{i,m}$, $c_{i,m}$). And it is given by the following equation,

$$\chi_f^2 = \frac{1}{N} \sum_{i=1}^{N} \left\{ \frac{(m_{i,o} - m_{i,m})^2}{{em_i}^2} + \frac{(c_{i,o} - c_{i,m})^2}{{ec_i}^2} \right\} \tag{1}$$

The prior assumptions for isochrone fitting for the LMC clusters include a metal fraction of $Z = 0.008$ (Piatti & Geisler 2013), a mean distance modulus of $D = 18.47$ (Pietrzyński et al. 2019) adjusted for an inclination of $i = 23.43°$ and position angle of the line of nodes, $\phi = 160.12°$ (Saroon & Subramanian 2022). Meanwhile, $Z = 0.004$ (Nayak et al. 2018) and $D = 18.97$ (Szewczyk et al. 2009) were chosen for the SMC. We used the models from PARSEC stellar evolutionary tracks (Bressan et al. 2012) to generate the theoretical isochrones as models for our study, with logarithmic ages ranging from 6 to 10 ($\log t$). The isochrone fits were obtained for the 1 to 5 stages of FSD for the same cluster. The final choice of the CMD is made after comparing age estimations from all the stages. For example, for the 3-FSD and 5-FSD, if the age difference lies within $\Delta \log t < 0.1$ and if $\Delta\chi^2 < 1$, 3-FSD is selected as it would have more member stars. For clusters falling outside the above range of $\Delta\chi^2 > 1$, fitted CMDs are selected from the

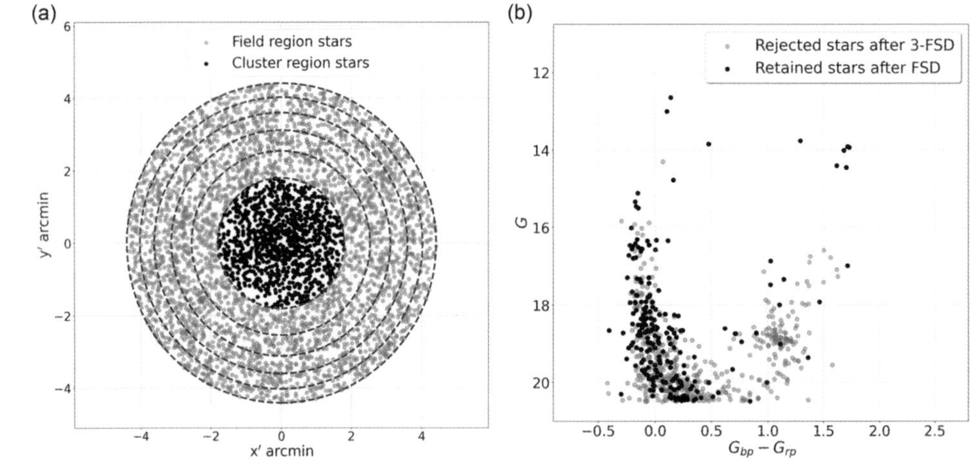

Cluster HW66: workspace

Cluster HW66: Cleaned CMD after 3-FSD

Figure 1. Cluster HW66: a) Cluster workspace. Black dots represent the cluster region stars, and red dots represent the field stars within the five annular field regions b) Cleaned cluster CMD, after 3-FSD. Prior selection criteria of rejecting stars fainter than $G = 20.5$, retaining stars with proper motion (pm) 2σ from mean pm of workspace and parallax < 2 mas is shown. The black and red dots represent the retained and rejected stars within the cluster region.

FSD stage with minimum χ^2. The remaining clusters with $\Delta \log t > 0.1$ are passed on to compare among 2-FSD and 4-FSD, then to 1-FSD and 2-FSD with the same above logic. This sequential filtering criterion ensures the identification of the best stage of FSD for parameter estimation, with ages coming close to the maximum likelihood domain given the prior information. Finally, we obtained a given cluster's maximum likelihood value of age and extinction. A sample cluster workspace and cleaned CMD of cluster HW66, in the LMC (located at, RA = 01h 12m 04s and DEC = -75° 11′ 51″, with radius, r = 1.8′) is shown in Fig 1.

2.3. *Estimation of cluster parameters*

The final parameters were estimated using the Bayesian approach by implementing a basic Markov Chain Monte Carlo (MCMC) sampling with the Metropolis-Hastings algorithm. The priors were obtained from the maximum likelihood estimate, and MCMC iterations were performed with a single walker for 20 thousand steps with a burn-in period of 5000 steps. The acceptance rate of sampling is kept to an optimum value between 0.25 to 0.32. The convergence of the parameters is found within the posterior distributions, and the estimated best parameters from the 50 percentile of the accepted samples are shown in Fig 2.

3. Preliminary results and future direction

We developed an automated algorithm to quantitatively estimate all the four parameters such as age, distance, extinction, and metallicity of star clusters in the MCs. Here we describe the initial and final parameters estimations method using a Bayesian approach. The initial results are compared with the literature values and show a satisfactory match.

We plan to apply this technique to all the isolated as well as clusters with one neighbor to make a comprehensive catalog of cluster parameters. We also plan to produce the spatiotemporal map and metallicity map of the clusters in the MCs. These will be

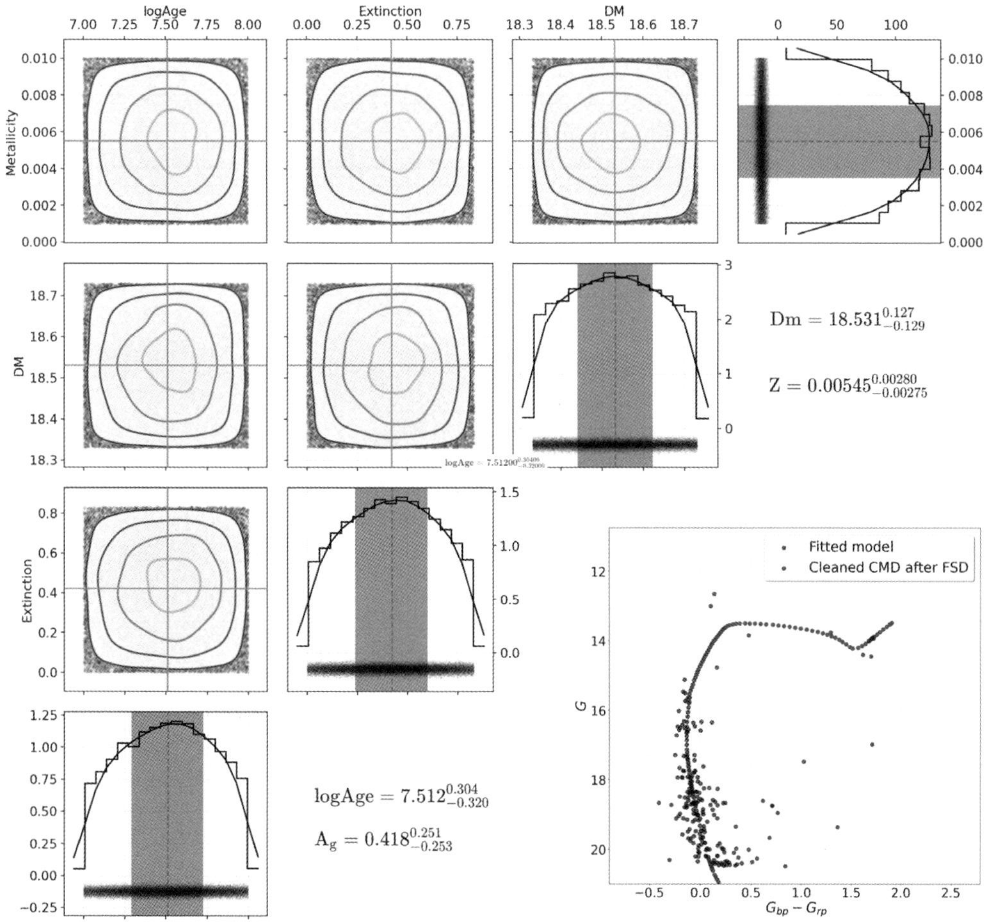

Figure 2. The estimated age, extinction, distance modulus, and metallicity for the cluster HW66 after MCMC is shown here. The 50 percentile median estimate is considered the best parameter from their posterior distribution. The cluster CMD with the fitted model is shown.

useful to identify correlated cluster formation episodes and look for evidence of the MCs interaction history.

References

Bica E., Bonatto C., Dutra C. M., Santos J. F. C., 2008, MNRAS, 389, 678.

Bressan A., Marigo P., Girardi L., Salasnich B., Dal Cero C., Rubele S., Nanni A., 2012, MNRAS, 427, 127.

Gaia Collaboration, Vallenari A., Brown A. G. A., Prusti T., de Bruijne J. H. J., Arenou F., Babusiaux C., et al., 2022, arXiv, arXiv:2208.00211.

Glatt K., Grebel E. K., Koch A., 2010, A&A, 517, A50.

Graczyk D., Pietrzyński G., Thompson I. B., Gieren W., Zgirski B., Villanova S., Górski M., et al., 2020, ApJ, 904, 13.

Piatti A. E., Geisler D., 2013, AJ, 145, 17.

Pietrzyński G., Graczyk D., Gallenne A., Gieren W., Thompson I. B., Pilecki B., Karczmarek P., et al., 2019, Natur, 567, 200.

Nayak P. K., Subramaniam A., Choudhury S., Indu G., Sagar R., 2016, MNRAS, 463, 1446.

Nayak P. K., Subramaniam A., Choudhury S., Sagar R., 2018, A&A, 616, A187.

Saroon S., Subramanian S., 2022, A&A, 666, A103.
Sitek M., Szymański M. K., Skowron D. M., Udalski A., Kostrzewa-Rutkowska Z., Skowron J., Karczmarek P., et al., 2016, AcA, 66, 255.
Sitek M., Szymański M. K., Udalski A., Skowron D. M., Kostrzewa-Rutkowska Z., Skowron J., Karczmarek P., et al., 2017, AcA, 67, 363.
Szewczyk O., Pietrzyński G., Gieren W., Ciechanowska A., Bresolin F., Kudritzki R.-P., 2009, AJ, 138, 1661.

Dynamical Masses of Local Group Galaxies
Proceedings IAU Symposium No. 379, 2025
P. Bonifacio, M.-R. Cioni, F. Hammer, M. S. Pawlowski & S. Taibi, eds.
doi:10.1017/S1743921323001849

The metallicity gradients of red giant branch stars in the Small Magellanic Cloud

Mahdieh Navabi[1], Noelia E. D. Noël[1], Ricardo Carrera[2,3] and Michele De Leo[1,4,5]

[1]Department of Physics, University of Surrey, Guildford, GU2 7XH, UK
email: m.navabi@surrey.ac.uk

[2]INAF-Osservatorio di Astrofisica e Scienza dello Spazio, via P. Gobetti 93/3, 40129, Bologna, Italy

[3]INAF-Osservatorio Astronomico di Padova, vicolo Osservatorio 5, 35122, Padova, Italy

[4]Instituto de Astrofísica, Pontificia Universidad Católica de Chile, Av. Vicuña Mackenna 4860, 7820436, Macul, Santiago, Chile

[5]Instituto Milenio de Astrofísica, Av. Vicuña Mackenna 4860, 7820436, Macul, Santiago, Chile

Abstract. The Small Magellanic Cloud (SMC), as one of the nearest galaxies to us, provides a superb laboratory for studying resolved stars in exquisite detail. We present here the largest sample of SMC Red Giant Branch (RGB) stars ($\sim$6000) observed with the AAOmega spectrograph fed by the Two Degree Field (2dF) multi-object system at the Anglo-Australian Telescope of the Siding Spring Observatory (Australia). Metallicities were recovered using a direct estimation of [Fe/H] from the equivalent widths of the Calcium triplet (CaT). We discuss the potential implications of the metallicity gradients and compare them with previous determinations of star formation history to assess the consequences of encounters between the SMC and the Large Magellanic Cloud.

Keywords. SMC, Magellanic Clouds, dwarf-galaxies, Calcium triplet (CaT)

1. Introduction

To unravel the evolution of nearby galaxies, we need to understand the processes that shape their internal kinematics and chemical composition. Because of their proximity, the Magellanic Clouds (MCs) provide an excellent laboratory for studying the chemical evolution of individual (resolved) stars. In this project we focus on studying the chemical enrichment history of the smaller of the two MCs, the Small Magellanic Cloud (SMC). The SMC is a fascinating object of study due to its proximity (~ 62 Kpc away from us; Scowcroft et al. 2016). Assessing the metallicity profile as a function of galactocentric distance is an important step to shed light on the evolution of the SMC. Previous spectroscopic studies of SMC Red Giant Branch (RGB) field stars show the presence of radial metallicity gradients in this intermediate-age population (Carrera et al. 2008; Dobbie et al. 2014) while other works (Parisi et al. 2010) found no convincing evidence of a gradient in the metallicity distribution of the intermediate-age stellar population. Given the perturbed nature of the SMC, having had several recent interactions with the Large Magellanic Cloud (LMC; e.g. Noël et al. 2013; De Leo et al. 2020), it could be

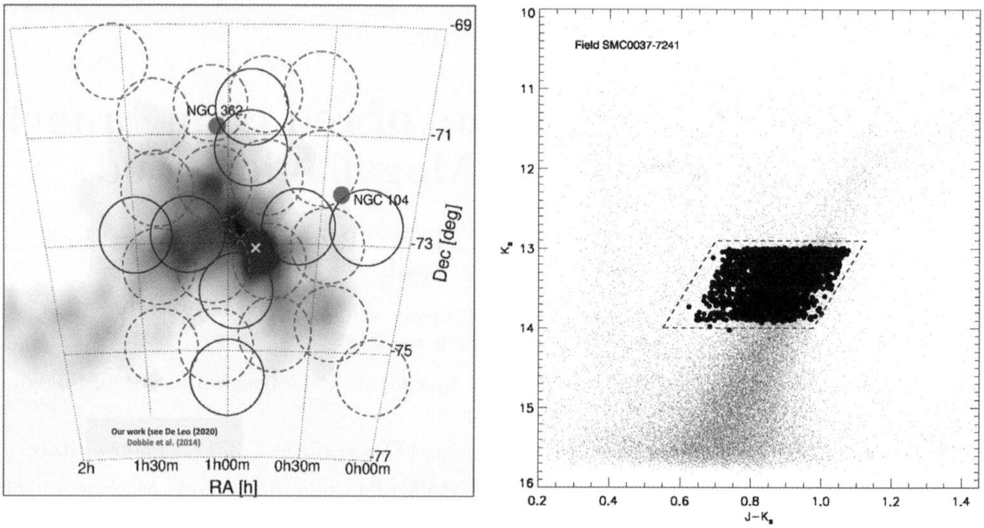

Figure 1. The left panel shows the Small Magellanic Cloud in HI with the fields observed for the present work (blue circles), sampling at different position angles. We created an 'extended' sample including Dobbie et al. (2014) data (red-dashed circles). The spectroscopic targets were selected using RGBs from the CMDs. The right panel shows the $(J - K_S, KS)$ CMD of one of the circles in the left panel with the targets overlapped by a blue dashed parallelogram.

possible that all the above studies are correct but there is much scatter depending on the location, i.e. whether the fields are located in the eastern SMC (facing the LMC) or in the western SMC (away from the LMC).

In order to find an answer to the above puzzle and, in turn, to further probe the evolution of the SMC, we present here a study consisting of an unprecedented sample of ~ 6000 RGB stars located in various fields of the SMC (see Figure 1) obtained using the Anglo-Australian Telescope at the Siding Spring Observatory. In section 2 we describe the observations and data. In section 3 we present the analysis. In section 4 we discuss the preliminary conclusions.

2. Data

The spectroscopic observations were obtained using the AAOmega spectrograph fed by the Two Degree Field (2dF) multi-object system at the Anglo-Australian Telescope at Siding Spring Observatory, Australia (De Leo et al. 2020). Our observations (the blue circles in the left panel of the figure 1) were performed during the nights of the 2nd and 3rd October 2011 and the 10th, 11th and 12th November 2014 with a spectral resolution of R ~ 8500 (Saunders et al. 2004). We combined our sample with that of Dobbie et al. (2014), who used the same instrument and setup, allowing us to compile the largest sample of SMC Red Giant Branch (RGB) stars, ~ 6000 in total located at distances of up to ~ 5 deg from the SMC photometric centre. Spectroscopy targets were selected based on the position of the upper RGB in the 2MASS colour-magnitude diagram (CMD, Skrutskie et al. 2006). The right panel of Figure 1 shows the location of the target star selection region overlaid on the $J - KS$ vs. K_S CMD for one of the fields in the left panel. Stars sampled with AAOmega are indicated by the black dots within the blue dashed parallelogram.

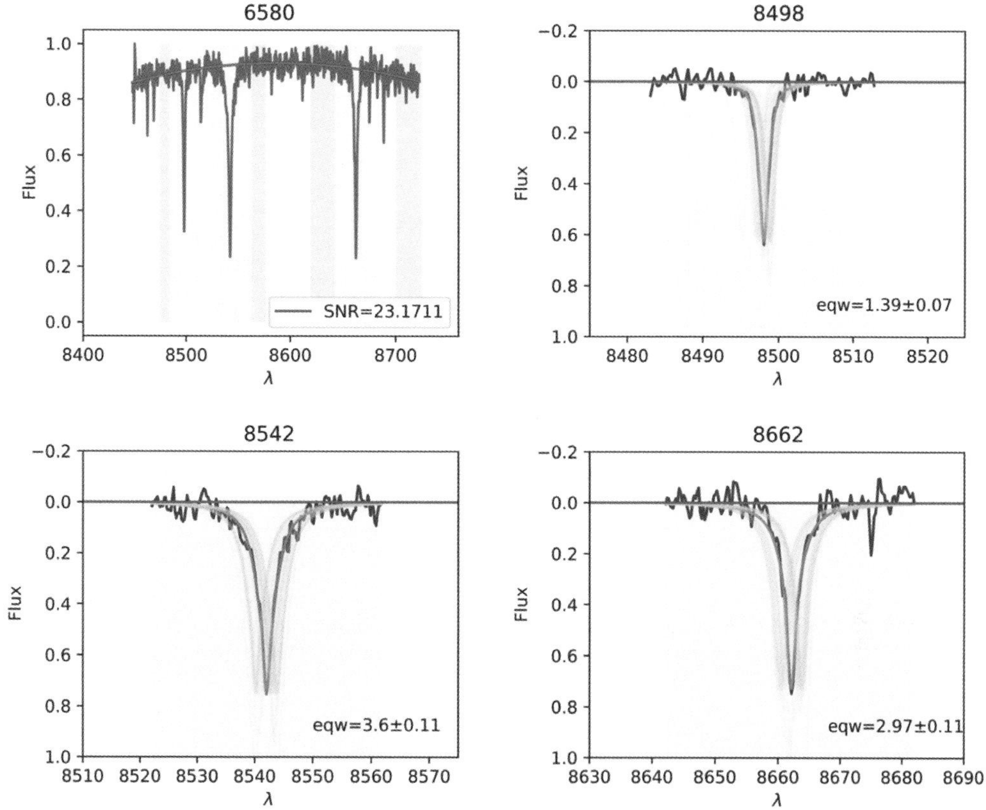

Figure 2. Example of the fitting technique used in the CaT method. Upper left panel: Full reduced stellar spectrum (in blue) with continuum fit (red line) and the line bands used for the line fit (pink contours). The ID of the star is at the top and the SNR as an inset at the bottom. Upper right panel: Line fit for the CaT line at 8498 Å (red line) superimposed on the spectrum within the line band. The green line from the continuum fit is also shown. Also, the grey lines are from the MCMC analysis to obtain the equivalent width error. Lower left panel: same as upper right panel, but for the line at 8542 Å. Bottom right panel: same as top right, but for the line at 8662 Å.

3. Analysis

The CaII triplet (CaT) lines at wavelengths 8498 Å, 8542 Å and 8662 Å (indicated by the faint yellow contours in Figure 2) have the advantage of being easily distinguishable even in low and medium resolution spectra (Carrera et al. 2009). The intensity of the CaT lines is determined by measuring the equivalent width, which is the area between the fitted profile and the continuum level. Since their intensities vary with metal content, CaT lines have been widely used as indicators of metallicity in various systems, including metal-poor and old Galactic systems such as globular clusters, dwarf spheroidal to open clusters, and irregular galaxies, as well as complex systems such as the Large Magellanic Cloud (Carrera et al. 2008; Starkenburg et al. 2010).

Figure 2 shows the continuum bandpasses (in the pink background) selected to avoid the inclusion of other spectral lines and molecular bands. The continuum position is calculated by fitting a two degree polynomial function to the means of each continuum bandpass. A combination of a Gaussian fit and a Lorentzian fit is used to determine the profile of each CaT line, as Cole et al. (2004) showed this method to be the best fit for

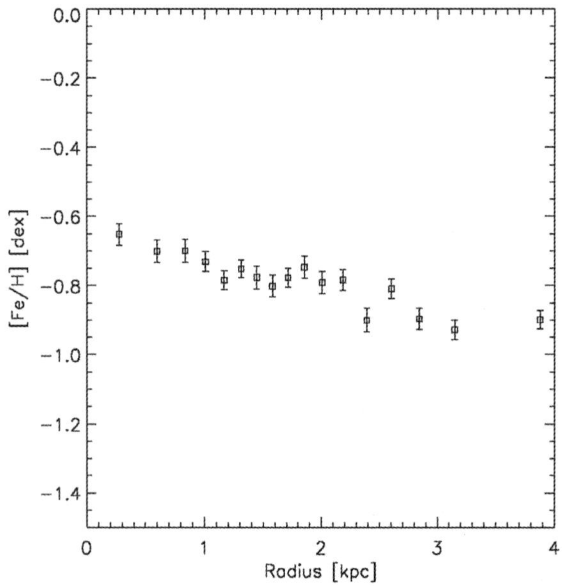

Figure 3. Estimation of the metallicity using the CaT index method for the "extended" sample versus the projected distance from the SMC centre.

both weak and strong lines (Carrera et al. 2013). This fitting function is represented by a red line superimposed on the spectrum in the upper right panel of Figure 2 and on the spectrum in the lower panels of Figure 2.

To obtain the metallicity from the CaT equivalent widths, we used the E.q. 1 from Carrera et al. (2013), which allows the direct estimation of [Fe/H] from the combination of the equivalent widths of three lines (ΣCaT) and the magnitude in the K-band (Mag_k).

$$[Fe/H] = a + bMag_k + c\Sigma CaT + d\Sigma CaT^{-1.5} + eMag_k\Sigma CaT \tag{1}$$

For the CaT calibration and determination of the coefficients in E.q. 1, we used two samples of cluster stars and metal-poor field stars. The first sample was collected by observing ~ 500 RGB stars in 29 different open and globular clusters of the Milky Way. This sample covered a wide range of metallicities from [Fe/H] = −2.3 to [Fe/H] = +0.4. The second sample consisted of stars with metallicities [Fe/H] $<$ −2.3 dex, observed in two runs in April 2011 and February 2012 with IDS at the Issac Newton Telescope (INT) at the Roque de los Muchachos Observatory in La Palma, Spain (Carrera et al. 2013).

4. Preliminary findings

After estimating the metallicity of all stars in the extended sample, we binned them according to their distance from the centre of the SMC, with each bin containing 300 stars. Figure 3 shows a linear trend of these stars, with increasing distance from the centre the metallicity of the bins decreases. The linear gradient of -0.071 ± 0.016 dex deg^{-1} obtained by our analysis is in very good agreement with other reports of metallicity gradients in this irregular dwarf galaxy, such as that found by Dobbie et al. 2014 (-0.075 ± 0.011 dex deg^{-1}) and roughly in agreement with the two estimates of

Choudhury et al. 2018 (-0.045 ± 0.004 dex deg^{-1} for the MCPS sample, -0.067 ± 0.006 dex deg^{-1} for the OGLE sample).

Although the metallicity gradient presented here is in good agreement with the literature, we find a slight overestimation of the sample metallicities compared to those reported in the literature (Dobbie et al. 2014; Choudhury et al. 2018). We are currently carrying out various tests and calibration checks to confirm our findings and proceed with the determination of alpha enhancements.

References

Carrera, R., Gallart, C., Aparicio, A., Costa, E., Méndez, R.A., and Noël, N.E.D.: 2008, AJ, 136, 1039
Carrera, R., Gallart, C., Aparicio, A., Costa, E., Hardy, E., Méndez, R.A., and, ...: 2009, AJ, 256, 275
Carrera, R., Pancino, E., Gallart, C., et al. 2013, MNRAS, 434, 1681
Choudhury, S., Subramaniam, A., Cole, A. A., et al. 2018, MNRAS, 475, 4279
Cole, A. A., Smecker-Hane, T. A., Tolstoy, E., et al. 2004, MNRAS, 347, 367
De Leo, M., Carrera, R., Noël, N. E. D., et al. 2020, MNRAS, 495, 98
Dobbie, P. D., Cole, A. A., Subramaniam, A., et al. 2014, MNRAS, 442, 1680
Noël, N. E. D., Greggio, L., Renzini, A., et al. 2013, ApJ, 772, 58
Noël, N. E. D., Conn, B. C., Read, J. I., et al. 2015, MNRAS, 452, 4222
Parisi M. C., Geisler D., Grocholski A. J., Clariá J. J., Sarajedini A., 2010, AJ, 139, 1168
Saunders, W., Bridges, T., Gillingham, P., et al. 2004, SPIE, 5492, 389
Scowcroft, V., Freedman, W. L., Madore, B. F., et al. 2016, ApJ, 816, 49
Skrutskie, M. F., Cutri, R. M., Stiening, R., et al. 2006, AJ, 131, 1163
Starkenburg, E., Hill, V., Tolstoy, E., et al. 2010, A&A, 513, A34

Dynamical Masses of Local Group Galaxies
Proceedings IAU Symposium No. 379, 2025
P. Bonifacio, M.-R. Cioni, F. Hammer, M. S. Pawlowski & S. Taibi, eds.
doi:10.1017/S1743921323001540

Dark Matter Halo Spin of the Dwarf Galaxy UGC 5288: Insights from Observations, N-body and Cosmological Simulations

Sioree Ansar[1,2*], Sandeep Kumar Kataria[3,4] and Mousumi Das[1]

[1]Indian Institute of Astrophysics, Bangalore 560034, India,
email: sioreeansar@gmail.com

[2]Pondicherry University, R.V. Nagar, Kalapet 605014, Puducherry, India

[3]Department of Astronomy, School of Physics and Astronomy, Shanghai Jiao Tong University, 800 Dongchuan Road, Shanghai 200240, China

[4]Key Laboratory for Particle Astrophysics and Cosmology (MOE) / Shanghai Key Laboratory for Particle Physics and Cosmology, Shanghai 200240, China

Abstract. Dark matter (DM) halo angular momentum is very challenging to determine from observations of galaxies. In this study, we present a new hybrid method of estimating the dimensionless halo angular momentum, halo spin of a gas-rich dwarf barred galaxy UGC5288 using N-Body/SPH simulations. We forward model the galaxy disk properties- stellar and gas mass, surface densities, disk scalelengths, bar length and bar ellipticity from observations. We use the HI rotation curve to constrain the DM halo density profile and further use the bar properties to determine the models that best represent the observed baryonic disk. We compare the halo spin profile from our models to the halo spin profiles of similar mass dwarf galaxy analogues of UGC5288 in the TNG50 simulations. The halo spin profile from our simulated models matches within ballpark values of the median spin profile of UGC5288 analogues in the TNG50 simulations, although there are some uncertainties due to the DM halo evolutionary history.

Keywords. Galaxies: bar, Dark Matter, Simulations, Galaxies: UGC 5288

1. Introduction

One of the current challenges in the determination of DM halo properties is to estimate the halo angular momentum. The DM halo angular momentum is difficult to measure for observed galaxies because of the lack of probes to measure the motion of the DM. In simulations the DM halo spin λ represents a dimensionless measure of the halo angular momentum (Peebles 1969).

$$\lambda = \frac{J|E|^{1/2}}{GM^{5/2}} \tag{1}$$

where J is the total angular momentum, E is the total energy and M is the mass within a radius r. Here, we investigate a new hybrid method combining semi-analytical techniques and N-Body/SPH simulations to forward model a dwarf Low Surface Brightness barred galaxy UGC 5288, to determine its DM halo density profile and spin profile. We consider two major selection criteria: (1) an isolated galaxy with no signatures of interaction with other galaxies in the recent past, as we want to model an isolated system and (2) the galaxy should have very low star formation, as we do not model star formation in our simulations.

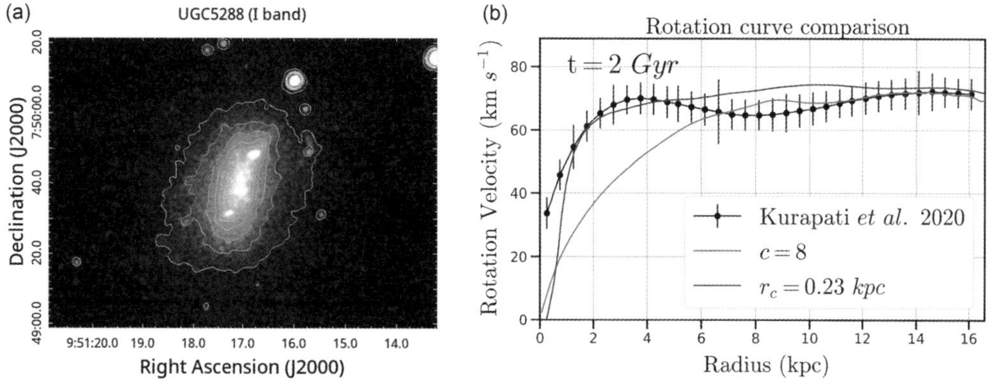

Figure 1. I-band image of UGC5288 (panel (a)). The contours show 0.6, 0.5, 0.4, 0.3, 0.2 and 0.1 times the peak flux in the image and highlight the barred structure. Panel (b) shows the comparison of the gas rotation curves of the best fit simulated models of Hernquist with halo concentration $c = 8$ (red line) and PISO halo profiles with core radius $r_c = 0.23$ kpc (blue line), with the HI rotation of UGC5288 (black line) from Kurapati et al. (2020). The PISO models show a better match to the observed HI rotation curve.

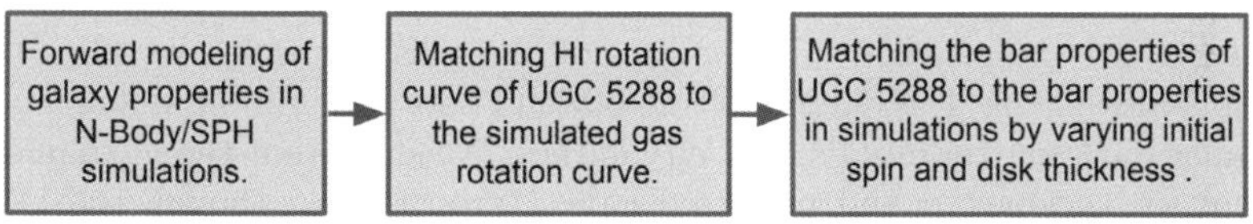

Figure 2. Flowchart of the method to estimate the halo spin profile of UGC5288.

2. Galaxy modelling

UGC5288 is a classical dwarf galaxy in the Lynx cancer void. It has an extended, undisturbed gas disk without signs of any interaction with other galaxies, a very low star formation rate (Werk et al. (2011)) and a well-studied HI rotation curve (Kurapati et al. 2020). We forward model UGC 5288's disk properties (stellar mass, gas mass, stellar and gas disk scalelengths and bar properties) and HI rotation curve in N-body/SPH simulations (see Figure 1). We model UGC 5288 with a stellar disk and gas disk with exponential profiles in radial direction and $sech^2$ profile in the direction perpendicular to the disk plane. We model the DM halo of the galaxy using two models- the cuspy Hernquist halo profile and the flat-core pseudo-isothermal (PISO) halo model. We use the N-body simulation code GalIC (Yurin & Springel 2014) to generate the initial conditions and GADGET2 (Springel 2005) to evolve the galaxy for a few Gyrs to ensure the stability of our simulated models. For more details about the modelling of the galaxy and the simulations see Ansar et al. (2023).

3. Essential features of Dark matter halo modeling

Here we discuss in detail the salient features in the galaxy DM halo modeling.

Importance of the steps of modelling. After matching the observed properties of UGC5288 with the stable simulated models, constraining the DM halo profile with the HI rotation curve is an essential step in our method to finally estimate the halo spin profile. Following the flowchart in Figure 2 (also see Ansar et al. (2023) for more details), first we need to constrain the dark matter halo profile (using the HI rotation curve) and in the next step, to compare the observed bar properties (bar position angle and ellipticity profiles) with the bar properties in the simulations. The reason is that the predicted spin profile is sensitive to the initial DM density profile (also pointed out in Fig. 11 of the Ansar et al. (2023)). This dependence is expected as the halo spin (Equation 1) has a strong dependence on the total DM mass inside a given radius. Hence we need to constrain the gas rotation curve in the simulations so that the underlying DM density profile is constrained. After the initial estimation of the DM profile from the rotation curve, the possibility of varied halo spin distribution is reduced a bit. We note that other assumptions of the velocity structure of the DM particles will also affect the final halo spin distribution. So we consider the most general velocity structure of the DM halo for our models with three integrals of motion in modelling. Very few of these models (based on the initial halo spin parameter and disk scale height) ultimately give a disk that hosts a weak bar. Now, what happens if we first constrain the bar properties? If we do not constrain the DM profile and in the first step try to constrain the bar properties, then there will be much more scatter in the values of possible halo spin. So after constraining the DM mass distribution, we compare the bar properties of our simulated models with the observed bar in UGC 5288. This reduces the scatter in the halo spin distribution. This is the best we could do at this moment as we lack information about the evolutionary history of the galaxy we are trying to model.

Presence of a stellar bar. We also test the impact of the presence of a bar in stellar disks hosted by DM halos with a similar type of mass distribution. We select barred and unbarred galaxies in the TNG50 and TNG100 data sets in different stellar mass ranges, having similar dark matter circular velocity curves. We show that the median spin values are different for the barred and unbarred galaxy groups even though the DM circular velocity curves are very similar (see Figure L.1. in Ansar et al. (2023)). This shows that the presence of a bar affects the halo spin irrespective of the effect of the dark matter density distribution. We are studying this in more detail in a following article (Ansar et al. in prep).

Can the method be used for unbarred galaxies? The barred structure will introduce an asymmetric potential and it will interact with the spherical dark matter halo by exchanging angular momentum. Hence in principle, it should significantly alter the final spin distribution and deviate it from that of an unbarred galaxy. In support of this argument, the connection between halo spin and bar properties has been already discussed in various idealized N-body simulations (Long et al. 2014; Collier et al. 2018; Kataria & Shen 2022), as well as TNG50 simulations (Rosas-Guevara et al. 2022), where the authors show that bars are more prominent in low-spin halos compared to high-spin dark matter halos. With a barred galaxy, we have a small range of values of halo spin that helps in recreating the baryonic disk properties. So it helps us narrow down on some specific models. If the galaxy is unbarred, we will only be using the constraints from the rotation curve for the DM mass distribution, but then we will have a different and wider range of values for halo spin and initial disk thickness that creates an unbarred galaxy. Hence, we will get more scatter in the values of halo spin for modelling an unbarred galaxy.

4. Comparison with cosmological simulations

We investigate halo spin distribution for barred and unbarred galaxies in the TNG50 (Nelson et al. 2019; Pillepich et al. 2019; Nelson et al. 2019) cosmological magneto hydrodynamical simulations at z=0. We find 24 similar stellar mass galaxy analogues of UGC5288 in the TNG50 simulations having similar flat rotation velocities, a small stellar disk with a weak bar ($A_2/A_0 > 0.2$) and an extended gas disk. 75% of the UGC5288

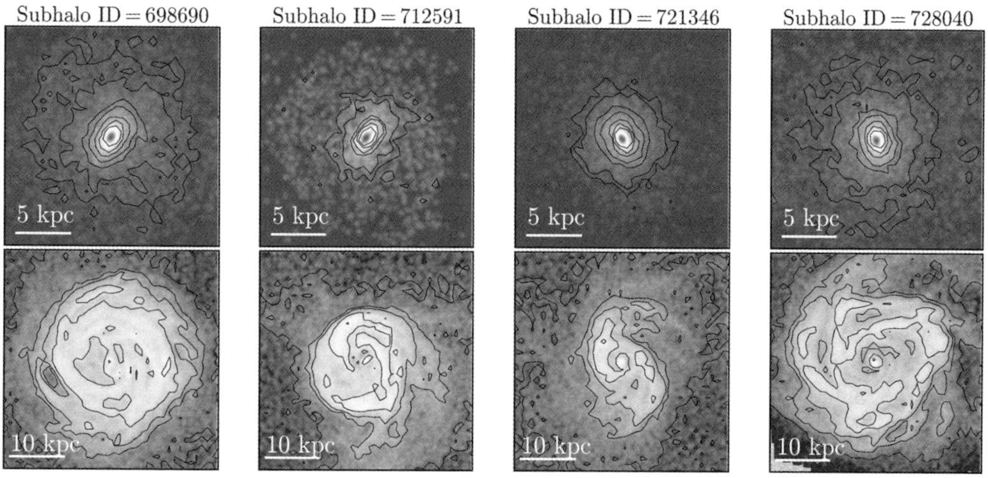

Figure 3. UGC5288 analogue galaxies in TNG50 with their Subhalo IDs (698690, 712591, 721346, 728040) undergoing ≤ 1 major merger all throughout their evolution. Among these galaxies, the first three galaxies (Subhalo ID- 698690, 721346 and 712591) have low halo spin similar to the halo spin of our predicted model with PISO halo profile and the last one (Subhalo ID-728040) have a higher halo spin.

analogue sample (~ 18 galaxies) have low halo spin similar to the halo spin prediction from our model and 4 out of the 24 galaxies undergo ≤ 1 major merger event throughout their evolution (see Figure 3). Hence a significant number of dwarf galaxies that undergo major mergers can still have low halo spin. Multiple merger properties, for example, satellite mass, number of pericenter passages, the inclination of orbit, and velocity of the incoming satellite will play important roles in determining the final halo spin of a galaxy at $z = 0$. Currently, we have not considered the effect of mergers in our analysis.

5. Discussion and conclusion

We discuss the salient features of a new hybrid method to estimate the halo spin profile of an observed galaxy UGC5288 by forward modelling of the galaxy properties in N-body simulation models having a stellar disk, gas disk and a dark matter halo. We first need to constrain the DM halo profile from the HI rotation curve and next to match the bar properties of the disk in the models. The presence of a bar helps in breaking the degeneracy between different models with similar DM halo profiles. Additionally, the scatter in the values of halo spin is lower for barred galaxies than the unbarred galaxies. In the process of comparing our model predictions with similar stellar mass barred galaxies in TNG50, we find that it is difficult to find well-resolved dwarf galaxies having similar stellar mass and gas mass in the data set. For a more accurate comparison with the simulated UGC5288 models, we need high-resolution dwarf galaxy data set from cosmological simulations. To validate our method we plan to apply the method to the virtual galaxy data from cosmological simulations. In future work, we will improve the methodology by including star formation and the effect of mergers that are not included in the current modelling.

Acknowledgement

SA thanks the IIA HPC facility NOVA where the simulations are run. We thank Denis Yurin and Volker Springel for providing the codes GalIC and GADGET2 publicly. We thank the IllustrisTNG team for making the data publicly available. MD acknowledges the support of the Science and Engineering Research Board (SERB) MATRICS grant MTR/2020/000266 for this research.

References

Peebles, P. J. E. 1969, ApJ, 155, 393.

Werk, J. K., Putman, M. E., Meurer, G. R., et al. 2011, ApJ, 735, 71.

Kurapati, S., Chengalur, J. N., Kamphuis, P., et al. 2020, MNRAS, 491, 4993.

Yurin, D. & Springel, V. 2014, MNRAS, 444, 62.

Springel, V. 2005, MNRAS, 364, 1105.

Ansar, S., Kataria, S. K., & Das, M. 2023, MNRAS, 522, 2967.

Long, S., Shlosman, I., & Heller, C. 2014, ApJL, 783, L18.

Collier, A., Shlosman, I., & Heller, C. 2018, MNRAS, 476, 1331.

Rosas-Guevara, Y., Bonoli, S., Dotti, M., et al. 2022, MNRAS, 512, 5339.

Kataria, S. K. & Shen, J. 2022, ApJ, 940, 175.

Nelson, D., Pillepich, A., Springel, V., et al. 2019, MNRAS, 490, 3234.

Pillepich, A., Nelson, D., Springel, V., et al. 2019, MNRAS, 490, 3196.

Nelson, D., Springel, V., Pillepich, A., et al. 2019, Computational Astrophysics and Cosmology, 6, 2.

Dynamical Masses of Local Group Galaxies
Proceedings IAU Symposium No. 379, 2025
P. Bonifacio, M.-R. Cioni, F. Hammer, M. S. Pawlowski & S. Taibi, eds.
doi:10.1017/S1743921323003435

Exploring the Mysteries of the Magellanic Stream: What We've Uncovered and What Remains Unknown

Elena D'Onghia

University of Wisconsin-Madison, 475 N Charter Str, Madison, USA
email: edonghia@astro.wisc.edu

Abstract. The Magellanic Stream is a lengthy, ribbon-like gas structure stretching 200 degrees across the sky and surrounding the Large and Small Magellanic Clouds. These two galaxies are the brightest dwarf galaxies orbiting the Milky Way (MW). The Stream is a major subject of study in galactic dynamics because it provides insights into the evolution of galaxies, including the MW and the Magellanic Clouds, its companion dwarf satellites, and the interstellar medium. Gas flows play a key role in galaxies' growth, evolution, and sustainability, but many questions related to the Stream remain unanswered. Here, I will review the main advance in this subject of the last decade and posit new questions that need to be addressed.

Keywords. Dynamics, Galaxies, Milky Way, Gas flows

1. Introduction

A fundamental problem in understanding galaxy formation is how gas accretion and large-scale outflows influence the evolution of galaxies. In particular, it is unclear whether the gas reservoir that maintains star formation in massive spiral galaxies like our MW is replenished predominantly by accreting filaments or via stripping from nearby dwarf galaxies. It is also unclear how the circum-galactic medium (CGM) gets enriched by metals. No studies have addressed the relative contribution of outflows versus gas tidally stripped by dwarf satellite galaxies to that enrichment. We know that the MW is accompanied by two dwarf galaxies, the Large and Small Magellanic Clouds (LMC and SMC, respectively, called the "Clouds" below), as well as a massive gaseous structure, the Magellanic Stream. This structure is an interwoven tail of filaments trailing the Clouds in their orbit around the Milky Way (Mathewson 1974; Bruns 2005; Nidever 2008, 2010). When considered in tandem with its Leading Arm (LA), the counterpart to the Stream on the opposite side of the Clouds leading their orbits, the Stream stretches over 200 degrees in the sky (as shown in Figure 1, colored in magenta).

Originally the Stream was detected in 21 cm observations of neutral hydrogen (Mathewson 1974) with a mass of $\approx 5 \times 10^8$ $M_\odot$. In addition to hundreds of head-tail clouds and intricate turbulent structures, data have shown that there are two spatially and kinematically distinct intertwined filaments in the Trailing Stream (Nidever 2008) (see Figure 2).

In early modeling (e.g. Gardiner 1996), the Clouds were thought to have completed multiple orbits around the Milky Way over a Hubble time. However, with the high astrometric precision of instruments such as the Advanced Camera for

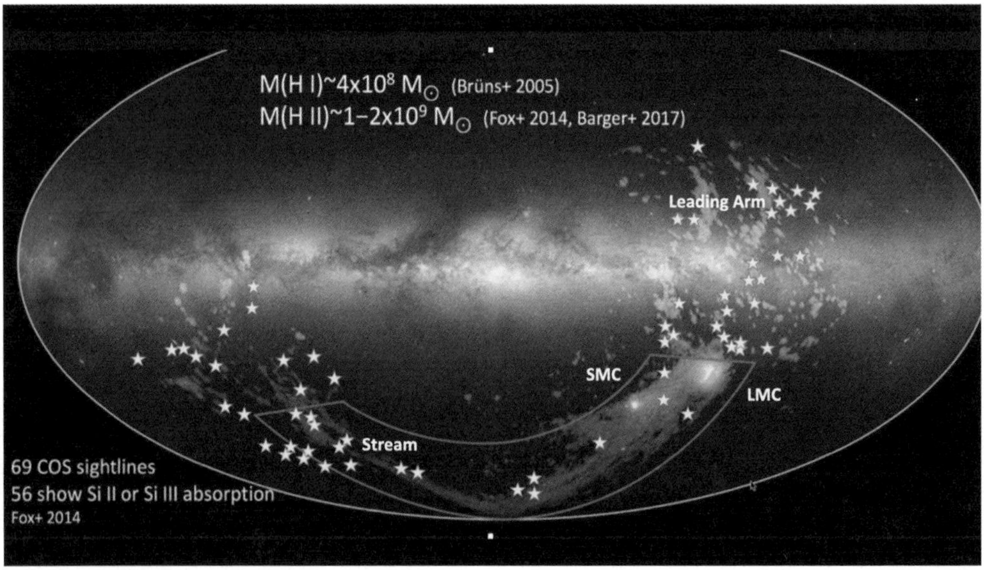

Figure 1. The Magellanic Stream in HI (magenta), credit (Nidever 2008). The bright blobs surrounded by the HI gas are the LMC and SMC, the HI gas ahead of the Clouds is the Leading Arm (LA) and the gas behind the Clouds is the Trailing Stream. The overlapping stars are the Cosmic Origins Spectrographs (COS) sightlines obtained with HST (Fox 2014). The total mass of the Stream is $\sim 5 \times 10^9$ $M_\odot$ consisting of HI ($\approx 5 \times 10^8$ $M_\odot$) and ionized gas ($\approx 5 \times 10^9$ $M_\odot$) (see D'Onghia & Fox 2016 for a recent review).

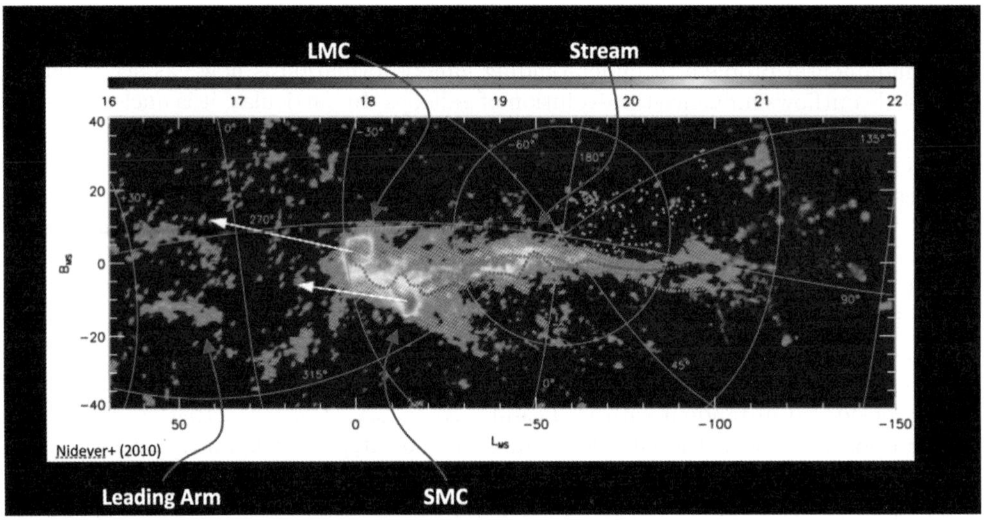

Figure 2. The Stream in Magellanic coordinates. There are two filaments, kinematically and chemically distinct indicating that both LMC and SMC gas are in the Stream (image adapted by Nidever (2010)).

Surveys (ACS) on the Hubble Space Telescope (HST) our knowledge of the Stream advanced significantly. Indeed, it was possible to measure accurate proper motions (Kallivayalil 2006a,b; Piatek 2008) and this showed that the Clouds have much higher orbital velocities than previously suspected, forcing a reconsideration of the conventional picture of their orbital history. Revised models thus had the LMC as a massive dwarf galaxy falling into the Milky Way's potential for the first time (Besla 2010). Further, the discovery of dwarf galaxies that may be associated with the Magellanic Clouds suggested the possibility that the Clouds are being accreted onto the Milky Way as part of a group of dwarf galaxies (D'Onghia 2008).

The second advance in the knowledge of the Magellanic Stream came again from the Hubble Space Telescope. In fact, absorption line spectroscopy studies allowed us to characterize the Stream's chemical composition and ionization state along dozens of sightlines (Lu 1994; Fox 2010, 2013, 2014; Richter 2013). By analyzing the spectrum of background quasars, we can detect very low-density gas between us and the source based on how much light has been absorbed since it was emitted. Spectroscopy gives the intensity of detected light as a function of frequency, allowing us to identify what chemical elements exist along the line of sight by cross-matching the frequencies of absorbed light with known atomic transitions. In the case of the Magellanic System, these measurements have resulted in two critical discoveries: (1) the two filaments are also chemically distinct, indicating that there is both LMC and SMC gas in the Stream (Fox 2013); and (2) the Stream is mostly ionized (Fox 2014).

The Stream also consists of a Leading Arm (LA), which is the counterpart to the Trailing Stream comprised of clumpy clouds of gas out ahead of the LMC and SMC in their orbits (seen in Figure 1). Due to their very high velocities, consistent with the LMC and SMC, it was proposed that these features are tidal material thrown out during the interactions between the Clouds (Putman 1998; Besla 2012; Pardy 2018). Absorption line spectroscopy studies have also investigated the chemical composition of the LA structure, but the results cannot conclusively point to a Magellanic origin (Lu 1994; Sembach 2001). An alternate explanation involves the LA being gas stripped from other dwarf galaxies that fell into the Milky Way ahead of the Magellanic Clouds (Yang 2014). However, dwarf galaxy candidates with positions and velocities to make this a plausible scenario have yet to be found (Tepper-Garcia 2019).

The Magellanic Bridge is also part of the Stream. It is a gaseous and stellar structure connecting the LMC and the SMC (see Figure 1). Interestingly, it contains stars while the Stream does not, and the Bridge and the Stream likely formed at different times (D'Onghia & Fox 2016). The stars and the gas in the Bridge have been well studied, indicating an SMC-like metallicity (Lehner 2009), a flow of stars from the SMC to the LMC, and intricate tidal structures such as a counter-bridge. Based on the existence of this Bridge and its proper motions, as well as the observed velocities of the galaxies, it is very likely that the LMC and SMC experienced a direct collision within the past several hundred million years.

2. Modeling

Early on, it was assumed that the LMC and SMC were long-lived satellites of the Milky Way, and the Stream had been formed through repeated tidal encounters between the LMC and our Galaxy (Fujimoto 1976). A significant computational effort went into determining a model that could reproduce the observations. However, Fujimoto (1977) found that no parameters for the Clouds' masses and orbits could reproduce the Stream's high negative radial velocity. Moreover, **?** discussed several alternative formation mechanisms and found that more than the proposed theories at that time were needed to explain the Stream. The next puzzle piece came with proper motion measurements for the Clouds. Using Hubble Space Telescope, Kallivayalil (2006a) determined that the LMC's velocity was very high (comparable with the Milky Way's escape velocity). Subsequent studies showed that these new data indicate that the Clouds are most likely on their first approach to the Milky Way(Besla 2007; Kallivayalil 2013). Therefore, previous tidal

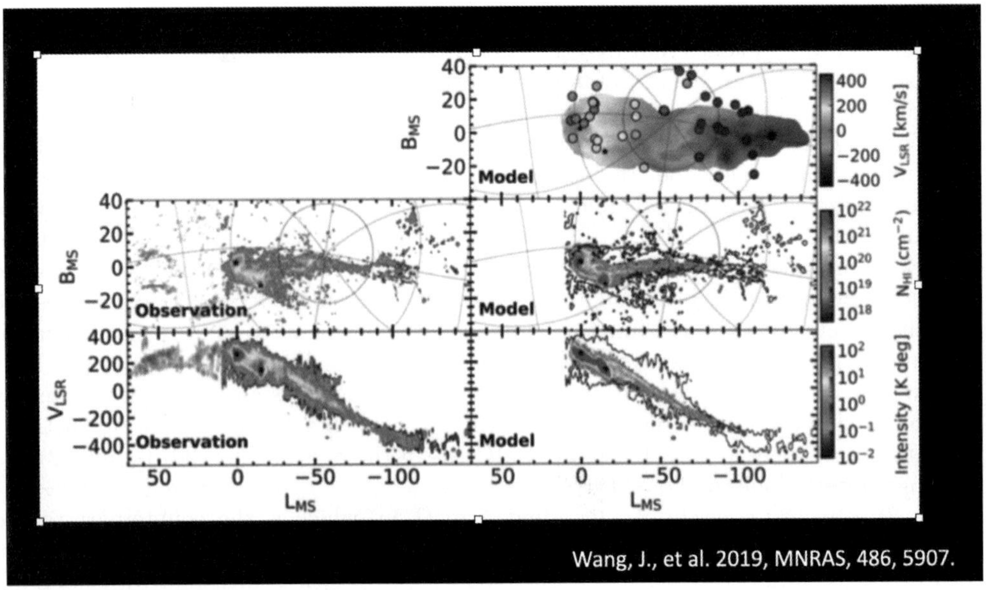

Figure 3. The ram pressure model proposes that the Stream material is pushed out of the disks of the Magellanic Clouds through hydrodynamical interactions with the Milky Way's hot gaseous halo. This model can also reproduce many of the features of the Trailing Stream (credits: Wang (2019)).

models were no longer viable. Besla (2010) proposed a new paradigm for the tidal formation of the Stream in which the material is stripped out of the SMC through interactions with the LMC before they approach our Galaxy (Besla 2012; Pardy 2018). This model has been widely considered a benchmark for forming the Stream.

Another significant development has been the inclusion of gas physics and hydrodynamic effects in the numerical experiments of the Stream. Several studies explored the effects of "ram-pressure" on the Stream's formation. In these models, the trailing material is pushed out of the Magellanic disks through interactions with an extended gaseous halo around the Milky Way (Moore 1994). The development of hydrodynamics simulation codes allowed for the whole, self-consistent exploration of both the tidal and ram pressure models for the Stream's formation. With advanced new codes and increases in computation power, modern simulations have come a long way (Besla 2012; Hammer 2015; Pardy 2018; Tepper-Garcia 2019; Wang 2019).

Currently, there are two leading theories for forming the Magellanic Stream. The original tidal model has evolved into a scenario in which the LMC and SMC are orbiting around each other, and the gravitational forces between the Clouds tidally strip material before they fall into the Milky Way (Besla 2012; Pardy 2018). In addition, the ram pressure model proposes that the Stream material is pushed out of the disks of the Magellanic Clouds through hydrodynamical interactions with the Milky Way's hot gaseous halo. This model is also able to reproduce many of the features of the Trailing Stream (Hammer 2015; Wang 2019) (see Figure 3). Just as the measurement of the proper motions of the Clouds ruled out multiple-passage models of the formation of the Stream, recent observations have been shown to conflict with these two existing models - (1) the Magellanic System is mostly ionized, and (2) the LMC has a total mass $> 10^{11}$ $M_{\odot}$. While the tidal model requires a high LMC mass, it alone cannot explain the ionized gas mass; and the

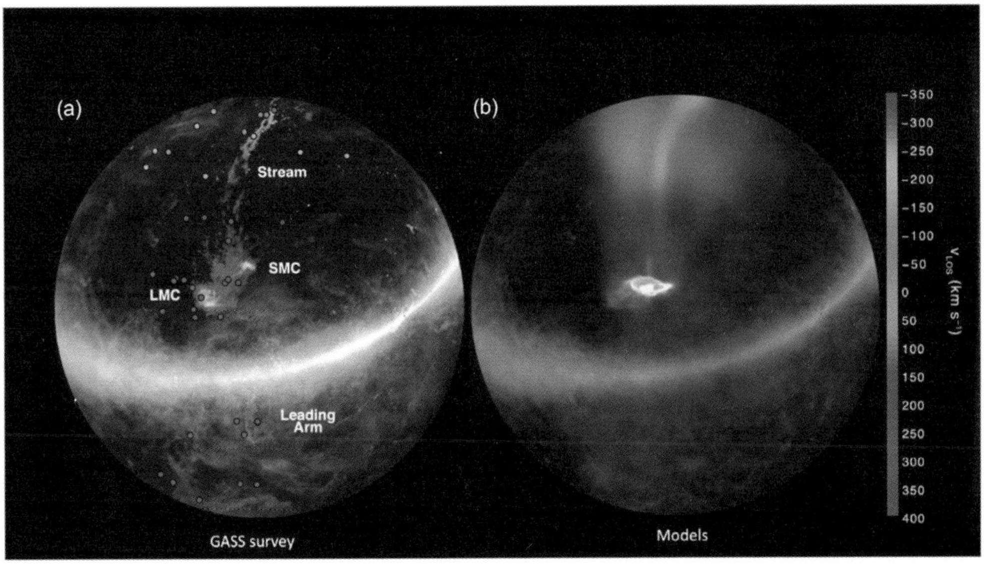

Figure 4. Hydrodynamic simulations of the Magellanic Stream with the inclusion of a Magellanic Corona (right panel) as compared to observations (McClure-Griffiths 2009)(left panel); credits: Lucchini (2020).

ram pressure model can form the ionized material, but it requires a very low mass for the LMC.

3. Understanding the Stream: the need for Galactic Coronae

A large fraction of all the ionized gas in the MW CGM is contained in the Stream (Fox 2014). All models of Stream formation via repeated encounters between the Magellanic Clouds underestimated its total mass, especially when considering the large ionized component. Even models that varied the gas fraction of the Clouds could not explain the large ionized mass (Pardy 2018). For the trailing Stream, ram-pressure stripping was suggested as the source of both the neutral material (Mathewson 1976; Moore 1994; Salem 2015; Hammer 2015; Wang 2019, 2022) and the ionized gas (Wang 2019; Bland-Howthorn 2017; Tepper-Garcia 2015). However, in a first-passage scenario, the Clouds are initially far from the MW where they encounter only low-density gas, which makes ram-pressure stripping inefficient.

The discovery of several ultra-faint dwarfs around the LMC (Bechtol 2015) suggests that the LMC and SMC approached the MW as part of a system of dwarf galaxies (the Magellanic Group), with the LMC as its largest member (D'Onghia 2008; Nichols 2011). Furthermore, cosmological simulations of MW-sized galaxies with LMC-like dwarf satellites (Pardy 2020) predict the existence of ionized halos surrounding those satellites (Hafen 2019; Jahn 2019). Thus, the LMC is expected to have an associated dark matter halo with a mass of $\sim 2 \times 10^{11}$ M$_\odot$ (Penarrubia 2016), and a virial temperature of $\sim 5 \times 10^5 K$, a factor of 4–10 lower than that of the MW halo. It is likely kept warm via energy input from stellar feedback and outflows. Recent work (Lucchini 2020) has shown that including a Magellanic Corona in hydrodynamic simulations of Stream formation solves many open questions, including the existence of the LA and the mass budget of the Stream (see Figure 4). The Corona has been directly observed in high-ion absorption toward quasars (Krishnarao 2022) (see Figure 5) and in "down-the-barrel" spectroscopic observations of hot stars in the LMC (Wakker 1998; Lehner 2009).

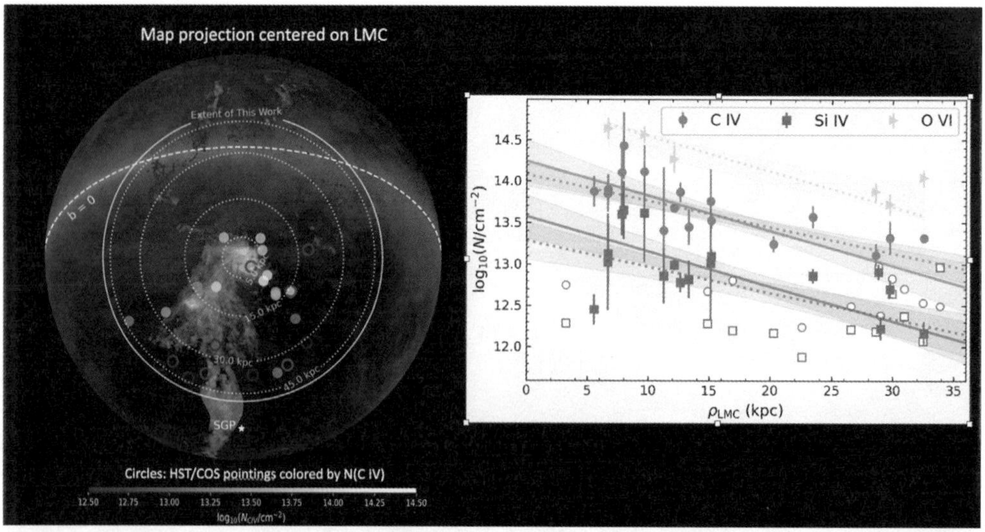

Figure 5. High-ion absorption toward quasars provides evidence for a corona around the LMC (pointings are shown in the left panel). The CIV and SiIV radial profile shows evidence of ionized coronal gas around the LMC. Credits: Krishnarao (2022)

The Magellanic Corona plays a dual role: it feeds the Stream and LA with ionized gas, thereby contributing to their mass budget, and it also *cocoons* them against disruption by the ram-pressure of the MW Corona (Lucchini 2020). The Corona therefore provides a shield around the Stream and LA allowing them to penetrate further into the MW Corona. While the MW Corona regulates the formation and morphology of the LA, the Magellanic Corona affects its spatial extent, so both Coronae have to be carefully modeled to fully explain the system. Even if the LA turns out to have a non-Magellanic origin, as recently suggested (Tepper-Garcia 2019), the inclusion of the Magellanic Corona still provides the bulk of the mass of the trailing Stream (Lucchini 2020).

4. A possible nearby Stream

The existence of a Magellanic Corona affects the orbital history of the Magellanic Clouds. As the SMC orbits around the LMC, it moves through the Corona, a warm gaseous medium, experiencing hydrodynamical drag. Hence, previous orbits of the Clouds are no longer consistent with their present-day positions and velocities. A new orbital model consistent with the locations and proper motions of the Clouds today was presented in Lucchini (2021). The model can reproduce the Trailing Stream, including the Magellanic Corona and a hot gas halo around the Milky Way. Intriguingly, these inclusions result in a Stream that is a factor of three to five closer to the Sun than previous models predicted.

5. Unsolved questions and validation tests

The existence of the LA poses a key challenge to models. If the MW corona is too dense, the infalling Magellanic gas stalls and no LA can form. Its formation is therefore tied to the properties of the MW and Magellanic Coronae. Recent models find that most (>50%) of the LA mass is contributed by the Magellanic Corona, with the remainder pulled out from both LMC and SMC (Lucchini 2020), but have not yet conducted a thorough parameter space study to determine what range of MW and Magellanic coronal parameters are consistent with the formation of the LA. Recent observational studies have constrained the metallicity of the LA using Ultraviolet (UV) spectroscopy from HST/COS found values of 5-10% solar, slightly lower than the current-day SMC metallicity (20% solar), consistent with the LA being formed from SMC material two Gyrs ago (Fox 2018; Richter 2018). At present we cannot yet rule out a scenario where the LA is unrelated to the Magellanic System, instead representing a shredded "forerunner" galaxy from the Magellanic Group (Tepper-Garcia 2019) or a complex of cool high-velocity clouds (HVCs) in the MW CGM. However, the LA origin remains uncertain.

While the recent models that include both Magellanic and Galactic Coronae break new ground and solve critical problems in Stream formation, they also pose new questions to address. If the Stream is so close, its stellar counterpart should be detectable. Furthermore, a key observational prediction of the nearby-Stream hypothesis is that the Stream should be detectable in absorption toward halo stars at known distances of 20–30 kpc (Lucchini 2021). A detection of Magellanic gas in absorption would provide the first direct distance constraint on the Stream since any gas seen in absorption in the stellar spectra must lie in front of the stars. Hence, this method offers a potential direct confirmation of the nearby-Stream hypothesis.

I am grateful for the contributions of my long-term collaborators, Andrew Fox and Scott Lucchini. Their hard work, and insightful perspectives have been invaluable. This work was supported by HST grant HST-AR-16363.001-A.

References

Bechtol, K., Drlica-Wagner, A., Balbinot, E. *et al.* 2022, *ApJ*, 807, 50
Besla, G., Kallivayalil, N., Hernquist, L. *et al.* 2007, *ApJ*, 668, 949
Besla, G., Kallivayalil, N., Hernquist, L. *et al.* 2010, *ApJ*, 721, 97
Besla, G., Kallivayalil, N., Hernquist, L. *et al.* 2012, *MNRAS*, 421, 2109
Bland-Hawthorn, J., Maloney, P.R., Stephens, A. 2017, *ApJ*, 849, 51
Brüns, C., Kerp, J., Staveley-Smith, L. *et al.* 2005, *A&A*, 432, 45
D'Onghia, E., Lake, G. 2008, *ApJ*, 686, 61
D'Onghia, E., Fox, A. 2016, *ARA&A*, 54, 363
Fox, A. *et al.* 2010, *ApJ*, 718, 1046
Fox, A. *et al.*, 2013, *ApJ*, 772, 110
Fox, A. *et al.* 2014, *ApJ*, 787, 147
Fox, A.J., Barger, K., Wakker, B.P. *et al.*, 2018, *ApJ* 854, 142
Fujimoto, M., Sofue, Y. 1976, *A&A*, 47, 263
Fujimoto, M., Sofue, Y. 1977, *A&A*, 61, 199
Gardiner, L. T., Noguchi, M. 1996, *MNRAS*, 278, 191
Jahn, E.D., Sales, L.V., Wetzel, A. *et al.* 2019, *MNRAS*, 489, 5348
Hafen, Z., Faucher-Giguère, C., Anglés-Alcázar, D. *et al.* 2019, *MNRAS*, 488, 1248
Hammer, F.,Yang, Y. B., Flores, H. *et al.* 2015, *ApJ* 813, 110
Kallivayalil, N. *et al.*, 2006a, *ApJ*, 638, 772
Kallivayalil, N. *et al.*, 2006b, *ApJ*, 652, 97
Kallivayalil, N., van der Marel, R.P. *et al.* 2013, *ApJ*, 764, 161
Krishnarao, D., Fox, A.J., D'Onghia, E. *et al.* 2022, *Nature*, 609, 915
Lehner, N., Staveley-Smith, L., Howk, J.C. 2009, *ApJ*, 702, 940
Lu, L., Savage, B.D., Sembach, K.R. 1994, *ApJ*, 437, 119
Lu, L., Savage, B.D., Sembach, K.R. 1994b, *ApJ* 426, 563
Lucchini, S., D'Onghia, E., Fox, A. J. *et al.* 2020, *Nature*, 585, 203
Lucchini, S., D'Onghia, E., Fox, A. 2021, *ApJ*, 921, 36
Mathewson, D. S., Cleary, M. N., Murray, J. D. 1974, *ApJ*, 190, 291
Mathewson, D. S., Schwarz, M. P. 1976, *MNRAS*, 176, 47
McClure-Griffiths, N. M., Pisano, D. J., Calabretta, M. R. *et al.* 2009, *ApJS*, 181, 398
Moore, B., Davis, M. 1994, *MNRAS*, 270, 209

Nichols, M., Colless, J., Colless, M., Bland-Hawthorn, J. 2011, *ApJ*, 742, 110
Nidever, David L., Majewski, Steven R., Butler Burton, W. 2008, *ApJ* 679, 432
Nidever, David L., *et al.* 2010, *ApJ*, 723, 1618
Pardy, S. A., D'Onghia, E., Fox, A.J. 2018, *ApJ*, 857, 101
Pardy, S.A., D'Onghia, E., Navarro, J.F. *et al.* 2020, *MNRAS*, 492, 1543
Peñarrubia, J., Gómez, F.A., Besla, G. *et al.* 2016, *MNRAS*, 456, 54
Piatek, S., Pryor, C., Olszewski, E.W. 2008, *AJ*, 135, 1024
Putman, M. *et al.* 1998, *Nature*, 394, 752
Richter, P. *et al.* 2013, *ApJ*, 772, 111
Richter, P., Fox, A.J., Wakker, B.P. *et al.* 2018, *ApJ*, 865, 145
Salem, M., Besla, G., Bryan, G. *et al.* 2015, *ApJ*, 815, 77
Sembach, K.R., Howk, J.C., Savage, B.D., Shull, J. M. 2001, *AJ* 121, 992
Tepper-García, T., Bland-Hawthorn, J., Sutherland, R.S. 2015, *ApJ*, 813, 94
Tepper-García, T., Bland-Hawthorn, J., Pawlowski, M.S., Fritz, T.K. 2019, *MNRAS*, 488, 918
Wang, J., Hammer, F., Yang, Y. *et al.* 2019, *MNRAS*, 486, 5907
Wang, J., Hammer, F., Yang, Y. 2022, *MNRAS*, 515, 940
Wakker, B., Howk, J.C., Chu, Y. *et al.* 1998, *ApJ*, 499, 87
Yang Y. *et al.* 2014, *MNRAS*, 442, 2419

Dynamical Masses of Local Group Galaxies
Proceedings IAU Symposium No. 379, 2025
P. Bonifacio, M.-R. Cioni, F. Hammer, M. S. Pawlowski & S. Taibi, eds.
doi:10.1017/S1743921324000310

The Dynamical Mass of the Large Magellanic Cloud

Laura L. Watkins[1], **Roeland P. van der Marel**[2] **and Paul Bennet**[2]

[1]AURA for the European Space Agency (ESA), Space Telescope Science Institute, 3700 San Martin Drive, Baltimore MD 21211, USA
email: lauralwatkins@gmail.com

[2]Space Telescope Science Institute, 3700 San Martin Drive, Baltimore MD 21211, USA

Abstract. In this contribution, we describe how we have used positions and velocities of 30 globular clusters in the Large Magellanic Cloud (LMC) to estimate its anisotropy and mass within 13 kpc, and then how we have used these estimates to extrapolate its virial mass. This is the first time that this family of mass estimation methods has been applied to the LMC. We also compare our estimate against other estimates of the LMC's mass via different methods and discuss the broader context of our results.

Keywords. Galaxy Dynamics, Galaxy Masses, Large Magellanic Cloud, Proper Motions

1. Introduction

Mass is a fundamental quantity that we use to understand and characterise astrophysical systems, and to compare one against another. The systems that we can study best are those in the Local Group (LG), and of those, the closer the system, the deeper our understanding.

At just 50 kpc away from us, the LMC is one of the best-studied galaxies in the Universe, and one that we can study in rare detail. It is also interesting, as the largest of the dwarf galaxies orbiting the Milky Way (MW). It sits in a very interesting niche within the LG, being both sufficiently different in mass than the MW to be an interesting comparison and sufficiently massive to be a major player in the MW's recent history and present state. In particular, it is massive enough to have a significant effect on the MW itself and on stellar streams in the MW halo, though the extent of its influence depends on its mass. Its mass is also interesting as a benchmark against which we can interpret observations of more distant objects.

Its mass has been well-studied in the last decade via a number of different techniques. Kallivayalil et al. (2013) set a lower limit on the LMC's mass by using simulations and identifying for what masses the LMC and its neighbour the Small Magellanic Cloud (SMC) are an infalling bound system. Peñarrubia et al. (2016) estimated the mass of the LMC by applying the Timing Argument to the M31-MW-LMC system. Laporte et al. (2018) estimated the mass of the LMC by measuring the response of the MW's disk to the LMC's infall, and Erkal et al. (2021) similarly estimated the mass via the response of the MW halo. Erkal et al. (2019) modeled the Orphan Stream and determined what mass LMC would produce the observed stream path, Vasiliev et al. (2021) similarly modeled the Sagittarius Stream, and Shipp et al. (2021) studied an ensemble of MW stellar streams to estimate the mass of the LMC. Abundance matching (e.g. Behroozi et al. 2013) can also be used to infer the mass of the LMC.

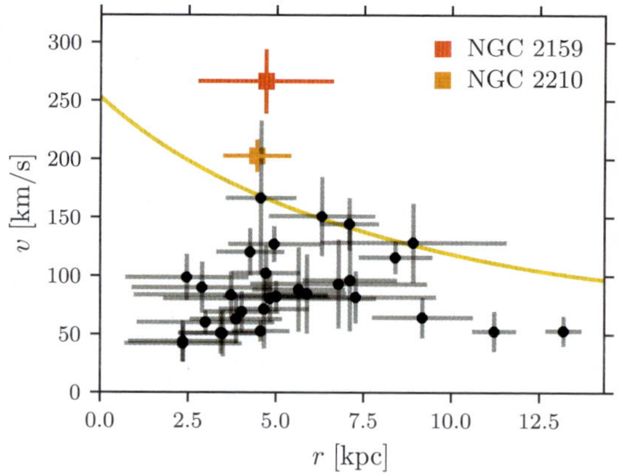

Figure 1. Positions r and velocities v relative to the LMC for 32 LMC GCs. The yellow line shows the escape velocity curve from Boubert et al. (2017). The red and orange points highlight GCs above the curve, which are excluded from our analysis. The black points are the GCs used for the analysis.

Another common approach to mass estimation is to use the motions of a population of tracer objects and to infer what potential must have given rise to such motions. This approach is the subject of this contribution.

2. Positions and Velocities of Globular Clusters in the LMC

Ideally, we require 6D phase space information for a set of tracers if we wish to use them to estimate the mass of their host. That is, positions on the plane of the sky, distances, line-of-sight velocities (LOSVs), and proper motions (PMs).

Both *Gaia* and *HST* are astrometric powerhouses, with *Gaia* measuring PMs of bright stars over the whole sky, and *HST* measuring PMs over more limited regions but to greater depth and often over longer baselines. Their complementary strengths mean that they can do more together than either can alone. That is, there are some objects for which PMs can be measured by both missions independently (with the better result depending on many factors related to both the objects under study and the observatories), but others that are only accessible by one and not the other. Although, despite all their strengths, there are some objects for which PMs cannot be measured with either observatory at present.

It is also possible to combine both missions more directly, such as with GaiaHub, a public tool that combines data from both *Gaia* and *HST* to measure PMs (del Pino et al. 2022). Recently, Bennet et al. (2022) used GaiaHub to measure PMs of globular clusters (GCs) in the LMC, in some cases measuring PMs for GCs that did not have an existing *HST* or *Gaia* measurement, and in some others improving on existing measurements. They supplemented these PMs with literature LOSVs and distances to compile a catalogue of 32 LMC GCs with 6D phase space information. For many years, the absence of PMs was a limiting factor in the accuracy and efficacy of tracer mass estimates, but now we are in a new and exciting era of having PMs readily available for many objects in the LG. The sample size of this catalogue was limited by the availability (or lack thereof) of distances and LOSVs.

We combine the GC catalogue with positions and motions of the LMC centre to transform the cluster properties into LMC-centric values. Figure 1 shows the clusters in the plane of distance r and velocity v relative to the LMC. The yellow line shows the escape velocity curve of the LMC from Boubert et al. (2017). Two clusters (NGC 2159 and

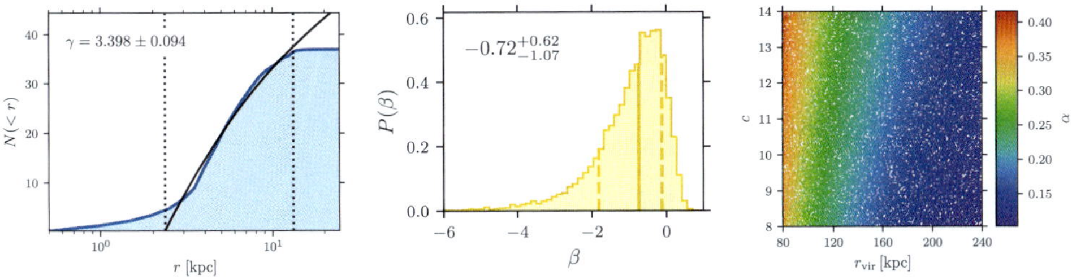

Figure 2. *Left:* Cumulative number profile of LMC GCs, with a power-law fit over the region spanned by the GCs with 6D position-velocity information. *Middle:* Anisotropy distribution of the LMC GCs within the region they span. *Right:* A grid of sample halos with the points coloured by the index of a power-law fit to each potential.

NGC 2210, shown in red and orange respectively) have velocities significantly above the escape velocity at their distance, indicating that they are unbound, so we remove these from the sample. This leaves us with 30 LMC GCs with positions and velocities with which to estimate the mass of the LMC.

3. Tracer Mass Estimates of the LMC

Watkins et al. (2010) introduced a tracer mass estimator (TME) under a set of simple assumptions. We start with a set of tracers with distances r and velocities v relative to the centre of their host, and spanning a region of interest $r_{\rm min}$ to $r_{\rm max}$. Over that region of interest, we assume that the tracers follow a power-law density profile with index γ, the potential is a power-law with index α, and the tracer velocity distribution is anisotropic with constant anisotropy β. Then the mass enclosed with $r_{\rm max}$ is given by

$$M_{\rm TME} = M\left(< r_{\rm max}\right) = \frac{\alpha + \gamma - 2\beta}{G\left(3 - 2\beta\right)}^{1-\alpha} \left\langle v^2 r^{\alpha} \right\rangle . \qquad (1)$$

For our LMC sample, $r_{\rm max} \sim 13$ kpc.

To estimate γ, we consider the cumulative number profile of all LMC GCs with 3D distance estimates (that is the GCs in our sample, and any others with distance estimates but which lack 3D velocity information), and ask which power-law fits best over the region of interest. The fit and γ estimate are shown in the left panel of Figure 2.

To estimate β, we monte-carlo sample the observed GC properties accounting for their uncertainties, and then calculate the LMC-centric second velocity moments in a spherical coordinate system $(\overline{v_r^2}, \overline{v_\theta^2}, \overline{v_\phi^2})$. From these, we calculate the anisotropy

$$\beta = 1 - \frac{\overline{v_\theta^2} + \overline{v_\phi^2}}{2\overline{v_r^2}} \qquad (2)$$

for each monte-carlo sample. The resulting distribution of β values is shown in the middle panel of Figure 2. The values are mostly negative, consistent with a flattened rotating system (Bennet et al. 2022).

To estimate α, we consider a range of possible potentials. We model the LMC as a disk, with fixed properties, and an NFW halo characterised by a virial radius $r_{\rm vir}$ and a concentration c. As the properties of the LMC halo are unknown – a driving force behind this work – we sample a grid of $r_{\rm vir}$ and c value, drawn at random. For each randomly selected potential, we fit a power-law over the region of interest to estimate α. The resulting grid of α values is shown in the right panel of Figure 2.

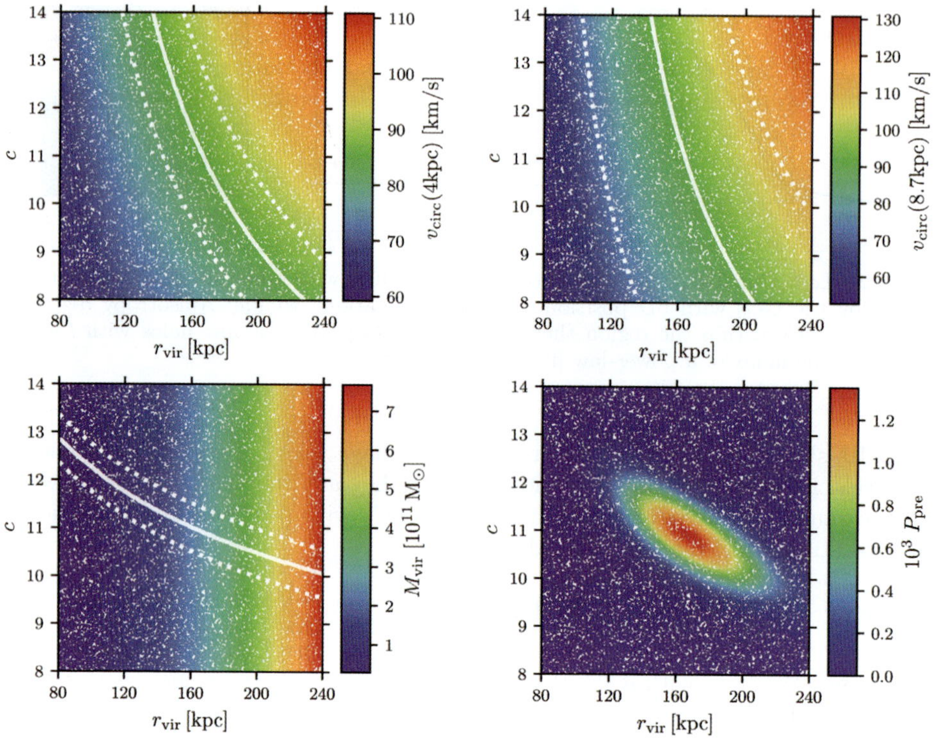

Figure 3. *Top-left:* Circular velocity at 4 kpc across the halo grid. The solid and dashed lines show the literature estimate and uncertainties. *Top-right:* Circular velocity at 8.7 kpc across the halo grid. The solid and dashed lines show the literature estimate and uncertainties. *Bottom-left:* Virial mass for each halo in the grid. The solid and dashed lines show a cosmological relation between concentration and virial mass and scatter amongst different relations. *Bottom-right:* Probability of each halo in the grid based on the circular velocities and the cosmological predictions shown in the other panels.

We can narrow down the grid of halos. We know the circular velocity of the LMC at 4 kpc from Gaia Collaboration et al. (2021) and at 8.7 kpc from van der Marel & Kallivayalil (2014). The top-left and top-right panels of Figure 3 show the circular velocities at 4 kpc and 8.7 kpc for each halo in the sample, with the literature values and their uncertainties marked as solid and dashed lines respectively.

Cosmological simulations also suggest a relationship between virial mass and concentration for NFW halos; at fixed concentration there is direct mapping from virial mass to virial radius, so this provides a cosmological prediction for virial radius and concentration as well. The bottom-left panel of Figure 3 shows virial masses for each halo in the sample with the cosmological relation from Klypin et al. (2011), transformed into $(r_{\rm vir}, c)$ space, shown as the solid white line. The dashed lines approximate the uncertainty among different simulations and relations.

Combining the observed circular velocity information and the cosmological simulation predictions, we can calculate the prior probability of each halo. Figure 3 shows the probability of each halo in the sample. We can see that this reduces the range of likely halos.

Finally, we can use Equation 1 to estimate the mass inside 13 kpc for each halo in the sample. We assume γ is fixed, we draw β from the monte-carlo distribution, and we have already calculated α directly for each halo. Then we weight $M_{\rm TME}$ for each halo by the

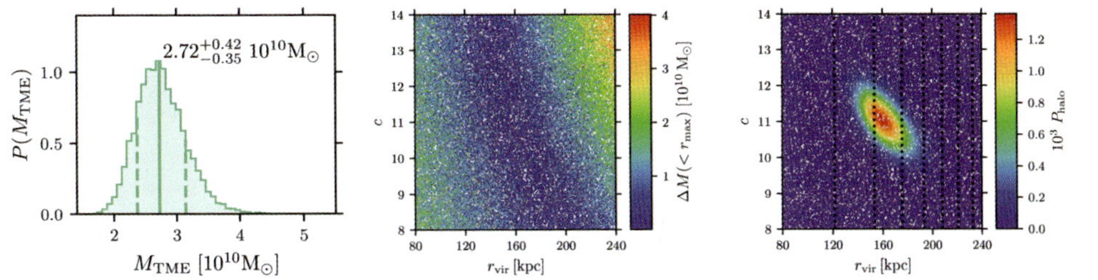

Figure 4. *Left:* Distribution of TME estimates for the mass of the LMC inside ~13 kpc. *Middle:* Difference between the TME mass and the model mass for each halo. *Right:* Revised probability for each halo based on the circular velocities, cosmological predictions and the TME mass estimates.

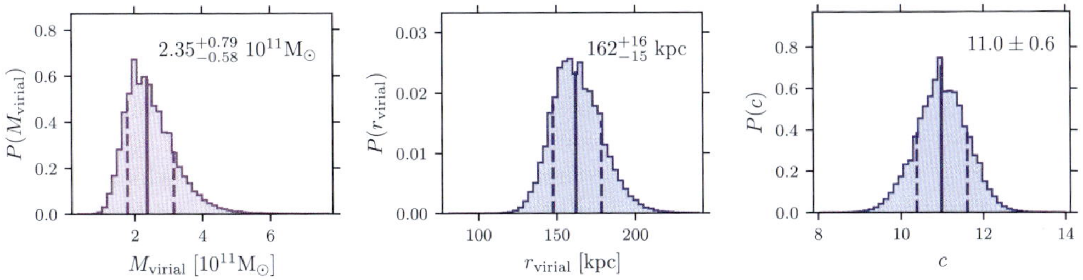

Figure 5. *Left:* Distribution of virial mass estimates for the LMC. *Middle:* Distribution of virial radius estimates for the LMC. *Right:* Distribution of halo concentration estimates for the LMC.

probability of that halo. The resulting distribution of masses is shown in the left panel of Figure 4, from which estimate $M(< 13\ \text{kpc}) = 2.72^{+0.42}_{-0.35} \times 10^{10}\text{M}_\odot$.

We can also compare these values for $M(< 13\text{kpc})$ calculated from the TME to the values predicted for each halo. These differences are shown in the middle panel of Figure 4, with purple showing regions where the calculated and predicted values agree best. We can fold this information into the halo probabilities we previously calculated to get a posterior halo probability. These probabilities are shown in the right panel of Figure 4, and we can immediately see that this added information tightens the range of likely halos. The black dotted lines show lines of constant virial mass from $1 - 7 \times 10^{11}\text{M}_\odot$.

So far the TME has given us the mass within ~13 kpc, which is itself useful. However, we are most interested in the LMC's total mass. So we take the distribution of virial masses across the grid weighted by the probability of each halo. The resulting distribution of virial masses is shown in the left panel of Figure 5, from which we estimate $M_{\text{vir}} = 2.35^{+0.79}_{-0.58} \times 10^{11}\text{M}_\odot$.

We can do likewise with the weighted distributions of virial radii and concentrations of our sample to estimate the best NFW halo properities. These distributions are shown in the middle and right panels of Figure 5, from which we estimate $r_{\text{vir}} = 162^{+16}_{-15}$ kpc and $c = 11.0 \pm 0.6$.

4. Comparison with Previous Measurements

Figure 6 shows how our estimate of the LMC's mass compares to those of the previous studies mentioned in 1. We can see that the estimate is on the high side of but broadly consistent with previous studies. The uncertainties are generous, thanks to extrapolating

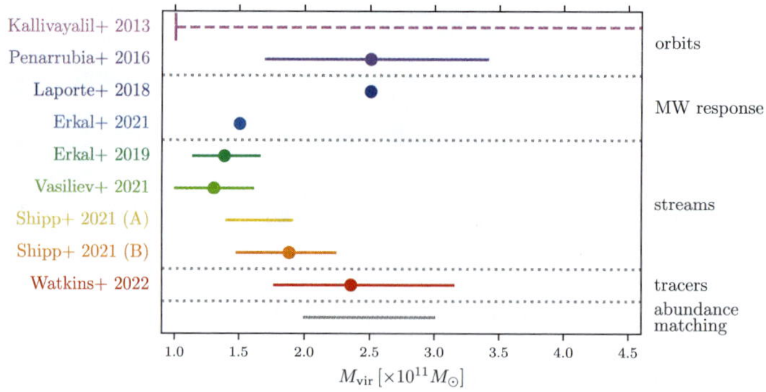

Figure 6. Comparison of our mass estimate for the LMC with previous studies. The left side of the figure gives the references for each study, the right side gives the methods used for each study.

out by over an order of magnitude to infer the virial mass at ∼160 kpc from the mass inside at ∼13 kpc, but are still very reasonable.

A driving factor in our virial mass estimate is the shape of the halo, which we have assumed to be NFW (inner slope -1, outer slope -3). The inner slope is the dominant contributor to the mass within 13 kpc, and the fact that the TME estimates and the model masses agree so well in this range (middle panel of Figure 4) suggests that an inner slope of -1 is a reasonable assumption. The extrapolation for the virial mass will be dominated by the adopted outer slope, with a steeper outer density profile resulting in a lower overall mass. If we were to adopt, for example, a Hernquist profile (inner slope -1, outer slope -4) we expect this would change the TME mass very little, but would reduce the virial mass. This work is still ongoing.

5. Conclusions

We have used a catalogue of 30 GCs in the LMC with 6D position-velocity phase space information to estimate the mass of the LMC within ∼13 kpc. We have used this estimate, combined with cosmological predictions and assumptions about the halo shape, to extrapolate the virial mass of the LMC. Our virial mass estimate is consistent with previous estimate of the LMC's mass using different dynamical mass estimation methods.

References

Behroozi, P. S., Wechsler, R. H., & Conroy, C. 2013, ApJ, 770, 57, doi: 10.1088/0004-637X/770/1/57
Bennet, P., Alfaro-Cuello, M., del Pino, A., et al. 2022, ApJ, 935, 149, doi: 10.3847/1538-4357/ac81c9
Boubert, D., Erkal, D., Evans, N. W., & Izzard, R. G. 2017, MNRAS, 469, 2151, doi: 10.1093/mnras/stx848
del Pino, A., Libralato, M., van der Marel, R. P., et al. 2022, ApJ, 933, 76, doi: 10.3847/1538-4357/ac70cf
Erkal, D., Belokurov, V., Laporte, C. F. P., et al. 2019, MNRAS, 487, 2685, doi: 10.1093/mnras/stz1371
Erkal, D., Deason, A. J., Belokurov, V., et al. 2021, MNRAS, 506, 2677, doi: 10.1093/mnras/stab1828
Gaia Collaboration, et al. 2021, A&A, 649, A7, doi: 10.1051/0004-6361/202039588
Kallivayalil, N., van der Marel, R. P., Besla, G., Anderson, J., & Alcock, C. 2013, ApJ, 764, 161, doi: 10.1088/0004-637X/764/2/161

Klypin, A. A., Trujillo-Gomez, S., & Primack, J. 2011, ApJ, 740, 102, doi: 10.1088/0004-637X/740/2/102

Laporte, C. F. P., Gómez, F. A., Besla, G., Johnston, K. V., & Garavito-Camargo, N. 2018, MNRAS, 473, 1218, doi: 10.1093/mnras/stx2146

Peñarrubia, J., Gómez, F. A., Besla, G., Erkal, D., & Ma, Y.-Z. 2016, MNRAS, 456, L54, doi: 10.1093/mnrasl/slv160

Shipp, N., Erkal, D., Drlica-Wagner, A., et al. 2021, ApJ, 923, 149, doi: 10.3847/1538-4357/ac2e93

van der Marel, R. P., & Kallivayalil, N. 2014, ApJ, 781, 121, doi: 10.1088/0004-637X/781/2/121

Vasiliev, E., Belokurov, V., & Erkal, D. 2021, MNRAS, 501, 2279, doi: 10.1093/mnras/staa3673

Watkins, L. L., Evans, N. W., & An, J. H. 2010, MNRAS, 406, 264, doi: 10.1111/j.1365-2966.2010.16708.x

Dynamical Masses of Local Group Galaxies
Proceedings IAU Symposium No. 379, 2025
P. Bonifacio, M.-R. Cioni, F. Hammer, M. S. Pawlowski & S. Taibi, eds.
doi:10.1017/S1743921323001321

The Formation of Magellanic System and the total mass of Large Magellanic Cloud

Jianling Wang[1], **Francois Hammer**[2], **Yanbin Yang**[2] **and Maria-Rosa L. Cioni**[3]

[1]CAS Key Laboratory of Optical Astronomy, National Astronomical Observatories, Beijing 100101, China
email: wjianl@bao.ac.cn

[2]GEPI, Observatoire de Paris, CNRS, Place Jules Janssen, F-92195 Meudon, France

[3]Leibniz-Institüt für Astrophysik Potsdam, An der Sternwarte 16, D-14482 Potsdam, Germany

Abstract. The Magellanic Stream is unique to sample the MW potential from ~50 kpc to 300 kpc, and is also unique in constraining the LMC mass, an increasingly important question for the Local Group/Milky Way modeling. Here we compare strengths and weaknesses of the two types of models (tidal and ram-pressure) of the Magellanic Stream formation. I will present our modeling for the formation of the Magellanic System, including those of the most recent discoveries in the Stream, in the Bridge and at the outskirts of Magellanic Clouds. This model has been successful in predicting most recent observations in both properties of stellar and gas phase. It appears that it is an over-constrained model and provides a good path to investigate the Stream properties. In particular, this model requires a LMC mass significantly smaller than 10^{11} $M_\odot$.

Keywords. Galaxies: Magellanic Clouds, Galaxies: interactions, Galaxy: halo, Galaxy: structure

1. Introduction

The Magellanic Stream (MS) and Leading Arm (LA) subtend an angle of 230°, which is identified to be anchored to the Magellanic Clouds in 1974 by Mathewson, Cleary & Murray (1974). The nature of its formation was considered still unknown in 2012 (Mathewson 2012). Modern observations of proper motion from both HST and GAIA are indicating that the the Clouds are presently at first passage to the Milky Way (Kallivayalil et al. 2006; Piatek, Pryor & Olszewski 2008; Kallivayalil et al. 2013). Besides large amount of neutral gas distributed along the Stream, there are mounting evidence that 3-4 times more ionized than neutral gas has been deposited along the Stream (Fox et al. 2014; Richter et al. 2017).

In the first infall frame, the explanations of the MS can be broadly classified into two schemes. One is the tidal tail model (Besla et al. 2012; Lucchini et al. 2020), the other is ram-pressure tails (Mastropietro 2010; Hammer et al. 2015; Wang et al. 2019; Wang, Hammer & Yang 2022). In the tidal tail model, the MS is generated by the mutual close interaction 1-2 Gyr ago before the MCs entering into the halo of MW. In this scenario, the SMC is assumed to be a long-lived satellite of LMC, which requires a LMC mass in excess of 10^{11} $M_\odot$.

There are several major limitations for the tidal model. First, it already lacks by a factor of 10 the amount of neutral gas observed in the Stream. Second, it is unable to

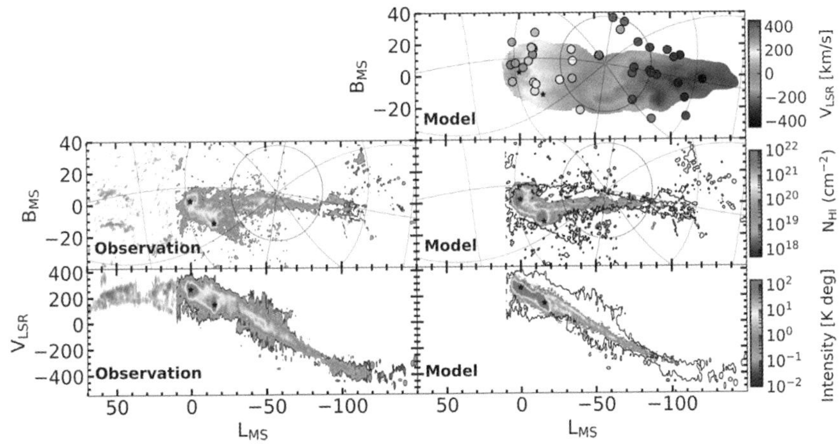

Figure 1. Comparing HI and ionized gas from simulation model (right panels) from Wang et al. (2019) with observations (left panels) of Nidever et al. (2010). The top-right panel shows the sky distribution of the simulated ionized gas with a color coding for the line-of-sight velocity. Circle points represent QSOs absorption line observations by HST/COS (Fox et al. 2014). The simulated ionized gas mass is consistent with that observed (D'Onghia & Fox 2016). The bottom two rows compare observed HI distributions of the Magellanic Stream with that of simulations. The black stars in each panel indicate the position of LMC and SMC. In the simulation panels the contours indicates the observations data. The LA is assumed to have another origin rather than the MC gas (Yang et al. 2014; Hammer et al. 2015; Tepper-García et al. 2019).

reproduce the huge amount of ionized gas that is observed along the Stream. Third, it can only produce a single stream filament, while the MS are made of two filaments, which have been clearly identified by chemical, kinematic, and morphological analyses (Nidever et al. 2010; Hammer et al. 2015). Fourth, no stars along the stream have been observed. Recent revised tidal model have been made by including hot corona of LMC to amend parts of above drawbacks (Lucchini et al. 2020; Lucchini, D'Onghia & Fox 2021). But these models required either a unreasonable massive corona of LMC that is even larger than that of MW, or a dramatic change of the Cloud orbits, which can not reproduce their observed proper motions within 3σ.

Conversely to the tidal model, the ram-pressure plus collision model (Hammer et al. 2015; Wang et al. 2019) naturally reproduce most observational properties associated with the Magellanic System, for instance, the dual filaments, huge amount of ionized gas, absence of stars in the Stream. Interestingly, several observations made after the elaboration of the model have been reproduced without fine tuning.

In this scenario, the Leading Arm are the trailing gas of front-runner dwarfs (Yang et al. 2014; Hammer et al. 2015; Tepper-García et al. 2019), which is well supported by the determination of low metal abundances in 3 part of this structure (see Philip Richter's contribution).

2. Two hydrodynamic filaments formed by ram-pressure plus collision

In the frame of the ram-pressure and collision model, we have built a stable model of Milky Way which include a hot gas corona. The progenitors of MCs are gas rich dwarf galaxies before entering the halo of MW. Figure 1 compare the observed neutral gas and ionized gas to this model. This model naturally generates two HI streams behind the MCs, and a huge amount of ionized gas deposited along the stream. The strong mutual interaction between the MCs totally stretched by gravitational tides the SMC

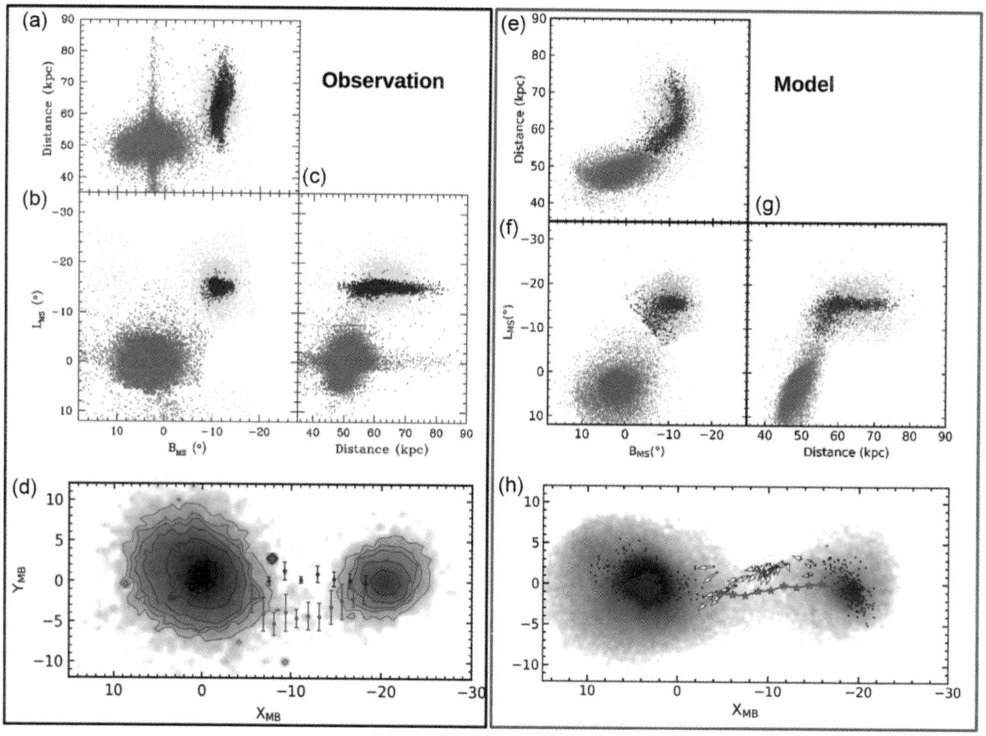

Figure 2. Comparing the stellar distribution of MCs between simulation model from Wang et al. (2019) (the right-hand panels enclosed within rectangle a red box) and observation data (the right-hand panels enclosed within a black box). In the observation panels (a,b,c), the green and cyan points represent ancient RR Lyrae stars from LMC and SMC, and blue points indicates the Classical Cepheids (Ripepi et al. 2017). In the simulation panels, particle numbers and sky distribution have been selected following the observations. The simulation model reproduces well the 'cigar' shape of SMC. Panel d shows the ancient star distribution on the sky, overlapping in the Bridge region RR Lyrae stars and young main sequence stars from Belokurov et al. (2017). Panel h shows simulation model from Wang et al. (2019) which show a similar offset of young stars with ancient stars in the Bridge region.

into a 'cigar' shape, which is well reproduced by this model as shown in Figure 1. Recent observation indicates that there is offset between the ancient stars and young stellar population in the Bridge region (Belokurov et al. 2017), which is also well reproduced in this model.

3. Many Predictions Are Confirmed by Observations

With the progress of observations, new data and findings provide essential test for any model aiming at reproducing the formation of Magellanic System. We will show that our ram-pressure plus collision model pass these tests with many predictions confirmed by recent observations.

3.1. *Two separated populations in the Bridge region*

Observations indicate that there are two different populations in the Bridge region, which are separated in both distance and kinematics space (e.g., Omkumar et al. 2021; James et al. 2021). Omkumar et al. (2021) found that two populations of red clump stars in the Bridge region starting from SMC to LMC, which show different brightnesses.

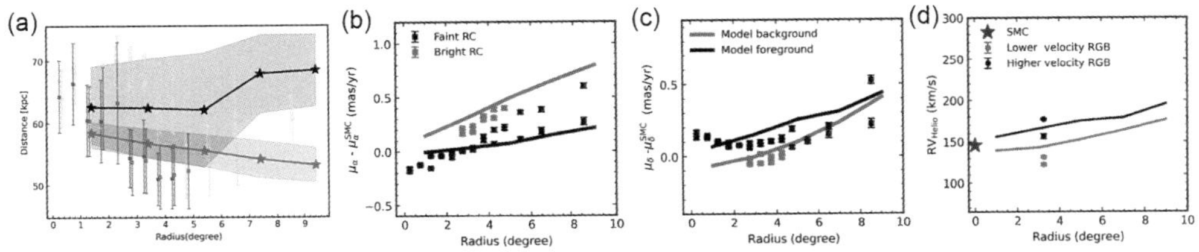

Figure 3. Comparing the observed distances (panel a), relative proper motion (panel b, c), and line-of-sight velocity (panel d) of foreground and background stellar population in the Bridge region as function radius to SMC between observation data and simulation model. In panel a, the green and cyan color points indicate bright and faint population of red clump stars from Omkumar et al. (2021) for north-east (solid square) and south-east region (open square). The proper motion of faint (black square) and bright (red square) RC in the left-hand and middle panel are from Omkumar et al. (2021). The radial velocity of the lower (red circles) and higher (black circles) velocity RGB stars are from James et al. (2021), which are corresponds to foreground components and main body of SMC. The blue star in the panel d indicates the SMC value. The observed proper motions and radial velocity of SMC are from Zivick et al. (2018).

The bright and faint red clump populations show different distance and kinematics, which consistent with the finding of James et al. (2021). In our model, the two populations are formed by SMC, which are tidally stripped by the LMC. The foreground population indicates the disk component of SMC, which is tidally stripped from SMC and showing debris of interaction stretching from SMC to LMC. While the background population come from the spheroid component which is less affected by the LMC tidal and distributed in the back of the Bridge region. This model naturally reproduces this new observation as shown in Figure 3.

3.2. *Periphery of the Clouds*

With deep observations, many faint features in the periphery of MCs have been discovered as shown in the left panel of Figure 4 from Gaia Collaboration et al. (2021), many of which have been well predicted by our model as shown in the right panel of Figure 4. The North Tidal Arm (NTA) is the largest tidal features rooted from the disk LMC, which is confirmed to originate from the LMC on the basis of its metallicity, distance, and kinematics (Cullinane et al. 2022a,b, 2020). Before observations, the model of Wang et al. (2019) predicted the existence of the NTA, which is formed by the Galactic tides exerted on the disk of LMC.

From Gaia EDR3, we have selected stars belonging to NTA according to its morphology position, proper motion. With this sample stars of NTA, we cross-matched with results of literature to get their distance and metallicity. The distance and metallicity variation as function of radius to the LMC are shown in Figure 5.

In order to compare simulation model with observations, we assigned metallicities to particles of the simulation model by painting that of the LMC with metallicities following the observational constraints (Grady, Belokurov & Evans 2021). Grady, Belokurov & Evans (2021) selected red giant stars of LMC from *Gaia* DR2, and used machine-learning method with data of Gaia+2MASS+WISE to estimate the photometric metallicity for these stars. With these data set, they estimated radial metallicity profile: [Fe/H] = α R + b. They found the $\alpha = -0.048 \pm 0.001$ dex kpc^{-1} and $b = -0.656 \pm 0.004$ dex. We use this relation to paint the initial metallicity of our modeled LMC. For simplicity, we have adopted the approximated relation [Fe/H] $\sim$ [M/H]. In Figure 5, the modeled data are shown with red points. The simulation model, can explain well the observational results

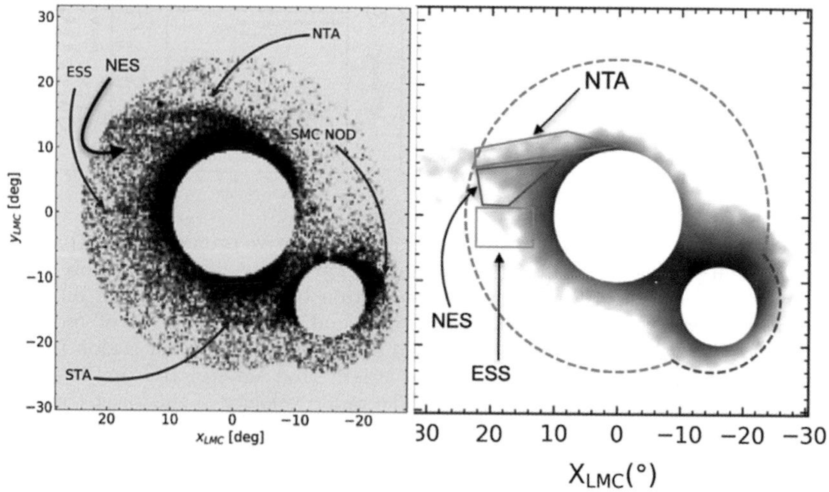

Figure 4. Comparing the morphology of MCs with Gaia EDR3 data (Gaia Collaboration et al. 2021) with simulation model (Wang, Hammer & Yang 2022). In the right panel, the red and blue dashed line indicate the sample selection region for Gaia data (Gaia Collaboration et al. 2021). In the right-hand panel, the colored polygon regions indicate different substructures associated with LMC detected by Gatto et al. (2022).

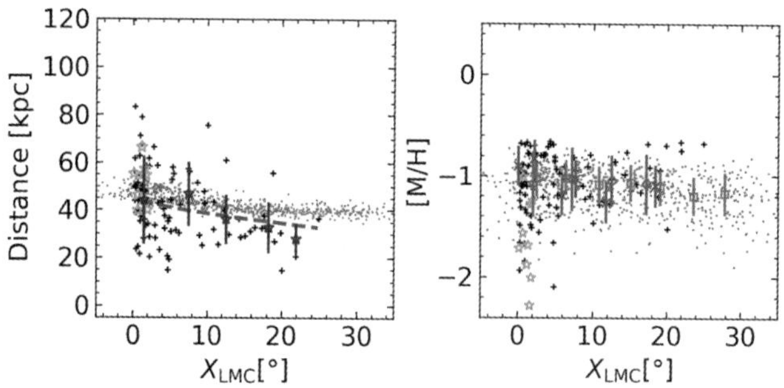

Figure 5. The distance and metallicity of NTA stars varied with X_{LMC}. The black crosses are selected with Gaia EDR3 (Gaia Collaboration et al. 2021) and StarHorse (Anders et al. 2022) (see the text for details), and the red points indicates data from our simulation model. The cyan stars indicate the data from Huang et al. (2022), who derived star distance and metallicity from SkyMapper DR2 (Onken et al. 2019) and Gaia EDR3. The blue stars and its associated error bars in the left panel indicate the mean and dispersion for the observed stars. The green line in the left panel denotes an inclined disk following van der Marel & Kallivayalil (2014). In the right panel, the magenta squares and its associated error bars show the mean and dispersion of observation data from Grady, Belokurov & Evans (2021). The green stars and its associated error bars indicate the value from observation data of MagES (Cullinane et al. 2022a). The red squares and its associated error bars are the mean and dispersion value of simulation model.

for both the distance and metallicity profiles, without fine tuning. The current observed metallicity profile of the LMC (Grady, Belokurov & Evans 2021) also reproduces the formation of NTA, which indicates that the LMC metallicity profile has been settled down before the formation of NTA, or the mutual interaction of MCs/gas loss have marginal influence on the metallicity structure of LMC.

4. Conclusion

The ram-pressure plus collision model (Hammer et al. 2015; Wang et al. 2019) can not only reproduce MS, but also succeed in predicting many observations that have been done in the meantime (see a description in Wang, Hammer & Yang 2022). This model naturally reproduces the two inter-twisted filaments of HI MS, as well as the huge amount of ionized gas associated with MS. This ability also validates that this model goes into the right direction to disentangle the mystery of Magellanic System formation (Mathewson, private communication). We conjecture that the LMC mass has to be small (a few times 10^{10} $M_{\odot}$) to form the Magellanic Stream, though further studies are needed to explore the exact mass range.

References

Anders F. et al., 2022, A&A, 658, A91
Belokurov V., Erkal D., Deason A. J., Koposov S. E., De Angeli F., Evans D. W., Fraternali F., Mackey D., 2017, MNRAS, 466, 4711
Besla G., Kallivayalil N., Hernquist L., van der Marel R. P., Cox T. J., Kereš D., 2012, MNRAS, 421, 2109
Cullinane L. R., Mackey A. D., Da Costa G. S., Erkal D., Koposov S. E., Belokurov V., 2022a, MNRAS, 510, 445
Cullinane L. R., Mackey A. D., Da Costa G. S., Erkal D., Koposov S. E., Belokurov V., 2022b, MNRAS, 512, 4798
Cullinane L. R. et al., 2020, MNRAS, 497, 3055
D'Onghia E., Fox A. J., 2016, ARA&A, 54, 363
Fox A. J. et al., 2014, ApJ, 787, 147
Gatto M., Ripepi V., Bellazzini M., Tortora C., Tosi M., Cignoni M., Longo G., 2022, ApJ, 931, 19
Gaia Collaboration et al., 2021, A&A, 649, A7
Grady J., Belokurov V., Evans N. W., 2021, ApJ, 909, 150
Hammer F., Yang Y. B., Flores H., Puech M., Fouquet S., 2015, ApJ, 813, 110
Huang Y. et al., 2022, ApJ, 925, 164
James D. et al., 2021, MNRAS, 508, 5854
Kallivayalil N., van der Marel R. P., Alcock C., Axelrod T., Cook K. H., Drake A. J., Geha M., 2006, The Astrophysical Journal, 638, 772
Kallivayalil N., van der Marel R. P., Besla G., Anderson J., Alcock C., 2013, The Astrophysical Journal, 764, 161
Lucchini S., D'Onghia E., Fox A. J., Bustard C., Bland-Hawthorn J., Zweibel E., 2020, Nature, 585, 203
Lucchini S., D'Onghia E., Fox A. J., 2021, ApJL, 921, L36
Mastropietro C., 2010, in American Institute of Physics Conference Series, Vol. 1240, Hunting for the Dark: the Hidden Side of Galaxy Formation, Debattista V. P., Popescu C. C., eds., pp. 150–153
Mathewson D., 2012, Journal of Astronomical History and Heritage, 15, 100
Mathewson D. S., Cleary M. N., Murray J. D., 1974, The Astrophysical Journal, 190, 291
Nidever D. L., Majewski S. R., Butler Burton W., Nigra L., 2010, ApJ, 723, 1618
Omkumar A. O. et al., 2021, MNRAS, 500, 2757
Onken C. A. et al., 2019, PASA, 36, e033
Piatek S., Pryor C., Olszewski E. W., 2008, The Astronomical Journal, 135, 1024
Richter P. et al., 2017, A&A, 607, A48
Ripepi V. et al., 2017, MNRAS, 472, 808
Tepper-García T., Bland-Hawthorn J., Pawlowski M. S., Fritz T. K., 2019, MNRAS, 488, 918
van der Marel R. P., Kallivayalil N., 2014, ApJ, 781, 121
Wang J., Hammer F., Yang Y., Ripepi V., Cioni M.-R. L., Puech M., Flores H., 2019, MNRAS, 486, 5907

Wang J., Hammer F., Yang Y., 2022, MNRAS, 515, 940
Yang Y., Hammer F., Fouquet S., Flores H., Puech M., Pawlowski M. S., Kroupa P., 2014, MNRAS, 442, 2419
Zivick P. et al., 2018, ApJ, 864, 55

Discussion

MÜLLER OLIVER: Do you think we can find other dwarfs showing 'cigar' shape as SMC in the extragalactic dwarfs ?

JIANLING WANG: This is difficult, since we need precise distance for individual star showing dwarf shapes in 3D.

Dynamical Masses of Local Group Galaxies
Proceedings IAU Symposium No. 379, 2025
P. Bonifacio, M.-R. Cioni, F. Hammer, M. S. Pawlowski & S. Taibi, eds.
doi:10.1017/S1743921323001527

Overdensities in the Milky Way stellar halo probe the LMC to MW mass ratio

Simon Rozier

University of Edinburgh
email: rozier@iap.fr

Abstract. In the last two decades, some arguments have accumulated for a more important mass ratio of the Large Magellanic Cloud (LMC) to the Milky Way (MW) than was previously thought, up to a value of 10% or more. This implies that the LMC has a measurable influence on the dynamics in the MW stellar halo, including both stellar densities and kinematics, as observed by Conroy et al. (2021) and Petersen and Peñarrubia (2021). While this merger has been previously reproduced using N-body simulations (see, e.g., Garavito-Camargo et al. 2019), I exploit here the results of a recent study (Rozier et al. 2022) which aimed at modelling the merger via linear response theory (LRT). More specifically, we integrated the linearized collisionless Boltzmann-Poisson system of partial differential equations using a methodology known as the matrix method. Our results display the same large scale behaviour as state-of-the-art simulations, with a dipolar over/underdense pattern related to the reflex motion of the MW, and an overdense wake trailing behind the LMC. Using LRT, I show that the response's self-gravity can be neglected, implying a direct proportionality between the LMC to MW mass ratio and the amplitude of the relative density variations of the MW stellar halo. However, these overdensities may also depend on other model parameters, such as the structure of the MW potential (including a dark matter component), the initial stellar halo density, as well as its internal kinematics. I focus on the latter source of degeneracy, showing how the stellar halo's velocity anisotropy impacts its response to the LMC. Interestingly enough, it appears that the density of the dipolar response is insensitive to the stellar halo's initial velocity anisotropy, and can therefore represent an efficient probe of the LMC to MW mass ratio.

Keywords. Galaxy: halo – galaxies: kinematics and dynamics – methods: analytical

1. Introduction

In the last 15 years, estimates of the Large Magellanic Cloud's (LMC) dynamical mass inflated by about an order of magnitude. While its stellar mass is estimated to $2.7 \times 10^9\ M_\odot$, early measurements of its internal kinematics within 8-9 kpc placed a lower limit on its dynamical mass at about $1.5 \times 10^{10}\ M_\odot$ (Schommer et al. 1992; van der Marel and Kallivayalil 2014). But when it later became probable that it is undergoing its first orbit around the Milky Way (MW; Besla et al. 2007), and that it is in a bound system with the Small Magellanic Cloud (Kallivayalil et al. 2013), a number of other probes coincided towards a much larger total mass of the LMC: from the first infall scenario, the total mass was estimated to $1.8 \times 10^{11}\ M_\odot$; abundance matching is in line with these values (Behroozi et al. 2013; Moster et al. 2013); the timing argument returns a high mass of $2.5 \times 10^{11}\ M_\odot$ (Peñarrubia et al. 2016); recent constraints considered the LMC's gravitational effect on MW stellar streams, yielding masses ranging from $1.3 \times 10^{11}\ M_\odot$ (Koposov et al. 2023) to $1.9 \times 10^{11}\ M_\odot$ (Shipp et al. 2021).

Such a high mass of the LMC, of about 10% or more that of the MW, means that the gravitational force applied by the dwarf on the galaxy as a whole can produce measurable effects. Early predictions were provided by Gómez et al. (2015), showing that such a massive satellite would induce a displacement of the MW centre w.r.t. its outer parts, an effect called reflex motion. Subsequent simulations (e.g., Garavito-Camargo et al. 2019) revealed the resulting structures in the MW halo's density, which can be separated in a dipolar part generated by the reflex motion, and an overdense wake trailing behind the satellite. Tentative evidence of both these effects (the motion and the density structure) were provided by Petersen and Peñarrubia (2021) and Erkal et al. (2021) on the one hand, and Conroy et al. (2021) on the other.

While most previous models of the merger were based on N-body simulations, here I exploit results from Rozier et al. (2022), relying on linear response theory (LRT). This analytical framework, based on the linearization of the collisionless Boltzmann – Poisson system of partial differential equations, can be used to compute the way the density of the MW halo (both stellar and DM) evolves when the LMC is introduced. This idealised approach is particularly useful because it allows us to analyse the merger in terms of simple fundamental dynamical ingredients. Indeed, in many of the existing constraints on the LMC's mass, the complexity of the models makes it hard to disentangle the degeneracies involved in the probes. Here, I will show that LRT reveals a direct relationship between the overdensities in the MW's stellar halo and the LMC to MW mass ratio.

The paper is organised as follows: I summarize the method in Sec. 2, I show its results in Sec. 3, I discuss degeneracies in Sec. 4 and I conclude in Sec. 5.

2. Method

Let me highlight some of the key aspects of LRT, referring the reader to Rozier et al. (2022) for more details. I start with a spherical equilibrium model for the MW, representing the DM and stellar halos. This potential ψ_0 (a Hernquist potential in the present case, see Binney and Tremaine 2008) fixes the orbits of halo stars and DM particles. I consider that the initial equilibrium distribution of orbits – the phase space distribution function (DF) – is given by F_0. I do not require a priori that F_0 self-consistently generates the Hernquist density ρ_0: if it does, then the system is fully self-consistent; if not, then I am only interested in the response of a fraction of the system (e.g., the stellar halo).

When the LMC is introduced, it adds a perturbation ψ^{e} to the equilibrium potential. This modifies the distribution of orbits in the MW, so the DF becomes $F_0 + f$. This results in a response potential ψ^{s}. Finally, I impose that the reference frame follows the density centre of the MW, which is accelerating in the presence of the LMC. I therefore have to compensate this motion with a pseudo-force that also acts on the orbits. All these perturbing quantities are related through the collisionless Boltzmann equation (CBE)

$$\frac{\partial f}{\partial t} + \boldsymbol{\Omega} \cdot \frac{\partial f}{\partial \boldsymbol{\theta}} - \frac{\partial F}{\partial \mathbf{J}} \cdot \frac{\partial \Delta H}{\partial \boldsymbol{\theta}} = 0, \tag{1}$$

where $\Delta H = \psi^{\mathrm{e}} + \psi^{\mathrm{s}} + \psi^{\mathrm{in}}$, with ψ^{in} the pseudo-potential associated with the acceleration of the reference frame. In eq. (1), I used action-angle coordinates to navigate phase space. In this set of canonical coordinates, the actions $\mathbf{J}$ are sets of three numbers that fully label orbits in the Hernquist potential, and the conjugate angles $\boldsymbol{\theta}$ identify the specific location of a particle on its orbit. When eq. (1) is coupled with the linearized Poisson equation

$$\Delta \psi^{\mathrm{s}} = 4\pi G \rho^{\mathrm{s}} = 4\pi G \int \mathrm{d}\mathbf{v}\, f(\mathbf{x}, \mathbf{v}), \tag{2}$$

they form a complete set of coupled partial differential equations, with the initial value given by the external perturber $\psi^{\mathrm{e}}(\mathbf{x},t)$. Our goal is to solve this system to get the response ρ^{s}.

Choosing phase space variables for solving this system is an important problem. While the CBE is best expressed in angle-action coordinates, the Poisson equation should be expressed in positions and velocities. Going from one set to the other is mathematically well defined, but can be very intricate numerically. This problem was resolved in Kalnajs (1977), where the Poisson equation was solved a priori by projecting every perturbing potential (resp. density) on a functional basis of potentials (resp. densities), with the potential and density bases related term by term through the Poisson equation. By such projections, the external perturber $\psi^{\mathrm{e}}(\mathbf{x},t)$ is transformed into a vector $\mathbf{b}(t)$, and the response potential into a vector $\mathbf{a}(t)$, so that

$$\psi^{\mathrm{s}}(\mathbf{x},t) = \sum_{p} a_p(t)\, \psi^{(p)}(\mathbf{x}) \quad ; \quad \psi^{\mathrm{e}}(\mathbf{x},t) = \sum_{p} b_p(t)\, \psi^{(p)}(\mathbf{x}), \tag{3}$$

where $\psi^{(p)}(\mathbf{x})$ is a potential basis function. Once these projections are injected in the CBE, it is transformed into a time integral matrix equation in a linear space

$$\mathbf{a}(t) = \int_0^t \mathrm{d}\tau\, \mathbf{M}(t-\tau)\, \big[\mathbf{a}(\tau) + \mathbf{b}(\tau)\big], \tag{4}$$

which gives its name to the method: the matrix method. This equation can be interpreted as follows: the response at the current time $\mathbf{a}(t)$ is equal to an integral over all past times $(0 < \tau < t)$ of all the perturbations occurring in the system $(\mathbf{a}(\tau) + \mathbf{b}(\tau))$, mediated by a kernel called the response matrix $(\mathbf{M})$. This matrix is an intrinsic property of the system (the background potential ψ_0 and the distribution of orbits F_0), and describes how it dynamically responds to any kind of perturbation. The method then goes in three steps: (i) take a model for the MW potential ψ_0, specify the collection of orbits that are allowed to respond F_0, and compute its response matrix $\mathbf{M}$; (ii) take a model for the LMC and project it on the basis to get $\mathbf{b}$; (iii) solve eq. (4) to get $\mathbf{a}(t)$, the response vector, and de-project it according to eq. (3).

3. Linear Response, Self-Gravity and the MW to LMC mass ratio

Let me first consider the response of the MW halo as a whole, including both stars and DM. I therefore take F_0 to self-consistently generate ψ_0. In my fiducial case, I assume an isotropic distribution of orbits, as given by Baes and van Hese (2007). Details on the LMC and MW models are given in Rozier et al. (2022) (masses, sizes, the LMC's orbit, etc.), as well as details on the matrix computation and time integration schemes. Here, let me highlight the resulting overdensity map of the MW halo, presented in the last panel of Fig. 3 of Rozier et al. (2022). Despite many differences in the model (different potentials, different MW DF, different LMC orbit, ...), LRT is consistent with results from N-body simulations by Garavito-Camargo et al. (2021) (last panel of their Fig. 3). In particular, we can identify the same features of the global response: the large scale dipole related to reflex motion, and the overdense wake trailing behind the LMC. These features have similar orientations and amplitudes. This implies that I can further exploit the results of LRT to gain insight that is generally inaccessible to N-body simulations.

Here, let me first consider the role of self-gravity. This is a natural step in the framework of LRT, because (i) self-gravity is the only way that two distinct populations of orbits can influence each other's response; (ii) it can be easily evaluated, by simply removing the $\mathbf{a}(\tau)$ within the time integral of eq. (4). Figure 4 of Rozier et al. (2022) shows the results of this procedure, i.e. the overdensity map of the MW halo when the response's self-gravity is neglected. This map is nearly identical to the self-gravitating case, the residuals remaining below 10% of the response itself. It means that self-gravity has hardly any role in the MW's response. Therefore, there is little interaction between the responses of different sub-components of the MW, in particular between the stellar and the DM halos. In the rest of the paper, I will consider the response of the stellar halo only.

I can model the stellar halo by a collection of orbits given by $F_1 = (M_*/M_{\rm MW})\, F_0$, where M_* is the mass of the stellar halo, $M_{\rm MW}$ the total MW mass, and F_0 is the initial self-consistent DF. Neglecting self-gravity, I can transform eq. (4) into $\underline{\mathbf{a}}_* = \underline{\mathbf{M}}_* \underline{\mathbf{b}}$, where $\underline{\mathbf{a}}_*$ is the stellar halo's response (i.e., its response density), $\underline{\mathbf{M}}_*$ is the response matrix of the stellar halo†, and $\underline{\mathbf{b}}$ is the LMC's density. Now, if I consider the scaling of these terms with the different physical quantities of interest, I have that $\underline{\mathbf{a}}_* \propto \rho_*^{\rm s}$, the stellar halo's response density; $\underline{\mathbf{M}}_* \propto M_*/M_{\rm MW}$; and $\underline{\mathbf{b}} \propto M_{\rm LMC}$. As a consequence, I can write the following scaling,

$$\boxed{\frac{\rho_*^{\rm s}}{\rho_*} \propto \frac{M_{\rm LMC}}{M_{\rm MW}}.} \tag{5}$$

This scaling is my crucial point here. It implies that the relative overdensities in the stellar halo (as measured in Conroy et al. 2021) directly inform us on the LMC to MW mass ratio. Of course, many complications are hidden in the proportionality, in particular degeneracies related to the background MW potential, the LMC's orbit, or the stellar halo's velocity distribution. But to leading order, the proportionality holds.

4. Degeneracies

Let me finally consider a particular source of degeneracy in the proportionality relation of eq. (5): that coming from the initial velocity distribution of the stellar halo. Baes and van Hese (2007) provide a variety of DFs for stellar halos with various initial velocity anisotropies. Here, I consider that the anisotropy β is fixed to a constant in the halo. I take two particular cases: one with tangential anisotropy, $\beta = -0.8$, and one with radial anisotropy, $\beta = 0.49$. Figure 5 of Rozier et al. (2022) shows the overdensity map of stellar halos with these initial anisotropies, keeping all other parameters identical to the fiducial isotropic halo. We can observe that both the shape and the amplitude of the overdensitites are sensitive to the initial halo's velocity distribution: a tangentially biased halo produces a shallower response, while a radial halo produces a stronger response. Hence, the initial anisotropy of the stellar halo is a source of degeneracy if one wishes to constrain the LMC to MW mass ratio using the amplitude of the halo overdensities.

However, leverage is provided by the following measurement: let me focus on the dipolar response of the halo, which results from the MW's reflex motion. When only this term ($\ell = 1$ in spherical harmonics) is extracted, we get the maps presented in Fig. 8 of Rozier et al. (2022), for the three values of the initial anisotropy. Now, this specific component of the response is insensitive to the initial halo's anisotropy. As a consequence, measuring the dipolar component of the stellar halo's response represents a way of constraining the LMC to MW mass ratio which is independent of the initial halo's velocity anisotropy.

If I now put the bits and pieces together, and do a non-rigorous, by eye comparison between the observations of Conroy et al. (2021), the simulations of Garavito-Camargo et al. (2021) and the present results, I can use the measured $\rho_*^{\rm s}/\rho_*$ to give an estimate of $M_{\rm LMC}/M_{\rm MW}$. From Fig. 2 of Conroy et al. (2021), I can estimate that the observed $\rho_*^{\rm s}/\rho_*$ is about 2.5 times as large as in the fiducial simulation of Garavito-Camargo et al. (2021), which assumes $M_{\rm LMC}/M_{\rm MW} = 15\%$. Now if I bluntly apply eq. (5), the estimate

† This transformation of eq. (4) is detailed in Rozier et al. (2022).

of the true mass ratio is

$$\frac{M_{\rm LMC}}{M_{\rm MW}} = 37.5\%. \tag{6}$$

Note that the comparison between observations and simulations in Conroy et al. (2021) considers the response as a whole, and is not restricted to the dipole (in particular, it is sensitive to the assumed halo velocity distribution). In addition, the measurements have large error bars, and many other sources of degeneracy remain. Varying the MW potential and the stellar halo's density profile are the next steps towards providing better constraints on $M_{\rm LMC}/M_{\rm MW}$.

5. Conclusion

Using results from Rozier et al. (2022), I showed that LRT is able to produce useful models of the MW's response to the LMC. This theory complements N-body simulations on two fronts. On the one hand, I could highlight the minor role of self-gravity, which implies that the MW stellar halo's response is mainly decoupled from the DM's response. From this finding, I could reveal the direct proportionality between the LMC to MW mass ratio and the amplitude of the overdensities of the stellar halo. On the other hand, I exploited the numerical efficiency of the matrix method to show that the proportionality relation is subject to degeneracies, focusing on the role of the stellar halo's initial velocity anisotropy. Yet, I could exhibit an observable that is insensitive to this source of degeneracy: the dipolar part of the response, related to the MW's reflex motion. Finally, applying these results to the observations by Conroy et al. (2021), I estimated a very high LMC to MW mass ratio of about 37.5%. This value should be taken with care, as it stems from the blunt use of results that still require refinement: the observational results are still debated, and the theory is subject to numerous degeneracies.

References

Baes, M., & van Hese, E. 2007, A & A, 471(2), 419–432.

Behroozi, P. S., Wechsler, R. H., & Conroy, C. 2013, Astrophys. J., 770(1), 57.

Besla, G., Kallivayalil, N., Hernquist, L., et al. 2007, Astrophys. J., 668(2), 949–967.

Binney, J., & Tremaine, S. 2008. *Galactic Dynamics: Second Edition.* Princeton University Press.

Conroy, C., Naidu, R. P., Garavito-Camargo, N., et al. 2021, Nature, 592(7855), 534–536.

Erkal, D., Deason, A. J., Belokurov, V., et al. 2021, Mon. Not. Royal Astron. Soc., 506(2), 2677–2684.

Garavito-Camargo, N., Besla, G., Laporte, C. F. P., et al. 2019, Astrophys. J., 884(1), 51.

Garavito-Camargo, N., Besla, G., Laporte, C. F. P., et al. 2021, Astrophys. J., 919(2), 109.

Gómez, F. A., Besla, G., Carpintero, D. D., et al. 2015, Astrophys. J., 802(2), 128.

Kallivayalil, N., van der Marel, R. P., Besla, G., Anderson, J., & Alcock, C. 2013, Astrophys. J., 764(2), 161.

Kalnajs, A. J. 1977, Astrophys. J., 212, 637–644.

Koposov, S. E., Erkal, D., Li, T. S., et al. 2023, Mon. Not. Royal Astron. Soc., 521(4), 4936–4962.

Moster, B. P., Naab, T., & White, S. D. M. 2013, Mon. Not. Royal Astron. Soc., 428(4), 3121–3138.

Peñarrubia, J., Gómez, F. A., Besla, G., Erkal, D., & Ma, Y.-Z. 2016, Mon. Not. Royal Astron. Soc., 456(1), L54–L58.

Petersen, M. S., & Peñarrubia, J. 2021, Nature Astronomy, 5, 251–255.

Rozier, S., Famaey, B., Siebert, A., et al. 2022, Astrophys. J., 933(1), 113.

Schommer, R. A., Suntzeff, N. B., Olszewski, E. W., & Harris, H. C. 1992, Astron. J., 103, 447.

Shipp, N., Erkal, D., Drlica-Wagner, A., et al. 2021, Astrophys. J., 923(2), 149.

van der Marel, R. P., & Kallivayalil, N. 2014, Astrophys. J., 781(2), 121.

Questions

Orlin Koop (Kapteyn Astronomical Institute): What are the prospects of applying this theory to what happens after the pericentric passage, and possibly on longer time scales? I am thinking about the response of the MW disc to the Sagittarius dwarf in particular.

SR: If I stick to an idealised (periodic) orbit of the LMC, then pushing the computation to larger times is not a problem, and I have looked at it. We see different behaviours of the wake and the dipole: the wake seems to get regenerated at each apocentric passage, in a periodic way. The dipole, on the other hand, seems to get partially phase mixed. This behaviour can also be interpreted in terms of transient and long-term features. In the few dynamical times that follow a perturbation, we expect transients to decay due to phase mixing. What remains is self-sustained modes. These phenomena are also expected to play a role in the MW disc's response to Sgr, especially because a disc usually produces a stronger response than a sphere. However, the linear theory of a thick axisymmetric body is much more computationally intensive than the spherical case, so we have to develop particularly efficient algorithmic tools to work out this problem.

Yanbin Yang (Paris Observatory): When you consider the LMC to MW mass ratio, the past orbit of the LMC will also be a source of degeneracy, which does depend on the MW mass independently on the LMC mass. Can you comment on that?

SR: There are a number of degeneracies that I did not mention, and a tricky one is related to the LMC's orbit. Luckily, most of the dipole builds up at late times, in particular in regions that are the closest to us and that we could hope to measure. At such late times, the trajectory of the LMC depends much less on the MW mass, because we are looking at a small fraction of the orbit. So, if we stick to the inner portion of the dipole, this degeneracy is not too worrying.

Dynamical Masses of Local Group Galaxies
Proceedings IAU Symposium No. 379, 2025
P. Bonifacio, M.-R. Cioni, F. Hammer, M. S. Pawlowski & S. Taibi, eds.
doi:10.1017/S1743921323001333

Visiting the North-Eastern outskirts of the Small Magellanic Cloud: Exploring star formation history and kinematics

Sipra Hota[1,2], Annapurni Subramaniam[1], S. R. Dhanush[1,2] and Maria-Rosa Cioni[3]

[1]Indian Institute of Astrophysics, Koramangala, Bangalore, India
email: siprahota1997@gmail.com

[2]Pondicherry University, R.V.Nagar, Kalapet, Puducherry, India

[3]Leibniz-Institut für Astrophysik Potsdam (AIP), Germany

Abstract. To investigate the impact of interaction in the North-Eastern Small Magellanic Cloud (SMC) shell region, we used far-UV images from eleven observed fields obtained from the Ultra Violet Imaging Telescope. Cleaned, science-ready images were created and point spread function photometry was performed to extract photometric data. The detected FUV stars were cross-matched with the Gaia EDR3 data to select SMC stars and to obtain kinematic information. The findings from the analysis of the FUV-optical CMDs suggest a history of episodic star formation with most of the stars with age $>$ 150 Myr, along with stars as young as 7 Myr. The spatial distribution of FUV stars shows a gradient decreasing radially outwards. We plan to perform a detailed analysis of the morphology, density, and kinematics of the young stars in this region.

Keywords. galaxies: kinematics and dynamics, galaxies: interactions, Magellanic Clouds

1. Introduction

The Small Magellanic Cloud (SMC), a unique nearest satellite galaxy of the Milky Way, is a dwarf irregular gas-rich galaxy located in the southern hemisphere. The SMC is known for its active star forming regions in a metal poor environment, which provide valuable insights into the star formation process and galaxies' dynamics. The outskirt of the SMC in the North-East direction, known as the Shell, shows features of a recent star formation and is expected to have experienced tidal forces due to its interaction with the Large Magellanic Cloud (LMC) and our Galaxy.

The previous studies by Brueck & Marsoglu (1978); Albers *et al.* (1987); Zaritsky et al. (2000) showed that the SMC shell region is associated with a fainter arm and contains the young stars of a few Myr. The recent work by Martínez-Delgado et al. (2019) using deep optical data from the Survey the MAgellanic Stellar History (SMASH) combined with Gaia, found that the SMC shell is comprised of younger stars and is formed due to the recent interactions between the LMC and SMC instead of tidal interaction. Piatti (2022), using the clusters in this region, noticed that the SMC shell region is stretched more along the line of sight.

This paper presents data from the Ultra Violet Imaging Telescope (UVIT) to study the SMC shell, combined with optical and proper motion (PM) data from the Gaia satellite.

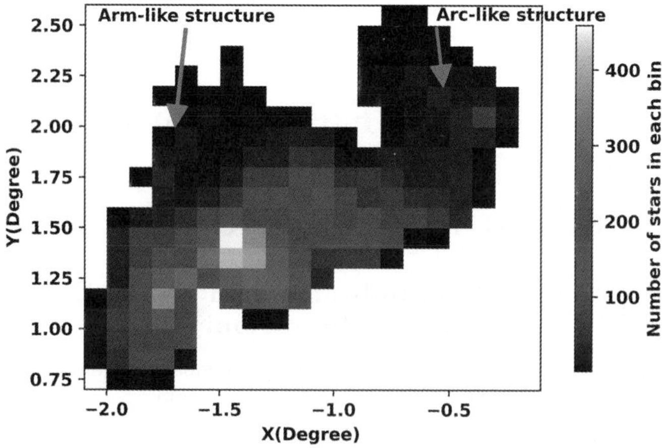

Figure 1. Density map of the FUV stars within the SMC shell. We have done 2d binning along X and Y - directions with the bin size of 0.1° and estimated the number of stars within each bin. FUV density drops radially outward.

We have created a spatio-temporal map of the SMC shell region using this combined data.

2. UVIT and Gaia Data

We have used the observations of the SMC shell region using the UVIT on board the AstroSat, the first Indian Space Observatory. The UVIT is a twin telescope with a spatial resolution of 1.4 arcseconds and a field of view of 0.5°. It has two channels, one for Far-UV (FUV) and the other for Near-UV (NUV) and Visible (VIS) observations. The FUV and NUV channels obtain photometric data, while the VIS channel is used for drift correction due to the spacecraft. Images of eleven fields of the SMC shell region using the FUV channel (F148W filter (1481±250 Å)) are used for this study. More information about UVIT calibration and instrumentation can be found in Tandon et al. (2017). Point spread function photometry has been carried out using the IRAF (Image Reduction And Analysis Facility), and we estimated FUV magnitudes in the F148W filter up to 22 mag with a maximum error of 0.2 mag. We have cross-matched these sources with the Gaia EDR3 data Gaia Collaboration et al. (2021) and obtained around 1.4×10^4 FUV sources with optical photometric and transverse motion data after a 3σ clipping to transverse motion and parallax.

3. Results

Using a bin size of 0.1°, the estimated spatial distribution of the FUV stars in the shell region is shown in Fig 1. It is important to note that the spatial distribution of young stars is not uniform and shows a gradient that decreases radially outward from the densest bin. We have marked the arm and arc-like structures, which are located on the outskirts of the SMC shell.

The FUV-optical color-magnitude diagram (CMD) shown in Fig 2 reveals important information about the star formation history in the SMC shell region. In the CMD, two prominent features are observed: the main-sequence (MS) branch and the sub-giant branch. The color difference between the blue loop stars and the sub-giants in the CMD is significant, indicating that the CMD is capable of discerning different stellar populations based on their ages. This makes the FUV-optical CMD a relevant tool for evaluating the age of the young population within the SMC shell region. From the FUV-optical

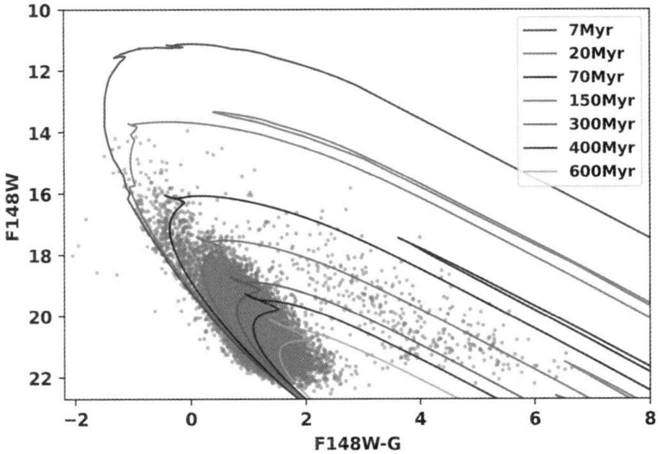

Figure 2. FUV-Optical color-magnitude diagram of the SMC shell. Two types of stellar features are in the CMD: Prominent main-sequence branch and sub-giant branch.

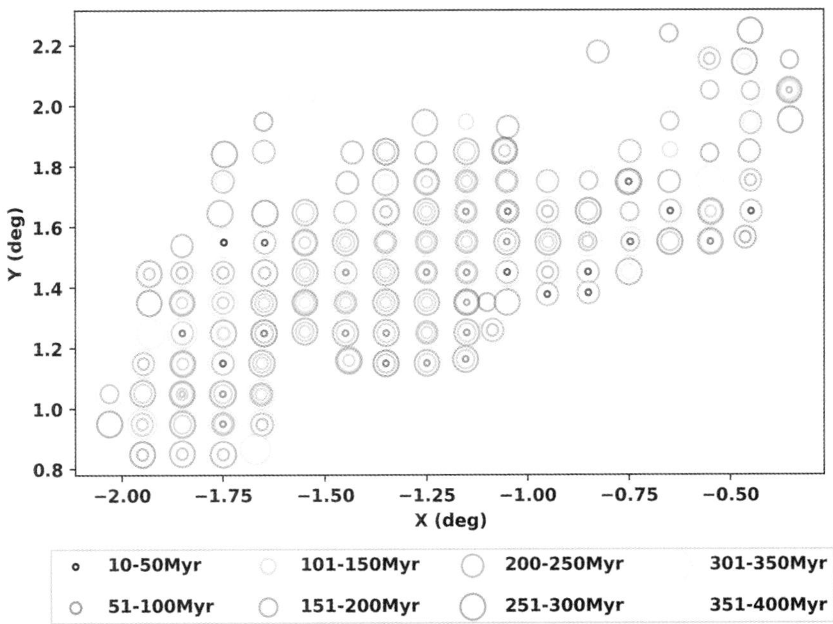

Figure 3. Spatial distribution star formation age in the SMC shell. The SMC shell region shows the episodic star formations.

CMD, we found that the SMC shell region shows significant star formation in between 150 to 300 Myr. This is in agreement with the results of Martínez-Delgado et al. (2019). There appears to be a recent burst of star formation at 70 Myr and contains FUV stars as young as 7 Myr.

The spatial age distribution within the SMC shell region is presented in Fig 3. We note that there is no noticeable star formation propagation within the SMC shell. Most of the SMC shell was formed between 200 to 250 Myr, as indicated by the peak in the age distribution during that time period. However, there is a noticeable decrease in star formation between 100–150 Myr.

4. Summary

This work presents the spatio-temporal distribution of FUV stars in the SMC shell. Using the UV data from UVIT, whose spatial resolution is three times better than the GALEX, we created the FUV-optical CMD, surface density map, and spatial map of the star formation age of the SMC shell. From the results, we can conclude that the SMC shell experienced recent star formation. The majority of the stars are older than 150 Myr, with some stars as young as 7 Myr. The spatial distribution shows a gradient decreasing radially outwards.

Supplementary material

To view supplementary material for this article, please visit https://doi.org/10.1017/S1743921323001333

References

Albers H., MacGillivray H. T., Beard S. M., Chromey F. R., 1987, in European Southern Observatory Conference and Workshop Proceedings. pp 251–257

Brueck M. T., Marsoglu A., 1978, A&A, 68, 193

de Grijs R., Bono G., 2015, AJ, 149, 179.

Gaia Collaboration, Brown A. G. A., Vallenari A., Prusti T., de Bruijne J. H. J., Babusiaux C., Biermann M., et al., 2021, A&A, 649, A1.

Haschke R., Grebel E. K., Duffau S., 2011, AJ, 141, 158.

Lemasle B., Groenewegen M. A. T., Grebel E. K., Bono G., Fiorentino G., François P., Inno L., et al., 2017, A&A, 608, A85.

Martínez-Delgado D., Vivas A. K., Grebel E. K., Gallart C., Pieres A., Bell C. P. M., Zivick P., et al., 2019, A&A, 631, A98.

Piatti A. E., 2022, MNRAS, 509, 3462.

Tandon S. N., Subramaniam A., Girish V., Postma J., Sankarasubramanian K., Sriram S., Stalin C. S., et al., 2017, AJ, 154, 128.

Zaritsky D., Harris J., Grebel E. K., Thompson I. B., 2000, ApJL, 534, L53.

Dynamical Masses of Local Group Galaxies
Proceedings IAU Symposium No. 379, 2025
P. Bonifacio, M.-R. Cioni, F. Hammer, M. S. Pawlowski & S. Taibi, eds.
doi:10.1017/S1743921323001898

Star Formation History of the Small Magellanic Cloud: the shell-like structure

J. D. Sakowska[1], N. E. D. Noël[1], T. Ruiz-Lara[2] and C. Gallart[3,4]

[1]Department of Physics, University of Surrey, Guildford, GU2 7JH, United Kingdom
email: j.sakowska@surrey.ac.uk

[2]Universidad de Granada (ugr), Departamento de Física Teórica y del Cosmos, Campus Fuente Nueva, Edificio Mecenas, 18071 Granada, Spain

[3]Instituto de Astrofísica de Canarias, Vía Láctea s/n, E-38205 La Laguna, Tenerife, Spain

[4]Departamento de Astrofísica, Universidad de La Laguna, E-38200 La Laguna, Tenerife, Spain

Abstract. We present the spatially resolved star formation history (SFH) of a shell-like structure located in the northeastern Small Magellanic Cloud (SMC). We quantitatively obtain the SFH using unprecedented deep photometric data ($g \sim 24$ magnitude) from the SMASH survey and colour-magnitude diagram (CMD) fitting techniques. We consider, for the first time, the SMC's line-of-sight depth and its optical effects on the CMDs. The SFH presents higher accuracy when a line-of-sight depth of ~ 3 kpc is simulated.

We find young star formation enhancements at ~150 Myr, ~200 Myr, ~450 Myr, ~650 Myr, and ~1 Gyr. Comparing the structure's SFH with the Large Magellanic Cloud's northern arm SFH we show strong evidence of synchronicity from at least the past ~2.8 Gyr, possibly ~3.5 Gyr. Our results place constraints on the orbital history of the Magellanic Clouds which, potentially, have implications on their dynamical mass estimates.

Keywords. galaxies: Magellanic Clouds, Local Group, formation, evolution, star formation

1. Introduction

The faint peripheries of galaxies contain stellar fossil records of a system's past events and, hence, hold the keys to understanding their formation and evolution (e.g. Elmegreen & Hunter 2017). The Small and Large Magellanic Cloud (SMC and LMC), located just ~50 kpc and ~60 kpc away from us (Pietrzyński et al. 2019; Graczyk et al. 2020), offer an outstanding natural laboratory to study the resolved stellar populations of two tidally disrupting galaxies in exquisite detail. Given the SMC's smaller total mass - estimated at ~1/10th of the LMC (De Leo et al., submitted to MNRAS) - it is likely the LMC's gravitational influence played a key role in tidally shaping the SMC (e.g. De Leo et al. 2020). In Fig. 1 (top) we see the SMC with a plethora of stellar structures around its main body, likely to have been produced during tidal interactions with the LMC. What we cannot see, however, are elongations along the line-of-sight which, at the extremes, could potentially be almost half of the distance to the SMC from us (~20 – 30 kpc in the east, e.g. Hatzidimitriou & Hawkins 1989; Nidever et al. 2013; Scowcroft et al. 2016; Ripepi et al. 2017; Tatton et al. 2021).

In this work, we focus on a shell-like structure of stars (hereafter, the shell) in the outskirts of the SMC, towards its north-east direction (Martínez-Delgado et al. 2019) seen in Fig. 1 (top panel). We derive a spatially resolved star formation history (SFH) of the shell using deep colour magnitude diagrams (CMDs) from the Survey of the MAgellanic

Stellar History (SMASH, Nidever et al. 2017, 2021; ugriz ∼24th mag). To achieve a quantitative determination of the SFH we have taken into account, for the first time, the puzzling line-of-sight depth of the SMC using the red clump (RC). In Section 2 we describe the analysis, including the derivation of the SFH. In Section 3 we discuss our results. We present the conclusions in Section 4.

2. Star formation history derivation

The bottom panel in Fig. 1 (left) shows the CMD of the shell. While the line-of-sight depth affects the entire CMD, this effect can be seen more conspicuously in the RC region (i.e. at $\sim 0.5 \leq (g-i)_0 \leq 1.0, 18.2 \leq i_0 \leq 19.5$). Fig. 1 (right) shows the observed RC luminosity function (blue) against a theoretical RC luminosity function for the SMC region (see caption for details). The line-of-sight depth in the shell region causes the observed RC to be more elongated in magnitude than the RC resulting from the models. We will discuss this in greater detail below.

To derive the SFH we simulate the observed distribution of stars into a 'solution' CMD by quantitatively comparing the observed and synthetic 'model' CMD. First, we quantified the photometric completeness of the observed CMD (affected by stellar crowding, blending, and other potential errors) by performing artificial star tests (ASTs) following the procedure described in Rusakov et al. (2021). For the CMD fitting, we only considered high-quality photometry above the 50% completeness threshold, which for the shell lies well below the main-sequence turn off at $M_g \sim 22$ mag for the SMC.

We generated the shell's model CMD using the solar-scaled BaSTI-IAC stellar evolution models (Hidalgo et al. 2018; Pietrinferni et al. 2004), with 2022 updates (S. Cassisi, private communication), containing 1.5×10^8 stars with uniform distributions in age ($0.03 \leq$ age ≤ 14 Gyr) and metallicity ($0.00001 \leq Z \leq 0.025$). We then simulated observational effects on the model CMD using `DisPar` (Sakowska et al. *in prep*). `DisPar` used the results from the ASTs to 'disperse' the stars from their actual positions on the model CMD in order to simulate the aforementioned observational effects seen in the observed CMD.

To derive the first version of the SFH, we used a Poisson adapted χ^2 minimisation algorithm (Bernard et al. 2015, 2018) that obtained the best combination of single stellar populations (SSP) from a synthetic CMD that fits the distribution of stars in the observed CMD. The best fitting combinations were reproduced in a 'solution' CMD from which we obtained the SFH.

In order to improve our SFH derivation, we must quantify the line-of-sight depth present in our observed CMD simulate it in the model CMD we fit it against. To do so, we considered the morphology of the RC region in the CMD (see Fig. 1 and description above). We illustrate this procedure in Fig. 2: the left panel shows the observed (blue) and the solution (orange) RC luminosity functions whereas the right panel shows the distribution with a line-of-sight depth which best reproduced the observed RC luminosity function (burgundy). We obtain the most accurate solution when simulating a ~ 3 kpc line-of-sight depth.

3. Results and discussion

In Fig. 3 we present the SFH of the shell-like structure (pink) after a ∼3 kpc line-of-sight depth has been simulated. We also include the SFH of the LMC's northern arm (Ruiz-Lara et al. 2020) in Fig. 3 (blue). The comparison is possible given that Ruiz-Lara et al. (2020) follow the same methodology we presented here but without including the line-of-sight depth since the effect is negligible in the LMC. As seen in Fig. 3, the SFH of the shell presents clear enhancements at young ages of ∼150 Myr, ∼200 Myr,

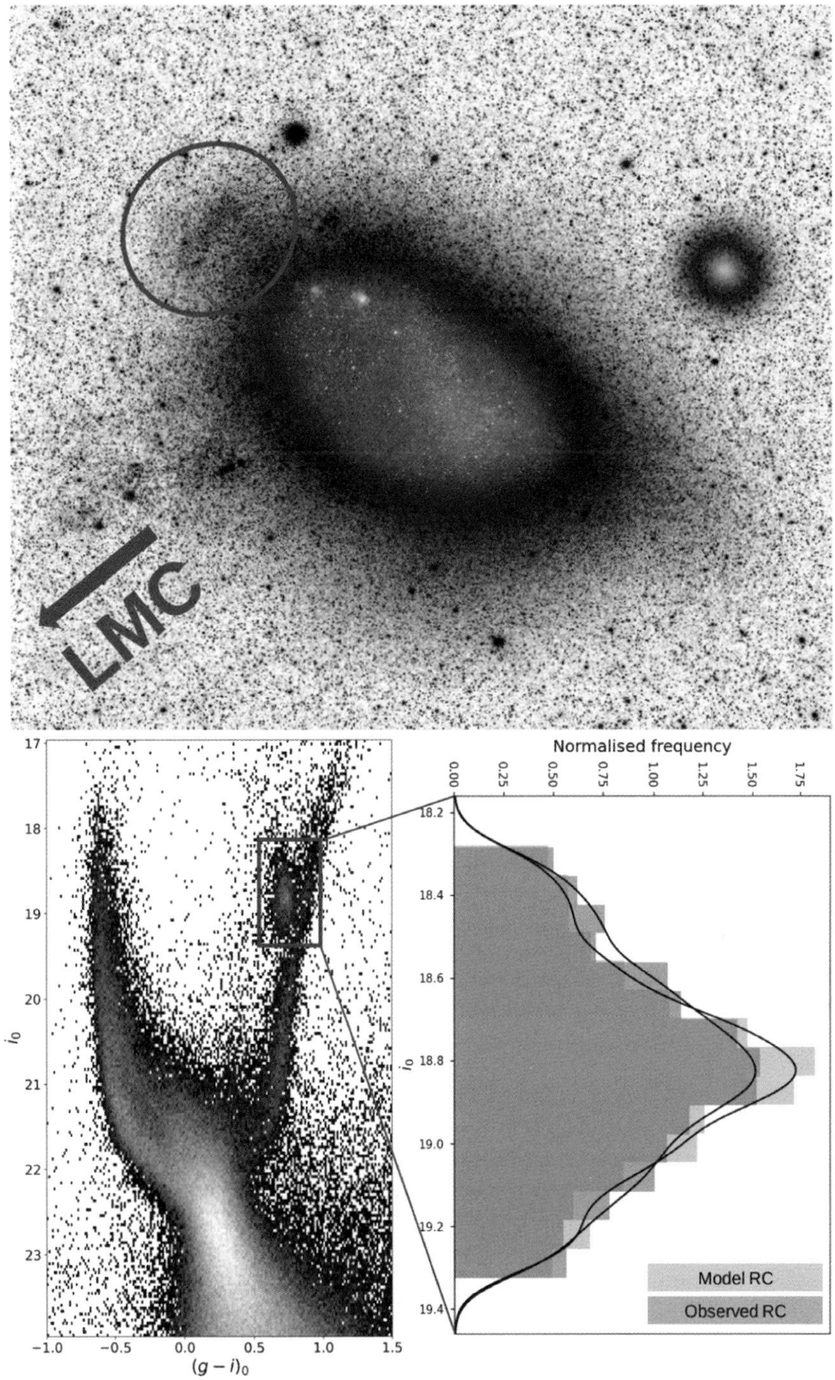

Figure 1. Top panel: SMC image taken from Martínez-Delgado et al. (2019). Bottom panel: CMD of the shell (left) and the observed vs. model luminosity function of the RC (right). The model RC shown here is taken from our first SFH (see Section 2).

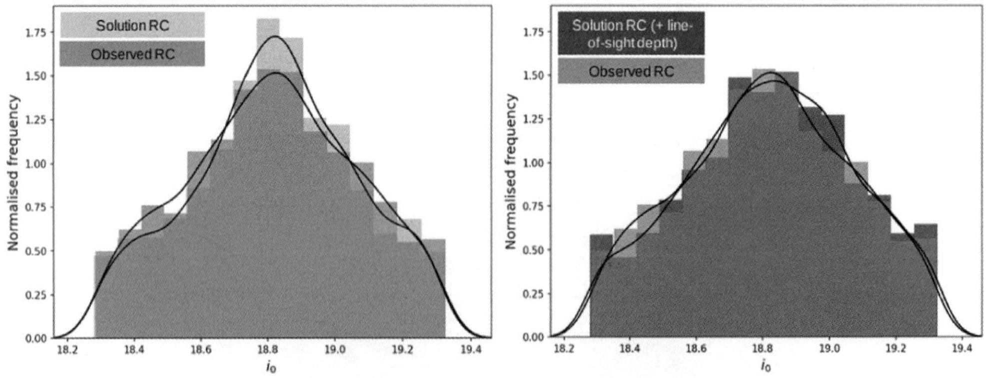

Figure 2. Left: luminosity function of the solution CMD RC (orange) versus observed (blue) for the shell. Right: luminosity function of the solution CMD RC with the best-fitting line-of-sight depth simulated (burgundy).

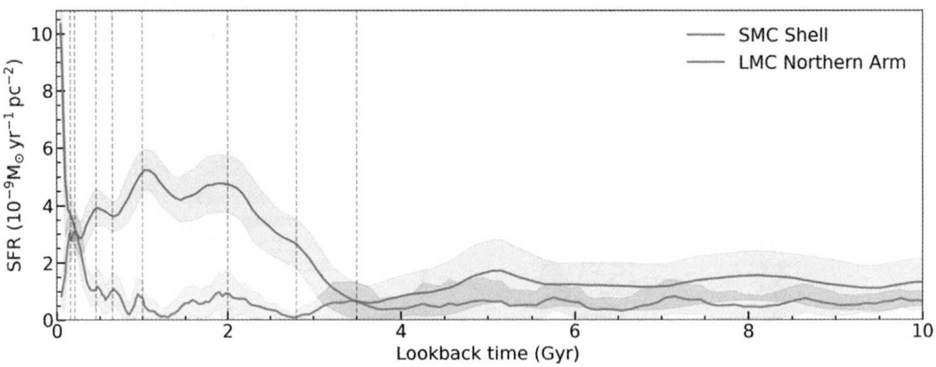

Figure 3. Star formation history of the shell with its line-of-sight depth accounted for (pink). Overlayed is the SFH of the LMC's northern arm also derived from SMASH (blue, Ruiz-Lara et al. 2020). We highlight SFR synchronicity at ~2.8 Gyr, ~2 Gyr, ~1 Gyr, ~450 Myr, ~200 Myr, ~150 Myr and the shell's SFR peak at ~650 Myr. Note the matching, potentially synchronous SFR at ~3.5 Gyr.

~450 Myr, ~650 Myr and ~1 Gyr, implying a possible synchronicity with the LMC's northern arm. This agrees with the results from Massana et al. (2022) who also followed the same methodology. Although the synchronicity in the SFHs is clear for at least the past ~2.8 Gyr, the shell and the LMC's northern arm SFHs seem to be very similar in the past ~3.5 Gyr. To probe the extent of the uncertainties associated with resolving SFH peaks at the intermediate ages, we ran tests using mock populations and 100 Myr wide bursts and demonstrated that the ~2.8 and ~3.5 Gyr times are sufficiently distinct from each other.

4. Conclusions

We present the first quantitative SFH of the SMC's shell-like structure which accounts for its line-of-sight depth. We found the best CMD fit to occur when simulating a ~3 kpc line-of-sight depth. The SFH presented here shows evidence for SMC/LMC co-evolution dating at least ~2.8 Gyr, but potentially even from ~3.5 Gyr, hinting that the LMC played a key role in tidally shaping the SMC's shell.

References

Bernard E. J., Ferguson A. M. N., Chapman S. C., et al., 2015, MNRAS, 453, L113
Bernard E. J., Schultheis M., Di Matteo P., et al., 2018, MNRAS, 477, 3507
De Leo M., Carrera R., Noël N. E. D., et al., 2020, MNRAS, 495, 98
Elmegreen B. G., Hunter D. A., 2017, ASSL, 434, 115
Graczyk D., Pietrzyński G., Thompson I. B., et al., 2020, ApJ, 904, 13
Hatzidimitriou D., Hawkins M. R. S., 1989, MNRAS, 241, 667
Hidalgo S. L., Pietrinferni A., Cassisi S., et al., 2018, ApJ, 856, 125
Martínez-Delgado D., Vivas A. K., Grebel E. K., et al., 2019, A&A, 631, A98
Massana P., Ruiz-Lara T., Noël N. E. D., et al., 2022, MNRAS, 513, L40
Nidever D. L., Monachesi A., Bell E. F., et al., 2013, ApJ, 779, 145
Nidever D. L., Olsen K., Walker A. R., et al., 2017, AJ, 154, 199
Nidever D. L., Olsen K., Choi Y., et al., 2021, AJ, 161, 74
Pietrinferni A., Cassisi S., Salaris M., et al., 2004, ApJ, 612, 168
Pietrzyński G., Graczyk D., Gallenne A., et al., 2019, Nature, 567, 200
Ripepi V., Cioni M.-R. L., Moretti M. I., et al., 2017, MNRAS, 472, 808
Ruiz-Lara T., Gallart C., Monelli M., et al., 2020, A&A, 639, L3
Rusakov V., Monelli M., Gallart C., et al, 2021, MNRAS, 502, 642
Scowcroft V., Freedman W. L., Madore B. F., et al., 2016, ApJ, 816, 49
Tatton B. L., van Loon J. T., Cioni M.-R. L.,et al., 2021, MNRAS, 504, 2983

Dynamical Masses of Local Group Galaxies
Proceedings IAU Symposium No. 379, 2025
P. Bonifacio, M.-R. Cioni, F. Hammer, M. S. Pawlowski & S. Taibi, eds.
doi:10.1017/S1743921323001886

The kinematic Structure of the LMC As Measured by its Field Stars and Star Clusters

Florian Niederhofer and Maria-Rosa Cioni

Leibniz-Institut für Astrophysik Potsdam, An der Sternwarte 16, D-14482 Potsdam, Germany
email: fniederhofer@aip.de

Abstract. In this contribution, we present the results from proper motion measurements across the Large Magellanic Cloud (LMC) using data from the VISTA survey of the Magellanic Cloud system (VMC). Using the derived proper motions, we modelled the structure of the LMC and analysed its internal kinematics. Within the central parts of the LMC, we found observational evidence for elongated orbits parallel to the bar's major axis, which are considered to provide the main contribution for the support of a bar structure. A peculiar kinematic structure in the outer regions of the LMC hints toward stripped material from the Small Magellanic Cloud. We further introduce an observational campaign utilising the *Hubble Space Telescope* to precisely measure the proper motions of star clusters within the LMC. These motions, combined with radial velocities and 3D positions will be used to trace the gravitational potential of the LMC.

Keywords. surveys, proper motion, stars: kinematics and dynamics, globular clusters, galaxies: individual: LMC, Magellanic Clouds

1. Introduction

The Large and Small Magellanic Clouds (LMC and SMC) are the largest and most luminous dwarf satellite companions of the Milky Way. Due to their close proximity they provide a unique opportunity to study the kinematics of resolved stellar populations within an interacting pair of galaxies in unparalleled detail. Within the last years, high-precision proper motion (PM) measurements of stars within the both galaxies had a tremendous impact on our understanding of the Magellanic system and its relation to our own Galaxy. In particular, the traditional view of the Magellanic Clouds being long-time companions to the Milky Way has been revolutionised based on space-based PM measurement (e.g. Kallivayalil et al. 2013, using *Hubble Space Telescope* (HST) observations). Further, the *Gaia* mission largely enhanced our understanding of the internal motions of both galaxies (e.g Gaia Collaboration et al. 2021). In terms of large-scale ground-based surveys, the Visible and Infrared Telescope for Astronomy (VISTA) survey of the Magellanic Cloud system (VMC; Cioni et al. 2011), provides the best multi-epoch survey for dynamical studies of the Magellanic Clouds (e.g. Cioni et al. 2014; Niederhofer et al. 2018, 2021; Schmidt et al. 2020).

2. VMC Observations and Proper Motion Measurements

The VMC survey provides deep and homogeneous photometric data of the Magellanic system in the near-infrared passbands Y, J, and K_s. In total, 68 slightly overlapping VMC tiles are distributed across the LMC, forming a continuous area on the sky of $\sim$105 deg^2. Each VMC tile has at least 13 epochs of observations in the K_s band that extend over a mean time baseline of ~2 years. Using these multi-epoch observations in the K_s filter, Niederhofer et al. (2022) and Schmidt et al. (2022) derived and analysed stellar PMs separately for the inner (the central 6×3 tiles) and outer regions (53 tiles, excluding the central 5×3 tiles) of the LMC. For the calculation of the PMs, the positions of all sources at every epoch were first transformed to a common astrometric reference frame. For each tile, the epoch with the best seeing conditions was selected as the reference epoch. For the tiles that cover the inner regions, the reference frame was defined by the position of likely LMC member star at the reference epoch. For the outer tiles, it was constructed using the positions of background galaxies at the reference epoch. PM were derived by a linear fit of the stellar positions as a function of observation epoch. The slope of the fit is a direct measure of the motion of the source in the reference frame. Since in the outer tiles, the motions are tied to background galaxies, which define a non-moving reference frame, the measured PMs are already in absolute units. For the inner tiles, the motions are measured relative to the bulk motion of the LMC. They have been calibrated to an absolute scale using PMs of stars from the *Gaia* eDR3 catalogue (Gaia Collaboration et al. 2021) that are in common with the VMC sample.

3. The Internal Kinematic Structure of the LMC

3.1. *The Dynamical Model*

To interpret and analyse the motions within the inner regions of the LMC, Niederhofer et al. (2022) described the kinematics of the galaxy employing a simple model, following the formalism described in van der Marel et al. (2002). In the adopted model, the LMC is considered as a flat, rotating disc which is inclined with respect to the plane of the sky. The galaxy itself moves with a three-dimensional centre of mass (COM) motion and its internal velocity field is represented by circular motions within the disc. The internal rotation curve $v_\phi(r)$ is chosen to consist of a linearly rising part and a flat part that approaches a constant terminal velocity, v_0.

To infer the best-fitting parameters, along with their uncertainties, a Bayesian approach was employed, using the Markov Chain Monte Carlo sampler EMCEE (Foreman-Mackey et al. 2013) to explore the posterior probability distribution of the model parameters. The constant rotational velocity v_0, the distance D_0 to the LMC as well as the systemic line-of-sight velocity v_{sys} of the LMC are set as fixed parameters in the fit. For the remaining parameters of the model the best-fitting values are found to be:

$$\alpha_0 = 79.95^{\circ}\,{}^{+0.22}_{-0.23}, \qquad \delta_0 = -69.31^{\circ}\,{}^{+0.12}_{-0.11}$$

$$\mu_{\mathrm{W},0} = -1.867\,{}^{+0.008}_{-0.008}\ \mathrm{mas\,yr^{-1}}, \quad \mu_{\mathrm{N},0} = 0.314\,{}^{+0.014}_{-0.014}\ \mathrm{mas\,yr^{-1}}$$

$$i = 33.5^{\circ}\,{}^{+1.2}_{-1.3}, \qquad \Theta = 129.8^{\circ}\,{}^{+1.9}_{-1.9}$$

(α_0, δ_0) are the RA and Dec. coordinates of the dynamical centre, ($\mu_{\mathrm{W},0}$, $\mu_{\mathrm{N},0}$) are the COM motion along the northern and western directions, (i) is the inclination angle of the LMC disc with respect to the plane of the sky and (Θ) is the position angle of the line of nodes.

3.2. *The COM Motion and Dynamical Centre of the LMC*

The values for the COM motion of the LMC as measured from VMC PMs are in general agreement with the range of measurements in the recent literature (e.g. Wan et al. 2020; Gaia Collaboration et al. 2021; Choi et al. 2022). However, the VMC results suggest a significantly faster motion in the northern direction, compared to what was measured using HST and *Gaia* DR2 (van der Marel & Kallivayalil 2014; Gaia Collaboration et al. 2018).

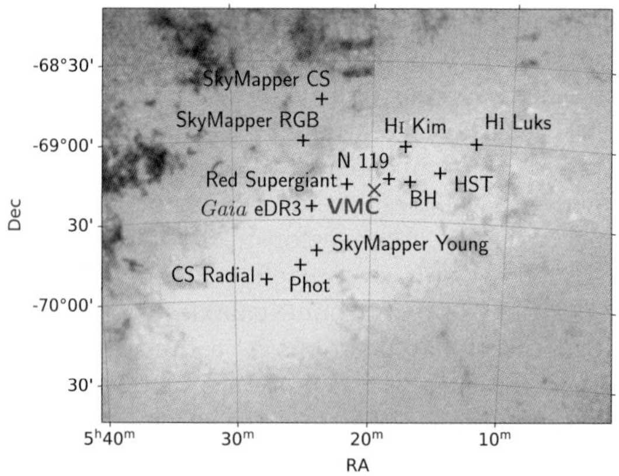

Figure 1. Overview of measured positions of the LMC's dynamical centre, resulting from VMC PMs and from literature studies. The labels in the figure refer to the following works: H I Luks and H I Kim: H I rotation curve from Luks & Rohlfs (1992) and Kim et al. (1998), respectively; HST: HST PMs from van der Marel & Kallivayalil (2014); Phot: photometric centre from van der Marel (2001); CS Radial: carbon stars radial velocities from van der Marel et al. (2002); Gaia eDR3: PMs from Gaia eDR3 from Gaia Collaboration et al. (2021); SkyMapper CS, RGB and Young: SkyMapper photometry combined with Gaia DR2 PMs of carbon, RGB and young stars from Wan et al. (2020); Red Supergiant: Gaia eDR3 PMs combined with line-of-sight velocities of red supergiants from Choi et al. (2022). We also show here the position of a possible central black hole (BH) as inferred by Boyce et al. (2017), as well as the position of the spiral-shaped H II region LHA 120-N 119 (N 119).

The location of the LMC's centre has been debated in the literature for many years without a conclusive answer. Different tracers that have been used to determine the centre have led to different results, reflecting the disturbed nature of the galaxy. The measured positions of the dynamical centre of the H I component of the disc as well as most derived centres of the stellar component reside to the western side of the bar, incompatible with the location of the photometric centre (see Fig. 1). The position of the dynamical centre as inferred from VMC PMs falls between the centres as derived using HST data (van der Marel & Kallivayalil 2014), one the one hand, and resulting from *Gaia* eDR3 PMs (Gaia Collaboration et al. 2021) on the other hand. It is consistent within the errors with the centres derived from HST data and from 3D velocities of young red supergiants (Choi et al. 2022).

Like for the centre location, various measurements of the viewing angles under which we see the LMC disc cover a wide range of values, suggesting that the various stellar populations and the structures they trace within the galaxy do not reside in the same plane but are warped and tilted with respect to each other. Thus, the results obtained depend on the studied region and sky coverage, as well as on the stellar tracer used. The inclination angle derived from VMC stellar kinematics within the central $\sim$7$\times$4 degrees of the LMC suggest that the bar is more inclined than the disc (see also Inno et al. 2016; Choi et al. 2022).

3.3. *The Rotation Curve of the LMC*

Within the inner regions of the LMC studied by Niederhofer et al. (2022) the intermediate-age/old ($\gtrsim$1 Gyr) stellar population follows the theoretical rotation curve resulting from the best-fitting parameters, whereas the young ($\lesssim$0.5 Gyr) stellar population follows a different curve with a steeper rising part and a higher constant rotational velocity (see Fig 2, left panel). Such behaviour is expected and can be explained by

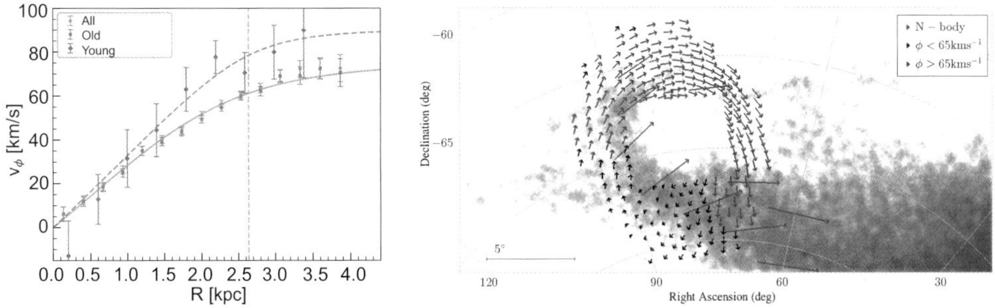

Figure 2. *Left:* Measured rotation curves of all stars (green dots), of the old population (orange squares) and of the young population (blue diamonds). The green solid line shows the rotation curve resulting from the best-fitting model parameters. The blue dashed line corresponds to the rotation curve that was fitted to only the young stars. *Right:* PMs of LMC stars within the outer VMC tiles. The background colour scale shows the density of stripped SMC as predicted by the model from Diaz & Bekki (2012). The red long-drawn arrows follow the median PMs of the model particles. Figure adapted from Schmidt et al. (2022).

an asymmetric drift, where the smaller velocity dispersion of the young population leads to higher rotation velocity (see, e.g., van der Marel & Kallivayalil 2014; Gaia Collaboration et al. 2021). Within the innermost ~0.5 kpc, the young stars further show negative mean rotation. This might indicate a population of stars that is counter-rotating with respect to the main sense of rotation of the galaxy. Alternatively, the young stars might move within a disc that is tilted and offset with respect to the one of the intermediate-age/old stellar population.

Within the outer VMC tiles, Schmidt et al. (2022) found that the south-eastern part of the LMC shows median rotational velocities that are systematically smaller than what would be expected from the overall rotation curve of the galaxy (see Fig. 2, right panel). *N*-body simulations of the LMC–SMC interaction history (Diaz & Bekki 2012) predict a population of stars that have been stripped from the SMC and move opposite to the main LMC rotation. Alternatively, this peculiar motion might be caused by perturbations induced by the LMC–SMC interaction (see contribution by L. Cullinane in this volume).

3.4. *Orbits within the Bar*

Resolved maps of the internal LMC velocity field reveal that intermediate-age/old stars in the region of the LMC bar preferentially move on elongated orbits that follow the orientation of stellar bar structure (see Fig 3). Such orbits which are part of the x_1 type family of periodic orbits, are predicted by theoretical models (e.g. Contopoulos & Papayannopoulos 1980) and are considered to provide the main support of a stellar bar structure. The velocity map based on VMC PMs provides direct observational evidence for this type of orbits within the LMC, also proving that they indeed are the prevailing one within the bar.

4. Proper Motions of the LMC Star Cluster System from HST

In the second part of this contribution, we introduce an observational campaign using HST (GO-16748, PI F.Niederhofer) to precisely measure PMs of a sample of 19 star clusters within the LMC. The programme adds an additional epoch of observations, resulting in time baselines for PM measurements of 6-12 years. The targeted clusters cover an age range between 40 Myr and 13 Gyr and have distances up to 13 kpc from the LMC centre (see Fig. 4 a).

Figure 3. Velocity streamlines (white solid lines) of the observed stellar motions within the inner parts of the LMC. In the regions along the bar, stars follow orbits that deviate from circular motions (indicated by dashed lines for comparison). The background image is constructed from VMC observations in the Y, J and K_s bands (Credit: ESO/VISTA VMC Survey).

For all clusters, the exposures from all epochs have first been processed by a single run of source detection and PSF fitting. The positions of the sources have subsequently been corrected for geometric distortion (Bellini et al. 2011) and put on a common master frame with is registered to the *Gaia* DR3 astrometric frame (Gaia Collaboration et al. 2022). This was then followed by a second-pass photometric run using the routine `KS2` (see Bellini et al. 2017) that simultaneously fits the stars of all exposures within one epoch (including neighbour subtraction) and produces output catalogue for individual exposures.

Relative PMs have been measured in an iterative way using the techniques described in Bellini et al. (2014) and Bellini et al. (2018). Briefly, within each iteration, the positions of the stars in the individual exposures have been transformed to the master frame using bright, unsaturated cluster members as reference stars. In the first iteration the reference stars have been selected based on their positions in the colour-magnitude diagram (see Fig. 4 c), in subsequent iterations based on their PMs. The positions of a given star on the master frame are then fit with a straight line, with the slope of the line giving the relative PM of the star. At the end of each iteration, the master frame positions are improved by using the measured PMs to shift them to a common reference epoch. As an example of the result of the method, the lower panels of Fig. 4 illustrate the quality of the measured PMs within NGC 1850 using four epochs of observations distributed over a time baseline of 12 years. For this cluster, the best-measured stars reach a precision of $\sim 20\ \mu\mathrm{as\,yr}^{-1}$. A posteriori correction have been applied to the measured PMs to mitigate the effects of small scale systematics that results from the different overlaps of the individual data sets (see e.g. Fig. 4 b) on the one hand and uncorrected geometric distortion and charge transfer efficient effects, on the other hand. Finally, the relative PMs have been calibrated to an absolute scale. For this the motions of well-measured

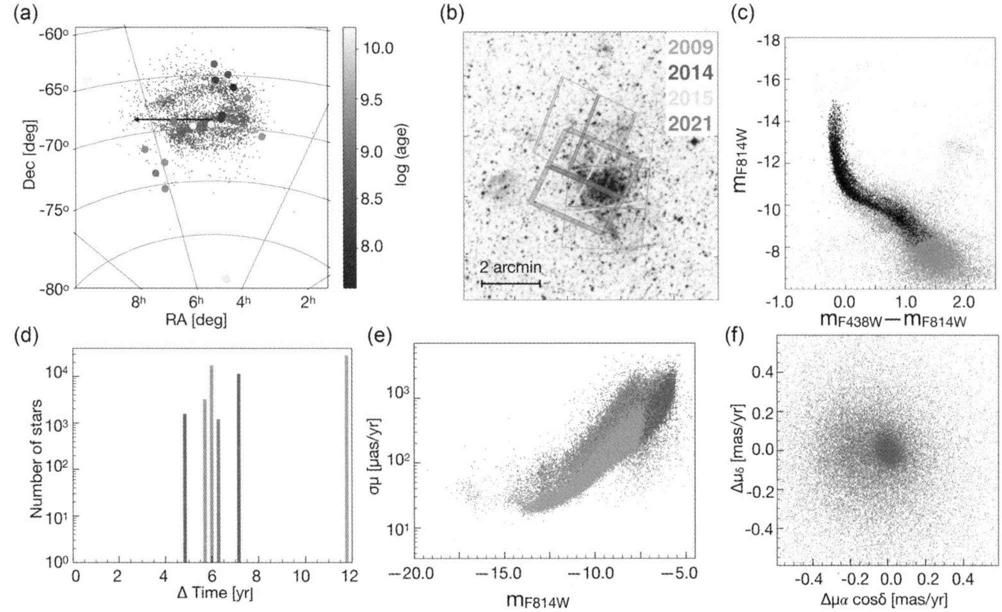

Figure 4. **(a)** On-sky distribution of the star clusters included in the HST programme, colour-coded by age. The small grey points denote Classical Cepheids taken from the OGLE-IV catalogue within the LMC. Also shown as a black arrow is the motion of the LMC on the sky over the next ~10 Myr. **(b)** Illustration of the HST fields used to measure the PMs within NGC 1850. **(c)** m_{F814W} vs $m_{F438W} - m_{F814W}$ colour-magnitude diagram (in instrumental magnitudes) of all stars in the combined field-of-view of NGC 1850 that have observations in at least two epochs. Reference stars used for master-frame transformations are coloured in black. **(d)** Histogram of the time baselines available for the PM calculations of the sources in NGC 1850. **(e)** PM uncertainties as a function of the m_{F814W} magnitude, colour-coded according to the time baseline as in panel (d). **(f)** Diagram of measured PM relative to the motion of the cluster.

Gaia DR3 stars that are in common with the HST catalogue have been used as reference stars to define the PM zero-point.

5. Future Work

For all VMC tiles, one additional epoch of VISTA observations have been acquired that we will use in future works to determine improved PMs across the VMC footprint. The additional data will increase the time baseline for PM measurements to ~ 12 years, resulting in significantly higher precisions.

We will combine the final HST PMs of all star clusters with literature measurements of radial velocities and distances to obtain their full 6D phase-space information to trace the gravitational potential of the LMC (see also the contribution by L. Watkins in this volume) and study in detail the kinematic structure of the star clusters within the LMC.

References

Bellini, A., Anderson, J., & Bedin, L. R. 2011, PASP, 123, 622
Bellini, A., Anderson, J., van der Marel, R. P., et al. 2014, ApJ, 797, 115
Bellini, A., Anderson, J., Bedin, L. R., et al. 2017, ApJ, 842, 6
Bellini, A., Libralato, M., Bedin, L. R., et al. 2018, ApJ, 853, 86
Boyce H., Lützgendorf N., van der Marel R. P., et al. 2017, ApJ, 846, 14
Cioni, M.-R. L., Clementini, G., Girardi, L., et al. 2011, A&A, 527, A116
Cioni, M.-R. L., Girardi, L., Moretti, M. I., et al. 2014, A&A, 562, A32

Choi, Y., Olsen, K. A. G., Besla, G., et al. 2022, ApJ, 927, 153
Contopoulos, G. & Papayannopoulos, T. 1980, A&A, 92, 33
Diaz, J. D. & Bekki, K. 2012, ApJ, 750, 36
Foreman-Mackey, D., Hogg, D. W., Lang, D., et al. 2013, PASP, 125, 306
Gaia Collaboration, Helmi, A., van Leeuwen, F., et al. 2018, A&A, 616, A12
Gaia Collaboration, Luri, X., Chemin, L., et al. 2021, A&A, 649, A7
Gaia Collaboration, Vallenari, A., Brown, A. G. A., et al. 2022, arXiv:2208.00211
Inno, L., Bono, G., Matsunaga, N., et al. 2016, ApJ, 832, 176
Kallivayalil, N., van der Marel, R. P., Besla, G., et al. 2013, ApJ, 764, 161
Kim, S., Staveley-Smith, L., Dopita, M. A., et al. 1998, ApJ, 503, 674
Luks, T. & Rohlfs, K. 1992, A&A, 263, 41
Niederhofer, F., Cioni, M.-R. L., Rubele, S., et al. 2018, A&A, 613, L8
Niederhofer, F., Cioni, M.-R. L., Rubele, S., et al. 2021, MNRAS, 502, 2859
Niederhofer, F., Cioni, M.-R. L., Schmidt, T., et al. 2022, MNRAS, 512, 5423
Schmidt, T., Cioni, M.-R. L., Niederhofer, F., et al. 2020, A&A, 641, A134
Schmidt, T., Cioni, M.-R. L., Niederhofer, F., et al. 2022, A&A, 663, A107
van der Marel, R. P. 2001, AJ, 122, 1827
van der Marel, R. P., Alves, D. R., Hardy, E., et al. 2002, AJ, 124, 2639
van der Marel, R. P. & Kallivayalil, N. 2014, ApJ, 781, 121
Wan, Z., Guglielmo, M., Lewis, G. F., et al. 2020, MNRAS, 492, 782

Dynamical Masses of Local Group Galaxies
Proceedings IAU Symposium No. 379, 2025
P. Bonifacio, M.-R. Cioni, F. Hammer, M. S. Pawlowski & S. Taibi, eds.
doi:10.1017/S174392132300131X

Circumgalactic gas flows in the Local Group

Philipp Richter

Institut für Physik und Astronomie, Universität Potsdam, Haus 28,
Karl-Liebknecht-Str. 24/25, 14476 Golm (Potsdam), Germany
email: prichter@astro.physik.uni-potsdam.de

Abstract. I discuss circumgalactic gas flows in the extended Milky Way halo in the context of the on-going formation and evolution of the Local Group. UV absorption-line measurements, in combination with H I 21 cm observations, provide detailed information on the chemical composition, dust content, physical conditions, and large-scale kinematics of the so-called high-velocity clouds, which are believed to trace neutral and ionized gas in the gaseous halos of the Milky Way, M31, and in the intragroup medium of the Local Group. Recent results in this field and ideas for future observational studies are being presented.

Keywords. Local Group, Milky Way, high-velocity clouds, CGM, UV absorption spectroscopy

1. Introduction

A considerable fraction of the gaseous material in spiral galaxies is situated outside the disk in an extended halo that reaches out to the virial radius and even beyond. We refer to this extraplanar gas component as the circumgalactic medium (CGM). The CGM connects to the large-scale cosmological envrionments in which galaxies reside (groups, clusters, filaments).

Also the Milky Way (MW) and M31 are surrounded by extended gaseous envelopes that originate in their CGM and in the intragroup medium of the Local Group (LG; see Richter et al. 2017, hereafter R17; Putman et al. 2012; Lehner et al. 2020). Studies that aim at characterizing the nature of the diffuse gas around Milky Way and M31 and its connection to the on-going assembly of the Local Group are very important for our general understanding of circumgalactic gas in the local Universe and its role in the hierarchical structure formation. This is because the gaseous outskirts of the Milky Way and M31 can be studied in great detail (in contrast to more distant galaxies) owing to the large amounts of emission and absorption-line data that are available from various instruments.

The Milky Way's CGM ususally is characterized by the so-called high-velocity clouds (HVCs), extraplanar gaseous structures that are observed at high radial velocities ($v_{\rm LSR} > 100$ km s^{-1}) in H I 21 cm radio emission or in line-absorption against extragalactic background sources, such as QSOs. The so-called intermediate-velocity clouds (IVCs) represent extraplanar clouds at lower radial velocities in the range $v_{\rm LSR} = 50 - 100$ km s^{-1}). IVCs are typically related to gas in the disk-halo interface at low vertical distances from the disk ($z \leq 2$ kpc; see Wakker 2001).

Substantial progress has been made over the last two decades in characterizing the role of the HVCs and IVCs in the on-going evolution of the Milky Way and the Local Group. By combining UV absorption-line data from HST/COS, HST/STIS, and FUSE with 21 cm observations, various research groups were able to put tight contraints on the chemical composition of the HVCs and IVCs, the physical conditions in the gas, and

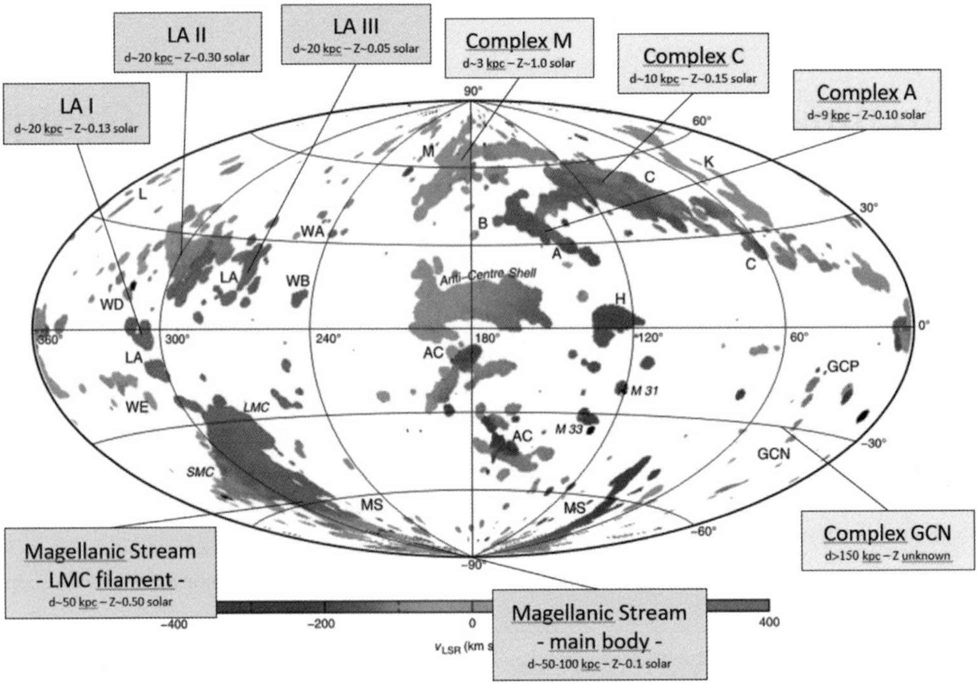

Figure 1. Properties of 21 cm HVCs in the Milky Way halo, as compiled from various studies (see R17 and references therein). The figure shows the 21 cm map of HVCs, with their radial velocities colour-coded, and further indicates position, metallicity, and distance for the most prominent HVCs.

even the distances to these clouds (see R17 and references therein). The Milky Way's CGM is metal-enriched, but typically has sub-solar abundances in the range 0.01 to 1.0 solar (e.g., Wakker et al. 1999; Richter et al. 2001a; Fox et al. 2013). The presence of low and high metal ions in the gas reflects an extreme multi-phase nature of the CGM in the temperature range $T=10^2$ to 10^7 K. Recent studies further imply that the bulk of the baryons in the Milky Way CGM reside in warm-hot, predominantly ionized gas phase that is detected in UV and X-ray absorption lines of intermediate and high ions against extragalactic background point sources (see R17). The main properties of the Milky Way's HVC population are summarized in Fig. 1.

2. Spatial distribution of 21 cm HVCs in the Milky Way halo

The HVC population of the Milky Way, as seen in neutral gas in H I 21 cm emission, has a total sky covering fraction of ~ 35 percent (for neutral gas column densities $N(\text{H I}) \geq 7 \times 10^{17}$ cm^{-2}; Wakker 2004 and references therein). This is substantially smaller than the covering fraction of HVCs seen in UV absorption (see Sect. 4.1). The most prominent Milky Way CGM feature is the Magellanic Stream (MS), an enormous stream of neutral and ionized gas in the southern hemisphere extending over more than 200 degrees (Nidever et al. 2010; see Fig. 1). The spatial extension of the MS into the northern sky is called Leading Arm (LA), a conglomerate of scattered clouds (labeled LA I, LA II, etc.) observed in 21 cm data and in UV absorption (e.g., Fox et al. 2018; R17; see also Fig. 1). The MS (including the LA) probably spans several 100 kpc in linear size and has an estimated distance in the range between 20 and 100 kpc (e.g., D'Onghia & Fox 2016). The stream is a result of the interaction between the Magellanic Clouds

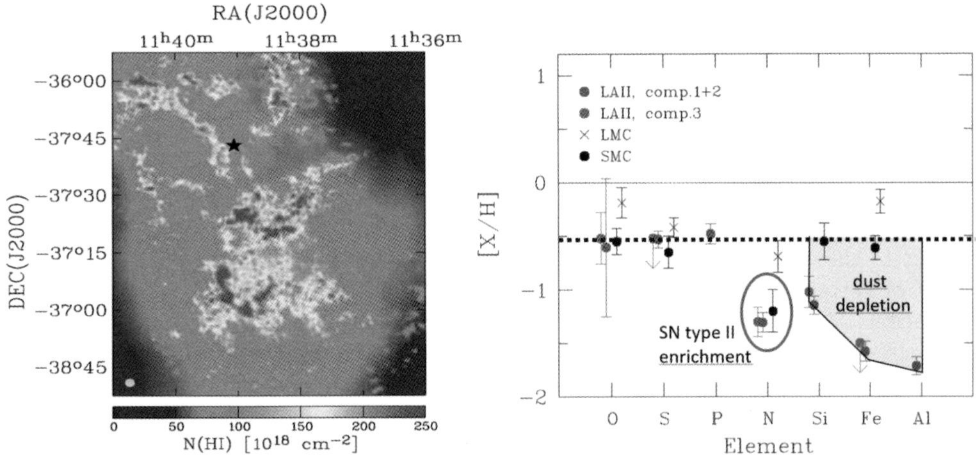

Figure 2. Precision measurements of metal abundances in the Leading Arm (cloud LA II) adopted from Richter et al. (2018). The left panel shows the H I 21 cm map of the LA with the position of NGC 3783, the background source used in this study, indicated by the black star. The right panel shows the abundances of various elements in the LA in the direction of NGC 3783.

and the Milky Way, but many details on *how* the MS was generated and what the exact origin of the individual components is, are still lacking. Various MS formation scenarios, motivated by observations as well as simulations (e.g., Lucchini et al. 2021; Hammer et al. 2015) were also discussed in the context of this IAU conference.

All other HVCs, for which direct distance information from the so-called bracketing method is available, are located within 20 kpc. The prominent HVC Complexes A & C have distances $d \approx 10$ kpc, the Cohen Stream and Complex GCP are at $d = 5 - 15$ kpc, and the HVC towards the LMC has $d \approx 9$ kpc (see R17 and references therein). Combining the 21 cm emission in Milky Way HVCs and in M31 halo clouds, Richter (2012) concludes that the H I covering fraction in HVCs around Milky Way-type galaxies declines exponentially with galactocentric distance. From deep 21 cm observations of the M31 halo, Westmeier et al. 2007 further conclude that the so-called compact high-velocity clouds (CHVCs), isolated high-velocity 21 cm gas clumps in the Milky Way halo with very small angular sizes, are located at $d < 50$ kpc, disproving a previous scenario in which CHVCs are regarded as gas-filled dark matter halos residing in the Local Group (e.g., Blitz et al. 1999).

3. Abundance measurements and origin of individual HVCs

3.1. *General trends*

Most of the IVCs have near-solar metallicities, supporting the idea that these clouds reflect the return flow of a Galactic Fountain (e.g., Richter et al. 2001b). In contrast, the metal abundance in many HVCs is substantially lower (by a factor $5 - 10$, typically). For Complex C, the metallicity has been constrained to ~ 0.15 solar (Wakker et al. 1999; Richter et al. 2001a). The metallicity of the main body of the Magellanic Stream has been constrained to 0.1 solar (Fox et al. 2010), but the MS also contains a filament that is somewhat more metal rich (0.3 solar; Richter et al. 2013). Complex A also has metallicity of only ~ 0.1 solar Wakker (2004). These low metal abundances support the scenario in which HVCs represent gas infalling from a pre-enriched intergalactic medium (or intragroup gas; Bouma et al. 2021). These low-metallicity halo clouds, on the other hand, could also trace material removed from satellite dwarf galaxies as they are being accreted by the Milky Way. Another possible scenario is that low-metallicity HVCs trace gas that has been part of the Milky Way a long time ago, then was ejected at relatively low metallicity by outflows/winds, and now is raining back towards the disk ("intergalactic fountain"). Even lower metallicities, down to ~ 0.01 solar, are probably seen in some CHVCs (Ashley et al., in prep.), pointing to an extragalactic origin for these clouds.

3.2. *The Leading Arm as a case study*

Next to the overall metal abundance (usually derived from O/H or S/H), abundance *ratios* of certain elements provide important information on the origin of HVCs and the various MS components. We recently have presented a detailed abundance study of elements such as O, S, P, N, Si, Fe, Al in the Leading Arm of the MS in the cloud LA II (Richter et al. 2018; see Fig. 2). We find that LA II has an α abundance and N/α ratio that mimics that of the current-day SMC, possibly indicating that the gas recently was removed from the SMC and pushed deep into the Milky Way halo. In this scenario, the age of LA II would be at most a few hundred Myr. On the other hand, if we consider only the chemical abundance pattern as age indicator, LA II could be older if the gas originates in the LMC, reflecting the LMC abundance pattern as it was 1-2 Gyr ago.

Interestingly, there is a substantial metallicity/age spread within the LA (Fox et al. 2018). The lowest metallicity ($\sim 0.04 - 0.06$ solar) is measured in LA III, followed by LA I, which has ~ 0.12 solar abundance, similar to that in the main body of the MS (Fox et al. 2010). The metallicity in LA II (~ 0.3 solar), as presented in Richter et al. (2018), is substantially higher (~ 0.3 solar) than for LA I and LA III, but lower than the LMC filament of the MS (Richter et al. 2018). This metallicity spread may hint at an episodic removal of gas from the SMC, with LA III being the oldest part stripped from the SMC together with the main body of the MS 1-2 Gyr ago, followed by the later removal of LA I and LA II due to more recent encounters of SMC and LMC (Fox et al. 2018). Another possible scenario is, that the LA (or parts of it) does *not at all* originates in the Magellanic Clouds, but instead represents gas stripped from *other* MW satellites, coincidentally being located in the region, where a leading arm would be expected. As it was discussed during this IAU conference, it would be extremely interesting to further investigate the latter scenario by combining MS formation simulations with new observations.

4. A UV absorption-line survey of HVCs in the Milky Way halo

4.1. *UV absorption characteristics*

In R17, we have presented the so-far largest survey of UV absorption the Milky Way's HVC population and in the LG using archival HST/COS data of 270 extragalactic background sources. The great advantage of such a survey in the UV is that it is sensitive to the various different gas phases (from cold/warm neutral to warm ionized gas) that are present in the CGM, and not only to the neutral gas phase (as for H I 21 cm-selected clouds). The aim of that survey was to characterize the absorption properties of the Milky Way's CGM and intragroup gas in the LG based on proper statistics, to explore the overall physical conditions and the distribution of the different CGM gas phases, and to estimate the total gas mass in HVCs and the Milky Way's gas-accretion rate.

The overall sky-covering fraction of high-velocity UV absorption is 77 percent for the most sensitive ion in the survey, Si III. This value is several times higher than the covering fraction of 21 cm neutral hydrogen (see above), demonstrating that the Milky Way's CGM is multi-phase and predominantly ionized. The equivalent-width ratios of low and intermediate metal ions (e.g., Si II, Si III, C II, C IV) are inhomogeneously distributed on large and small angular scales. This implies a complex spatial distribution of multi-phase gas that surrounds the neutral 21 cm HVCs. From our survey we estimate that the total

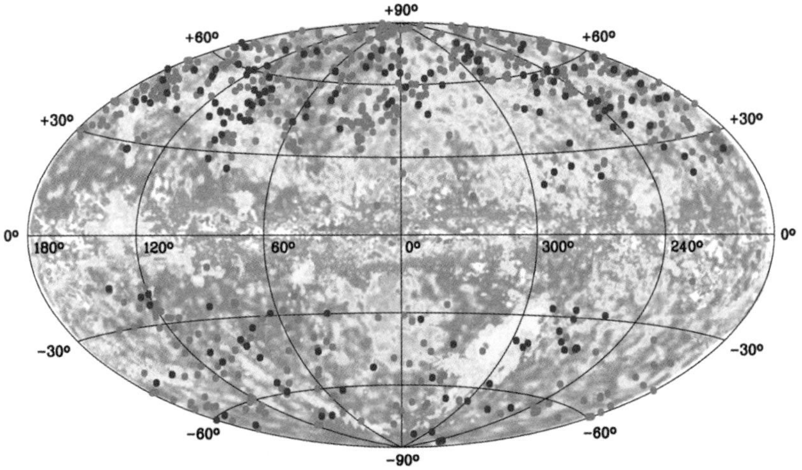

Figure 3. Sky coverage of QSO sightlines to study the Milky Way CGM and gas in the Local Group overlaid on a CGM heat map. The blue dots indicate the positions of the QSOs used in R17, the red dots show the positions of additional QSO sightlines for which archival UV data are available.

mass and accretion rate of neutral and warm ionized gas in the Milky Way CGM is $M_{\rm HVC} > 3 \times 10^9\, M_\odot$ and $\mathrm{d}M_{\rm HVC}/\mathrm{d}t > 6 M_\odot\,\mathrm{yr}^{-1}$, where the MS contributes with more than 90 percent to this mass/mass-flow.

4.2. *Relation between HVCs and other LG galaxies*

As part of the UV absorption survey, we have also systematically investigated sightlines that pass the outskirts of Local Group galaxies located outside the MW's virial radius to search for absorption signatures that might be related to the CGM of these galaxies. Along seven sightlines, high-velocity absorption from gas in the extended M31 halo is present in the LSR velocity range between -300 and -150 km s^{-1}, in line with the more recent AMIGA survey of the M31 CGM presented by Lehner et al. (2020). Interestingly, no CGM absorption is found at impact parameters $\rho \leq 2R_{\rm vir}$ and at velocities close to the systemic velocities of the galaxies NGC 3109, Antlia, Sextans A, Aquarius (one sightline per galaxy, respectively), and Leo I (12 sightlines). Absorption *is* found along a sightline that passes IC 1613 at $\sim 1.5R_{\rm vir}$ near the systemic velocity of IC 1613, but this absorption most likely is associated with gas from the MS. Therefore, only for M31 and the Milky Way the COS data provide evidence for the presence of an extended CGM, while for the LG dwarfs/satellite population outside the Milky Way halo there is no compelling hint for the existence of a circumgalactic gas component from UV observations. This apparent lack of CGM gas in LG dwarfs deserves more attention in future LG absorption-line surveys and in simulations that study gas stripping/gas removal processes in group environments.

4.3. *Local Group gas flows*

Another interesting aspect of our UV survey is the possible detection of a velocity signature of intragroup gas that is moving towards the LG barycenter at a speed that is lower than that of the LG member galaxies. High-velocity absorbers in two distinct regions in the northern sky (northern ionized region, NIR) and in the southern sky (SIR), that are roughly aligned with the LG's major axis, show a particularly high degree of ionization, in line with the idea that they could be located at large distances beyond the virial radius of the MW. In R17, we demonstrate that the NIR and SIR form a dipole on the sky with opposite signs for the observed absorption velocities. In view of the observed high radial velocities and the high absolute Galactic latitudes of these antipodal regions, the observed dipole pattern cannot be produced by the rotation of the Galactic disk. Galaxies outside the virial radius of the Milky Way form a similar dipole-velocity pattern on the sky, which reflects the general flow of galaxies towards the barycenter of the cosmological filament that forms the LG. The Milky Way, being part of that flow, leaves behind LG gas that moves at smaller speeds towards the LG barycenter (leading to red-shifted absorbers in the anti-barycenter direction), while it is moving towards gas that falls in from the other side of the LG barycenter (leading to blue-shifted absorbers in the barycenter direction).

This scenario is supported by constrained cosmological simulations of the LG and its gaseous environment from the CLUES project (Nuza et al. 2014) and, more recently, from the HESTIA project (Biaus et al. 2022).

5. Summary and outlook

As we have discussed here, UV absorption-line measurements of the Milky Way CGM and high-velocity absorbers possibly related to gas in the Local Group provide a detailed insight into the on-going assembly of the Local Group and the evolution of its member galaxies. Since the R17 survey, several hundred *additional* QSO sightlines have been observed with HST/COS (see Fig. 3), providing a rich data basis for another, more detailed all-sky UV survey of large-scale gas flows in the Local Group - a project, that we are currently working on.

References

Biaus, L., Nuza, S.E., Richter, P., et al. 2022, MNRAS 517, 6170
Blitz, L., Spergel, D.N., Teuben, P.J., Hartmann, D., & Burton, W.B. 1999, ApJ, 514, 818
Bouma, S.J.D., Richter, P. & Fechner, C., 2019, A&A, 627, A20
Casetti-Dinescu, D.I., Moni Bidin, C., Girard, T.M., et al. 2014, ApJL, 784, L37
D'Onghia, E. & Fox, A.J. 2016, ARA&A, 54, 363
Fox, A.J., Wakker, B.P., Smoker, J.V., et al. 2010, ApJ, 718, 1046
Fox, A.J., Richter, P., Wakker, B.P. et al. 2013, ApJ, 772, 110
Fox, A.J., Barger, K.A., Wakker, B.P., et al. 2018, ApJ, 854, 142
Hammer, F., Yang, Y.B., Flores, H., Puech, M., Fouquet, S. 2015, ApJ, 813, 110
Lehner, N., Berek, S.C., Howk, J.C., et al. 2020, ApJ, 900, 9
Lucchini, S., D'Onghia, E., & Fox, A.J. 2021, ApJL, 921, L36
Nidever, D.L., Majewski, S.R., Butler Burton, W., & Nigra, L. 2010, ApJ, 723, 1618
Nuza, S.E., Parisi, F., & Scannapieco, C., et al. 2014, MNRAS, 441, 2593
Putman, M. E., Peek, J. E. G., & Joung, M. R. 2012, ARA&A, 50, 491
Richter, P., Wakker, B.P., Sembach, K.R., et al. 2001a, ApJ, 559, 318
Richter, P., Savage, B. D., Wakker, B. P., Sembach, K. R., & Kalberla, P. M. W. 2001b, ApJ, 549, 281
Richter, P. 2012, ApJ, 750, 165
Richter, P., Fox, A.J., Wakker, B.P., et al. 2013, ApJ, 772, 111
Richter, P. 2017, ASSL, 430, 15
Richter, P., Nuza, S. E., Fox, A. J., et al. 2017, A&A, 607, A48
Wakker, B.P., Howk, J.C., Savage, B.D. 1999, Nature, 402, 388
Wakker, B.P. 2001, ApJS, 136, 463
Wakker, B.P. 2004, in High Velocity Clouds, ed. H. van Woerden, B.P. Wakker, U.J. Schwarz, & K.S. de Boer (Dordrecht: Kluwer), 25

Westmeier, T., Braun, R., Brüns, C., Kerp, J., & Thilker, D.A. 2007, New. Astron. Rev., 51, 108

Zhang, L., Moni Bidin, C., Casetti-Dinescu, D.I., et al. 2017, ApJ, 835, 285

Discussion

JIANLING WANG: Excellent talk! From the metallicity of the LA it is very likely that the LA has a different origin than the Magellanic Clouds. You have shown distances of the LA in your talk. I am curious: how do you get these distances? And what are their errors?

PHILIPP RICHTER: The cited distance of $d = 20$ kpc for the LA is based on the studies by Casetti-Dinescu et al. (2014) and Zhang et al. (2017), who claimed that there is a young stellar population in the LA. There are also some more recent indirect measurements using H α fluxes. All these results are, however, afflicted with systematic uncertainties, so there is also a substantial uncertainty in this distance estimate.

Session 5:
Near-field cosmology and galaxy masses beyond the Local Group

Dynamical Masses of Local Group Galaxies
Proceedings IAU Symposium No. 379, 2025
P. Bonifacio, M.-R. Cioni, F. Hammer, M. S. Pawlowski & S. Taibi, eds.
doi:10.1017/S1743921323002600

Stepping beyond our backyard: the era of Near-Field Cosmology

Denija Crnojević

Department of Physics & Astronomy, University of Tampa, 401 West Kennedy Boulevard, Tampa, FL 33606, USA
email: dcrnojevic@ut.edu

Abstract. The study of resolved stellar populations in the nearest galaxies, or "near-field cosmology", provides key constraints on the physics underlying galaxy formation and evolution. Deep, wide-field surveys of nearby groups of galaxies allow us to characterize the past and ongoing accretion processes shaping the halos of Milky Way-mass galaxies. This field is set to experience significant advancements with the current and future generations of state-of-the-art telescopes.

Keywords. galaxies: evolution, galaxies: halos, galaxies: resolved stellar populations, galaxies: luminosity function

1. Introduction

Much of this conference was focused on dynamical mass measurements within the Local Group. Most notably, the Milky Way (MW) mass has to date been estimated with tens of different methods (see, e.g., Wang et al. 2020b), which still struggle to converge to a conclusive value; similarly, we have been able to obtain increasingly accurate estimates for our closest giant neighbour, M31, which is approximately twice as massive as the MW. Besides the giant Local Group hosts, the Gaia proper motions in conjunction with current and upcoming wide-field spectrographs have dramatically increased our knowledge of the internal kinematics, dynamical masses, chemo-dynamical properties, and orbital properties of the MW and M31 satellites. Streams and halo substructures are continuously discovered and characterized, opening new avenues for studies of infall histories and group assembly, dark matter, and much more.

Beyond the Local Group, the amount of information we can extract is inevitably limited by the increasing distance of the targets, making it challenging to focus on the same scientific goals as for the galaxies in our neighbourhood. In contrast to the Local Group, most galaxies only have a couple of robust mass estimates each. Nevertheless, dynamical masses have been routinely estimated for almost a century, mainly focusing on rotation curves when possible, and more recently on tracer mass estimates (satellite galaxies and/or globular clusters, GCs) as well as Jeans modelling. Such measurements must carefully include caveats stemming from the uncertainties in distance, inclination, completeness of the sample, symmetry and equilibrium assumptions, etc. Nevertheless, they can give important insights into a number of open questions: the masses of the target galaxies are necessary in order to accurately compare the observational constraints (such as number and spatial distribution of observable satellites, galaxy stellar luminosity/mass profiles, metallicity gradients, accretion features) to theoretical predictions on the mass and shape of galaxy haloes (giants and dwarfs), the satellite luminosity function, the stellar halo mass fraction, the accretion history of a galaxy, environmental effects on dwarf galaxy evolution, planes of satellites, star formation at the smallest scales, to name a few.

Given the notable differences between the two main galaxies in the Local Group (assembly history, halo substructure, stellar profile, chemical content), despite a mere factor of two difference in mass, the need to step beyond our backyard is only natural. While increasing the sample of targets is necessary, tackling the above mentioned questions is arguably best done when exploiting the power of resolved stellar studies. The current state-of-the-art telescopes now allow us to obtain increasingly detailed views of galaxies within the Local Volume ($\lesssim 10-15$ Mpc), thus shifting our focus from the quantity of the galaxy targets to the quality of the available observational constraints.

2. Near-field cosmology: resolved surveys of nearby galaxies

The resolved study of nearby galaxies is dubbed *near-field cosmology*. The advantages of resolved stellar photometry are multiple, compared to integrated light photometry; on the other hand, obtaining data at such faint limiting magnitudes for individual stars beyond the Local Group translates into a very time-consuming task (even when utilizing wide-field cameras), and the number of possible target galaxies is intrinsically limited by the available number of objects within the Local Volume.

In the past few years, a considerable amount of effort has been put into surveying galaxies in the Local Volume. The main approaches are two: going deep by using the power of space-based facilities, or going wide by exploiting wide-field imagers from the ground. Here I summarize a few recent highlights from each approach.

Space-based studies. While the first resolved HST studies of nearby galaxies have appeared as long as two decades ago, they mainly focused on the inner regions of the target galaxies rather than their extended halos. Such studies were capable of providing accurate tip of the red giant branch measurements and, to some extent, metallicity and metallicity gradients information from the old stellar populations in their targets (for an overview of such early studies, the reader is referred to Crnojević 2017). One of the first and most comprehensive studies of nearby galaxy halos and their features is the Galaxy Halos, Outer disks, Substructure, Thick disks, and Star clusters (GHOSTS) survey. This systematic effort targets 14 nearby disk galaxies within 17 Mpc, and reaches surface brightness values as low as $\mu_V \sim 30\,\mathrm{mag\,arcsec^{-2}}$ in their halos. The deep HST CMDs reveal a huge diversity in the derived halo mass fractions, metallicity gradients and stellar density slopes (Harmsen et al. 2017). A detailed comparison to targeted cosmological magneto-hydrodynamical simulations confirms that the observed trends in these quantities can be reasonably reproduced by models, and that they likely arise due to a diversity in merger history (Monachesi et al. 2019), even though further investigations are warranted. An interesting result is the apparent correlation between metallicity and mass of the galaxies' extended halos, hinting at the crucial role played by the most massive and/or recent accretion event (e.g., D'Souza and Bell 2018).

Ground-based, wide-field studies. On the ground-based side, a number of groups have recently obtained extended resolved stellar maps of Local Volume MW-mass galaxies with the Magellan/Megacam, Subaru/HyperSuprimeCam, CTIO/DECam imagers. Examples include: the early map of NGC 891 by Mouhcine et al. (2010), showcasing the first giant stream detected beyond the LG; the recent surveys of the highly interacting M81 group (Okamoto et al. 2015; Smercina et al. 2020); the Panoramic Imaging Survey of Centaurus and Sculptor (PISCeS) of the spiral NGC 253 and the elliptical NGC 5128 (Centaurus A; Sand et al. 2014; Crnojević et al. 2016), reaching some ~ 150 kpc in the outer halos of these galaxies and down to $\mu_V \sim 32\,\mathrm{mag\,arcsec^{-2}}$ in surface brightness. The key advantage of such surveys is the global view they provide, comparable to the one of PAndAS for M31: individual, albeit deeper, HST pointings randomly placed in halo regions carry the risk of targeting non-representative regions, such as a an overdensity associated with an unknown stellar stream. All the mentioned surveys highlight the large number and variety of features in the halos of the target galaxies, from recent low surface brightness merger/accretion events, to star formation occurring as a result of galaxy interaction, to extended stellar clouds, to previously unknown GCs at large radii (see Figs. 1 and 2).

Particularly important clues can also be derived from the satellite galaxy populations, and specifically from their structural properties, metallicities, spatial distribution and luminosity functions (Bullock and Boylan-Kolchin 2017). The resolved surveys mentioned above have been very successful at uncovering faint dwarf galaxies ($M_V \lesssim -12$) in the halos of the considered targets, even reaching into the ultra-faint regime (e.g., Chiboucas et al. 2009, 2013; Crnojević et al. 2014, 2016; Smercina et al. 2018; Okamoto et al. 2019; Mutlu-Pakdil et al. 2022; Bell et al. 2022), with tip of the red giant branch distances confirming their group membership. Several other studies have simultaneously taken advantage of wide-field, integrated light surveys of nearby galaxies, where candidate dwarfs can be recognized as faint surface brightness enhancements (Merritt et al. 2014; Bennet et al. 2017; Taylor et al. 2018; Müller et al. 2015, 2017, 2018a). Once these candidates are followed-up with resolved imaging to confirm their group membership, the obtained luminosity functions can be compared among Local Volume galaxies (e.g., Smercina et al. 2018; Bennet et al. 2019; Crnojević et al. 2019; Carlsten et al. 2021). The recent Exploration of Local VolumE Satellites (ELVES) Survey is a systematic effort to survey 30 hosts in the Local Volume by analyzing ground-based data for wide-field imagers with a semi-automated algorithm, in search for their satellites down to $M_V \lesssim -9$ (Carlsten et al. 2022). They find about 400 candidates, of which more than 300 are confirmed to be at the distance of their host. These exquisite dataset is being studied to obtain constraints on satellite quenching, satellite stellar-to-halo mass relation, satellite radial distribution, GCs and nuclear star clusters, and more. One of the early results confirms and extends the surprisingly high scatter found by previous studies among the luminosity functions, which is only partly explained by theoretical predictions; the observed hosts furthermore have more bright and fewer faint satellites with respect to model predictions. The closer we look at the faintest among galaxies, the more they challenge our current understanding of baryonic processes and their dependence on the surrounding environment.

For the first time, the lower-mass range of galaxy hosts has started to be considered as well: surveys such as the SOlitary LOcal dwarfs survey (Higgs et al. 2021), the Magellanic Analog Dwarf Companions And Stellar Halos survey (Carlin et al. 2016, 2019), and the DECam Local Volume Exploration Survey (Drlica-Wagner et al. 2022) target the halos of nearby, relatively isolated galaxies in the mass range of the Small Magellanic Cloud to the Large Magellanic Cloud. Simulations are now achieving the necessary resolution to characterize the tiny satellites of such low-mass dwarfs (e.g., Wheeler et al. 2015; Dooley et al. 2017; Munshi et al. 2019), and observational constraints on such objects are particularly useful at these scales, where baryonic processes are highly stochastic and not well understood. Preliminary results from these surveys show that dwarf galaxies can indeed host and disrupt satellites of their own (see Fig. 1).

Detailed follow-up studies of individual systems. Beyond the detection of tidal features, satellites, and GCs, wide-field surveys can be complemented by targeted follow-up (deeper imaging and, if possible, spectroscopy; see, e.g., Toloba et al. 2016) in order to characterize the new discoveries' stellar populations, and reconstruct their origin and orbital history. Importantly, these halo inhabitants have been used as tracers to derive mass estimates of nearby systems, including recently discovered satellite galaxies, as in e.g., Karachentsev et al. (2022). Dwarf satellites have also been found to often be arranged in planes around nearby galaxies (which might or might not be at odds with theoretical

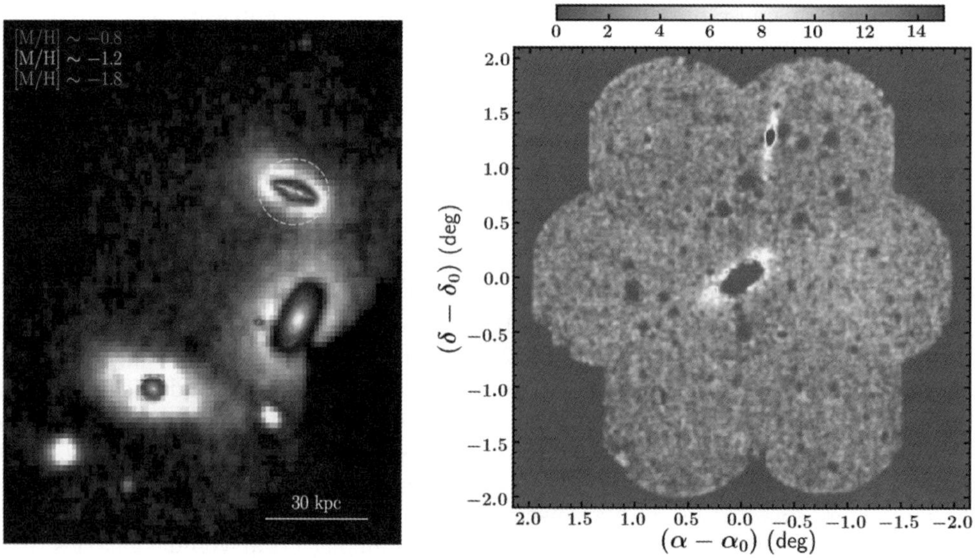

Figure 1. Red giant branch stellar density maps derived from Subaru/HyperSuprimeCam wide-field surveys of two nearby galaxies (at $\sim 3 - 3.5$ Mpc). *Left.* Smoothed density image of the highly interacting M81 group, with intensity mapped to stellar density, where each channel represents stars in three bins of metallicity (as indicated). The interiors of the three main galaxies (M81, M82, and NGC 3077) are filled with to-scale images from HST. Figure from Smercina et al. (2020), their Fig. 14; reproduced by permission of the ©AAS. *Right.* Smoothed density map of the Large Magellanic Cloud-mass galaxy NGC 2403. The field is centered on NGC 2403 (the hole in the center is due to extreme crowding), and its disrupting satellite DDO 44 is to the north. The colorbar encodes the number of stars per 0.75 arcmin bin. Figure from Carlin et al. (2019), their Fig. 3.

predictions), even though the existence of such structures needs careful distance confirmation and kinematic measurements (see, e.g., Müller et al. 2018b). One further step can be taken by matching the photometric maps to the kinematic information derived for GCs, as previously done for M31 (Veljanoski et al. 2014): this powerful method allows the identification of kinematically cold GC samples that are associated with tidal features (Hughes et al. 2023; see Fig. 2), and can further help constrain the host galaxy's mass, reconstruct its past assembly history, as well as the progenitors' properties. The elliptical Cen A has been a benchmark for such studies, given its proximity and active interaction history, and models are starting to answer questions about its past evolution (e.g., Wang et al. 2020a; Pearson et al. 2022). Some of the most stringent constraints come from a combination of the photometric, kinematic, and structural properties of stellar streams (see Fig. 2): beyond the Local Group, we are just beginning to explore these tracers (Pearson et al. 2019; Shipp et al. 2023), and novel methods are being developed to fully exploit the plethora of data that upcoming missions and surveys will deliver in the next decade.

3. Future Perspectives

Near-field cosmology is just starting to be exploited to its fullest, as increasingly sophisticated theoretical predictions go hand-in-hand with the latest observational strategies in order to explore the halos of nearby galaxies and their inhabitants. To advance the field, in the near future we need to be able to survey galaxies of different morphologies and in different environments to a level of detail comparable to the one achieved to date in the Local Group, and we have to perform meaningful comparisons between observations

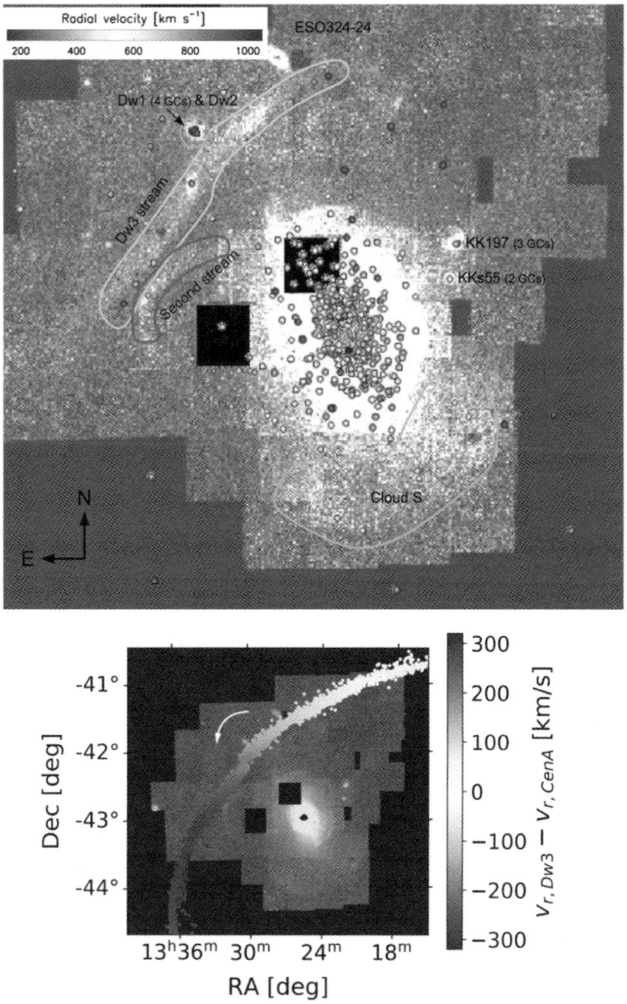

Figure 2. Stellar density maps of red giant branch stars in the halo of Cen A (~ 3.5 Mpc), obtained with Magellan/Megacam as part of the PISCeS survey. Several tidal features trace both ongoing and past accretion events. *Top.* Dots are confirmed globular clusters, color-coded by their radial velocity (Cen A has a velocity of 541 km/s). The colored contours mark the cold stellar features corresponding to substructures (e.g., disrupting dwarfs and a sparse cloud). Figure from Hughes et al. (2023), their Fig. 10. *Bottom.* The best-fit model stream for the disrupting dwarf CenA-MM-Dw3 (seen also in the top panel), colored by radial velocity in Cen A's rest frame, overplotted on the Cen A image data. The white arrow indicates the predicted direction of motion of the dwarf. Adapted from Pearson et al. (2022), their Fig. 5.

and theoretical predictions, which necessitates accurate mass estimates for the target galaxies.

In the upcoming decade, the next generation of telescopes and instruments will open a new window on our knowledge of galaxy evolution: the ground-based 30-m telescopes, the Vera Rubin Observatory (previously known as LSST), the JWST, Euclid, and WFIRST space-borne missions will allow us to achieve a significantly larger number of target galaxies (about 100 within ~ 10 Mpc) for which detailed stellar density maps will be derived out to large galactocentric radii and down to very faint surface brightness limits (see, e.g., Mutlu-Pakdil et al. 2021). Many currently open questions will be answered by such exquisite datasets, and new ones will certainly arise: what are the relative fractions of in-situ vs ex-situ stellar components in galaxy halos, and what is their physical extent and metallicity content? What are the properties of the accreted vs the surviving satellites of galaxies with different morphologies and environments? Are planes of satellites ubiquitous, and what governs the (scatter in the) satellite luminosity function? What are the environmental effects on the build-up of galaxy halos, and are the assembly mechanisms the same at all galaxy masses? What is the shape and mass of the underlying dark matter halos? What can stellar streams teach us about their host galaxies? In preparation for the golden era of near-field cosmology, we need to coordinate systematic surveys of galaxies within the Local Volume and meaningfully compare diverse datasets, learn to fully exploit the synergy between ground-based wide-field surveys and deep, pencil-beam follow-up studies, as well as translate these observational constraints into "simulation language", and viceversa.

Discussion

ANGELES PEREZ-VILLEGAS: How complete is the GC sample around CenA?

DC: Different surveys have recently discovered hundreds of candidate GCs, even fainter than the GC luminosity function peak, out to large galactocentric radii, but many still need to be spectroscopically confirmed.

LAURA WATKINS: Cen A has some 1500 GCs, some associated with substructure. Can you flip the problem and search for patterns or groups in GCs to identify substructures that are too faint or diffuse for us to see?

DC: Absolutely, that is a promising strategy, and the candidate substructures could then be followed-up with deeper observations.

KYLE OMAN: Are all of the external ultra-faint dwarfs discovered so far around more massive hosts?

DC: Most of them are, but the recently discovered Tucana B (Sand et al. 2022), beyond the Local Group boundaries, is very isolated. In the near future, many such faint and isolated galaxies will be discovered.

MARCEL PAWLOWSKI: What do you think is the way forward in near-field cosmology?

DC: The answer to this question is now elaborated in the "Future Perspectives" section above.

Acknowledgements

I would like to thank the organizers for an enjoyable and very stimulating conference. I acknowledge support from a Faculty International Professional Development Grant from the University of Tampa, and from NSF grant AST-1814208.

References

Bell, E. F., et al. 2022. *ApJL*, 937(1), L3.
Bennet, P., et al. 2019. *ApJ*, 885(2), 153.
Bennet, P., et al. 2017. *ApJ*, 850, 109.
Bullock, J. S. & Boylan-Kolchin, M. 2017. *Annual Review of Astronomy and Astrophysics*, 55, 343–387.
Carlin, J. L.,et al. 2019. *ApJ*, 886(2), 109.

Carlin, J. L., et al. 2016. *ApJL*, 828, L5.
Carlsten, S. G., Greene, J. E., Beaton, R. L., Danieli, S., & Greco, J. P. 2022. *ApJ*, 933(1), 47.
Carlsten, S. G., Greene, J. E., Peter, A. H. G., Beaton, R. L., & Greco, J. P. 2021. *ApJ*, 908(1), 109.
Chiboucas, K., Jacobs, B. A., Tully, R. B., & Karachentsev, I. D. 2013. *AJ*, 146, 126.
Chiboucas, K., Karachentsev, I. D., & Tully, R. B. 2009. *AJ*, 137, 3009–3037.
Crnojević, D. 2017. volume 434 of*Astrophysics and Space Science Library*, 31.
Crnojević, D., et al. 2019. *ApJ*, 872(1), 80.
Crnojević, D., et al. 2014. *ApJL*, 795, L35.
Crnojević, D., et al. 2016. *ApJ*, 823, 19.
Dooley, G. A., et al. 2017. *MNRAS*, 471(4), 4894–4909.
Drlica-Wagner, A., et al. 2022. *ApJS*, 261(2), 38.
D'Souza, R. & Bell, E. F. 2018. *MNRAS*, 474, 5300–5318.
Harmsen, B., et al. 2017. *MNRAS*, 466(2), 1491–1512.
Higgs, C. R., et al. 2021. *MNRAS*, 503(1), 176–199.
Hughes, A. K., et al. 2023. *ApJ*, 947(1), 34.
Karachentsev, I. D., Makarova, L. N., Anand, G. S., & Tully, R. B. 2022. *AJ*, 163(5), 234.
Merritt, A., van Dokkum, P., & Abraham, R. 2014. *ApJL*, 787, L37.
Monachesi, A., et al. 2019. *MNRAS*, 485(2), 2589–2616.
Mouhcine, M., Ibata, R., & Rejkuba, M. 2010. *ApJL*, 714, L12–L15.
Müller, O., Jerjen, H., & Binggeli, B. 2015. *A&A*, 583, A79.
Müller, O., Jerjen, H., & Binggeli, B. 2018,a. *A&A*, 615a, A105.
Müller, O., Pawlowski, M. S., Jerjen, H., & Lelli, F. 2018,b. *Science*, 359b(6375), 534–537.
Müller, O., Scalera, R., Binggeli, B., & Jerjen, H. 2017. *A&A*, 602, A119.
Munshi, F., et al. 2019. *ApJ*, 874(1), 40.
Mutlu-Pakdil, B., et al. 2021. *ApJ*, 918(2), 88.
Mutlu-Pakdil, B., et al. 2022. *ApJ*, 926(1), 77.
Okamoto, S., et al. 2015. *ApJL*, 809, L1.
Okamoto, S., et al. 2019. *ApJ*, 884(2), 128.
Pearson, S., et al. 2022. *ApJ*, 941(1), 19.
Pearson, S., et al. 2019. *ApJ*, 883(1), 87.
Sand, D. J., et al. 2014. *ApJL*, 793, L7.
Sand, D. J., et al. 2022. *ApJL*, 935(1), L17.
Shipp, N., et al. 2023. *ApJ*, 949(2), 44.
Smercina, A., et al. 2018. *ApJ*, 863(2), 152.
Smercina, A., et al. 2020. *ApJ*, 905(1), 60.
Taylor, M. A., et al. 2018. *ApJL*, 867(1), L15.
Toloba, E., et al. 2016. *ApJL*, 830, L21.
Veljanoski, J., et al. 2014. *MNRAS*, 442, 2929–2950.
Wang, J., Hammer, F., Rejkuba, M., Crnojević, D., & Yang, Y. 2020,a. *MNRAS*, 498a(2), 2766–2777.
Wang, W., Han, J., Cautun, M., Li, Z., & Ishigaki, M. N. 2020,b. *Science China Physics, Mechanics, and Astronomy*, 63b(10), 109801.
Wheeler, C., et al. 2015. *MNRAS*, 453, 1305–1316.

Dynamical Masses of Local Group Galaxies
Proceedings IAU Symposium No. 379, 2025
P. Bonifacio, M.-R. Cioni, F. Hammer, M. S. Pawlowski & S. Taibi, eds.
doi:10.1017/S1743921323001989

Dwarf early-type galaxies in the Coma cluster: internal dynamics, stellar populations

Igor V. Chilingarian[1,2], Kirill A. Grishin[3,2], Anton V. Afanasiev[4,2], Anton Mironov[5], Daniel Fabricant[1], Sean Moran[1], Nelson Caldwell[1], Ivan Yu. Katkov[6,2] and Irina Ershova[5]

[1]Center for Astrophysics – Harvard and Smithsonian, 60 Garden St. Cambridge, MA, 02138 USA
email: igor.chilingarian@cfa.harvard.edu

[2]Sternberg Astronomical Institute, Moscow State University, 13 Universitetsky prospect, Moscow, Russia

[3]Université Paris Cité, CNRS, Astroparticule et Cosmologie, F-75013 Paris, France

[4]LESIA, Observatoire de Paris, 5 place Jules Janssen, 92195, Meudon, France

[5]Faculty of Space Research, Moscow State University, 1 Leninskie Gory, bld. 52, Moscow, Russia

[6]New York University Abu Dhabi, Saadiyat Island, PO Box 129188, Abu Dhabi, UAE

Abstract. We present preliminary results from our spectroscopic survey of low-luminosity early-type galaxies in the Coma cluster conducted with the Binospec spectrograph at the 6.5 m MMT. From spatially-resolved profiles of internal kinematics and stellar population properties complemented with high-resolution images, we placed several low-luminosity dEs on the fundamental plane in the low-luminosity extension of the available literature data. We also discovered unusual kpc-sized kinematically-decoupled cores in several dwarf galaxies, which had been probably formed before these galaxies entered the cluster.

Keywords. dwarf elliptical galaxies, dynamics of galaxies, stellar populations

---

1. Introduction

Dwarf elliptical (dE) and lenticular (dS0) galaxies are low-stellar-mass ($M_* \lesssim 5 \cdot 10^9 M_\odot$) quiescent stellar systems commonly found in galaxy clusters and massive groups (Sandage & Binggeli 1984). The formation and evolution of dE/dS0s are thought to be influenced by a combination of internal processes, such as stellar feedback (Dekel & Silk 1986) and environmental effects, ram pressure stripping of gas by the hot intracluster medium (Gunn & Gott 1972) and tidal interactions with other cluster members as well as the cluster potential (Moore et al. 1996). Dwarf early-type galaxies are not as simple as they look. Faint low-contrast spiral arms and bars were found in several 'luminous' dE/dS0s in the Virgo and Abell 496 clusters (Jerjen et al. 2000; Chilingarian et al. 2008b). The diversity of rotational support in dE/dS0s (Geha et al. 2002), residual central star formation (Lisker et al. 2006), and young metal-rich embedded substructures and kinematically decoupled cores (Chilingarian et al. 2007b,c; Chilingarian 2009) strengthened their evolutionary connection to more massive 'normal' galaxies (Boselli et al. 2008). However, which evolutionary mechanism dominates the dE/dS0 formation is still not understood. In 2017–2022 we carried out a spectroscopic campaign to characterise dE/dS0 populations in three nearby clusters ($0.023 < z < 0.045$) and assess their evolutionary paths by obtaining spatially resolved internal kinematics and stellar population characteristics of several hundred dE/dS0 galaxies.

2. Observations, Data Reduction and Analysis

We observed galaxies in the three low-redshift clusters, Coma=Abell 1656 ($z = 0.023$), Abell 2147 ($z = 0.035$), and Abell 168 ($z = 0.045$) using the high-throughput multi-object optical spectrograph Binospec (Fabricant et al. 2019) operated at the f/5 focus of the 6.5 m MMT at Mt. Hopkins, Arizona. The field-of-view of 16'×15' at this redshift range allows us to observe up-to 100 galaxies in each multi-slit mask at the spatial resolution of 0.4–0.7 kpc (corresponding to the typical 0.8" seeing). Our sample includes > 200 dEs/dS0s in the Coma cluster (8 masks), 30 in Abell 2147 (1 mask), and 50 in Abell 168 (2 masks). Using tilted slits, we observed a subsample of ~ 100 Coma and Abell 168 dEs/dS0s along their major and minor axes. We integrated for 2–4 h on each mask and reached a surface brightness $\mu_B \sim 25.5$ mag arcsec^{-2} at S/N $= 3$ pix^{-1} in the wavelength range $3700 < \lambda < 5300$ Å. The 1000 gpm grating provides a spectral resolution $R = 4800$. This setup is well suited for internal kinematic studies and stellar population analysis of faint absorption-line spectra (Chilingarian & Grishin 2020; Afanasiev et al. 2023). In addition to the spectroscopic data, we used high-resolution Hubble Space Telescope images.

We reduced spectroscopic data using the Binospec data reduction pipeline (Kansky et al. 2019) optimized for low-surface brightness data. We used the NBursts full spectum fitting code (Chilingarian et al. 2007a) with $R = 10000$ simple stellar population models computed with the PEGASE.HR code (Le Borgne et al. 2004) and a dedicated grid of synthetic spectra, which model a process of a ram-pressure-induced starburst (Grishin et al. 2019) for a subsample of objects with signs of recent quenching. We analyzed images using GALFIT (Peng et al. 2010) by using up-to three light profile components and computed an equivalent half-light radius from several components.

3. Preliminary Results

3.1. *Coma dEs in κ-space*

A good proxy for a dynamical mass of a virialized stellar system is its position in the Fundamental Plane (FP, Djorgovski & Davis 1987), which requires the knowledge of R_e, $\langle\mu_e\rangle$ and σ_*. Here we use a so-called κ-space FP projection (Bender et al. 1992), where one of the axes, κ_3, is related to the dynamical mass-to-light ratio and another one, κ_1 can be expressed as a logarithm of the stellar mass. In Fig. 1 we present a literature compilation of structural and dynamical measurements for dwarf and giant early-type galaxies, ultradiffuse galaxies (UDGs), bulges, and compact stellar systems. New measurements from Binospec spectra and HST images are shown as purple stars. The purple stars fall on the extension of the dE locus towards lower stellar masses and lower velocity dispersions. Interestingly, these objects seem to have lower dynamical mass-to-light ratios than Virgo cluster dEs of the same stellar mass. All fall below the M/L rise in the dwarf spheroidal regime, where KDG 64, one of the most massive dSphs in the Local Volume, resides (Afanasiev et al. 2023). Interestingly, UDGs have higher M/L ratios for the same stellar mass.

3.2. *Kinematically decoupled cores*

Another intriguing result of our project is a discovery of several kinematically decoupled cores (KDCs) in brighter ($M_V < -17$ mag) Coma cluster dEs. In the Coma cluster,

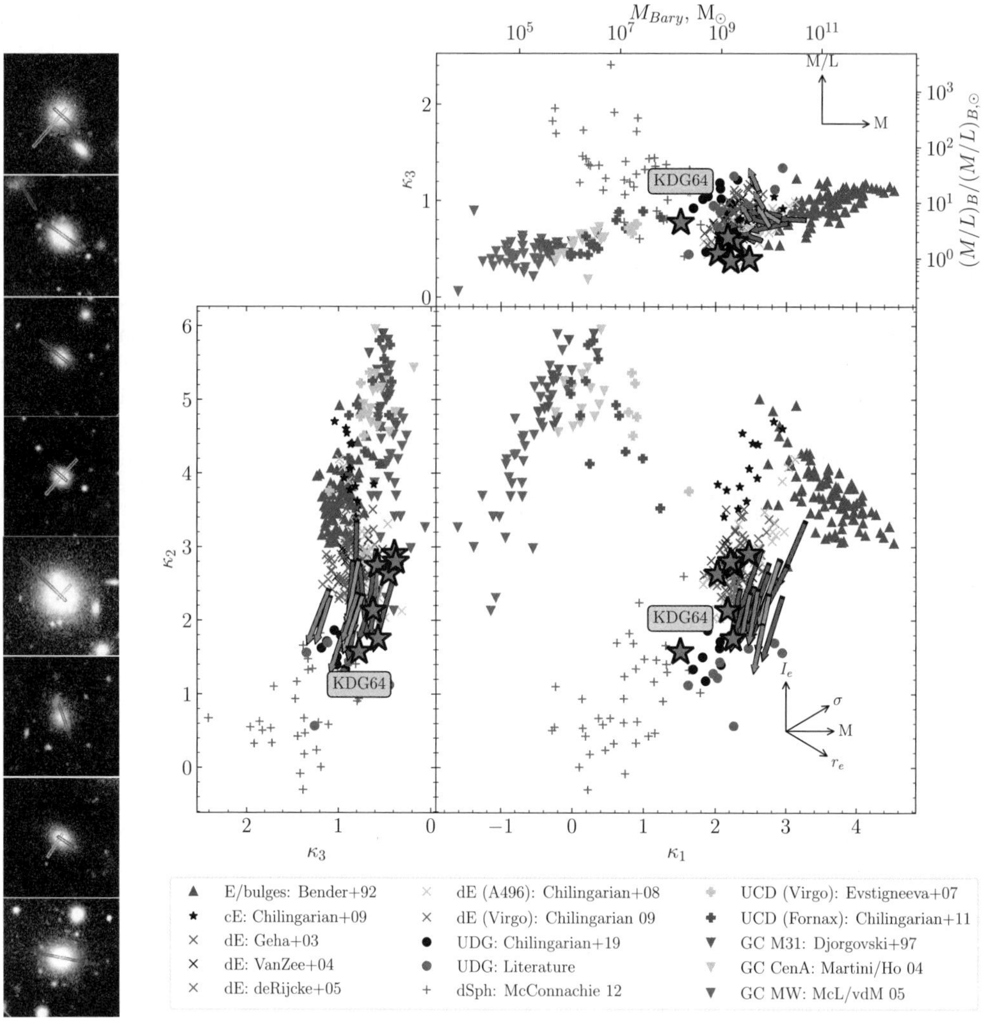

Figure 1. Left: examples of Coma cluster dEs with positions of Binospec slits overplotted. Four galaxies were observed along two axes (green and red slits). **Right:** κ-space FP projection, where the results of our analysis for Coma cluster dEs are shown by purple stars. KDG 64, a large dwarf spheroidal or a small UDG in the M 81 group (Afanasiev et al. 2023) is marked: its position is clearly distinct from the remaining dEs. The arrow represent predicted passive evolution of diffuse post-starburst galaxies.

to our surprise, we found KDCs which are much more extended (> 1 kpc) than those in the Virgo cluster (see Fig. 2). Because the distance is 6 times greater, we cannot detect KDCs as compact as those in the Virgo cluster ($200 \ldots 300$ pc) in ground-based seeing-limited observations, however, no 'large' KDCs have been found in the Virgo dEs to date (Toloba et al. 2014). Both 'small' KDCs in Virgo dEs and 'large' KDCs in Coma dEs are usually distinct in stellar age and metallicity; they are younger and more metal-rich. KDCs are thought to form via minor mergers, cold gas accretion, or galaxy flybys at a low relative velocity. While mergers can happen in clusters, despite a low probability of dwarf-dwarf interactions, the two latter channels are practically unfeasible due to the dynamically hot environment. However, a pre-existing KDC in a disk galaxy can survive its morphological transformation into a dE during infall. The environmental phenomena

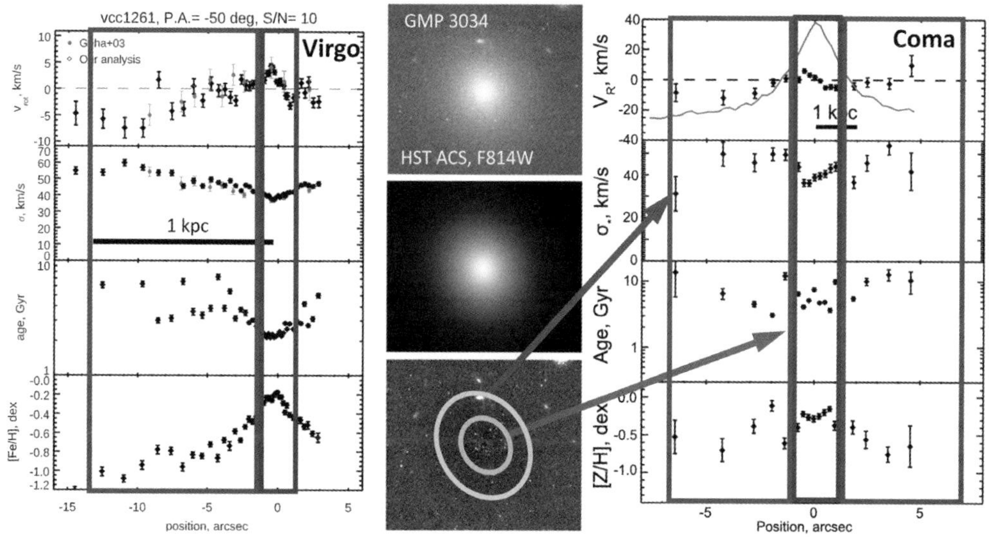

Figure 2. Examples of kinematically decoupled cores found in VCC 1261 in the Virgo cluster (re-analysis of archival observations from Keck ESI presented in Geha et al. 2003) and GMP 3034 (Binospec data). Note the difference of spatial scales (0.08 and 0.5 kpc arcsec^{-2} for Virgo and Coma respectively).

in the massive Coma cluster are expected to be more violent than in Virgo, therefore more massive galaxies with larger pre-existing KDCs can be efficiently quenched and transformed into 'oversized' dEs with large KDCs.

References

Afanasiev, A. V., Chilingarian, I. V., Grishin, K. A., et al. 2023, MNRAS, 520, 6312

Bender, R., Burstein, D., & Faber, S. M. 1992, ApJ, 399, 462

Boselli, A., Boissier, S., Cortese, L., & Gavazzi, G. 2008, ApJ, 674, 742

Chilingarian, I., Cayatte, V., Revaz, Y., et al. 2009, Science, 326, 1379

Chilingarian, I., Prugniel, P., Sil'chenko, O., & Koleva, M. 2007a, in IAU Symposium, Vol. 241, Stellar Populations as Building Blocks of Galaxies, ed. A. Vazdekis & R. R. Peletier (Cambridge, UK: Cambridge University Press), 175–176

Chilingarian, I. V. 2009, MNRAS, 394, 1229

Chilingarian, I. V., Afanasiev, A. V., Grishin, K. A., Fabricant, D., & Moran, S. 2019, ApJ, 884, 79

Chilingarian, I. V., Cayatte, V., & Bergond, G. 2008a, MNRAS, 390, 906

Chilingarian, I. V., Cayatte, V., Durret, F., et al. 2008b, Astronomy & Astrophysics, 486, 85

Chilingarian, I. V., & Grishin, K. A. 2020, PASP, 132, 064503

Chilingarian, I. V., Mieske, S., Hilker, M., & Infante, L. 2011, MNRAS, 412, 1627

Chilingarian, I. V., Prugniel, P., Sil'Chenko, O. K., & Afanasiev, V. L. 2007b, MNRAS, 376, 1033

Chilingarian, I. V., Sil'Chenko, O. K., Afanasiev, V. L., & Prugniel, P. 2007c, Astronomy Letters, 33, 292

de Rijcke, S., Michielsen, D., Dejonghe, H., Zeilinger, W. W., & Hau, G. K. T. 2005, Astronomy & Astrophysics, 438, 491

Dekel, A., & Silk, J. 1986, ApJ, 303, 39

Djorgovski, S., & Davis, M. 1987, ApJ, 313, 59

Djorgovski, S. G., Gal, R. R., McCarthy, J. K., et al. 1997, ApJL, 474, L19

Evstigneeva, E. A., Gregg, M. D., Drinkwater, M. J., & Hilker, M. 2007, AJ, 133, 1722

Fabricant, D., Fata, R., Epps, H., et al. 2019, PASP, 131, 075004

Geha, M., Guhathakurta, P., & van der Marel, R. P. 2002, AJ, 124, 3073
—. 2003, AJ, 126, 1794
Grishin, K. A., Chilingarian, I. V., Afanasiev, A. V., et al. 2021, Nature Astronomy, 5, 1308
Grishin, K. A., Chilingarian, I. V., Afanasiev, A. V., & Katkov, I. Y. 2019, arXiv e-prints, arXiv:1909.13460
Gunn, J. E., & Gott, III, J. R. 1972, ApJ, 176, 1
Jerjen, H., Kalnajs, A., & Binggeli, B. 2000, Astronomy & Astrophysics, 358, 845
Kansky, J., Chilingarian, I., Fabricant, D., et al. 2019, PASP, 131, 075005
Le Borgne, D., Rocca-Volmerange, B., Prugniel, P., et al. 2004, Astronomy & Astrophysics, 425, 881
Lisker, T., Glatt, K., Westera, P., & Grebel, E. K. 2006, AJ, 132, 2432
Martini, P., & Ho, L. C. 2004, ApJ, 610, 233
McConnachie, A. W. 2012, AJ, 144, 4
McLaughlin, D. E., & van der Marel, R. P. 2005, APJS, 161, 304
Moore, B., Katz, N., Lake, G., Dressler, A., & Oemler, A. 1996, NAT, 379, 613
Peng, C. Y., Ho, L. C., Impey, C. D., & Rix, H.-W. 2010, AJ, 139, 2097
Sandage, A., & Binggeli, B. 1984, AJ, 89, 919
Toloba, E., Guhathakurta, P., van de Ven, G., et al. 2014, ApJ, 783, 120
van Zee, L., Barton, E. J., & Skillman, E. D. 2004, AJ, 128, 2797

Dynamical Masses of Local Group Galaxies
Proceedings IAU Symposium No. 379, 2025
P. Bonifacio, M.-R. Cioni, F. Hammer, M. S. Pawlowski & S. Taibi, eds.
doi:10.1017/S1743921323001990

Dark matter content and dynamical masses of ultra-diffuse galaxies in the Coma cluster

Igor V. Chilingarian[1,2], Kirill A. Grishin[3,2], Anton V. Afanasiev[4,2], Anton Mironov[5], Daniel Fabricant[1], Sean Moran[1], Nelson Caldwell[1], Ivan Yu. Katkov[6,2] and Irina Ershova[5]

[1]Center for Astrophysics – Harvard and Smithsonian, 60 Garden St. Cambridge, MA, 02138 USA
email: igor.chilingarian@cfa.harvard.edu

[2]Sternberg Astronomical Institute, Moscow State University, 13 Universitetsky pr., Moscow, Russia

[3]Université Paris Cité, CNRS, Astroparticule et Cosmologie, F-75013 Paris, France

[4]LESIA, Observatoire de Paris, 5 place Jules Janssen, 92195, Meudon, France

[5]Faculty of Space Research, Moscow State University, 1 Leninskie Gory, bld. 52, Moscow, Russia

[6]New York University Abu Dhabi, Saadiyat Island, PO Box 129188, Abu Dhabi, UAE

Abstract. Ultra-diffuse galaxies (UDGs) are spatially extended, low surface brightness stellar systems with regular elliptical-like morphology found in large numbers in galaxy clusters and groups. Studies of the internal dynamics and dark matter content of UDGs have been hampered by their low surface brightnesses. We identified a sample of low-mass early-type post-starburst galaxies, 'future UDGs' in the Coma cluster still populated with young stars, which will passively evolve into UDGs in the next 5–10 Gyr. We collected deep observations for a sample of low-mass early-type galaxies in the Coma cluster using MMT Binospec, which includes present-day and future UDGs. We derived their dark matter content within a half-light radius (70–95 %) and total dynamical masses ($M_{200} = 5.5 \cdot 10^9 - 1.4 \cdot 10^{11} M_\odot$) assuming the Burkert density profile and assess how different proposed evolutionary channels affect dark and visible matter in UDGs. We also discuss observational methodology of present and future UDG studies.

Keywords. ultradiffuse galaxies, dwarf galaxies, dynamics of galaxies, stellar populations

1. Introduction

Ultra-diffuse galaxies (UDGs) are low-surface brightness ($\langle \mu_r \rangle > 24$ mag arcsec^{-2}) extended (R_e up-to 10 kpc) stellar systems. Initially discovered in the Virgo cluster in the 80s (Sandage & Binggeli 1984), they were nearly forgotten for almost two decades until Caldwell (2006) resolved red giant branch stars in two of them. Another decade later, UDGs attracted a lot of attention, when hudreds of them were discovered in the Coma cluster (van Dokkum et al. 2015; Koda et al. 2015) and progress in astronomical instrumentation enabled their spectroscopic studies in integrated light. The main unanswered question is whether UDGs represent an extension of dwarf elliptical (dE) galaxy population to larger sizes and lower surface brightnesses (Conselice 2018) and share common evolutionary paths, or whether they form a separate galaxy class with different formation scenarios (see e.g. Yozin & Bekki 2015).

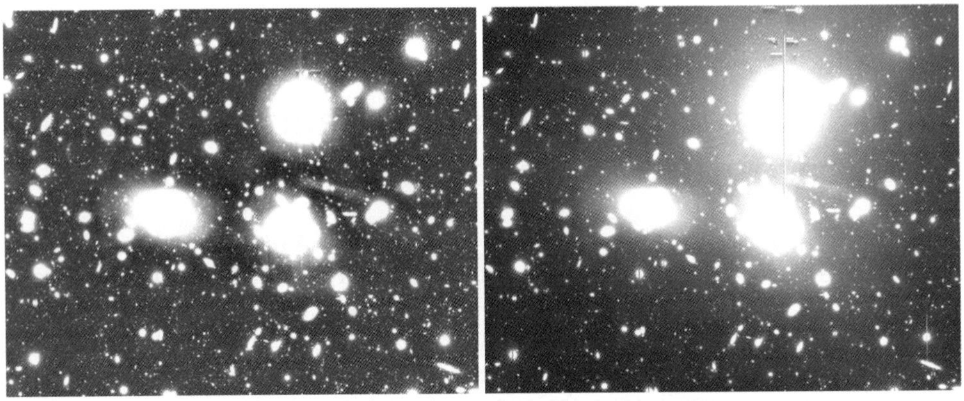

Figure 1. Improvement in sky background subtraction for HyperSuprimeCam data for the central region of the Coma cluster. We show contrast-enhanced images obtained from the HyperSuprimeCam Legacy Archive (left) and using our own data reduction (right). The background over-subtraction in the left panel reults in dark rings around the two dominant cluster galaxies.

The proposed UDG formation scenarios can be confirmed or ruled out by detailed assessment of their internal dynamics, dark matter content, and star formation histories. Here we review our recent results on dynamical modelling of UDGs and put them into the context of their formation and evolution.

2. UDG Observations and Data Reduction: Methodological Considerations

Dynamical modelling of collisionless stellar systems requires knowledge of the stellar density distribution and internal kinematics (stellar velocities, velocity dispersions, possibly higher order moments of the line-of-sight velocity distribution (LOSVD), and proper motions of individual stars for nearby galaxies and star clusters). Therefore, an input observational dataset usually contains images and spectra, or, in rare cases, a fully calibrated spectral data cube, which can be used for stellar density measurements, even though direct images are typically more reliable in the areas of low surface brightness.

2.1. *Imaging and Light Profile Decomposition*

When observed from the ground, UDG central surface brightnesses fall below that of the dark moonless sky, and one typically needs to trace light profiles to $\mu_r >$ 26 mag arcsec^{-2} to confidently measure an effective radius and Sérsic concentration. Therefore, the quality of sky subtraction is the key quality factor for light profiles. Unfortunately, standard mosaic imaging data processing pipelines use local sky background modelling, leading to background oversubtraction near extended targets and to artefacts in galaxies close to the edges of individual detectors in the array. For this project, we modified the Subaru HyperSuprimeCam data reduction pipeline (Bosch et al. 2018) to construct a global sky background model (see Fig. 1), which allowed us to mitigate these problems and confidently reach the surface brightnesses $\mu_g = 28.5 \ldots 29.0$ mag arcsec^{-2}.

Relatively nearby galaxies (to a distance of ~20 Mpc) can be resolved into stars using Hubble and the James Webb Space telescopes. Star counts allow a more precise determination of the outer light profiles. Because resolved images don't reach the faintest stars we need to renormalize the star count density profile using ground-based data

(Afanasiev et al. 2023). Even though crowding might become an issue in the inner parts of a galaxy we are able to define a region where both unresolved and resolved profiles overlap for renomalization. At the Virgo cluster distance (~16 Mpc) at a surface brightness of $\mu_r = 23.5 \ldots 26.0$ mag arcsec^{-2} confusion does not affect the population of red giant stars.

2.2. *Optical Spectroscopy*

Obtaining integrated-light spectra of UDGs and using them to extract stellar kinematics, ages, and metallcities remains challenging. However, even relatively low signal-to-noise spectra (S/N~3 pix^{-1}) can yield good measurements of internal dynamics with an appropriate observing strategy (Chilingarian et al. 2019).

The key factors are spectral resolution, choice of wavelength range, and slit width for long-slit or multi-slit observations. Three main rules apply. (1) The slit width should not exceed the seeing FWHM; otherwise the galaxy light profile will define the spectral line shape making internal kinematics measurements impossible. (2) The spectral resolution ($\sigma_{\rm inst}$) should match the expected velocity dispersion of the galaxy ($R \approx 5000$ for UDGs). If the resolution is too low measuring velocity dispersions and radial velocities will be difficult or impossible. If the resolution is too high, the LOSVD will be oversampled, leading to no improvement in v_* and σ_* uncertainties. (3) The chosen wavelength range must contain many strong absorption lines. This selection affects the quality of kinematics more than any other factor (Chilingarian & Grishin 2020). Additionally, to make stellar population measurements possible, one should include age- (e.g. Balmer lines) and metallicity-sensitive features in the spectral range, even though many more spectral lines can be used to determine stellar ages (Chilingarian 2009) than typically understood. The best (optical) wavelength range for studies of internal kinematics and stellar populations of faint quiescent galaxies is between 3900–5400 Å.

3. Observations, Data Reduction and Analysis

We observed UDGs and diffuse post-starburst galaxies (PSGs) or 'future UDGs' in the Coma=Abell 1656 ($z = 0.023$) and Abell 2147 ($z = 0.035$) galaxy clusters using the high-throughput multi-object optical spectrograph Binospec (Fabricant et al. 2019) at the f/5 focus of the 6.5 m MMT. The wavelength range $3700 < \lambda < 5300$ Å at a spectral resolving power $R = 4800$ was chosen to target low-surface brightness absorption-line galaxies. We used deep archival imaging of the Coma cluster from HyperSuprimeCam at the 8-m Subaru telescope.

We reduced spectroscopic data using the standard Binospec data reduction pipeline (Kansky et al. 2019). We reduced HSC images using a customized version of the HSC data reduction pipeline were we added a block to construct a global sky model.

We use the NBursts full spectum fitting code (Chilingarian et al. 2007) with $R = 10000$ simple stellar population models computed with the PEGASE.HR code (Le Borgne et al. 2004) and a dedicated grid of synthetic spectra to model a ram-pressure-induced starburst (Grishin et al. 2019) for diffuse PSGs (Fig. 2). We analyze images with GALFIT (Peng et al. 2010) using up to three light profile components. We perform Jeans axisymmetric modelling including a spherical dark matter halo in addition to the stellar component (Afanasiev et al. 2023).

4. Dynamical Masses of UDGs

Full details of the Binospec data analysis for the 'bona fide UDGs' and for diffuse PSGs are presented in Chilingarian et al. (2019) and Grishin et al. (2021) respectively. Here we discuss the derived dark matter content within R_e, how it is related to full

Figure 2. A sample of diffuse PSGs (8 'future UDGs' and 2 'future dEs' in the Coma and Abell 2147 galaxy clusters, for which we performed full dynamical modelling using internal kinematics out to $1.5 - 3R_e$. Color images are assembled from archival Canada-France-Hawaiian Telescope images. For GMP 2640 we show a Subaru Suprimecam R-band image.

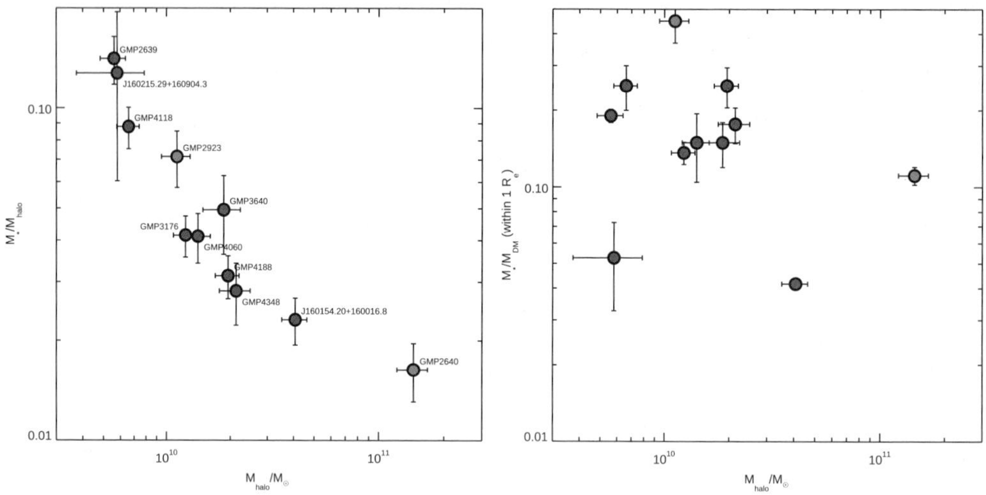

Figure 3. *Left:* Total stellar-to-halo mass ratio vs total extrapolated halo mass for a sample of diffuse PSGs in Coma and Abell 2147. *Right:* Stellar-to-dark matter mass ratio within R_e vs total extrapolated halo mass. Red and blue dots mark galaxies, which will passively evolve into UDGs and dEs respectively.

(extrapolated) dynamical masses, and how assumptions about the dark matter profiles affect this extrapolation.

The derived dark matter content within 1 R_e (70–95 %) derived for UDGs using sparse measurements with σ_* profiles including between 1 and 5 data points and extending barely to 1 R_e, agree remarkably well with the measurements of 'future UDGs' from well sampled kinematics extending to 1.5-3 R_e. There seems to be a trend of increased DM for smaller stellar surface densities but this trend should be verified with a larger sample. The agreement between the two approaches validates the low-S/N measurements and proves that diffuse PSGs are indeed 'future UDGs' despite their current high surface brightness arising from abundant young stars.

At the same time, the total virial masses M_{200} derived under assumptions of NFW (Navarro et al. 1997) and Burkert (1995) DM profiles derived from stellar kinematics within the central 1–3 R_e are different by up-to an order of magnitude (Burkert profiles

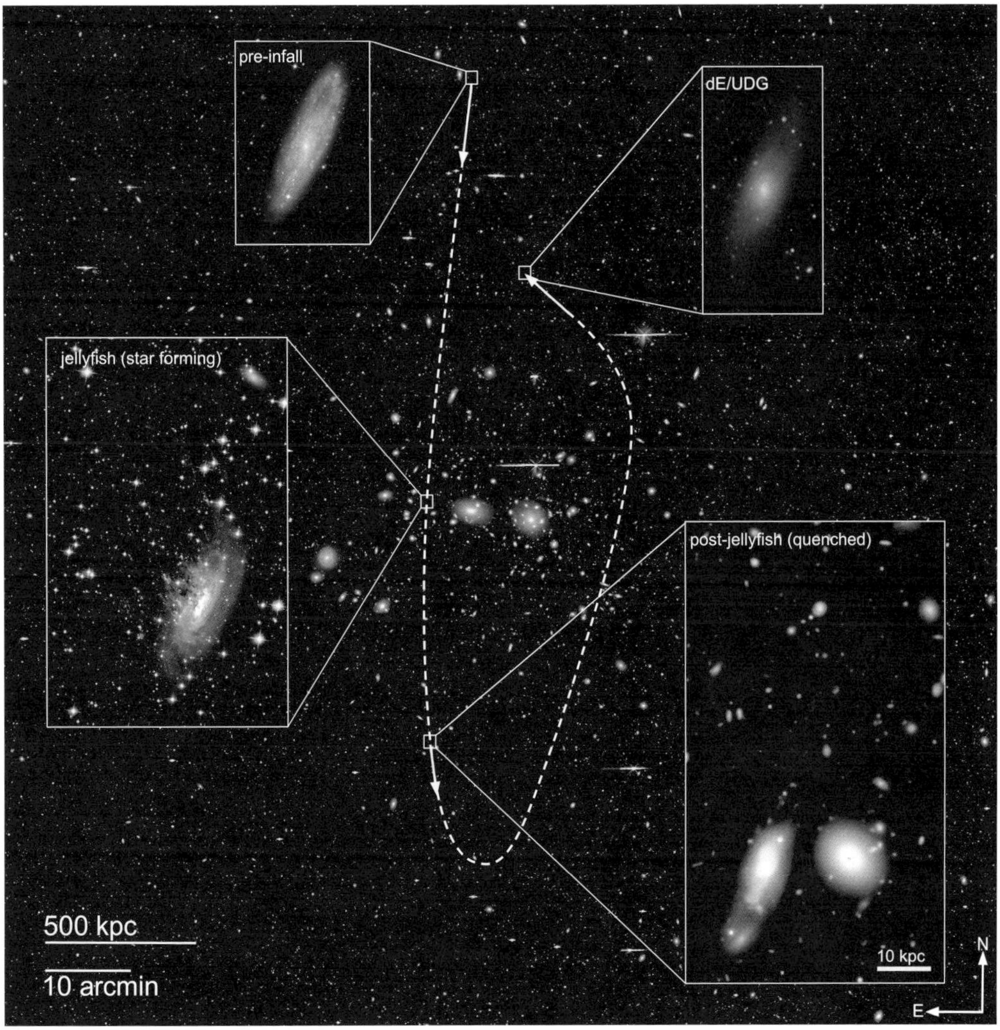

Figure 4. An artist's impression of an evolutionary path of a normal low-mass disk galaxy into a UDG/dE in a cluster. (a) An unperturbed galaxy falls into the cluster following an orbit with a substantial tangential component (shown by a dashed line). (b) Interaction with the intergalactic medium induces a burst of star formation and causes gas stripping, creating a 'jellyfish' galaxy. (c) Star formation and stripping remove all of the gas, quenching further star formation and causing the disk expansion. (d) The galaxy evolves into a UDG or dE inside the cluster for billions of years.

being lighter). This phenomenon is best illustrated by the case of KDG 64, a small UDG or a large dwarf spheroidal galaxy in the nearby M 81 group where despite very low surface brightness, the profiles of internal kinematics are well sampled within 1 R_e (Afanasiev et al. 2023).

Using the measurements presented in Grishin et al. (2021) we can compare total stellar mass M_* to a total dark matter halo mass M_{200} computed for the Burkert profile, and perform the same comparison for the quantities within R_e. Interestingly, there seems to be a clear trend of the total M_*/M_{200} ratio as a function of M_{200} while no correlation is seen for the M_*/M_{DM} within R_e (see Fig. 3). The total mass ratio trend goes in the opposite direction from expectation for the low-mass end of the galaxy mass function in the 'star formation efficiency' diagram from Behroozi et al. (2019). For our sample of 'future UDGs' the stellar-to-halo mass ratio increases when the halo mass decreases. We also notice that our values of $M_*/M_{\rm halo}$ are over an order of magnitude higher than those from Behroozi et al. (2019). This discrepancy may be reduced if we use a NFW halo density profile.

5. Discussion

Which phenomena are responsible for the trend of M_*/M_{200} vs M_{200}? Does this result from assumptions made during dynamical modelling? Grishin et al. (2021) demonstrated statistically that almost half the brightest UDGs in the Coma cluster were likely formed by ram pressure stripping of gas by the hot intracluster medium (Gunn & Gott 1972) from low-mass disky progenitors. Moreover, prior to stripping, a burst of star formation is induced by compression of the gas inside an infalling galaxy, and up-to 50% of the total stellar mass is formed during this event. Later on, gas from a galaxy gets stripped entirely, and star formation quenches. In a gas-rich galaxy, a substantial fraction of the mass can be stripped in the form of gas, so that the gravitational potential in the galactic disk changes, and quasi-circular orbits of stars become elliptical leading to the disk expansion. Later, the disk slowly expands even further as the galaxy passively evolves, losing stellar mass over time as the galaxy interacts with other members of the cluster. This process is illustrated in Fig. 4.

The dark matter halo helps to stabilize the process of ram pressure stripping preventing a galaxy from a complete dissipation as a result of substantial mass loss. The distributions of both dark matter and baryons can change as a result of mass loss, potentially affecting the dynamical modelling results from our spectroscopic observations ($< 2 \ldots 3R_e$), where we have assumed a fixed halo density profile.

Another possibility is halo truncation from tidal interactions with other galaxies in the cluster. In this case the extrapolation of the dynamical properties derived from the 'dense' baryonic part of the galaxy becomes invalid at R_{200}. However, we do not expect this process to be important for galaxies infalling into the cluster for the first time, such as the diffuse PSGs we observe.

References

Afanasiev, A. V., Chilingarian, I. V., Grishin, K. A., et al. 2023, MNRAS, 520, 6312
Behroozi, P., Wechsler, R. H., Hearin, A. P., & Conroy, C. 2019, MNRAS, 488, 3143
Bosch, J., Armstrong, R., Bickerton, S., et al. 2018, PASJ, 70, S5
Burkert, A. 1995, ApJL, 447, L25
Caldwell, N. 2006, ApJ, 651, 822
Chilingarian, I., Prugniel, P., Sil'chenko, O., & Koleva, M. 2007, in IAU Symposium, Vol. 241, Stellar Populations as Building Blocks of Galaxies, ed. A. Vazdekis & R. R. Peletier (Cambridge, UK: Cambridge University Press), 175–176
Chilingarian, I. V. 2009, MNRAS, 394, 1229
Chilingarian, I. V., Afanasiev, A. V., Grishin, K. A., Fabricant, D., & Moran, S. 2019, ApJ, 884, 79
Chilingarian, I. V., & Grishin, K. A. 2020, PASP, 132, 064503
Conselice, C. J. 2018, Research Notes of the American Astronomical Society, 2, 43
Fabricant, D., Fata, R., Epps, H., et al. 2019, PASP, 131, 075004
Grishin, K. A., Chilingarian, I. V., Afanasiev, A. V., et al. 2021, Nature Astronomy, 5, 1308
Grishin, K. A., Chilingarian, I. V., Afanasiev, A. V., & Katkov, I. Y. 2019, arXiv e-prints, arXiv:1909.13460
Gunn, J. E., & Gott, III, J. R. 1972, ApJ, 176, 1
Kansky, J., Chilingarian, I., Fabricant, D., et al. 2019, PASP, 131, 075005

Koda, J., Yagi, M., Yamanoi, H., & Komiyama, Y. 2015, ApJL, 807, L2
Le Borgne, D., Rocca-Volmerange, B., Prugniel, P., et al. 2004, Astronomy & Astrophysics, 425, 881
Navarro, J. F., Frenk, C. S., & White, S. D. M. 1997, ApJ, 490, 493
Peng, C. Y., Ho, L. C., Impey, C. D., & Rix, H.-W. 2010, AJ, 139, 2097
Sandage, A., & Binggeli, B. 1984, AJ, 89, 919
van Dokkum, P. G., Abraham, R., Merritt, A., et al. 2015, ApJL, 798, L45
Yozin, C., & Bekki, K. 2015, MNRAS, 452, 937

Dynamical Masses of Local Group Galaxies
Proceedings IAU Symposium No. 379, 2025
P. Bonifacio, M.-R. Cioni, F. Hammer, M. S. Pawlowski & S. Taibi, eds.
doi:10.1017/S1743921323001850

Dynamical modelling of ATLAS^{3D} galaxies

S. Thater[1], **P. Jethwa**[1], **E. J. Lilley**[1], **A. Zocchi**[1], **G. Santucci**[2,3] **and G. van de Ven**[1]

[1]Department of Astrophysics, University of Vienna, Türkenschanzstraße 17, 1180 Vienna
email: sabine.thater@univie.ac.at

[2]International Centre for Radio Astronomy Research (ICRAR), M468, University of Western Australia, 35 Stirling Hwy, Crawley, WA 6009, Australia

[3]ARC Centre of Excellence for All Sky Astrophysics in 3 Dimensions (ASTRO 3D), Australia

Abstract. Triaxial dynamical models of massive galaxies observed in the ATLAS^{3D} project can provide new insights into the complex evolutionary processes that shape galaxies. The ATLAS^{3D} survey is ideal as the sample comprises a good mix of fast and slow rotators with vastly different mass assembly histories. We present a detailed dynamical study with our triaxial modelling code DYNAMITE, which models galaxies as a superposition of their stellar orbits. The models allow us to constrain the intrinsic shape of the stellar component, the distributions of the visible and invisible matter and the orbit distribution in these nearby early-type galaxies and to relate it with different evolutionary scenarios. Triaxial modelling is essential for these galaxies to understand their complex kinematical features.

Keywords. Galaxies: kinematics and dynamics, Galaxies: structure

1. Introduction

Early-type galaxies (ETGs) are distinguished into two classes based on their apparent angular momentum: fast rotators and slow rotators (e.g. Emsellem et al. 2007; Cappellari et al. 2007; Emsellem et al. 2011). While fast rotators are nearly axisymmetric and often have an oblate shape, slow rotators are weakly triaxial (but not far from isotropic). These two classes likely depict two different channels of galaxy formation. Slow rotators are thought to have assembled in the centres of massive halos and after an intense star formation period at high redshift have evolved from gas-poor major mergers. In contrast, fast rotators have likely formed out of star-forming discs and their evolution is dominated by gas accretion, bulge growth and quenching (Cappellari 2016). In this study we investigate the evolutionary histories of ATLAS^{3D} galaxies by searching for imprints in their dynamically inferred intrinsic shapes and stellar orbit distributions.

2. The ATLAS^{3D} sample

ATLAS^{3D} (Cappellari et al. 2011) is a multiwavelength survey that includes 260 ETGs with stellar masses $M_* > 6 \times 10^9$ $M_\odot$ within the local volume (42 Mpc). The sample was deduced from a parent sample which was carefully selected to be statistically representative of the nearby galaxy population. About 25% of the ATLAS^{3D} ETGs were classified as elliptical galaxies and 75% as lenticular galaxies. The survey was carried out with the SAURON integral field unit on the William Herschel Telescope. From these observations,

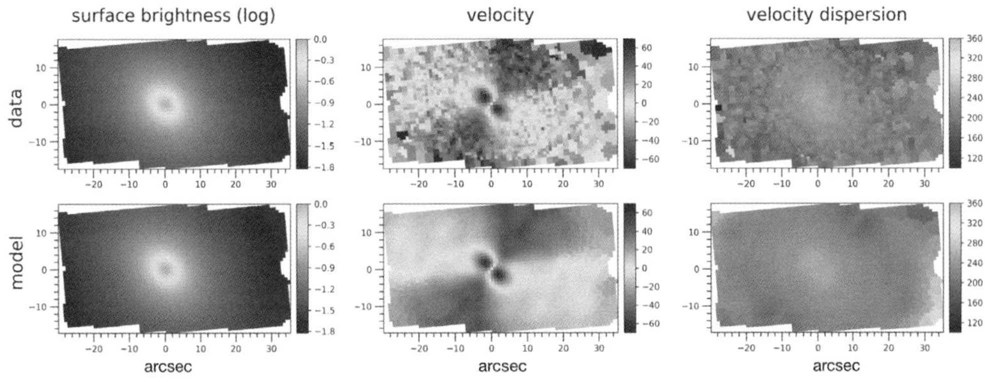

Figure 1. DYNAMITE makes it possible to dynamically model complex triaxial features such as kinematically decoupled components. We show here the surface brightness, mean velocity and velocity dispersion of the SAURON data (top) and of the Schwarzschild model (bottom) of the ATLAS^{3D} galaxy NGC 4365.

detailed 2-dimensional stellar kinematic maps of mean velocity, velocity dispersion, and h_3 and h_4 Gauss-Hermite polynomials with high S/N are available, which are ideal for our dynamical study. The kinematics (e.g. Krajnović et al. 2011; Emsellem et al. 2011), star formation history (e.g. McDermid et al. 2015) and environment (e.g. Cappellari et al. 2011; Serra et al. 2012) of the ATLAS^{3D} galaxy sample have been extensively studied in the past decade, allowing us to put our dynamical results in the context of galaxy assembly histories.

3. First dynamical results with DYNAMITE

We obtained the data products from the ATLAS^{3D} webpage† and for the first time constructed triaxial Schwarzschild orbit-superposition models for these galaxies using DYNAMITE‡ (van den Bosch et al. 2008; Jethwa et al. 2020; Thater et al. 2022). DYNAMITE allows us to recover the enclosed mass, the intrinsic shape of the stellar component, the dark matter fraction and the orbit distribution of the modelled galaxies. So far, we have finished modelling 63 galaxies, about a quarter of the full ATLAS^{3D} sample. Figure 1 shows a comparison between the observed SAURON kinematics and the best-fit DYNAMITE model of NGC 4365. NGC 4365 harbours a kinematically decoupled core (KDC) in its centre that is remarkably well recovered by our models. In the ATLAS^{3D} survey, many galaxies are exhibiting complex kinematic features which require triaxial Schwarzschild modelling, and not only regularly rotating galaxies (that can be well modelled assuming axisymmetry).

We modelled the ATLAS^{3D} galaxies with the same modelling setup as described in Santucci et al. (2022) for SAMI galaxies. Our gravitational potential consists of a stellar component, a central black hole, and dark matter parametrised as a spherical halo with a Navarro-Frenk-White (NFW; Navarro et al. 1996) radial profile. In total, we consider five free parameters: a constant stellar mass-to-light ratio M_*/L, intrinsic stellar axis-length ratios p (intermediate-to-long) and q (short-to-long), the stellar projected-to-intrinsic scale-length ratio u and the dark matter fraction $f_{\rm DM} = M_{200}/M_*$ within the virial radius r_{200}. The dark matter concentration c was fixed using the $M_{200} - c$ relation by

† `http://www-astro.physics.ox.ac.uk/atlas3d/`
‡ `https://dynamics.univie.ac.at/dynamite_docs/`

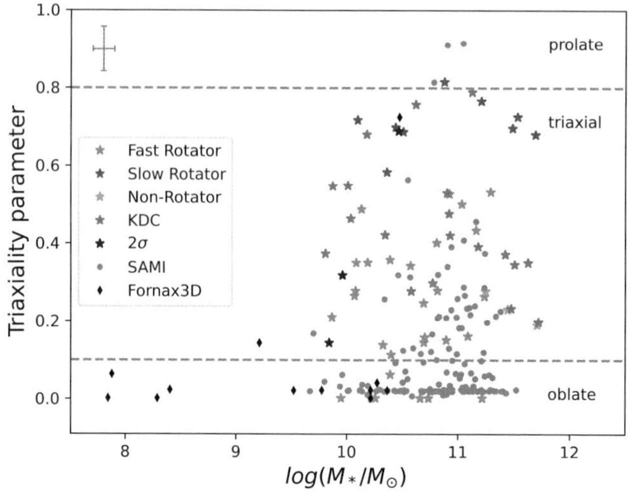

Figure 2. Triaxiality parameter T_{Re} as a function of stellar mass. We colour-code the ATLAS^{3D} galaxies based on their kinematic classification from Krajnović et al. (2011): regular fast-rotating galaxies are represented here in blue, featureless slow-rotators in red, non-rotators in orange, KDCs in light purple, and 2σ galaxies in dark purple. Typical uncertainties at 1σ confidence level are shown in the top left corner of the plot. For comparison, we also show two other samples of galaxies that were modelled with DYNAMITE, from the SAMI (Santucci et al. 2022) and Fornax3D (Ding et al. 2023) surveys, represented here with grey dots and black diamonds, respectively.

Dutton & Macciò (2014) to get a better handle on the degeneracy between M_*/L and dark matter. The ATLAS^{3D} kinematics do not have the spatial resolution to constrain the central black hole mass (M_{BH}), therefore we fixed black hole masses using the empirical $M_{\mathrm{BH}} - \sigma_{\mathrm{e}}$ relation by van den Bosch (2016). From our modelling approach, we obtain dynamical masses that are consistent with those obtained by Cappellari et al. (2013) and Poci et al. (2017) using Jeans modelling.

In this study, we aim at constraining the intrinsic shape of the stellar component of massive ETGs. The triaxiality parameter at one effective radius (R_{e}) is defined as $T_{\mathrm{Re}} = (1 - p^2_{\mathrm{Re}})/(1 - q^2_{\mathrm{Re}})$. Galaxies with $T_{\mathrm{Re}} = 0$ are classified as oblate, galaxies with $T_{\mathrm{Re}} = 1$ as prolate and galaxies with values of T_{Re} between 0.1 and 0.8 as triaxial. In Figure 2, we show the triaxiality parameter as a function of stellar mass; the ATLAS^{3D} galaxies we modelled are shown as stars in the plot, with different colours indicating the kinematic classification by Krajnović et al. (2011). While most of our modelled galaxies have triaxial intrinsic shapes, we find trends that depend on their kinematic classification. Most of the modelled galaxies are regularly fast-rotating galaxies (blue) and have a mildly triaxial shape ($\overline{T_{\mathrm{Re}}} = 0.25$). However, the scatter is quite strong and some galaxies have oblate or strongly triaxial shapes. Featureless slow-rotators ($\overline{T_{\mathrm{Re}}} = 0.71$, red) on the other hand are strongly triaxial or even prolate (which is also seen in the kinematics). Interestingly, the modelled non-rotating galaxies (orange; only 3 galaxies) have a mildly triaxial shape ($\overline{T_{\mathrm{Re}}} = 0.23$) similar to fast-rotators. The ATLAS^{3D} survey contains about 20 galaxies with KDCs. For these galaxies we find on average a strongly triaxial intrinsic shape ($\overline{T_{\mathrm{Re}}} = 0.45$) which is consistent with these galaxies having mainly formed through mergers. Surprisingly, we also find strong triaxiality for the three 2σ galaxies (galaxies exhibiting two peaks in the velocity dispersion). As this feature is typically found in galaxies with a pair of extended, counter-rotating discs, we would have expected these

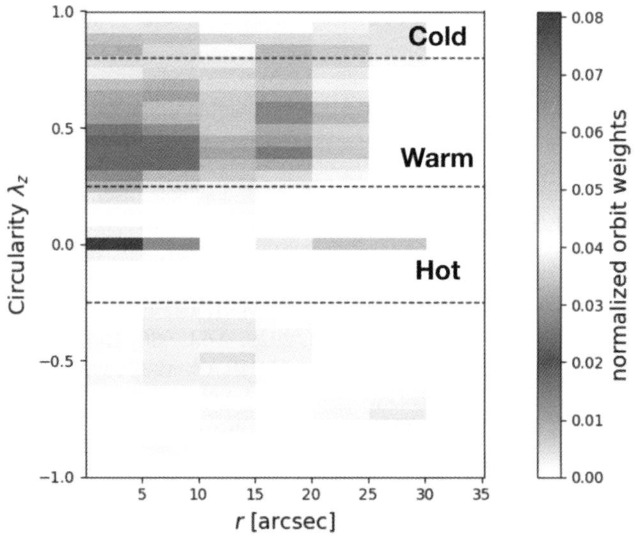

Figure 3. Stellar orbit distribution of the fast-rotating galaxy NGC 821. Stellar orbits are characterized by two main properties: the time-averaged radius r, representing the size of each orbit, and the circularity λ_z, which compares the angular momentum of the orbit with the one of a circular orbit. The density of the orbits in phase space is normalized to unity within the data coverage ($\sim$ 30 arcsec) and indicated by the different colours. The three horizontal black dashed lines indicate λ_z = 0.8, λ_z = 0.25 and λ_z = -0.25, dividing the orbits into four regions (cold, warm, hot and counter-rotating orbits).

galaxies to be more oblate. The triaxial shape might be a remnant signature of their merger history. More galaxies and a deeper analysis are essential to confirm our first results. In contrast to our modelled ATLAS^{3D} galaxies, the majority of modelled galaxies in the SAMI (Santucci et al. 2022) and Fornax3D (Ding et al. 2023) surveys have oblate shapes, as shown in Fig. 2 (with grey dots and black diamonds, respectively).

DYNAMITE models also allow a detailed look at the distribution of the stellar orbits of the galaxies. In Fig 3, we show the orbit distribution of NGC 821, one of our fast-rotating ETGs. As expected for bulge-dominated galaxies, we see a high fraction of hot orbits with a circularity $\lambda_z \sim 0$ (consistent with a pressure-supported bulge). However, there is also a large fraction of warm and cold orbits, meaning that the ETG exhibits significant net angular momentum. We will compare the orbit distribution for galaxies belonging to the various kinematic classes and search for signatures of accretion as in Zhu et al. (2020, 2022).

In the future, we will extend our dynamical study to the remaining galaxies of the ATLAS^{3D} survey and search for correlations between the intrinsic shapes/orbit distributions and kinematic properties, environment and star formation histories, in order to better understand the formation and evolutionary processes of early-type galaxies.

This research was supported by the European Union's Horizon 2020 research and innovation programme under grant agreement NO 724857 (Consolidator Grand ArcheoDyn).

References

Cappellari, M., Emsellem, E., Bacon, R., et al. 2007, MNRAS, 379, 418
Cappellari, M., Emsellem, E., Krajnović, D., et al. 2011, MNRAS, 413, 813

Cappellari, M., Scott, N., Alatalo, K., et al. 2013, MNRAS, 432, 1709
Cappellari, M. 2016, ARAA, 54, 597
Ding, Y., Zhu, L., van de Ven, G., et al. 2023, A&A, 672, A84
Dutton, A. A. & Macciò, A. V. 2014, MNRAS, 441, 3359
Emsellem, E., Cappellari, M., Krajnović, D., et al. 2007, MNRAS, 379, 401
Emsellem, E., Cappellari, M., Krajnović, D., et al. 2011, MNRAS, 414, 888
Jethwa P., Thater S., Maindl T., van de Ven G., 2020, ascl.soft. ascl:2011.007
Krajnović, D., Emsellem, E., Cappellari, M., et al. 2011, MNRAS, 414, 2923
McDermid, R. M., Alatalo, K., Blitz, L., et al. 2015, MNRAS, 448, 3484
Navarro, J. F., Frenk, C. S., & White, S. D. M. 1996, ApJ, 462, 563
Poci, A., Cappellari, M., & McDermid, R. M. 2017, MNRAS, 467, 1397
Santucci, G., Brough, S., van de Sande, J., et al. 2022, ApJ, 930, 153
Serra, P., Oosterloo, T., Morganti, R., et al. 2012, MNRAS, 422, 1835
Thater S., Jethwa P., Tahmasebzadeh B., et al., 2022, A&A, 667, A51
van den Bosch, R. C. E., van de Ven, G., Verolme, E. K., et al. 2008, MNRAS, 385, 647
van den Bosch, R. C. E. 2016, ApJ, 831, 134
Zhu L., van de Ven G., Leaman R., et al., 2020, MNRAS, 496, 1579
Zhu L., van de Ven G., Leaman R., et al., 2022, A&A, 664, A115

Dynamical Masses of Local Group Galaxies
Proceedings IAU Symposium No. 379, 2025
P. Bonifacio, M.-R. Cioni, F. Hammer, M. S. Pawlowski & S. Taibi, eds.
doi:10.1017/S1743921323001539

Dwarf Properties and Satellite Planes Beyond the Local Volume

N. Heesters[1], O. Müller[1], R. Habas[2], F. Marleau[3], P.-A. Duc[2], M. Poulain[4], R. Sánchez-Janssen[5], P. Durrell[6], S. Lim[7] and S. Paudel[7]

[1]École Polytechnique Fédéale de Lausanne (EPFL), 1290 Sauverny, Switzerland
email: nick.heesters@epfl.ch

[2]Observatoire Astronomique de Strasbourg (ObAS), Universite de Strasbourg - CNRS, UMR 7550 Strasbourg, France

[3]Universität Innsbruck, Technikerstraße 25/8, Innsbruck, A-6020, Austria

[4]University of Oulu, P.O. Box 3000, FI-90014, Oulu, Finland

[5]UK Astronomy Technology Centre, Royal Observatory, Blackford Hill, Edinburgh, EH9 3HJ, UK

[6]Youngstown State University, One University Plaza, Youngstown, OH 44555 USA

[7]Yonsei University, Seoul 03722

Abstract. Dwarf satellite galaxies in our Milky Way and different galaxy systems in the Local Volume appear to be arranged in thin, vast planes. It has been argued that these phase-space correlations cannot be explained to a satisfactory degree by the ΛCDM paradigm but it is unclear whether these planes in our neighborhood are statistical outliers, or if they are perhaps a common phenomenon in the Universe. Recent deep imaging surveys have significantly increased the number of known dwarf galaxies and allow us to advance such small-scale tensions beyond the Local Volume. We present our study analyzing the spatial distribution of 2210 dwarf galaxies identified in the MATLAS survey as well as results from follow-up observations with the MUSE instrument on the VLT. Spectral information for 56 of these dwarf galaxies allow for a deeper dive into their properties and for a comparison to the Local Volume dwarfs.

Keywords. Cosmology: dark matter, Cosmology: observation, Galaxies: dwarf

1. Introduction

Dwarf galaxies are thought to be the fundamental building blocks of more massive galaxies in a ΛCDM context (Frenk & White 2012). As such they are ideal test beds for the study of formation and evolution of galaxies. Furthermore, they are not only the most dark matter dominated type of galaxy but also the most abundant one in the Universe (e.g., Mateo et al. 1991). Even though dwarfs are numerous, they are also on average small and faint, with magnitudes > -17 (Tammann 1994). These properties make them difficult to observe, in particular at larger distances. Therefore, until recently instrumental limits have constrained the highly detailed study of dwarf galaxies mostly to the Local Group (LG) (Mateo 1998; McConnachie 2012; Drlica-Wagner et al. 2020), a few nearby groups (e.g., Chiboucas et al. 2013; Danieli et al. 2017; Carlsten et al. 2019; Byun et al. 2020; Müller et al. 2021) and a hand-full of galaxy clusters (e.g., Ferrarese et al. 2012; Eigenthaler et al. 2018; Venhola et al. 2019). Thisraises the important question of whether we have been able to paint a complete picture of dwarf galaxies via the investigation of these mostly nearby environments. It is plausible that dwarfs show a range of different properties in various environments and in more distant regions of the Universe.

Recently there have been great efforts to push observations of dwarf galaxies beyond the Local Volume ($D \leq 10$ Mpc). A number of surveys have been conducted and are currently underway (e.g., Duc et al. 2015; Geha et al. 2017; La Marca et al. 2022). These new observations will enable us to further address the so-called small-scale challenges of ΛCDM which arise on dwarf galaxy scales and have been discussed for decades (see e.g., Bullock & Boylan-Kolchin 2017, for a review). The so-called planes-of-satellites problem is interesting in particular, as it cannot be solved by the careful treatment of baryonic physics in cosmological simulations and is dominated by gravity (e.g., Müller et al. 2018). In several nearby galaxy systems dwarf satellites are orbiting their host galaxy in thin, co-rotating structures (e.g., Kunkel & Demers 1976; Pawlowski et al. 2012; Ibata et al. 2013; Müller et al. 2018). This is unexpected in a ΛCDM Universe, since the model of hierarchical structure formation predicts that satellites should be distributed in close to isotropic fashion and to move with random motions (Kroupa et al. 2005).

Another avenue to further improve our understanding of dwarf galaxies are the so-called scaling relations. Much can be learned by investigating if they persist in different environments or if there are deviations. Although galaxy formation and evolution is thought to depend on the complex interplay of various factors, galaxies have shown some remarkably tight correlations or scaling relations between their basic dynamical and stellar properties (see e.g., D'Onofrio et al. 2021, for a recent review). One such relation is the stellar mass-metallicity relation which shows a tight correlation for the dwarf galaxies in the LG (Kirby et al. 2013). In this work we report on our studies investigating the spatial distribution, the radial velocities and stellar population properties of dwarfs in the MATLAS survey beyond the Local Volume. In particular, we study the dwarfs in light of the planes-of-satellites problem as well as the universal stellar mass-metallicity relation for dwarf galaxies.

2. Flattened structures of dwarf satellites in the MATLAS survey

2.1. *MATLAS: data*

The "Mass Assembly of early-Type GaLAxies with their fine Structures "(MATLAS)" survey (Duc et al. 2015) is a deep optical imaging survey which was conducted using the MegaCam on the Canada-France-Hawaii Telescope (CFHT) from 2010 until 2015. The survey is an extension of the larger ATLAS^{3D} project (Cappellari et al. 2011) which has the scientific goal of characterizing the morphology and kinematics of 260 early-type galaxies in distances between 10 and 45 Mpc. In Habas et al. (2020) 2210 dwarf galaxies were identified in the 1 deg^2 fields around the targeted ETGs and their structural and morphological parameters extracted and discussed in Poulain et al. (2021). Around 3% of this sample were classified as Ultra-Diffuse Galaxies (UDGs) which were discussed in Marleau et al. (2021). In Heesters et al. (2021) we present the 2D spatial distribution of the dwarf galaxies identified in the MATLAS fields in the context of the planes-of-satellites problem. We summarize the methods and results of this work in the following.

2.2. *MATLAS: methods*

In order to characterize the spatial distribution of the dwarfs, we utilize the so-called Hough transform (Hough 1959, 1962). This mathematical technique was originally developed to find lines and other simple shapes in digital image processing. The principle of this approach is that every pixel is transformed into a parameter space and all possible slopes and intercepts for a given pixel or data point are calculated. In this 'Hough space' a voting procedure determines the best fit that is suitable for the majority of data points. The Hough method allows for an educated guess on whether or not a dwarf is part of a given flattened structure. The use of this method is also motivated by the fact that only about half of the dwarf population of M31 appear to lie in a thin co-rotating plane, the Great Plane of Andromeda (GPoA; Ibata et al. 2013).

2.3. *MATLAS: results*

We identify 31 (26%) statistically significant (p-value ≤ 0.05) flattened dwarf structures in 119 MATLAS fields. Due to the absence of distance estimates, we are only able to observe these structures edge-on and are missing all flattened face-on dwarf configurations. The number of flattened structures we observe should therefore be considered a lower limit. The majority of these structures show physical dimensions between the LG planes (VPOS & GPoA) and the Centaurus A satellite plane (CASP). We find no correlation between the presence of a statistically significant flattened dwarf structure and the properties of the assumed host such as stellar mass or photometric and kinematic position angle. In fields featuring multiple massive host galaxies we perform a simple total least square fit to the massive galaxies and use it as a proxy estimation to the larger scale structure. We find a preference (50%) for flattened structures to be aligned with the orientation of these massive hosts. Furthermore, we find a small excess of flattened structures around slow rotators (40%) when compared to fast rotators (16%). Such a correlation is expected if these planes-of-satellites candidates originate from galaxy mergers, since slow rotators have been shown to host more tidal features and other signs of recent mergers (Duc 2019; Bílek et al. 2020).

3. Spectroscopic follow-up of 56 dwarf galaxies with MUSE

3.1. *MUSE follow-up: data*

We followed up 56 dwarf galaxies which were idenitfied in the MATLAS fields with the Multi Unit Spectroscopic Explorer (MUSE; Bacon et al. 2010, 2012) at the Very Large Telescope (VLT) of the European Southern Observatory (ESO). The observations were part of different filler programmes (PI Marleau, F.) which were conducted with the aim to generate a reference sample of dwarf galaxies with spectroscopic data beyond the LV. Due to the nature of the filler programmes, the dwarf galaxies are located in different galaxy systems and do not form complete systems. For this reason we cannot study potential co-rotational trends in the identified flattened dwarf structures to further investigate phase-space correlations in light of the planes-of-satellites problem. Instead we use the data to confirm the dwarf and satellite nature of these dwarfs through line-of-sight velocity measurements and study their stellar population properties through full spectrum fitting with single stellar population models.

3.2. *MUSE follow-up: methods*

We use full spectrum fitting with the Penalized Pixel-Fitting algorithm (pPXF; Cappellari & Emsellem 2004; Cappellari 2017) in order to extract the radial velocities, ages and metallicities of the dwarfs in our sample. We fit the spectra using linear combinations of SSP models from the E-MILES library (Vazdekis et al. 2016) ranging from 14 Gyr - 70 Myr in age and from solar to -2.27 dex in metallicity. By varying the aperture around the dwarf from which the spectra are extracted we find the aperture which maximizes the signal-to-noise ratio (SNR) in the resulting spectrum.

We estimate the errors in the measured parameters by first calculating the residuals between the input galaxy spectrum and the best fit at every wavelength. The signs of the residuals are randomized at each wavelength, added back to the best fit and re-fitted giving slightly different values than the best fit parameters. This procedure is repeated 400 times for every galaxy. The standard deviation of the resulting MC distribution of parameter values around the best fit values gives the errors.

3.3. *MUSE follow-up: results*

Using the recessional velocities returned by pPXF we match them with the velocities of the massive galaxies observed in the ATLAS^{3D} survey and can determine their satellite nature. In this way we find that 42/56 (75%) dwarfs are consistent with being satellites of nearby massive ATLAS^{3D} galaxies. Almost a third (30%) of these matched dwarfs show star formation activity. 10 dwarfs (18%) are located further in the background based on their radial velocities. 70% of these are star forming which can be explained by the specific semi-automatic approach used to identify the dwarf galaxies (see Habas et al. 2020). We cannot detect any spectral lines for 4 (7%) of the dwarfs due to low SNR.

We find that the dwarfs in our sample are systematically offset towards lower metallicities when compared to the universal stellar mass-metallicity relation presented in Kirby et al. (2013). This discrepancy might be attributed to the difference in methodologies for measuring the metallicity. Kirby et al. (2013) analyze and average spectra from resolved individual red giant branch (RGB) stars in the MW dwarfs and present their results in the form of iron metallicities [Fe/H]. This work and other studies used to compare our results to use full spectrum fitting and thus report total metallicities [M/H]. We note that the environment the dwarfs reside in, an insufficient SNR or the inclusion or exclusion of the nucleus in nucleated dwarfs in our sample are unlikely to cause the observed systematic offset.

3.4. *Conclusions*

- We find 31 (26%) flattened dwarf structures around massive ETGs in 119 MATLAS fields beyond the Local Volume.
- These planes-of-satellites candidates are based on a 2D analysis and co-rotation and thus phase-space correlation remains to be investigated via follow-up spectroscopic observations.
- We followed up 56 MATLAS dwarf galaxies with the MUSE integral field spectrograph on the VLT.
- We confirm their dwarf nature and find that 75% of the dwarfs are satellites of nearby massive hosts in the ATLAS^{3D} survey volume. 10% are background dwarfs and 7% show no spectral lines due to insufficient SNR.
- We find that our dwarf metallicities are offset towards lower metallicities when compared to the LG dwarfs on the universal stellar mass-metallicity relation. This shift might be attributed to the different methodologies used to extract the metallicity in this work (full spectrum fitting) vs Kirby et al. (2013) (resolved RGB stars).

Acknowledgements

N.H. and O.M. are grateful to the Swiss National Science Foundation for financial support under the grant number PZ00P2_202104. M.P. is supported by the Academy of Finland grant n:o 347089. S.L. acknowledges the support from the Sejong Science Fellowship Program by the National Research Foundation of Korea (NRF) grand funded by the Korea goverment (MSIT) (No. NRF-2021R1C1C2006790). N.H. and O.M. thank Guisseppina Battaglia for pointing out the work by Boecker et al. (2020). We thank the International Space Science Institute (ISSI) in Bern (Switzerland) for hosting our international team for a workshop on "Space Observations of Dwarf Galaxies from Deep Large Scale Surveys: The MATLAS Experience".

Supplementary material

To view supplementary material for this article, please visit https://doi.org/10.1017/S1743921323001539

References

Bacon, R., Accardo, M., Adjali, L., et al. 2010, Proceedings of the SPIE, 7735, 773508.
Bacon, R., Accardo, M., Adjali, L., et al. 2012, The Messenger, 147, 4
Bílek, M., Duc, P.-A., Cuillandre, J.-C., et al. 2020, MNRAS, 498, 2138.
Bullock, J. S. & Boylan-Kolchin, M. 2017, Annu. Rev. Astron. Astrophys., 55, 343.
Byun, W., Sheen, Y.-K., Park, H. S., et al. 2020, APJ, 891, 18.
Cappellari, M. 2017, MNRAS, 466, 798.
Cappellari, M. & Emsellem, E. 2004, Publ. Astron. Soc. Pac., 116, 138.
Cappellari, M., Emsellem, E., Krajnović, D., et al. 2011, MNRAS, 413, 813.
Carlsten, S. G., Beaton, R. L., Greco, J. P., et al. 2019, APJl, 878, L16.
Chiboucas, K., Jacobs, B. A., Tully, R. B., et al. 2013, AJ, 146, 126.
Danieli, S., van Dokkum, P., Merritt, A., et al. 2017, APJ, 837, 136.
Drlica-Wagner, A., Bechtol, K., Mau, S., et al. 2020, APJ, 893, 47.
Duc, P.-A. 2020, Proceedings IAUS355, arXiv:2007.13874.
Duc, P.-A., Cuillandre, J.-C., Karabal, E., et al. 2015, MNRAS, 446, 120.
D'Onofrio, M., Marziani, P., & Chiosi, C. 2021, Frontiers in Astronomy and Space Sciences, 8, 157.
Eigenthaler, P., Puzia, T. H., Taylor, M. A., et al. 2018, APJ, 855, 142.
Ferrarese, L., Côté, P., Cuillandre, J.-C., et al. 2012, APJs, 200, 4.
Frenk, C. S. & White, S. D. M. 2012, Annalen der Physik, 524, 507.
Geha, M., Wechsler, R. H., Mao, Y.-Y., et al. 2017, APJ, 847, 4.
Habas, R., Marleau, F. R., Duc, P.-A., et al. 2020, MNRAS, 491, 1901.
Heesters, N., Habas, R., Marleau, F. R., et al. 2021, A&A, 654, A161.
Hough PV. 1959, International Conference on High Energy Accelerators and Instrumentation, 554–556
Hough PV. 1962, US Patent, 3,069,654
Ibata, R. A., Lewis, G. F., Conn, A. R., et al. 2013, Nature, 493, 62.
Kirby, E. N., Cohen, J. G., Guhathakurta, P., et al. 2013, APJ, 779, 102.
Kroupa, P., Theis, C., & Boily, C. M. 2005, A&A, 431, 517.
Kunkel, W. E. & Demers, S. 1976, The Galaxy and the Local Group, 182, 241
La Marca, A., Peletier, R., Iodice, E., et al. 2022, A&A, 659, A92.
Marleau, F. R., Habas, R., Poulain, M., et al. 2021, A&A, 654, A105.
Mateo, M., Olszewski, E., Welch, D. L., et al. 1991, AJ, 102, 914.
Mateo, M. L. 1998, Annu. Rev. Astron. Astrophys., 36, 435.
McConnachie, A. W. 2012, AJ, 144, 4.
Müller, O., Fahrion, K., Rejkuba, M., et al. 2021, A&A, 645, A92.
Müller, O., Pawlowski, M. S., Jerjen, H., et al. 2018, Science, 359, 534.
Pawlowski, M. S., Pflamm-Altenburg, J., & Kroupa, P. 2012, MNRAS, 423, 1109.
Poulain, M., Marleau, F. R., Habas, R., et al. 2021, MNRAS, 506, 5494.
Tammann, G. A. 1994, European Southern Observatory Conference and Workshop Proceedings, 49, 3
Vazdekis, A., Koleva, M., Ricciardelli, E., et al. 2016, MNRAS, 463, 3409.
Venhola, A., Peletier, R., Laurikainen, E., et al. 2019, A&A, 625, A143.

Dynamical Masses of Local Group Galaxies
Proceedings IAU Symposium No. 379, 2025
P. Bonifacio, M.-R. Cioni, F. Hammer, M. S. Pawlowski & S. Taibi, eds.
doi:10.1017/S1743921323001345

Origins of the north-south asymmetry in the ALFALFA H i velocity width function

Richard A. N. Brooks[1,2,3], Kyle A. Oman[2,3] and Carlos. S. Frenk[2,3]

[1]Department of Physics and Astronomy, University College London, London WC1E 6BT, UK
email: richard.brooks.22@ucl.ac.uk

[2]Institute for Computational Cosmology, Durham University, South Road, Durham DH1 3LE, UK

[3]Department of Physics, Durham University, South Road, Durham DH1 3LE, UK

Abstract. The number density of extragalactic 21-cm radio sources as a function of their spectral line-widths – the H I width function (H I WF) – is a tracer of the dark matter halo mass function. The ALFALFA 21-cm survey measured the H I WF in northern and southern Galactic fields finding a systematically higher number density in the north; an asymmetry which is in tension with Λ cold dark matter models which predicts the H I WF should be identical everywhere if sampled in sufficiently large volumes. We use the *Sibelius-DARK* N-body simulation and semi-analytical galaxy formation model *GALFORM* to create mock ALFALFA surveys to investigate survey systematics. We find the asymmetry has two origins: the sensitivity of the survey is different in the two fields, and the algorithm used for completeness corrections does not fully account for biases arising from spatial galaxy clustering. Once survey systematics are corrected, cosmological models can be tested against the H I WF.

Keywords. galaxies: abundances – galaxies: luminosity function, mass function – radio lines: galaxies - dark matter

1. Introduction

The kinematics of observable tracers in a galaxy are linked to the HMF because the maximum circular velocity of a halo, $v_{\rm max}$, is correlated to its mass (Navarro et al. 1997). Connecting a kinematic tracer to $v_{\rm max}$ often requires additional modelling, e.g., 21-cm spectral line widths. The 21-cm velocity spectrum is the H I mass-weighted line-of-sight velocity distribution, the full width at half maximum of the spectrum thus parameterises the line-width, w_{50}. Instead of attempting to infer $v_{\rm max}$ from line-width measurements, a more straightforward technique is to predict the number density of extragalactic 21-cm sources as a function of w_{50} – the H I width function (H I WF) – and then compare with observational measurements.

Zavala *et al.* (2009), and more recently Oman (2022), found differences in the normalisation of the ALFALFA H I WF when measured in its northern and southern Galactic field. The driver of this asymmetry has been previously only speculated (e.g., see Oman 2022). We use the *Sibelius-DARK* N-body simulation (McAlpine *et al.* 2022) populated with galaxies using the *GALFORM* semi-analytical model (Lacey *et al.* 2016) to create mock ALFALFA surveys. Our mock surveys provide a footing for an investigation into survey systematics that may be responsible for driving the asymmetry; our approach enables us to provide the first quantitative estimates for the magnitude of these effects and assess whether they can explain the observed asymmetry. A full account of this work is detailed in Brooks *et al.* (2023).

2. Mock 21-cm survey: *Sibelius-DARK* and *GALFORM*

The ALFALFA survey (Giovanelli *et al.* 2005) mapped $\sim 7,000\,\mathrm{deg}^2$ of the sky visible from Arecibo at 21-cm wavelengths out to $\sim 250\,\mathrm{Mpc}$. The survey was completed in 2012, and is composed of two separate fields on the sky; one in the northern Galactic hemisphere, visible during the spring, and the other in the southern Galactic hemisphere, visible during the autumn. By convention, these fields are labelled 'spring' and 'fall', respectively (Jones *et al.* 2018). We define a selection of ALFALFA sources from which we measure the H I WF in a similar way to Oman (2022), but we impose a distance cut $\mathrm{d_{mw}} \leq 200\,\mathrm{Mpc}$ in order to facilitate comparison with the *Sibelius-DARK* simulation which is contaminated by low-resolution particles from outside of the zoom-in region beyond this distance (McAlpine *et al.* 2022). Only sources above the 50 per cent completeness limit (CL) of the survey are selected. There are 20,857 sources above the 50 per cent CL in the ALFALFA catalogue, of which 13,006 are in the spring field and 7,851 are in the fall field.

'Simulations Beyond The Local Universe' (*Sibelius*; Sawala *et al.* 2022) aims to connect the Local Group with its cosmological environment using ΛCDM initial conditions that are constrained such that the large-scale structure is accurately reproduced. The first simulation from the *Sibelius* project is *Sibelius-DARK* (McAlpine *et al.* 2022), a realisation of a volume constrained within 200 Mpc of the Milky Way, making it ideal to compare with the ALFALFA survey. The N-body *Sibelius-DARK* simulation has its galaxy population inferred using the Lacey *et al.* (2016) variant of the semi-analytical model of galaxy formation *GALFORM*.

Four steps are required to construct a mock 21-cm survey from the simulation: calculation of the galactic circular velocity curve, determination of the amount of H I gas as a function of line-of-sight velocity to produce a 21-cm spectrum, convolution with a kernel to model the thermal broadening of the 21-cm line, and application of the selection criteria consistent with the ALFALFA survey. The result is a mock survey containing 21,031 sources, of which 12,408 are in the spring field and 8,623 are in the fall field.

Finally, to account for the undetected galaxy population we use the $1/V_{\mathrm{eff}}$ maximum-likelihood estimator. The $1/V_{\mathrm{eff}}$ estimator is 2D because the CL of the ALFALFA survey depends both on the integrated 21-cm flux and line-width of the source. From the summation of the values of the $1/V_{\mathrm{eff}}$ weights in 2D bins in M_{HI} and w_{50} we compute the 2D H I mass-width function. The sum along the w_{50} axis gives the H I MF, while the same along the mass axis gives the H I WF.

3. ALFALFA and *Sibelius-DARK* + *GALFORM* H I WFs

Understanding the origins of the asymmetric H I WFs is crucially important in the context of using it as a constraint on cosmology. The ΛCDM cosmological model predicts that the dark matter HMF should be universal (in shape and normalisation, e.g. Frenk *et al.* 1988; Mo *et al.* 1996; Sheth *et al.* 2002; Crain *et al.* 2009) and therefore similar in the two fields surveyed in ALFALFA because the volumes sampled are sufficiently large. We have checked this explicitly in the *Sibelius-DARK* simulation: the HMFs in the spring and fall volumes differ by no more than 8 per cent (within their uncertainties) at any halo mass $10^8 < M_{\mathrm{vir}}/\mathrm{M}_\odot < 10^{14}$. The most straightforward prediction for the H I WF is that it should also have the same shape and normalisation (within about 8 per cent) in the two fields. In the right panels of Fig. 1 we show the

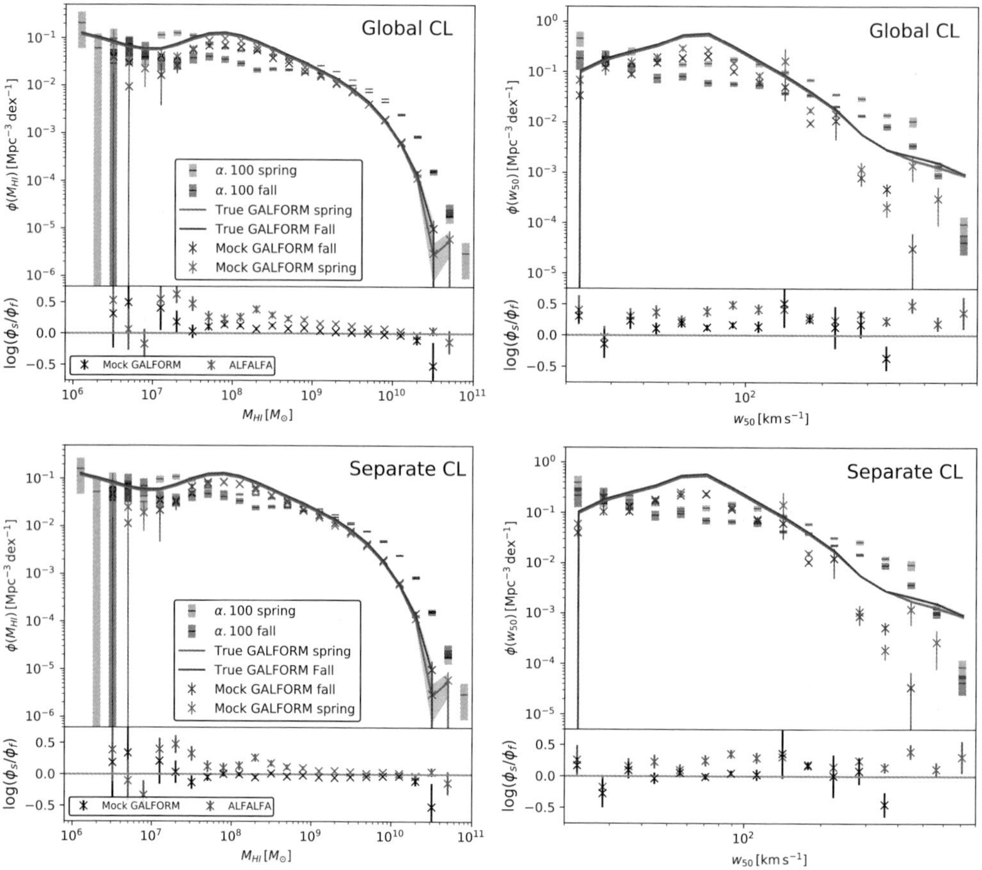

Figure 1. *Upper left panel:* the H I mass function (H I MF) calculated by adopting the same completeness limit in the spring and fall fields. The main panel shows the H I MF measured from the ALFALFA (α.100, dashes with shaded 1σ uncertainties) and *GALFORM* catalogues (crosses with 1σ uncertainties shown with error bars), separately for the two ALFALFA fields, spring (green) and fall (purple). Additionally, the 'true' *GALFORM* H I MF (solid lines with 1σ uncertainty shown with shaded band) is shown for the spring and fall fields. The lower sub-panel shows the ratio of the spring and fall *GALFORM* (black crosses) and ALFALFA (red crosses) H I MFs with 1σ uncertainties. *Upper right panel:* The H I width function (H I WF) calculated by adopting the same completeness limit in the spring and fall fields. Lines and symbols are as in upper left panel. *Lower left and right panels:* similar to upper panels, but showing the H I MF and H I WF calculated by adopting separate completeness limits in the spring and fall fields.

simulation H I WF of all galaxies with $M_{\rm HI} > 10^6\,{\rm M}_\odot$ in the spring and fall survey fields (regardless of whether they would be detected) with the green and purple solid lines, respectively. The two H I WF curves differ by no more than 8 per cent. If asymmetry of the H I WFs cannot otherwise be explained, then the ΛCDM model could be called into question.

First, we make the measurement of the H I MF and H I WF for the spring and fall mock catalogues separately, following the procedure outlined in Oman (2022) where the globally derived CL for the ALFALFA survey is assumed in the $1/V_{\rm eff}$ estimator; top-left and right panels of Fig. 1 respectively show the H I MFs and H I WFs measured using this approach. The measurements of Oman (2022), now retaining only sources within 200 Mpc, are also shown in these panels for comparison. We repeat the measurement

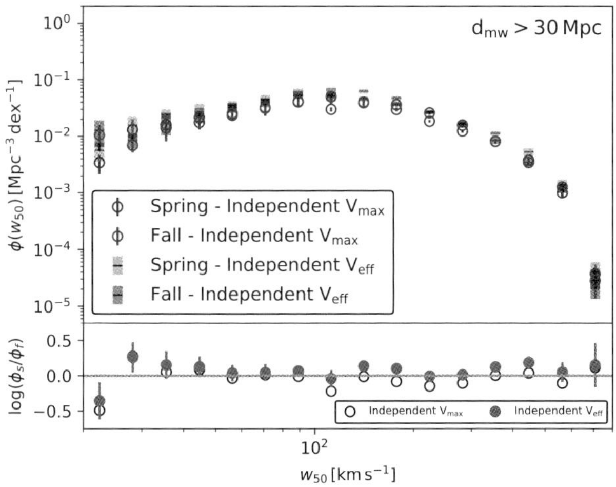

Figure 2. The spring (green) and fall (purple) H I WF derived using the 1/V_{eff} (dashes) and 1/V_{max} (open circles) weights assigned independently but including only those sources detected at distances $d_{mw} > 30$ Mpc. The lower sub panel again shows the spring-to-fall ratio for the independent 1/V_{eff} (red points) and independent 1/V_{max} (open black circles) 1/V_{eff} weights.

of the H I MF and H I WF for the spring and fall ALFALFA and mock catalogues, but this time assuming CLs derived separately for the spring and fall ALFALFA fields in the calculation of the $1/V_{eff}$ weights; bottom panels of Fig. 1. By adopting the separately derived CLS, the spring and fall mock survey H I WFs are no longer systematically offset from each other. Therefore, when the 'correct' CLs used to construct the mock catalogues are also used to correct them for incompleteness, the $1/V_{eff}$ algorithm correctly recovers the fact that the H I WFs in the two fields are indistinguishable. For ALFALFA, the systematic offset between the spring and fall H I WFs is reduced when the 'correct' CLs are used in the $1/V_{eff}$ estimator, but does not disappear: the median ratio between the spring to fall H I WFs is $\log_{10}(\phi_s \,/\, \phi_f) = 0.25 \pm 0.09$. The persisting asymmetry suggests that this is not the only systematic effect influencing the measurement.

The origin of the remaining asymmetry stems from the number of detected sources in the foreground of the survey ($d_{mw} \lesssim 30$ Mpc). The 1/V_{eff} estimator incorrectly extrapolates the foreground overdensity in the spring field through the entire survey volume, biasing the spring field to higher number densities. This effect is similar to what would occur if we used the $1/V_{max}$ estimator (Schmidt 1968), which assumes that galaxies are uniformly distributed in space, but less severe – the $1/V_{eff}$ algorithm is intended to compensate for non-uniformity in the galaxy distribution, but does so imperfectly. In Fig. 2, we show the H I WFs in the two fields measured with both (i) a conservative CL imposed to remove bias from the differing sensitivity in the two survey field and (ii) sources within 30 Mpc removed. In this case the H I WFs in the two fields are very close to agreement; the median ratio between the spring to fall H I WFs derived using the independent V_{eff} weights is $\log_{10}(\phi_s \,/\, \phi_f) = 0.07 \pm 0.12$.

In summary, we attribute the asymmetry between the spring and fall ALFALFA H I WFs to: (i) the adopted completeness limit for the survey; and (ii) the $1/V_{eff}$ estimator incorrectly extrapolating the foreground overdensity in the spring field through the entire survey volume. Accounting for these systematic effects leads to H I WFs in the spring and fall fields that are consistent with being identical, as expected from ΛCDM cosmologies.

4. Outlook – the H i WF as a cosmological constraint

The influence of individual over/underdense regions on the calculation of $1/V_{\rm eff}$ weights can be mitigated by simply surveying a larger area on the sky; the ongoing WALLABY survey (Koribalski *et al.* 2022) will cover an area about four times wider than that covered by ALFALFA. As the abundance of sources at low line-widths is sensitive to the adopted dark matter model, the H i WF has the potential to become a stringent test of cosmological models. Realising this potential will require a deeper understanding of the systematic biases influencing measurements, such as those due to spatial (or temporal) variability in survey sensitivity. Progress on theoretical issues is also needed. We look forward to the prospect of using the H i WF as a constraint on the nature of dark matter.

References

Brooks, Richard A. N. and Oman, Kyle A. and Frenk, Carlos S, 2023, MNRAS
Crain, Robert A., Giuseppe, 2009, MNRAS, 399, 4
Frenk, Carlos S. and White, Simon D. M., 1988, ApJ, 327, 507
Giovanelli, Riccardo *et al.*, 2005, Astrophysical Journal, 130, 6
Jones, Michael G. *et al.*, 2018, MNRAS, 477, 1
Koribalski, Bärbel S. *et al.*, 2022, APSS, 365, 7, 118
Lacey, Cedric G. *et al.*, 2016, MNRAS, 462, 4
McAlpine, Stuart *et al.*, 2022, MNRAS, 512, 4
Mo, H. J. and White, Simon D. M., 1996, MNRAS, 282, 2
Navarro, Julio F. and Frenk, Carlos S. and White, Simon D. M., 1997, ApJ, 490, 2
Oman, Kyle A., 2022, MNRAS, 509, 3
Sawala, Till *et al.*, 2022, MNRAS, 509, 1
Schmidt, Maarten, 1968, ApJ, 151, 393
Sheth, Ravi K. and Tormen, Giuseppe, 2002, MNRAS, 329, 1
Zavala, J. *et al.*, 2009, ApJ, 700, 2

Dynamical Masses of Local Group Galaxies
Proceedings IAU Symposium No. 379, 2025
P. Bonifacio, M.-R. Cioni, F. Hammer, M. S. Pawlowski & S. Taibi, eds.
doi:10.1017/S1743921323001126

Low-mass galaxy rotation curves that fail as dynamical mass tracers

K. A. Oman[1,2] and E. R. Downing[1,2]

[1]Institute for Computational Cosmology, Durham University, South Road, Durham, DH1 3LE, United Kingdom
email: kyle.a.oman@durham.ac.uk

[2]Department of Physics, Durham University, South Road, Durham, DH1 3LE, United Kingdom

Abstract. It is routinely assumed that galaxy rotation curves are equal to their circular velocity curves (modulo some corrections) such that they are good dynamical mass tracers. We analysed 33 low-mass galaxies from the APOSTLE simulation suite to explore the limits of validity of this assumption. Only 3 galaxies have rotation curves similar to their circular velocity curves; the rest are undergoing a wide variety of dynamical perturbations. We assessed how many galaxies are likely to be strongly perturbed by processes in several categories: mergers/interactions, bulk gas flows, non-spherical DM halo, warps, and IGM ram pressure. Most galaxies fall into more than one of these categories; only 5/33 are not in any of them. The sum of these effects leads to an underestimation of the low-velocity slope of the baryonic Tully-Fisher relation that is difficult to avoid, and could contribute to the observed diversity in low-mass galaxy rotation curve shapes.

Keywords. Galaxy kinematics and dynamics, Dwarf galaxies, Dark matter

1. Introduction

The present contribution summarizes the content of the poster presented at IAU Symposium 379. A more detailed and complete account of the work presented is published in Downing & Oman (2023).

2. The APOSTLE simulations

The APOSTLE simulations (Sawala *et al.* 2016; Fattahi *et al.* 2016) are a suite of fully cosmological hydrodynamical zoom-in simulations of 12 volumes approximately 5 Mpc in diameter. The simulations used in this work have a mass resolution of about $10^4\,M_\odot$ (gas) or $10^5\,M_\odot$ (dark matter), with a minimum force softening of about 130 pc. The APOSTLE project uses an identical galaxy formation model as was used in the EAGLE simulation project, and the 'REFERENCE' model calibration (Schaye *et al.* 2015; Crain *et al.* 2015). APOSTLE galaxies are labelled with a unique identifier according to: their resolution level (in this work always the highest resolution 'L1' used in the project); the identifier of the volume from 1 to 12 (e.g. 'V6'); their group identifier composed of a friends-of-friends group identifier (indexed from 1) and a subhalo identifier (indexed from 1 – a value of 0 indicates a 'central' as opposed to a 'satellite' galaxy). Fig. 1 shows visualisations of the gas discs of the low-mass galaxies analysed in this work.

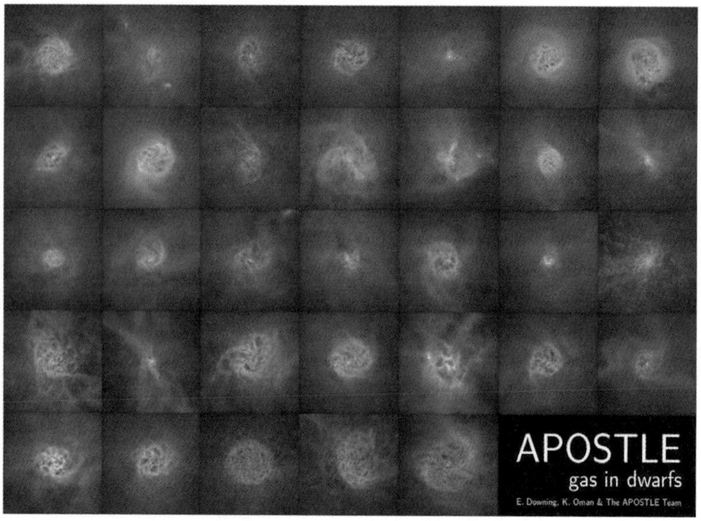

Figure 1. Gas discs at $z = 0$ seen face-on of the 33 low-mass ($60 < v_{\rm max}/{\rm km\,s^{-1}} < 120$) central (not satellite) galaxies from the highest-resolution APOSTLE simulation runs used in this work.

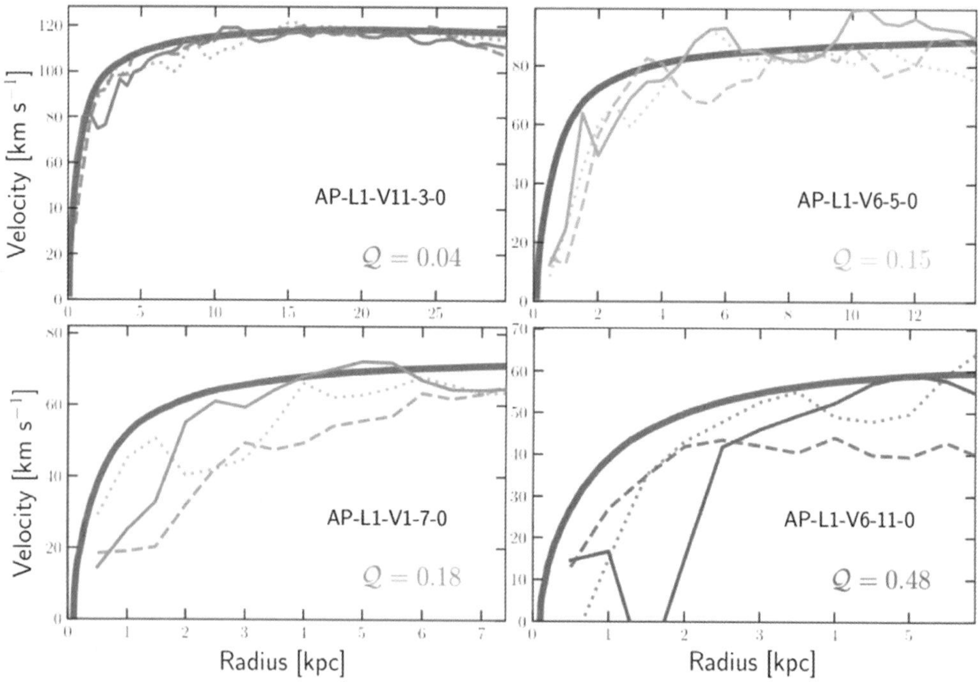

Figure 2. Rotation curves for 4 example galaxies at the end of the simulation (solid lines) and 300 and 600 Myr earlier (dashed and dotted lines). The heavy purple lines show the circular velocity curves at the end of the simulation (these evolve only very slowly).

3. Rotation curves and circular velocity curves

For every galaxy we calculated the orbital velocity expected from the potential: the circular velocity. We also measured the rotation curve (azimuthalvelocity of the gas)

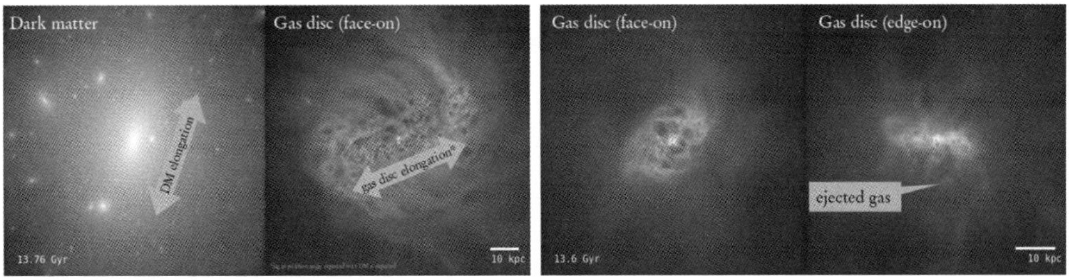

Figure 3. *Left panel pair:* Face-on view of the gas disc (right) and the dark matter (left, same perspective) of galaxy AP-L1-V11-6-0. The DM halo is visibly elongated, driving transient lopsided ($m = 1$ harmonic) and bisymmetric ($m = 2$) deformations of the gas disc. *Right panel pair:* Face-on (left) and edge-on (right) views of the gas disc of galaxy AP-L1-V4-13-0. Several prominent whisps of vertically ejected gas are visible both above and below the disc in the interval from about 13.0 – 13.8 Gyr.

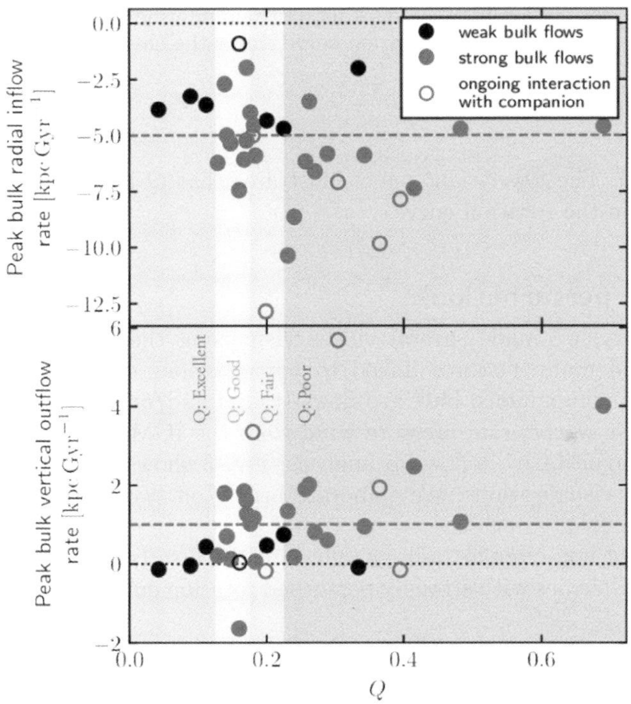

Figure 4. *Upper panel:* Anti-correlation between peak (cylindrical) radial flow rate in the last 600 Myr of the simulation and the degree to which the rotation curve traces the circular velocity curve, Q. *Lower panel:* Correlation between peak vertical (out)flow rate and Q. Points which exceed a threshold for 'strong' flows (dashed line) in either panel are coloured red.

at 300 Myr intervals. We defined a measure Q of the agreement between the rotation and circular velocity curves. A higher Q value indicates a rotation curve that is a poorer tracer of the circular velocity curve. We qualitatively classified galaxies as 'excellent'/'good'/'fair'/'poor' based on their Q. Fig. 2 shows one example from each of these categories. The lower-left panel illustrates that Q is penalized for discrepancies between the present-day rotation curve and circular velocity curve (over at least 25% of

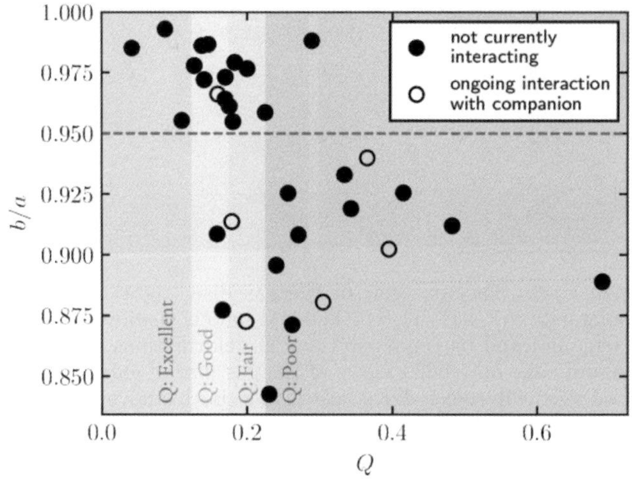

Figure 5. Anti-correlation between dark matter halo intermediate-to-major axis ratio b/a and the degree to which the rotation curve traces the circular velocity curve, Q.

its radial extent). The lower-right panel illustrates that Q is futher penalized for more time-variability in the rotation curve.

4. Kinematic perturbations

For each galaxy, we made several videos visualising the gas and dark matter. We visually identified many features linked to perturbations of the gas disc: mergers or interactions with companions; bulk gas flows (radial and/or vertical); elongation of the DM halo; gas disc warps; ram-pressure wind from the IGM. Our initial visual analysis guided a more quantitative follow-up analysis. Fig. 3 shows two examples drawn from our collection of visualisations. We elaborate briefly on two types of perturbations, as examples.

The strength of non-circular bulk gas flows in the disc (radial or vertical) correlates with Q (Fig. 4). Galaxies with strongly perturbed rotation curves tend to also have strong bulk gas flows.

DM halo intermediate-to-major axis ratio (b/a) anti-correlates with Q (Fig. 5). Even a mildly prolate or triaxial DM halo shape in the disc region ($b/a \lesssim 0.95$) can strongly perturb the gas.

5. Conclusions

Only about 30% (12/33) of galaxies in our sample have rotation curves that trace their circular velocity curves reasonably well, and only about 10% (3/33) are close matches. Causes include: mergers/interactions with companions (6/33); accretion-driven gas inflows (19/33); supernova-driven gas outflows (15/33); prolate/triaxial DM haloes (17/33); gas disc warps (8/33); and IGM ram pressure (5/33). Most galaxies fall in more than one category.

As a consequence, the baryonic Tully-Fisher relation (McGaugh 2000) and other kinematic scaling relations may suffer from severe biases, as illustrated in Fig. 6.

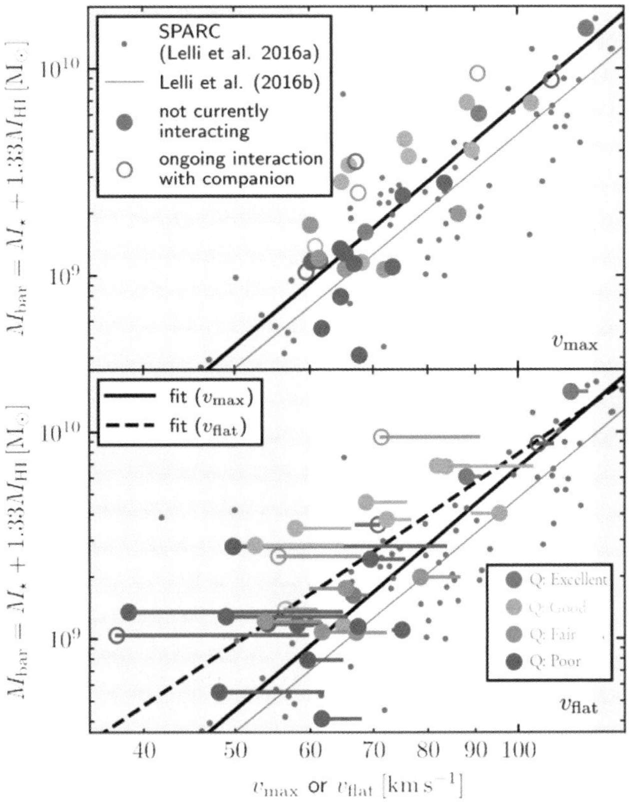

Figure 6. *Upper panel:* The baryonic Tully-Fisher relation (BTFR) for our sample of low-mass galaxies where v_{max} is calculated from their circular velocity curves – the 'true' BTFR. The heavy line is a fit to the filled coloured points. *Lower panel:* The BTFR for the same sample of galaxies, but where v_{flat} is measured from their rotation curves. The points are joined to their positions in the upper panel by horizontal line segments. The maximum gas rotation velocity (v_{flat}) systematically underestimates the maximum circular velocity (v_{max}), and the underestimates get systematically worse at lower v_{max}; this has the potential to bias both the normalisation and slope of the BTFR, as illustrated by the dashed fit line.

Acknowledgements

KAO: European Research Council (ERC) Advanced Investigator grant to C.S. Frenk, DMIDAS (GA 786910), and STFC grant ST/T000244/1. ERD: a Durham Physics Developing Talent Award to KAO. This work used the DiRAC@Durham facility managed by the Institute for Computational Cosmology on behalf of the STFC DiRAC HPC Facility (www.dirac.ac.uk). The equipment was funded by BEIS capital funding via STFC capital grants ST/K00042X/1, ST/P002293/1, ST/R002371/1 and ST/S002502/1, Durham University and STFC operations grant ST/R000832/1. DiRAC is part of the National e-Infrastructure. Elements of this work, in particular Figs. 2, 4, 5 & 6, are adapted from Downing & Oman (2023) as permitted under the terms of that work's license (CC BY 4.0; `https://creativecommons.org/licenses/by/4.0/`).

References

Crain, R. A., Bahé, Y. M., Lagos, C. del P., *et al.* 2015, MNRAS, 450, 1937
Downing, E. R., Oman, K. A. 2023, MNRAS, in press.

Fattahi, A., Navarro, J. F., Sawala, T., *et al.* 2016, MNRAS, 457, 844
McGaugh, S. S. 2000, ApJ, 533, L99
Sawala, T., Frenk, C. S., Fattahi, A., *et al.* 2016, MNRAS, 457, 1931
Schaye, J., Crain, R. A., Bower, R. G., *et al.* 2015, MNRAS, 446, 521

Dynamical Masses of Local Group Galaxies
Proceedings IAU Symposium No. 379, 2025
P. Bonifacio, M.-R. Cioni, F. Hammer, M. S. Pawlowski & S. Taibi, eds.
doi:10.1017/S1743921323001643

Mass models of disk galaxies from gas dynamics

Federico Lelli

INAF - Arcetri Astrophysical Observatory, Largo Enrico Fermi 5, 50125, Florence, Italy
email: federico.lelli@inaf.it

Abstract. I review methods and techniques to build mass models of disk galaxies from gas dynamics. I focus on two key steps: (1) the derivation of rotation curves using 3D emission-line datacubes from H I, CO, and/or Hα observations, and (2) the calculation of the gravitational field from near-infrared images and emission-line maps, tracing the stellar and gas mass distributions, respectively. Mass models of nearby galaxies led to the establishment of the radial acceleration relation (RAR): the observed centripetal acceleration from rotation curves closely correlates with that predicted from the baryonic distribution at each galaxy radius, even when dark matter supposedly dominates the gravitational field. I conclude by discussing the (uncertain) location of Local Group dwarf spheroidal galaxies on the RAR defined by more massive disk galaxies.

Keywords. Dark matter — galaxies: kinematics and dynamics — galaxies: spiral — galaxies: dwarf

1. Introduction

Rotation curves from gas kinematics are one of the most direct methods to measure the gravitational potential (Φ) of galaxies and, therefore, constrain their baryonic and dark matter (DM) content. The reason is simple. In gas disks, the rotation velocity ($V_{\rm rot}$) is typically much larger than the velocity dispersion ($\sigma_{\rm V}$), so pressure support is negligible ($V_{\rm rot}/\sigma_{\rm V} \gg 1$) and $V_{\rm rot}$ is a direct proxy of the circular velocity of a test particle ($V_c^2 = -R\,\nabla\Phi$). For example, this is not the case for the stellar components of galaxies because both rotation and pressure support can be important, so $V_{\rm c}$ must be estimated by solving the Jeans' equations, which require either full 6D phase-space information (as for the Milky Way thanks to the Gaia mission) or assumptions on the velocity dispersion tensor (as for external galaxies with stellar spectroscopy).

Over the past 20 years, the study of gas dynamics made enormous progress thanks to dedicated surveys of different emission lines, probing different gas phases: the H I line at 21 cm tracing atomic gas, CO lines in the submm tracing molecular gas, and Balmer lines in the optical (mostly the Hα line) tracing warm ionized gas. Another key advancement has been the availability of near infrared (NIR) images from space telescopes, such as Spitzer and WISE. The NIR light is only marginally affected by dust extinction and represents the best tracer of the stellar mass distribution, so it allows us to accurately compute the stellar gravitational field.

For galaxies in groups and in the field, H I disks are typically more extended than the stellar components (on average by $\sim$4 times the NIR half-light radii $R_{\rm h}$, Lelli et al. 2016), so they trace the gravitational potential out to the most DM-dominated regions. To date, the main disadvantage of H I observations is the low spatial resolution (typically $5'' - 30''$) but the situation will drastically improve with the upcoming Square Kilometer

Array (SKA). On the other hand, CO and Hα disks are typically confined to the inner galaxy regions, but can be routinely observed at spatial resolutions of $\sim 1'' - 2''$, which can be pushed further down to sub-arcsec resolutions thanks to adaptive optics for Hα data and ALMA long baselines for CO data. The best approach to study gas kinematics is a multiwavelength one, combing CO and/or Hα data at high resolution in the inner parts with H I data at lower resolutions in the outer regions.

In this brief review, I focus on methods and techniques to measure circular-velocity curves (Sect. 2) and build mass models of galaxies (Sect. 3). The scientific implications of these observational and modeling efforts have been extensively reviewed elsewhere (de Blok 2010; McGaugh et al. 2020; Lelli 2022) and will not be repeated here. I will only mention the radial acceleration relation (RAR) in context with Local Group (LG) dwarf galaxies (Sect. 4).

2. Measuring the circular-velocity curve

2.1. *Circular motions*

For a geometrically and optically thin disk with nearly circular orbits, the projected line-of-sight velocity $V_{\rm l.o.s.}$ at a sky position (x, y) is given by

$$V_{\rm los}(x, y) = V_{\rm sys} + V_{\rm rot}(R)\sin(i)\cos(\theta), \tag{1}$$

where $V_{\rm sys}$ is the systemic velocity (due to the Hubble expansion and peculiar motions), $V_{\rm rot}$ is the rotational velocity at radius R in the galaxy plane, i is the inclination angle between the normal to the disk and the line of sight, and θ is the azimuthal angle in the disk plane given by

$$\cos(\theta) = \frac{-(x - x_0)\sin(P.A.) + (y - y_0)\cos(P.A.)}{R}, \tag{2}$$

where $(x_0,\ y_0)$ are the center coordinates and $P.A.$ is the position angle taken in anti-clockwise direction between the North direction and the major axis of the projected disk. Thus, the rotation curve $V_{\rm rot}(R)$ can be measured knowing six parameters: $V_{\rm sys}$, x_0, y_0, $P.A.$, and i.

The most basic approach to measure $V_{\rm rot}$ is to obtain slit spectroscopy along the disk major axis ($\theta = 0°$), so that $V_{\rm rot}(R) = [V_{\rm los}(R) - V_{\rm sys}]/\sin(i)$. This 1D approach must assume (x_0, y_0), $P.A.$, and i from independent observations (e.g., optical images) and cannot account for warped disks in which $P.A.$ and i can vay with R. A better strategy is to obtain velocity maps ($V_{\rm l.o.s.}$ at each sky position) and fit them with Eq. 1 (Begeman 1989). This 2D approach is robust for well-resolved disks with hundreds of resolution elements, but becomes unreliable in less resolved disks (e.g., dwarf and/or distant galaxies) because of beam-smearing effects, which systematically underestimate $V_{\rm rot}$ and overestimate $\sigma_{\rm V}$ (e.g., Di Teodoro and Fraternali 2015).

The best approach is to fit directly the 3D emission-line cube (with two spatial axes and one velocity axis) to reproduce the full shape of the line profiles at each (x, y). In this 3D approach, disk models are built starting from some rotation curve $V_{\rm rot}(R)$, intrinsic velocity dispersion profile $\sigma_{\rm V}(R)$, gas surface density profile $\Sigma_{\rm gas}(R)$, and disk thickness $z_{\rm gas}$. The disk models are projected on the sky to produce a mock cube, which is smoothed to the same spectral and spatial resolution of the data (reproducing beam smearing effects) and compared with the observed cube until a best-fit is found (Swaters et al. 2009). The 3D approach exploits the full information available in the data, but requires more free parameters than the 2D approach to model $\sigma_{\rm V}$, $\Sigma_{\rm gas}$, and $z_{\rm gas}$. In most cases, $\Sigma_{\rm gas}$ can be inferred from the observed intensity map (with the exception of edge-on galaxies), while $z_{\rm gas}$ is generally unmeasurable because is below the angular resolution, so it is fixed to sensible values. Over the past years, several public 3D fitting softwares became available: ^{3D}Barolo (Di Teodoro and Fraternali 2015) and FAT (Kamphuis et al. 2015) adopt a so-called *tilted-ring modeling*, while KinMS (Davis et al. 2013) and GalPak3D (Bouché et al. 2015) use a *parametric modeling*.

The tilted-ring modeling (Rogstad et al. 1974; Begeman 1989) divides the gas disk into N rings (whose separation is typically set by the spatial resolution) and fits Eq. 1 independently in each ring. It is common to perform multiple, iterative fits. After a first fit, $V_{\rm sys}$ and (x_0, y_0) are measured as the mean (or median) values across the rings and fixed in subsequent fits with less free parameters. The angles $P.A.$ and i can be estimated in the same way unless the gas disk is warped: in such cases the radial variations of $P.A.$ and i are fitted with appropriate smooth functions, which are then imposed in a final fit to infer $V_{\rm rot}$ and $\sigma_{\rm V}$ in each ring.

The parametric modeling (e.g., Courteau 1997) uses parameteric functions to describe $V_{\rm rot}(R)$ and fits a global model to the whole disk. Warps can be accounted for, but require a "trial and fail" procedure to choose appropriate functions to model $P.A.(R)$ and $i(R)$. The parametric modeling has less free parameters than the tilted-ring modeling; this facilitates the use of Markov Chain Monte Carlo algorithms (Bouché et al. 2015) and neural networks (Dawson et al. 2021). In the parametric modeling, $V_{\rm rot}$ may be described by an empirical fitting function, such as $V_{\rm t}\arctan(R/R_{\rm t})$ (Courteau 1997), or by a mass model that includes different velocity components (see Sect. 3). The parametric modeling, however, does not provide an actual, empirical derivation of the rotation curve because a smooth shape is imposed, neglecting possible real features in $V_{\rm rot}(R)$. Moreover the intrinsic $\sigma_{\rm V}$ is often described by a single, radially averaged value, which may be dominated by the inner bright regions during the 3D fit. Alternatively, one must assume a parametric functional form for $\sigma_{\rm V}(R)$ as well.

2.2. *Noncircular motions*

Noncircular motions may affect the measured $V_{\rm rot} \simeq V_{\rm c}$. The simplest form of non-circular motion is a radial flow ($V_{\rm rad}$) in the disk plane, giving an additional term $V_{\rm rad}(R)\sin(i)\sin(\theta)$ in the right-end side of Eq. 1. More generally, $V_{\rm rot}$ and $V_{\rm rad}$ are the first-order terms of an harmonic expansion of the line-of-sight velocity (Schoenmakers et al. 1997):

$$V_{\rm l.o.s.}(x,y) = V_{\rm sys} + \sin(i) \sum_{m=1} \left[c_m(R)\cos(m\theta) + s_m(R)\sin(m\theta)\right], \tag{3}$$

where the harmonic order $m=1$ gives $c_1 = V_{\rm rot}$ and $s_1 = V_{\rm rad}$. A perturbation of the gravitational potential of order m causes harmonics of order $m-1$ and $m+1$ in projection (Schoenmakers et al. 1997). For example, bar-like and oval distortions with $m=2$ give $m=1$ and $m=3$ terms in $V_{\rm l.o.s}$ (Spekkens and Sellwood 2007). The harmonic decomposition further increases the number of free parameters in tilted-ring fits, so it can only be applied to high-resolution, high-sensitivity data. To date, this method has been applied up to $m=3$ with 2D fits, but further progress may be done using a 3D approach that models the full line profiles.

H I studies find that noncircular motions are typically smaller than 10 km s^{-1}, corresponding to ∼1% to ∼10% of $V_{\rm rot}$ (Gentile et al. 2005; Trachternach et al. 2008; Oh et al. 2015; Marasco et al. 2018). They may be larger in barred and starburst galaxies, and become progressively more important in dwarf galaxies with small $V_{\rm rot}$ (e.g., Lelli et al. 2012). Nevertheless, nonciruclar motions are usually comparable to uncertainties in $V_{\rm rot}$, which are computed considering the difference in $V_{\rm rot}$ from the approaching and receding sides of the disk (Swaters et al. 2009). These uncertainties are not formal errors, but quantify global kinematic asymmetries, effectively capturing noncircular motions in most cases. Noncircular motions could be larger in Hα disks (Simon et al. 2005; Spekkens and Sellwood 2007), possibly because ionized gas is more closely related to the sites of star formation activity than atomic gas.

2.3. *Random motions*

Pressure support due to random motions is often negligible in gas disks. For atomic and molecular gas, the intrinsic σ_V is typically between $5 - 15$ km s^{-1}, so effectively $V_c = V_{rot}$ for galaxies with $V_{rot} \gtrsim 50$ km s^{-1} (Swaters et al. 2009) and pressure support become important only in the tiniest gas-bearing dwarfs (Iorio et al. 2017). The situation can be different for ionized gas because the intrinsic σ_V can be higher ($\sim 15 - 30$ km s^{-1}), so pressure support can be important in galaxies with $V_{rot} \lesssim 100$ km s^{-1} (Barat et al. 2020). For both atomic and ionized gas, the velocity dispersion is largely driven by turbulent gas motions rather than thermal motions, so the pressure support is also referred to as turbulence support.

For gas disks with $V_{rot}/\sigma_V \lesssim 4$, pressure support can be accounted for using the so-called asymmetric-drift correction (ADC, Binney and Tremaine 1994). Assuming that σ_V is isotropic (as expected for collisional gas due to frequent energy exchanges), we obtain

$$V_c^2 = V_{rot}^2 - \sigma_V^2 \left(\frac{\partial \ln \rho_{gas}}{\partial \ln R} + \frac{\partial \ln \sigma_V^2}{\partial \ln R} \right), \tag{4}$$

where $\rho_{gas}(R, z)$ is the gas volume density, which is not directly observable.

A common approach (Meurer et al. 1996) is to assume that the vertical density distribution does not vary with radius, so $\partial \ln \rho_{gas}/\partial \ln R = \partial \ln \Sigma_{gas}/\partial \ln R$, where the gas surface density profile $\Sigma_{gas}(R)$ can be traced from emission-line maps. Then, one may either estimate the radial derivatives of Σ_{gas} and σ_V separately, or fit the product $\Sigma_{gas}\sigma_V^2$ with a smooth parametric function with a trivial radial derivative (Iorio et al. 2017). The main caveat with this approach is that the thickness of H I disks may actually increase with radius (Bacchini et al. 2020).

Another approach to solve Eq. 4 (Burkert et al. 2010) is to consider a self-gravitating disk where σ_V is independent of z, so $\rho_{gas}(R, z) = \rho_0(R)\text{sech}^2(z/z_d)$ (Spitzer 1942) where $\rho_0(R) = \pi G_N \Sigma_{gas}^2(R)/2\sigma_V^2(R)$ is the volume density at $z = 0$ and G_N is Newton's constant. The gas disks of galaxies, however, are not self-gravitating. A self-consistent ADC should use an iterative approach: (i) measure $\Phi(R, z)$ fitting some initial $V_c(R)$ with a 3D mass model (Sect. 3), (ii) measure the vertical structure as a function of R assuming hydrostatic equilibrium for the given $\Phi(R, z)$, (iii) measure again V_{rot}, σ_V, and V_c for the new vertical structure, and (iv) iterate.

At any rate, the dominant uncertainty in the ADC is driven by $\sigma_V(R)$. In the relevant cases (dwarf galaxies), σ_V is often poorly measured and assumed to be constant with radius. Then, for an exponential gas disk with scale length R_d, Eq. 4 simplifies to $V_c^2 = V_{rot}^2 + \sigma_V^2(R/R_d)$. This equation can be used as a zeroth-order approximation in poorly resolved galaxies.

3. Mass Models

3.1. *Measuring the baryonic gravitational field*

To study the relative distribution of baryons and DM in galaxies from the measured $V_c(R)$, it is necessary to compute the Newtonian gravitational acceleration from various mass components (van Albada and Sancisi 1986). In cylindrical coordinates (R, z), a razor-thin exponential disk (Freeman 1970) gives the following velocity contribution in the disk plane ($z = 0$):

$$V_c^2(R) = \frac{G_N M_d}{R_d} 2y^2 \left[I_0(y)K_0(y) - I_1(y)K_1(y) \right] \tag{5}$$

where M_d the disk mass, R_d the disk scale length, $y = R/(2R_d)$, and I_n and K_n are modified Bessel functions of the first and second kind, respectively. A razor-thin exponetial disk is just a zeroth-order representation of the mass distribution in galaxies because it neglects the finite thickness of stellar and gas disks, as well as relevant features in their mass distribution such as inner concentrations (bulges, pseudobulges, nuclei), inner depressions, outer breaks, bumps, wiggles, and so on. For a more realistic model, we need to numerically solve $\nabla^2\Phi_i = 4\pi G_N \rho_i$ for each baryonic component i. There are two main approaches: (1) to measure the projected radial density profile $\Sigma(R)$ and make assumptions on the intrinsic 3D geometry, or (2) to fit 2D images with the multi-Gaussian expansion (MGE) method (Emsellem et al. 1994).

In the former approach, in cylindrical symmetry, we have $\rho(R, z) = \Sigma(R)Z(z)$ where $\Sigma(R)$ is the observed radial density profile (from optical/NIR images for stars and emission-line maps for gas) and $Z(z)$ is an assumed vertical density profile. The $Z(z)$ profile can be directly studied only in edge-on disks; common parametrizations are $\mathrm{sech}^2(z/z_d)$, $\exp(-z/z_d)$, or $\exp(-z^2/z_d^2)$, where the scale height z_d is found to correlate with the scale length R_d (van der Kruit and Freeman 2011). The velocity contribution at $z = 0$ is given by (Casertano 1983):

$$V_c^2(R) = -8G_N R \int_0^\infty \frac{\partial\Sigma(\tilde{R})}{\partial\tilde{R}} \left[\int_0^\infty Z(\tilde{z}) \frac{\mathscr{K}(p) - \mathscr{E}(p)}{\sqrt{pR\tilde{R}}} \, d\tilde{z} \right] \tilde{R} d\tilde{R}, \tag{6}$$

where $\mathscr{K}$ and $\mathscr{E}$ are complete elliptic integrals of the first and second kind, respectively, and $p = x - \sqrt{x^2 - 1}$ with $x = (R^2 + \tilde{R}^2 + z^2)/(2\tilde{R}R)$. Notably, $V_c(R)$ depends on $\partial\Sigma(R)/\partial R$ so bumps and wiggles in the mass profile have a relevant effect. In addition, V_c at R is given by a double integral in $d\tilde{R}$ and $d\tilde{z}$ from zero to infinity, so depends on the entire mass distribution. This occurs because *Newton's shell theorem does not apply in disks*: mass at $R > R_0$ does contribute to the gravitational field at R_0. As a result, V_c^2 can sometimes be "negative" at some radii, in the sense that the gravitational field V_c^2/R is directed towards the outer galaxy regions because the mass at $R > R_0$ pulls more than the mass at $R < R_0$. This effect often occurs in the gravitational contribution of H I disks with central holes or strong depressions (see Fig. 1).

If there is a central mass concentration (a "bulge"), its contribution can be treated separately from the disk. In spherical symmetry, the velocity contribution at $z = 0$ is given by (Kent 1986):

$$V_c^2(R) = \frac{2\pi G_N}{R} \int_0^R \tilde{R}\Sigma(\tilde{R})d\tilde{R} + \frac{4G_N}{R} \int_R^\infty \left[\sin^{-1}(R/\tilde{R}) - R(\tilde{R}^2 - R^2)^{-1/2} \right] \tilde{R}\Sigma(\tilde{R})d\tilde{R}. \tag{7}$$

Eq. 7 is simply $V_c^2(r) = G_N M(r)/r$ considering the deprojection of spherical shells of radius r.

For an oblate spheroid with intrinsic axial ratio q observed at an inclination i, the velocity contribution at $z = 0$ is given by (Noordermeer 2008):

$$V_c^2(R) = -4G_N \sqrt{q^2 \sin^2(i) + \cos^2(i)} \int_{T=0}^{R} \left[\int_{\tilde{R}=T}^{\infty} \frac{\partial\Sigma(\tilde{R})}{\partial\tilde{R}} \frac{d\tilde{R}}{\sqrt{\tilde{R}^2 - T^2}} \right] \frac{T^2 dT}{\sqrt{R^2 - T^2 + q^2T^2)}}. \tag{8}$$

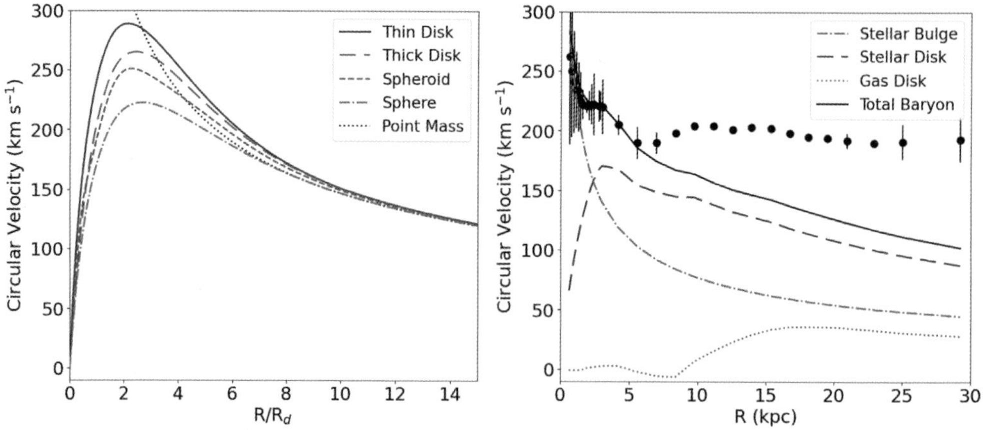

Figure 1. *Left panel*: circular-velocity curves for a model galaxy with a projected exponential density profile and total mass of 5×10^{10} M$_\odot$, assuming different geometries: razor-thin disk (Eq. 5), thick exponential disk with $z_{\rm d} = 1/3R_{\rm d}$ (Eq. 6), oblate spheroid with $q = 0.5$ (Eq. 8 or Eq. 9), and sphere (Eq. 7). The $V_{\rm c}$ of a corresponding point mass is also shown. Flattened mass distributions reach the Keplerian decline at $R \gtrsim 10R_{\rm d}$. *Right panel*: Mass model for the spiral galaxy UGC 3546. The Hα+HI rotation curve (black points with errorbars, Noordermeer et al. 2007) is well reproduced at $R \lesssim 6$ kpc by the baryonic gravitational field computed from NIR photometry (Lelli et al. 2016). Note that $V_{\rm gas}$ is negative at $5 \lesssim R/{\rm kpc} \lesssim 10$.

Assuming that the spheroid's inclination i is the same as that of the gas disk (from kinematic fits, see Sect. 2), the mean axial ratio $\tilde{q}$ of the observed isophotes (from optical or NIR images) can be used to infer the intrinsic axial ratio as $q^2 = [\tilde{q}^2 - \cos^2(i)]/\sin^2(i)$.

For a spheroid where q varies with R, the MGE method (Emsellem et al. 1994) is most effective. The 2D image is fitted by a sum of N 2D Gaussian functions j with luminosity L_j, standard deviation σ_j, axial ratio q_j, and $P.A._j$. The potential $\Phi_j(R, z)$ of a Gaussian component stratified on oblate concentric ellipsoids is given by (Chandrasekhar 1969; Cappellari 2002):

$$\Phi_j(R,z) = \frac{2G_{\rm N}\Upsilon_j L_j}{\sqrt{2\pi}\sigma_j}\int_0^1 \exp\left[\frac{-T^2}{2\sigma_j^2}\left(R^2 + \frac{z^2}{1-T^2+q_j^2T^2}\right)\right](1-T^2+q_j^2T^2)^{-1/2}dT \tag{9}$$

where Υ_j is the mass-to-light ratio of the component j. Thus, one can compute the total potential Φ and infer the circular velocity $V_c^2(R, z=0) = -R\nabla\Phi$. The MGE approach infers the intrinsic 3D distribution from fitting observed isophotes in the 2D image. This is robust for early-type galaxies with regular isophotes, but becomes more uncertain for late-type galaxies (with spiral arms, star formation, dust lanes, etc.) and for gas maps with irregular distributions.

Figure 1 shows $V_{\rm c}(R)$ for a model galaxy with $M = 5 \times 10^{10} M_\odot$ and a projected surface density profile $\Sigma(R) = \Sigma_0 \exp(-R/R_{\rm d})$, using the various equations described above. In regions probed by gas kinematics ($R \lesssim 6R_{\rm d} - 7R_{\rm d}$), differences in $V_{\rm c}$ due to geometry can be up to $\sim 25\%$. Importantly, the Keplerian decline of the corresponding point mass is reached only at $R \gtrsim 10R_{\rm d}$ or equivalently $R \gtrsim 6R_{\rm h}$. Thus, the concept of "dynamical mass" can be ill-defined unless (1) we model the entire mass distribution for $R \to \infty$, and/or (2) we have measurements out to very large R, where the monopole term of $\Phi(R, z)$ dominates and $M_{\rm dyn}(R) \simeq V_{\rm c}^2 R/G_{\rm N}$.

3.2. *Measuring the baryonic masses*

For a typical galaxy, the expected circular velocity from baryons is given by:

$$V_{\rm bar}^2 = \Upsilon_{\rm bul}|V_{\rm bul}|V_{\rm bul} + \Upsilon_{\rm disk}|V_{\rm disk}|V_{\rm disk} + \Upsilon_{\rm gas}|V_{\rm gas}|V_{\rm gas} \tag{10}$$

where $V_{\rm bul}$, $V_{\rm disk}$, and $V_{\rm gas}$ are the contributions from stellar bulge, stellar disk, and gas disk, respectively. $\Upsilon_{\rm bul}$, $\Upsilon_{\rm disk}$, and $\Upsilon_{\rm gas}$ are dimensionless factors (mass-to-light ratios) that scale the contributions according to the total mass of each component, defined for $R \to \infty$. These factors can be either fixed to some fiducial values, or used as free parameters when fitting the rotation curve, ideally imposing a Gaussian prior in a Bayesian context (e.g., Li et al. 2020).

The stellar contribution is best traced using NIR images. For late-type disks (Sc-dI), the bulge contribution is often negligible and inner mass concentrations ("pseudobulges") can be modeled together with the disk. For early-type disks (S0-Sb), instead, it is appropriate to separate bulge & disk components because they have different geometries and stellar populations. The corresponding mass-to-light ratios can be estimated in several ways (e.g., McGaugh and Schombert 2015). Stellar population models typically give $\Upsilon_{\rm bul} \simeq 0.7-1.0$ and $\Upsilon_{\rm disk} \simeq 0.3-0.7$ at Spizer 3.6 μm, depending on the assumed star-formation history and chemical enrichment (Schombert et al. 2022). For galaxies on the star-forming main sequence, it is sensible to assume $\Upsilon_{\rm disk} \simeq 0.5$ with a 1σ variation of $\sim$25% (McGaugh and Schombert 2015). Naturally, $\Upsilon_{\rm disk}$ is expected to vary with radius because of metallicity and stellar population gradients, but the effect is somewhat degenerate with geometry (Fig. 1) and subdominant with respect to the absolute calibration in $\Upsilon_{\rm disk}$. Notably, to implement a radially variable mass-to-light ratio, we cannot simply use some $\Upsilon_{\rm disk}(R)$ in Eq. 10 but have to recalculate $V_{\rm disk}$ using the integrals in Sect. 3.1 for a rescaled surface density profiles $\Upsilon_{\rm disk}(R)\Sigma_{\rm disk}(R)$.

The gas contribution is usually dominated by atomic gas, so H I maps are used to compute $V_{\rm gas}$. The conversion from H I luminosity to H I mass is known from atomic physics, so $\Upsilon_{\rm gas}$ must only account for Helium and heavier elements. Considering big bang nucleosynthesis and stellar chemical enrichment, one has $\Upsilon_{\rm gas} \simeq 1.34-1.41$ depending on the gas metallicity (McGaugh et al. 2020). The uncertainty on $V_{\rm gas}$ is dominated by the absolute H I flux calibration, which is typically $\sim$10%. The smaller contribution of molecular gas can be separately computed using CO maps (Frank et al. 2016), but CO & H I data are rarely available for the same galaxy samples. Luckily, molecular gas is distributed in a similar way as the stellar disk (similar $R_{\rm d}$), so its contribution in star-forming galaxies can be roughly taken into account with a systematic correction to $V_{\rm disk}^2$ of the order of 7% (McGaugh et al. 2020), or simply included in the error budget of $\Upsilon_{\rm disk}$. Warm ionized gas ($T \simeq 10^4$ K) typically gives a negligible mass contribution, while hot ionized gas ($T \simeq 10^{6-7}$ K) is thought to form low-density halos extending over hundreds of kpc, so its gravitational contribution is negligible within the H I disk.

The DM contribution can be added to Eq. 10 assuming a spherical halo with a given volume density profile. Equations for common halo models are summarized in Li et al. (2020).

4. Radial acceleration relation

Thanks to large H I and NIR surveys, mass models have been built for hundreds of nearby galaxies, spanning over $\sim$5 dex in stellar mass and $\sim$3 dex in effective surface brightness (e.g., Lelli et al. 2016; Iorio et al. 2017). The scientific implications of these mass models have been recently reviewed in McGaugh et al. (2020) and Lelli (2022). Here we only recall some basic facts. In high-mass ($M_\star > 3 \times 10^9$ M$_\odot$) and high-surface-brightness (HSB) galaxies, baryonic matter can generallyexplain the inner

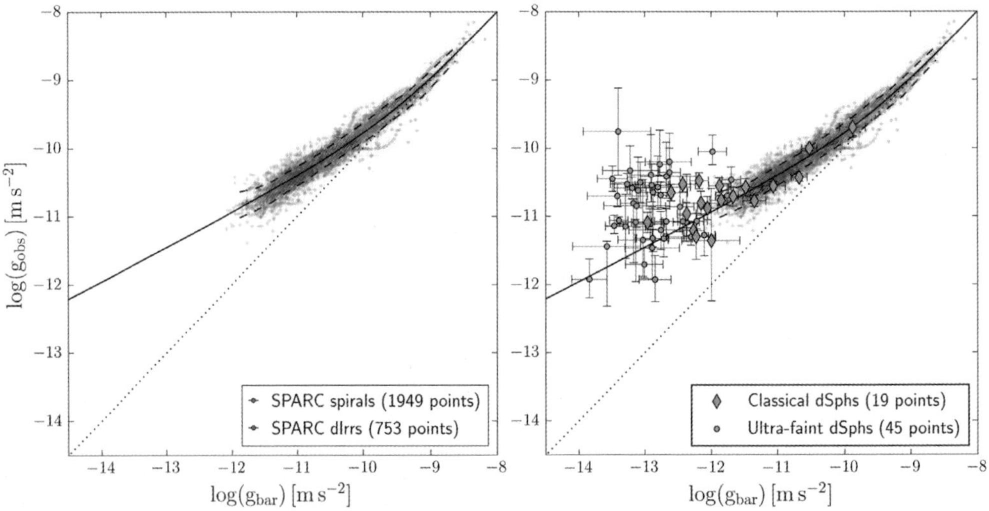

Figure 2. The Radial Acceleration Relation (RAR): the observed centripetal acceleration at each radius is plotted against that expected from the baryons' distribution. The left panel shows the RAR from 153 disk galaxies from the SPARC database; blue and red dots correspond to data from spirals and dIrrs, respectively. The right panel adds 62 dSphs in the LG; orange diamonds and green circles show "classical" dSphs and ultra-faint dSphs. In both panels, the dotted line is the line of unity; the solid line shows a fit to the SPARC data; dashed lines correspond to 1 standard deviation from the mean. Adapted from Lelli et al. (2017).

dynamics ($R \lesssim 1-2R_{\rm h}$) while the DM effect appears in the outer regions. In low-mass ($M_\star \leq 3\times10^9$ M$_\odot$) and low-surface-brightness (LSB) galaxies, instead, the DM effect is already important at small radii. In addition, the baryonic distribution and the rotation-curve shapes appear tightly coupled in galaxies at a local level (Sancisi 2004).

The RAR (Fig. 2) is an effective way to quantify the *local* baryon-dynamics coupling in galaxies (McGaugh et al. 2016; Lelli et al. 2017). At each radii, the observed acceleration from rotation curves ($g_{\rm obs} = V_{\rm rot}^2/R$) correlates with that expected from the distribution of baryons ($g_{\rm bar} = V_{\rm bar}^2/R = -\nabla\Phi_{\rm bar}$). At high accelerations, $g_{\rm obs} = g_{\rm bar}$ so there is no need of DM. At low accelerations, below a characteristic acceleration scale of $\sim 10^{-10}$ m s^{-2}, $g_{\rm obs} > g_{\rm bar}$ and the DM effect emerges. Intriguingly, in the RAR plane, the outer DM-dominated regions of spiral galaxies smoothly overlap with the inner DM-dominated regions of dwarf irregulars (dIrrs) as if the two regions "know" about each other. The observed scatter of the RAR is just $\sim 25\% - 30\%$, so its intrinsic scatter must be tiny, if not zero (Li et al. 2018; Desmond 2023).

Remarkably, the existence and properties of the RAR were predicted *a-priori* by Milgrom (1983) using his Modified Newtonian Dynamics (MOND). It is thus important to test whether different galaxy types follow the same RAR. Early-type galaxies (ellipticals and lenticulars) lie on the same RAR of spirals and dIrrs (Lelli et al. 2017; Shelest and Lelli 2020). The situation is more uncertain for dwarf spheroidal galaxies (dSphs) in the LG. Given their low surface densities, dSphs are truly unique systems to study the low RAR down to very low accelerations, but they lack a rotating gas disk, so $g_{\rm obs}$ must be inferred from stellar kinematics.

Lelli et al. (2017) compiled a sample of 62 dSphs with stellar velocity dispersion $\sigma_\star$ from single-star spectroscopy. Assuming that dSphs are fully pressure-supported and spherically symmetric, we can compute $g_{\rm obs} = 3\sigma_\star^2/r_{1/2}$ and $g_{\rm bar} = G_{\rm N}M_\star/(2r_{1/2}^2)$, where $r_{1/2}$ is the 3D half-light radius. This choice is motivated by the fact that the mass-anisotropy degeneracy in spherical, pressure-supported systems is nearly broken at $r_{1/2}$ (Wolf et al. 2010). For dSphs, therefore, we have a single point per galaxy rather than a spatially resolved analysis. Figure 2 (right panel) show that "classical" dSphs roughly follow the same RAR as disk galaxies, whereas ultra-faint dwarfs (UFDs) display a much larger scatter and are systematically shifted towards higher $g_{\rm obs}$ than expected from the RAR extrapolation. Several observational uncertainties may affect the location of UFDs on the RAR: (1) $M_\star$ and $r_{1/2}$ are not inferred from NIR surface photometry but from star counts in optical images, after candidate stars are selected using color-magnitude diagrams and template isochrones, so there could be systematics between different datasets, (2) $\sigma_\star$ may be inflated by undetected stellar binaries and/or small number statistics, especially when only a dozen of bright stars are available, and (3) both $\sigma_\star$ and $r_{1/2}$ may be inflated by tidal forces from the host galaxies (the Milky Way and Andromeda), possibly driving the systems out of dynamical equilibrium. Major observational and theoretical efforts are truly needed to clarify the location of UFDs on the RAR.

Discussion

MARINA REJKUBA: What are the prospects to use the stellar $V_{\rm rot}$ in low surface brightness outskirts? Maybe in resolved galaxies?

FEDERICO LELLI: It is hard to study stellar kinematics beyond 1-2 $R_{\rm h}$ with existing integral-field spectrographs due to the low surface brightness. Resolved individual-star spectroscopy is currently limited to LG galaxies and require large spectrophotometric campaigns. Future 40-meters telescopes, such as the ESO ELT, will surely push stellar kinematics further. However, in addition to the observational challenges, there is a theoretical limitation. To infer the circular velocity tracing the gravitational field, we need to correct the stellar $V_{\rm rot}$ for pressure support, which requires assumptions on the shape of the velocity dispersion tensor.

MARINA REJKUBA: You brought up uncertainties in $R_{\rm h}$ as possible explanation for the scatter in the RAR for LG dwarfs. What did you use for $R_{\rm h}$? In particular, for ultra-faint dwarfs, this can indeed be uncertain as they may be out of equilibrium or affected by the Milky Way.

FEDERICO LELLI: We compiled $R_{\rm h}$ from various literature sources, then we calculated the deprojected 3D half-light radius $r_{1/2}$ that is needed to compute centripetal accelerations. In addition to the technical challenges in measuring $R_{\rm h}$ in ultra-faint dwarfs, I fully agree that out-of-equilibrium dynamics is a major concern because it could artificially increase both $R_{\rm h}$ and the measured velocity dispersion $\sigma_\star$. As far as I can tell, the observed scatter in the RAR may be entirely driven by observational uncertainties.

YABIN YANG: About the dSphs scattered distribution on the RAR, perhaps taking their distances to be an extra parameter could help to understand their large dispersion?

FEDERICO LELLI: We used the distances of dSphs from their host galaxies (Milky Way or Andromeda) to compute the expected gravitational acceleration ($g_{\rm host}$) and tidal acceleration ($g_{\rm tides}$) at their location. The scatter in the RAR decreases imposing a cut in $g_{\rm obs}/g_{\rm tides}$, but there is no simple trend, nor obvious cut to use. The observed scatter is probably driven by multiple effects and uncertainties, so considering a single quantity does not tell the full story.

Ting Li: What should the velocity dispersion be for the ultra-faint dwarfs if they follow the same RAR as disk galaxies?

Federico Lelli: On average, the velocity dispersion of ultra-faint dwarfs should be a factor of 2 smaller to have them on the low-acceleration extrapolation of the RAR of disk galaxies.

References

Bacchini, C., Fraternali, F., Pezzulli, G., & Marasco, A. 2020, The volumetric star formation law for nearby galaxies. Extension to dwarf galaxies and low-density regions. *A&A*, 644, A125.

Barat, D., D'Eugenio, F., Colless, M., Sweet, S. M., Groves, B., & Cortese, L. 2020, SHαDE: survey description and mass-kinematics scaling relations for dwarf galaxies. *MNRAS*, 498(4), 5885.

Begeman, K. G. 1989, HI rotation curves of spiral galaxies. I. NGC 3198. *A&A*, 223, 47.

Binney, J. & Tremaine, S. 1994,. *Galactic Dynamics.* Princeton University Press, Princeton, USA.

Bouché, N., Carfantan, H., Schroetter, I., Michel-Dansac, L., & Contini, T. 2015, GalPak^{3D}: A Bayesian Parametric Tool for Extracting Morphokinematics of Galaxies from 3D Data. *AJ*, 150(3), 92.

Burkert, A., Genzel, R., Bouché, N., & 13 co-authors 2010, High-redshift Star-forming Galaxies: Angular Momentum and Baryon Fraction, Turbulent Pressure Effects, and the Origin of Turbulence. *ApJ*, 725(2), 2324.

Cappellari, M. 2002, Efficient multi-Gaussian expansion of galaxies. *MNRAS*, 333(2), 400–410.

Casertano, S. 1983, Rotation curve of the edge-on spiral galaxy NGC 5907: disc and halo masses. *MNRAS*, 203, 735.

Chandrasekhar, S. 1969,. *Ellipsoidal figures of equilibrium.* Yale University Press, New Haven, USA.

Courteau, S. 1997, Optical Rotation Curves and Linewidths for Tully-Fisher Applications. *AJ*, 114, 2402.

Davis, T. A., Alatalo, K., Bureau, M., & 23 co-authors 2013, The ATLAS^{3D} Project - XIV. The extent and kinematics of the molecular gas in early-type galaxies. *MNRAS*, 429(1), 534.

Dawson, J. M., Davis, T. A., Gomez, E. L., & Schock, J. 2021, A self-supervised, physics-aware, Bayesian neural network architecture for modelling galaxy emission-line kinematics. *MNRAS*, 503(1), 574.

de Blok, W. J. G. 2010, The Core-Cusp Problem. *Advances in Astronomy*, 2010, 789293.

Desmond, H. 2023, The underlying radial acceleration relation. *arXiv e-prints*, arXiv:2303.11314.

Di Teodoro, E. M. & Fraternali, F. 2015, ^{3D}BAROLO: a new 3D algorithm to derive rotation curves of galaxies. *MNRAS*, 451(3), 3021–3033.

Emsellem, E., Monnet, G., & Bacon, R. 1994, The multi-gaussian expansion method: a tool for building realistic photometric and kinematical models of stellar systems I. The formalism. *A&A*, 285, 723.

Frank, B. S., de Blok, W. J. G., Walter, F., Leroy, A., & Carignan, C. 2016, The Impact of Molecular Gas on Mass Models of Nearby Galaxies. *AJ*, 151(4), 94.

Freeman, K. C. 1970, On the Disks of Spiral and so Galaxies. *ApJ*, 160, 811.

Gentile, G., Burkert, A., Salucci, P., Klein, U., & Walter, F. 2005, The Dwarf Galaxy DDO 47 as a Dark Matter Laboratory: Testing Cusps Hiding in Triaxial Halos. *ApJ*, 634(2), L145.

Iorio, G., Fraternali, F., Nipoti, C., Di Teodoro, E., Read, J. I., & Battaglia, G. 2017, LITTLE THINGS in 3D: robust determination of the circular velocity of dwarf irregular galaxies. *MNRAS*, 466(4), 4159.

Kamphuis, P., Józsa, G. I. G., Oh, S. H., Spekkens, K., Urbancic, N., Serra, P., Koribalski, B. S., & Dettmar, R. J. 2015, Automated kinematic modelling of warped galaxy discs in large H I surveys: 3D tilted-ring fitting of H I emission cubes. *MNRAS*, 452(3), 3139–3158.

Kent, S. M. 1986, Dark matter in spiral galaxies. I. Galaxies with optical rotation curves. *AJ*, 91, 1301.

Lelli, F. 2022, Gas dynamics in dwarf galaxies as testbeds for dark matter and galaxy evolution. *Nature Astronomy*, 6, 35.

Lelli, F., McGaugh, S. S., & Schombert, J. M. 2016, SPARC: Mass Models for 175 Disk Galaxies with Spitzer Photometry and Accurate Rotation Curves. *AJ*, 152(6), 157.

Lelli, F., McGaugh, S. S., Schombert, J. M., & Pawlowski, M. S. 2017, One Law to Rule Them All: The Radial Acceleration Relation of Galaxies. *ApJ*, 836(2), 152.

Lelli, F., Verheijen, M., Fraternali, F., & Sancisi, R. 2012, Dynamics of starbursting dwarf galaxies. II. UGC 4483. *A&A*, 544, A145.

Li, P., Lelli, F., McGaugh, S., & Schombert, J. 2018, Fitting the radial acceleration relation to individual SPARC galaxies. *A&A*, 615, A3.

Li, P., Lelli, F., McGaugh, S., & Schombert, J. 2020, A Comprehensive Catalog of Dark Matter Halo Models for SPARC Galaxies. *ApJS*, 247(1), 31.

Marasco, A., Oman, K. A., Navarro, J. F., Frenk, C. S., & Oosterloo, T. 2018, Bars in dark-matter-dominated dwarf galaxy discs. *MNRAS*, 476(2), 2168.

McGaugh, S. S., Lelli, F., & Schombert, J. M. 2016, Radial Acceleration Relation in Rotationally Supported Galaxies. *PRL*, 117(20), 201101.

McGaugh, S. S., Lelli, F., & Schombert, J. M. 2020, Scaling Relations for Molecular Gas and Metallicity: Impact on the Baryonic Tully-Fisher Relation. *RNAAS*, 4(4), 45.

McGaugh, S. S. & Schombert, J. M. 2015, Weighing Galaxy Disks With the Baryonic Tully-Fisher Relation. *ApJ*, 802(1), 18.

Meurer, G. R., Carignan, C., Beaulieu, S. F., & Freeman, K. C. 1996, NGC 2915.II.A Dark Spiral Galaxy With a Blue Compact Dwarf Core. *AJ*, 111, 1551.

Milgrom, M. 1983, A modification of the Newtonian dynamics - Implications for galaxies. *ApJ*, 270, 371.

Noordermeer, E. 2008, The rotation curves of flattened Sérsic bulges. *MNRAS*, 385(3), 1359.

Noordermeer, E., van der Hulst, J. M., Sancisi, R., Swaters, R. S., & van Albada, T. S. 2007, The mass distribution in early-type disc galaxies: declining rotation curves and correlations with optical properties. *MNRAS*, 376(4), 1513.

Oh, S.-H., Hunter, D. A., Brinks, E., & 13 co-authors 2015, High-resolution Mass Models of Dwarf Galaxies from LITTLE THINGS. *AJ*, 149(6), 180.

Rogstad, D. H., Lockhart, I. A., & Wright, M. C. H. 1974, Aperture-synthesis observations of H I in the galaxy M83. *ApJ*, 193, 309.

Sancisi, R. The visible matter – dark matter coupling. In *Dark Matter in Galaxies* 2004, volume 220 of *IAU Symposia*, 233.

Schoenmakers, R. H. M., Franx, M., & de Zeeuw, P. T. 1997, Measuring non-axisymmetry in spiral galaxies. *MNRAS*, 292(2), 349–364.

Schombert, J., McGaugh, S., & Lelli, F. 2022, Stellar Mass-to-light Ratios: Composite Bulge+Disk Models and the Baryonic Tully-Fisher Relation. *AJ*, 163(4), 154.

Shelest, A. & Lelli, F. 2020, From spirals to lenticulars: Evidence from the rotation curves and mass models of three early-type galaxies. *A&A*, 641, A31.

Simon, J. D., Bolatto, A. D., Leroy, A., Blitz, L., & Gates, E. L. 2005, High-Resolution Measurements of the Halos of Four Dark Matter-Dominated Galaxies: Deviations from a Universal Density Profile. *ApJ*, 621(2), 757.

Spekkens, K. & Sellwood, J. A. 2007, Modeling Noncircular Motions in Disk Galaxies: Application to NGC 2976. *ApJ*, 664(1), 204.

Spitzer, Lyman, J. 1942, The Dynamics of the Interstellar Medium. III. Galactic Distribution. *ApJ*, 95, 329.

Swaters, R. A., Sancisi, R., van Albada, T. S., & van der Hulst, J. M. 2009, The rotation curves shapes of late-type dwarf galaxies. *A&A*, 493, 871.

Trachternach, C., de Blok, W. J. G., Walter, F., Brinks, E., & Kennicutt, R. C., J. 2008, Dynamical Centers and Noncircular Motions in THINGS Galaxies: Implications for Dark Matter Halos. *AJ*, 136(6), 2720.

van Albada, T. S. & Sancisi, R. 1986, Dark matter in spiral galaxies. *Royal Society of London Philosophical Transactions Series A*, 320, 447.

van der Kruit, P. C. & Freeman, K. C. 2011, Galaxy Disks. *ARA&A*, 49, 301–371.

Wolf, J., Martinez, G. D., Bullock, J. S., Kaplinghat, M., Geha, M., Muñoz, R. R., Simon, J. D., & Avedo, F. F. 2010, Accurate masses for dispersion-supported galaxies. *MNRAS*, 406, 1220.

Dynamical Masses of Local Group Galaxies
Proceedings IAU Symposium No. 379, 2025
P. Bonifacio, M.-R. Cioni, F. Hammer, M. S. Pawlowski & S. Taibi, eds.
doi:10.1017/S1743921323001667

Super cuspy dark matter halos of massive galaxies due to baryon-driven contraction

Pengfei Li

Leibniz-Institute for Astrophysics, An der Sternwarte 16, 14482 Potsdam, Germany
email: pli@aip.de

Abstract. The interplay between dark matter (DM) and baryons has long been ignored when building galaxies semi-empirically and observationally. Here I show that baryonic gravity leads to an adiabatic contraction of DM halos, which is most significant in massive galaxies. Ignoring this effect, the derived DM halos are not guaranteed in dynamical equilibrium. I present a new approach to deriving DM halos from rotation curves, which incorporates the adiabatic contraction. The compressed halos turn out super cuspy with respect to NFW halos, which require smaller baryonic contributions and less concentrated primordial halos. I also examine the semi-empirical approach to building galaxies, and find the adiabatic contraction can shift massive galaxies from the observed radial acceleration relation dramatically. Both approaches lead to super cuspy DM halos for massive galaxies, demonstrating the importance of the baryon-driven contraction, which has to be taken into account in order to make an apple-to-apple comparison with simulations.

Keywords. dark matter — galaxies: kinematics and dynamics — galaxies: dwarf — galaxies: spiral — galaxies: irregular

1. Introduction

Galaxy formation in the standard ΛCDM model relies on a critical component, dark matter halos, which are invisible but set the very initial seeds for all the observed galaxies. To understand their structures, dark matter-only simulations have been employed, which suggest a Navarro-Frenk-White model (NFW, Navarro et al. 1996). To test these simulations, the properties of dark matter halos have been extracted from rotation curves for a variety of late-type galaxies. The comparison shows that while massive galaxies are consistent with simulations, dwarf galaxies present a cusp-core problem (e.g. see de Blok & Bosma 2002): the NFW model predicts a halo that is too cuspy to fit observed rotation curves. Another way to test simulations is to build galaxies semi-empirically by assembling simulated dark matter halos and observed baryonic distributions into galaxies (e.g. see Di Cintio & Lelli 2016; Desmond 2017). One then tests whether the assembled galaxies are able to reproduce observed scaling relations established from dynamical studies of galaxies (e.g. Navarro et al. 2017).

In both tests, the interplay between dark matter and baryons has generally been ignored, i.e. the two components are treated separately and independently, though they are closely coupled through mutual gravitational interaction. Baryonic gravity could possibly change the structure of dark matter halos, causing a net contraction. As a result, the built galaxies are not guaranteed in dynamicalequilibrium. Therefore, whether the

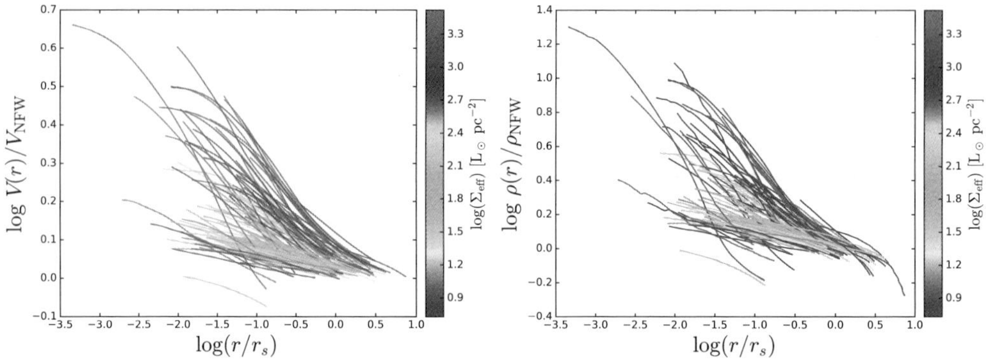

Figure 1. Left: Ratios of velocities contributed by compressed halos to primordial NFW halos; Right: density ratios between compressed halos and NFW halos. Lines are color-coded by their effective surface brightness. The initial NFW halos are from the best fits in Li et al. (2020).

previous comparisons between observations and simulations are robust remains an open question to be tested.

2. Significance of baryon-driven contraction

To quantify the significance of the baryon-driven contraction, we can model the evolution of dark matter halos through an adiabatic process (Choi et al. 2006). Young (1980) introduced this approach by conserving three adiabatic actions and Sellwood & McGaugh (2005) adopted it into the "COMPRESS" code. Using this code, one can compute stabilized dark matter density profiles for any given baryonic distributions and dark matter halos.

The Spitzer Photometry & Accurate Rotation Curves (SPARC, Lelli et al. 2016) database presents a representative sample of rotationally supported galaxies, from dwarfs to large spirals. Their NFW dark matter halos have been derived and presented in Li et al. (2020). To test the dynamical state of the best-fit halos, we compute the adiabatic changes using the "COMPRESS" code (Li et al. 2022).

Figure 1 shows the comparison between the best-fit NFW halos and the stabilized halos. Galaxies are color-coded by their effective surface brightness. For low-surface-brightness galaxies, their dark matter halos contract very little, so the best-fit NFW halos in Li et al. (2020) are approximately in dynamical equilibrium. However, for galaxies with a surface brightness higher than 100 $L_\odot$ pc^{-2}, the dark matter halos experience strong adiabatic contraction. Their density profiles are significantly compressed by baryonic gravity. The inner density increases more significantly than the outer density, given the surface brightness is higher at small radii than at large radii. As a result, their contributions to rotation velocity increase dramatically. This breaks the obtained rotation curve fits, and it brings a dilemma for massive galaxies: the best-fit dark matter halos are not in dynamical equilibrium with embedded baryons, and the stabilized dark matter halos cannot describe observed rotation curves.

3. Deriving dark matter halos from rotation curves

Since the traditional approach to deriving dark matter halos from rotation curves cannot guarantee the dynamical equilibrium of the whole system, it requires an improvement in the technique. This is made possible by combining evolving dark matter halos and fitting rotation curves (Li et al. 2022). The new approach starts by assuming an initial dark matter halo and evolve it. One then calculates the stabilized halo using the

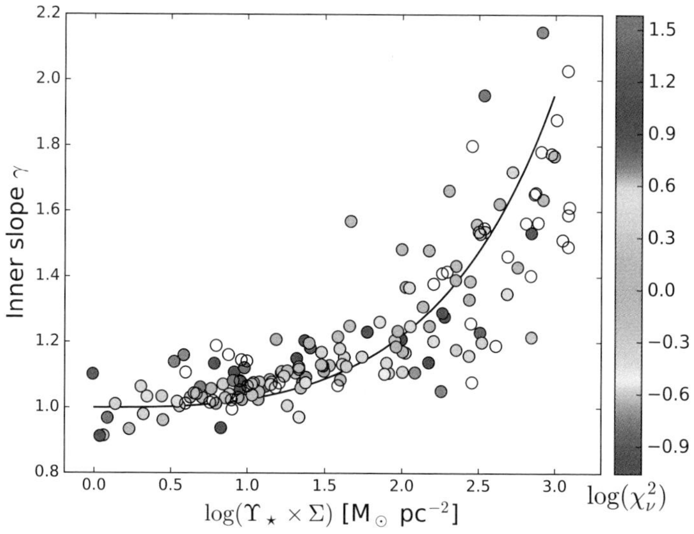

Figure 2. Best-fit values of the inner slope γ as a function of surface mass density (Li et al. 2022). Galaxies are color-coded by the fit quality. The solid line presents the selected exponential function that can roughly describe the mean relation.

"COMPRESS" code and compare it with observed rotation curves. The process is iterated multiple times until a satisfactory fit is obtained. Since we fit stabilized halos to data, it guarantees the dynamical equilibrium for the coupled system.

The initial halos can be parameterized by the NFW model, given simulations suggest that halos prior to galaxy formation are in the shape of NFW. The properties of the primordial NFW halos can be determined from the abundance matching relation by Moster et al. (2013) and halo mass-concentration relation by Dutton & Macciò (2014). However, there is no guarantee that these properties will be preserved after baryonic compression, even if they are imposed as priors.

Though baryonic compression can significantly increase the inner density of dark mater halos, it is generally possible to achieve satisfactory rotation curve fits. This is because the strength of baryonic compression is quite sensitive to stellar mass-to-light ratios $\Upsilon_\star$. Reducing the value of $\Upsilon_\star$ leads to a lower surface mass density and thereby less significant adiabatic contraction. It also reduces the baryonic contribution to rotation curves, so that there is larger room for the contribution from dark matter halos. Since the net effect of baryonic compression is that halos become more concentrated, the concentrations for the initial halos are supposed to be smaller. In contrast, total halo mass is relatively less affected, as baryonic compression only changes the structure. Li et al. (2022) indeed obtain good rotation curve fits for most of the SPARC galaxies, but the price is that both stellar mass-to-light ratios and initial halo concentrations are systematically turned down.

In fact, the compressed halos are not even in NFW shape. To describe the stabilized halos, we can fit them with the generalized (α, β, γ) model (Zhao 1996). The fits suggest that the inner density slope γ strongly correlates with surface mass density, as Figure 2 shows. The correlation can be roughly described by an exponential function,

$$\gamma = \exp\left(0.025(\log \Sigma_m)^3\right), \tag{1}$$

where $\Sigma_m = \Upsilon_\star \times \Sigma$ is the surface mass density. When the surface mass density is below $10\ M_\odot\ \mathrm{pc}^{-2}$, the inner slope is the same as NFW halos, suggesting that the baryonic compression is negligible. At higher surface mass density, the inner slope increases and the halos deviate from the NFW shape. For massive galaxies with $\Sigma_m > 100\ M_\odot\ \mathrm{pc}^{-2}$, the inner slope increases dramatically, forming super cuspy halos, which generally require relatively smaller stellar mass-to-light ratios to achieve satisfactory fits.

In the above calculation, I did not consider the effect of baryonic feedback. Feedback is believed to help convert cuspy halos to cores. The currently introduced mechanisms, such as supernova (e.g. see Di Cintio et al. 2014a,b), stellar winds, star formation (e.g. Read et al. 2016), only work in dwarf galaxies. They are inefficient for massive galaxies, so the super cuspy halos we found for massive galaxies remain invariant. Other feedback mechanisms might work. Since the new approach considers the end product of the coupled system, it would be interesting to see whether feedback can keep counteracting baryonic compression all the time.

4. Building galaxies semi-empirically

The interplay between baryons and dark matter halos also plays an important role when building galaxies in a semi-empirical way. This approach sets up each component of galaxies separately and independently, such as stellar disks, gas disks, central bulges (if any) and dark matter halos. The detailed properties of each component can be set using results from observations (for baryonic components) and simulations (for dark matter halos).

Many studies have been carried out using this approach (e.g. see Di Cintio & Lelli 2016; Desmond 2017; Navarro et al. 2017) to reproduce observed scaling relations in galaxies, such as the radial acceleration relation (RAR, McGaugh et al. 2016; Lelli et al. 2017), which is a tight correlation that holds in individual galaxies once the uncertainties on observational variables are taken into account (Li et al. 2018). Navarro et al. (2017) used this approach to setup galaxies and claimed that the RAR can be naturally explained by abundance matching relations. However, they did not consider the interplay between the baryonic components and the dark matter halos. So the galaxies they built may not be in dynamical equilibrium.

To investigate how significant the interplay could affect the structure of dark matter halos, we build 80 model galaxies in Li et al. (2022). Stellar disks and gas disks are modeled with exponential profiles, and their properties (scale length and total mass) are chosen to match the SPARC galaxies (Lelli et al. 2016). For galaxies with $M_\star > 10^9\ M_\odot$, a central bulge is included using the Hernquist profile (Hernquist 1990). Similar to Navarro et al. (2017), dark matter halos are parameterized using the NFW model. One can employ abundance matching relation to determine the halo mass from their stellar mass. We find that the relation by Kravtsov et al. (2018) can generate the RAR that is most comparable to the observed one. Halo concentrations are determined using the mass-concentration relation from Dutton & Macciò (2014).

The left panel of Figure 3 shows that the 80 modeled galaxies nicely reproduce the observed RAR except for dwarf galaxies at small radii, which present some up-bended hooks. This is consistent with Navarro et al. (2017), but they cut off the inner parts, so their results do not present hooks. In contrast, we cut the modeled rotation curves based on the radial range of observed galaxies. The up-bended hooks are due to the cuspy NFW model, which is known to overestimate the inner density of dark matter halos. They can be removed if we model dark matter halos using cored halo profiles such as the DC14 model (Di Cintio et al. 2014b). Therefore, it is possible to build a sample of galaxies using the semi-empirical approach to reproduce the observed scaling relation by carefully choosing halo models and the abundance matching relation.

To examine the dynamical state of the model galaxies, we compute the stabilized halos by evolving the parameterized halos underthe baryonic gravity. The right panel

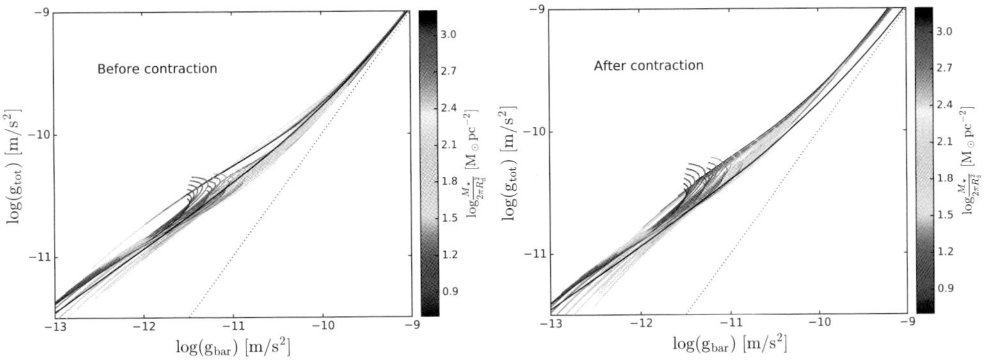

Figure 3. Radial acceleration relation from 80 galaxies built by setting up stellar disks, gas disks, central bulges (for massive galaxies) and NFW halos presented in Li et al. (2022). Galaxies are color-coded by their surface mass density. The left panel shows the results without incorporating the adiabatic contraction of dark matter halos, while the right panel shows the relation after adiabatic contraction.

of Figure 3 shows the RAR after adiabatic contraction. While low-mass galaxies remain almost invariant, high-mass galaxies systematically shift upwards and significantly deviate from the observed RAR. This demonstrates that their dark matter halos experienced strong baryonic compression, and the inner mass density increases dramatically. As such, the total accelerations outweigh the baryonic ones at small radii, resulting in a net upward shift in the RAR panel. This suggests that the modeled NFW halos for massive galaxies turn to be super cuspy. Accommodating these super cuspy halos poses a challange for the CDM model.

5. Conclusion and outlook

Dynamical tests demonstrate that the baryon-driven contraction is a quite significant effect for massive galaxies. To fairly compare observations with simulations, the adiabatic contraction has to be taken into account, and it leads to super cuspy dark matter halos that are difficult to be accommodated by observed rotation curves.

There are two possible ways to get rid of these super cuspy halos. First, baryonic feedback is known to be able to reduce the inner density of dark matter halos, which is however most efficient in dwarf galaxies. For massive galaxies, one might need to introduce more powerful feedback processes. Second, the primordial halos may be less cuspy. Since we assume the primordial halos are NFW, which is already quite cuspy, it might be able to reproduce a normally cuspy halo if we start with core halos. Whether these two possibilities work out is worth further studies.

References

Choi, J.-H., Lu, Y., Mo, H. J., et al. 2006, Mon. Not. Royal Astron. Soc., 372, 1869. doi:10.1111/j.1365-2966.2006.10990.x

de Blok, W. J. G. & Bosma, A. 2002, A & A, 385, 816. doi:10.1051/0004-6361:20020080

Desmond, H. 2017, Mon. Not. Royal Astron. Soc., 464, 4160. doi:10.1093/mnras/stw2571

Di Cintio, A., Brook, C. B., Macciò, A. V., et al. 2014, Mon. Not. Royal Astron. Soc., 437, 415. doi:10.1093/mnras/stt1891

Di Cintio, A., Brook, C. B., Dutton, A. A., et al. 2014, Mon. Not. Royal Astron. Soc., 441, 2986. doi:10.1093/mnras/stu729

Di Cintio, A. & Lelli, F. 2016, Mon. Not. Royal Astron. Soc., 456, L127. doi:10.1093/mnrasl/slv185

Dutton, A. A. & Macciò, A. V. 2014, Mon. Not. Royal Astron. Soc., 441, 3359. doi:10.1093/mnras/stu742
Hernquist, L. 1990, Astrophys. J., 356, 359. doi:10.1086/168845
Kravtsov, A. V., Vikhlinin, A. A., & Meshcheryakov, A. V. 2018, Astronomy Letters, 44, 8. doi:10.1134/S1063773717120015
Lelli, F., McGaugh, S. S., & Schombert, J. M. 2016, Astron. J., 152, 157. doi:10.3847/0004-6256/152/6/157
Lelli, F., McGaugh, S. S., Schombert, J. M., et al. 2017, Astrophys. J., 836, 152. doi:10.3847/1538-4357/836/2/152
Li, P., Lelli, F., McGaugh, S., et al. 2018, A & A, 615, A3. doi:10.1051/0004-6361/201732547
Li, P., Lelli, F., McGaugh, S., et al. 2020, Astrophys. J. Supplement, 247, 31. doi:10.3847/1538-4365/ab700e
Li, P., McGaugh, S. S., Lelli, F., et al. 2022, Astrophys. J., 927, 198. doi:10.3847/1538-4357/ac52aa
Li, P., McGaugh, S. S., Lelli, F., et al. 2022, A & A, 665, A143. doi:10.1051/0004-6361/202243916
McGaugh, S. S., Lelli, F., & Schombert, J. M. 2016, Phys. Rev. Lett., 117, 201101. doi:10.1103/PhysRevLett.117.201101
Moster, B. P., Naab, T., & White, S. D. M. 2013, Mon. Not. Royal Astron. Soc., 428, 3121. doi:10.1093/mnras/sts261
Navarro, J. F., Frenk, C. S., & White, S. D. M. 1996, Astrophys. J., 462, 563. doi:10.1086/177173
Navarro, J. F., Benítez-Llambay, A., Fattahi, A., et al. 2017, Mon. Not. Royal Astron. Soc., 471, 1841. doi:10.1093/mnras/stx1705
Read, J. I., Agertz, O., & Collins, M. L. M. 2016, Mon. Not. Royal Astron. Soc., 459, 2573. doi:10.1093/mnras/stw713
Sellwood, J. A. & McGaugh, S. S. 2005, Astrophys. J., 634, 70. doi:10.1086/491731
Young, P. 1980, Astrophys. J., 242, 1232. doi:10.1086/158553
Zhao, H. 1996, Mon. Not. Royal Astron. Soc., 278, 488. doi:10.1093/mnras/278.2.488

Discussion

HAMMER: With this contracted halo for the Milky Way, what would be the ratio of the dark matter density to that of the baryons, at says, 6 kpc from the centre?

LI: The ratio of the dark matter denisty to that of baryons at 6 kpc or the solar position should be higher than what we thought.

Dynamical Masses of Local Group Galaxies
Proceedings IAU Symposium No. 379, 2025
P. Bonifacio, M.-R. Cioni, F. Hammer, M. S. Pawlowski & S. Taibi, eds.
doi:10.1017/S1743921323001369

The oblateness of dark matter halos of nearby galaxies and its correlation with gas mass fractions

Mousumi Das[1], **Roger Ianjamasimanana**[2], **Stacy McGaugh**[3], **James Schombert**[4] **and K. S. Dwarakanath**[5]

[1]Indian Institute of Astrophysics, Bengaluru 560034, India
email: mousumi@iiap.res.in

[2]Instituto de Astrofísica de Andalucía (CSIC), Glorieta de la Astronomía, E-18008 Granada, Spain

[3]Department of Astronomy, Case Western Reserve University, 10900 Euclid Avenue, Cleveland, OH 44106, USA

[4]Department of Physics, University of Oregon, 120 Willamette Hall, 1371 E 13th Avenue, Eugene, OR 97403461, USA

[5]Raman Research Institute, C.V. Raman Avenue, Sadashivanagar, Bengaluru, Karnataka 560080, India

Abstract. We present a method to measure the the oblateness parameter q of the dark matter halo of gas rich galaxies that have extended HI disks. We have applied our model to a sample of 20 nearby galaxies that are gas rich and close to face-on, of which 6 are large disk galaxies, 8 have moderate stellar masses and 6 are low surface brightness (LSB) dwarf galaxies. We have used the stacked HI velocity dispersion and HI surface densities to derive q in the outer disk regions. Our most important result is that gas dominated galaxies (such as LSB dwarfs) that have M(gas)/M(baryons)>0.5 have oblate halos (q< 0.55), whereas stellar dominated galaxies have a range of q values from 0.2 to 1.3. We also find a significant positive correlation between q and stellar mass, which indicates that galaxies with massive stellar disks have a higher probability of having halos that are spherical or slightly prolate, whereas low mass galaxies preferably have oblate halos. We briefly also discuss how the halo shape affects the disks of galaxies, especially the oblate halos.

Keywords. Galaxies: spirals, Galaxies : dwarfs, Dark Matter halos, HI line emission, Galaxy Structure

1. Introduction

Galaxy disks are known to be embedded in large dark matter halos, the main evidence being the flat rotation curves of disk galaxies (Kent 1987). The halos play a dominant role in galaxy evolution, as they affect disk stability and hence star formation (Yurin & Springel 2015; Sellwood 2016). Although we know that halos are important for disk evolution and galaxy rotation curves indicate that they contain most of the galaxy mass, their detailed properties are not well constrained. Halo angular momentum or spin, halo shape or even halo masses are not well understood. We are also not certain where the halos really ends, as there is no clear tracer beyond the outer limits of the HI rotation curve. In this paper we focus on the determination of halo shape. We present a new method to determine the shape of dark matter halos using the neutral hydrogen (HI) velocity dispersion measured in face-on galaxies.

The earliest observational studies of halo shapes used the polar rings around galaxies to model halo shapes (Combes 2014). But such rings are quite rare and so the method cannot be applied to a large sample of galaxies. Later studies used the flaring of HI disks in edge-on galaxies (Banerjee & Jog 2008), the kinematics of stars in our Milky way and other edge on galaxies (Olling 1996), the distribution of globular clusters in galaxies (Posti & Helmi 2019) and more recently the distribution and kinematics of stellar streams around galaxies (Pearson et al. 2022).

Numerical simulations have also helped our understanding of halo shapes. Early studies showed that the gas infall, star formation and the resulting build-up of stellar mass within galaxy halos affects the triaxial nature of halos and makes them rounder (Dubinski 1994). This is also seen in recent cosmological simulations where the halo axes ratio in the disk plane (b/a) is often close to 1 (Prada et al. 2019). The halos vary from prolate to oblate, but the Milky Way mass halos are close to spherical (Chua et al. 2019). However, one disadvantage of these studies is that they do not resolve the smaller dwarf galaxies with stellar masses $< 10^8 M_{\odot}$. So the variation of halo shape with stellar or halo masses is not not well understood.

In the following sections we briefly present a new way to determine the shapes of galaxy halos using the HI velocity dispersion in their extended disks. As is well known, the dark matter halo dominates the mass in the outer disks of galaxies, so our method focuses on the outer disk regions, where the stellar disk mass declines. Our method is applied to face-on disk galaxies rather than edge-on galaxies, as previously investigated in the literature (Banerjee & Jog 2011). Our basic method is described in Das et al. (2020), where it has been shown that halo dark matter is important for the vertical support of extended HI disks and the method has been applied to a galaxy sample in Das et al. (2023). In this paper we first describe need for halo dark matter for the vertical support of extended gas disks, the method to determine the halo vertical axis ratio q (where $q = c/a$) and the its importance for understanding disk dynamics in galaxies.

2. The vertical support of galaxy disks

Galaxy disks are supported by rotation in the radial direction where the gravitational force at any disk radius is balanced by the centrifugal force, so that $\mathrm{v}^2 = s\ \mathrm{GM(dyn)/R}$. This simple equality can be used to determine approximately the total mass within a radius R in a galaxy, but holds only if the dark matter halo is spherical. The vertical equilibrium in the disk is different. The gravitational binding force provided by the disk stars is balanced by the pressure in the gas and the velocity dispersion of the stars. As shown in Das et al. (2020), the disk may also contain a significant amount of dark matter that is generally associated with the halo.

In the outer parts of galaxy disks where there is often HI gas but very little stellar mass, the role of the dark matter becomes much more important for maintaining the vertical disk equilibrium. Some good examples are galaxies such as NGC628 which has a very extended HI disk (see Das et al. 2020 for the image). Also, dwarf galaxies such as DDO187 (Figure 1) where the comparison of the near-infrared (NIR) and HI image clearly shows that the stellar disk is much less extended. Extreme examples of this are the LSB dwarfs lying in isolated voids such as KK246 (Kreckel et al. 2011) and dwarfs lying along cosmic filaments such as NGC4701 (Das et al. 2019).

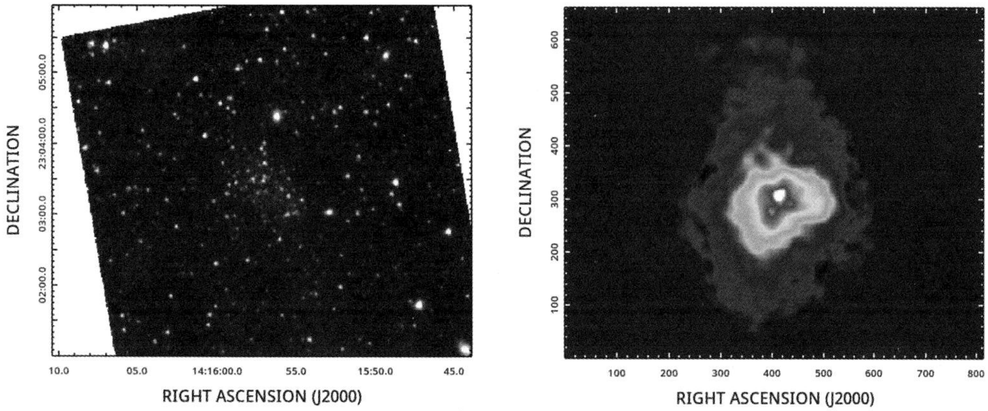

Figure 1. Comparison of the stellar and gas distribution in the dwarf galaxy DDO187. The image on the left is the Spitzer 3.6 μm image and that on the right is the THINGS VLA HI 21 cm emission map. Both Right Ascension (RA) and Declination are in J2000, and the images are matched in WCS. Note that the HI is more extended than the stellar disk as traced by the 3.6 μm emission.

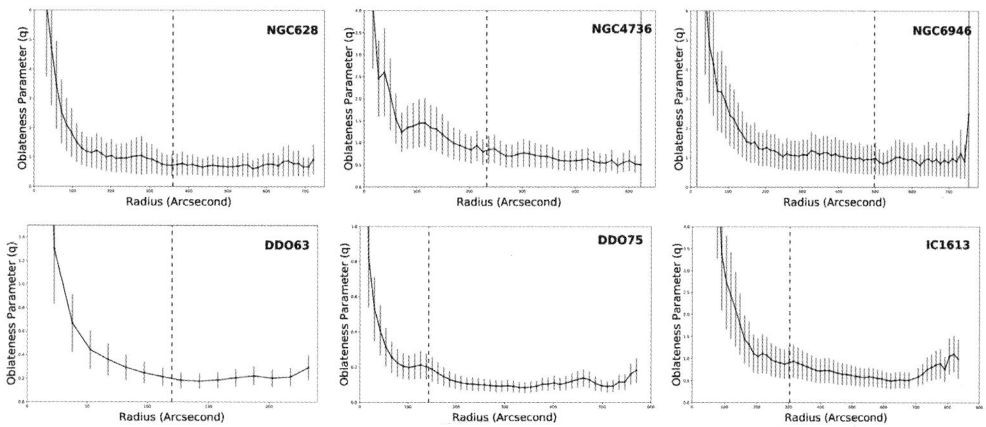

Figure 2. The top panel shows how q varies with radius for three relatively massive galaxies whereas the bottom panel shows the same for less massive dwarf galaxies. The dashed vertical line marks the R_{25} radius where the stellar disk ends.

3. Our methods, galaxy sample and analysis

We have assumed hydrostatic equilibrium for the gas disk in the vertical direction. The potential is composed of the stellar disk, molecular hydrogen gas (H_2), HI and dark matter. The dark matter halo has been modeled using the logarithmic potential; the stellar and gas disks have been assumed to have an exponential form. The detailed calculations are described in Das et al. (2023). The final expression used in the determination of the halo axis ratio $q = c/a$ is given by,

$$q^2 = \frac{{v_0}^2 {z_{g0}}^2}{R^2} \frac{1}{({\sigma_{z\mathrm{H\,I}}}^2 - \pi G z_{g0} \Sigma(HI)} \qquad (1)$$

where v_0 is the flat rotation velocity in units of km s^{-1}, the vertical disk scale height z_{g0} and the radius are in kpc, the vertical HI velocity dispersion $\sigma_{z\mathrm{H\,I}}$ is in m s^{-1}, the gas surface density in $M_{\odot}pc^{-2}$ and the gravitational constant is $G = 6.67 \times 10^{-8}$. The final expression after including all the constants becomes,

$$q = \frac{v_0 z_{g0}}{R} \times \frac{10^3}{[\sigma_{\mathrm{zH\,I}}{}^2 - 13.97 \times 10^6 z_{g0} \Sigma(HI)]^{1/2}} \tag{2}$$

where $\Sigma(HI)$ includes the correction for helium in the HI gas surface density.

We have applied this expression to a sample of 20 quiescent, nearly face-on galaxies, where the HI velocity dispersion over the cube represents $\sigma_{\mathrm{zH\,I}}$ and the HI surface density $\Sigma(HI)$ can be determined from the moment0 map. The galaxy sample was derived from the The H I Nearby galaxy Survey (THINGS; Walter et al. 2008), and the Local Irregulars That Trace Luminosity Extremes; The H I Nearby galaxy Survey (LITTLE THINGS; Hunter et al. 2012) and two galaxies from the VLA ACS Nearby galaxy Survey Treasury (VLA ANGST) survey (Ott et al. 2012). Of the 20 galaxies, 8 had stellar masses $M(*) > 10^9 M_{\odot}$, and the remaining 12 galaxies are low mass dwarf galaxies.

The disk flat rotation velocity was determined using the baryonic Tully-Fisher relation. The method for determining H I velocity dispersion $\sigma_{\mathrm{zH\,I}}$ variation with radius is summarised in (Ianjamasimanana et al. 2017) and Das et al. (2020), and involves stacking the individual velocity profiles over radial bins in the data cube. This helps to have a good signal-to-noise ratio (S/N) for the $\sigma_{\mathrm{zH\,I}}$ values. The plots are shown in the Appendix of Das et al. 2023.

4. Main Results

1. An important outcome of this study is that the q does not vary much in the outer disks of galaxies at radii $R > R_{25}$ (see Figure 1). Of the sample galaxies, only the gas-rich, LSB dwarf galaxy DDO46 shows a significant decrease in q from q = 0.5 at R_{25} to 0.3 at $2R_{25}$. However, as shown in equation [1] and [2], q depends on the thickness of the gas disk, z_{g0}, and this may vary with radius, especially in cases where the HI disk flares in the outer parts. We discuss this further in the discussion.

2. The value of q shows a significant dependence on galaxy mass, with q increasing for increasing stellar mass. This suggests that the more massive galaxies with $M(*) > 10^9 M_{\odot}$ have rounder halos that are close to spherical and some even slightly prolate ($q \sim 1.4$). However, there is some scatter and error; hence the Pearsons weighted correlation coefficient is 0.78. A larger sample will reveal whether this trend is real or not.

3. Since a large fraction of our sample are gas dominated galaxies, we also plotted q against the gas to baryon mass fraction (Figure 3). There is a clear trend for gas dominated galaxies to have lower q values compared to the stellar dominated ones. All the gas dominated galaxies lying to the right of the dashed line in Figure 3 are LSB dwarf galaxies and have halo axis ratios varying from $q = 0.11$ to 0.55. whereas the stellar dominated ones have a wide range of q values, from $q = 0.21$ to 1.25.

5. Implications of our results

1. Halo Spin : One of the most interesting results from this study is that gas dominated galaxies have more flattened or oblate dark matter halos compared to stellar dominated galaxies. Such galaxies are also dwarfs with low stellar masses. One reason could be that these galaxies have higher halo spin compared to massive galaxies. This has been suggested in previous studies (Kim & Lee 2013) and is shown also shown by the modeling of the halo spin of the LSB dwarf galaxy UGC5288 by Ansar et al. (2023), and is discussed in a poster presentation by Sioree Ansar in this conference. The higher spin could be due to tidal torques experienced by the dwarfs at early epochs. Such galaxies are usually found in ioslated environments. On the other hand, the more massive stellar dominated

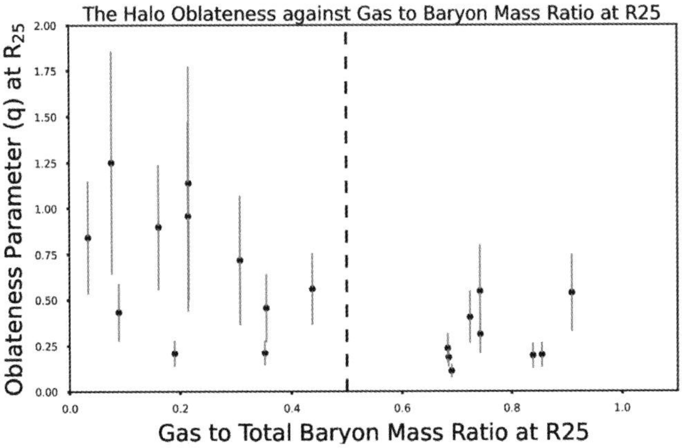

Figure 3. The above plot shows the oblateness parameter q plotted against the gas to baryon mass fraction in galaxies. the dashed line is just the 0.5 mark, so that galaxies to the right of it are gas mass dominated and those to the left are stellar mass dominated. It is interesting to note that the gas dominated galaxies (which are all LSB dwarfs, have oblate halos.

galaxies may have experienced mergers and so they have a distribution of halo shape and spin.

2. Disk dynamical evolution: One of the results of having oblate halos is that there is more halo dark matter associated with the disks. This can affect the disk dynamics. For example Kumar et al. (2022) have shown that oblate halos delay bar formation in galaxy disks and consequently delay bar buckling. The formation and buckling of bars are important for triggering star formation and bulge growth, which are important parts of disk evolution. It is possible that spiral arm formation may also be curbed, as it is also a global disk instability like bars. The underlying reason for the slow growth of disk instabilities in oblate halos is that the disk becomes dynamically hotter. This is clearly noted in Kumar et al. (2022), where the velocity dispersion of the disk stars increases as the halo becomes more oblate.

3. The HI disk height : In our study we have derived the HI gas disk height from the stellar disk radius using the ratio of $R_d/z_{g0} \sim 0.85$ (Kregel et al. 2002). This relation has been derived for stellar disks and may not be accurate for HI disks. This was adopted for simplicity, since there is no published study linking the disk scale length with HI thickness in galaxies. However, we do not think that including the flaring will change our basic result i.e. gas rich dwarfs have more oblate halos compared to massive, stellar dominated galaxies. To test this hypothesis we changed the disk thickness by 10% for DDO187. For R = 58.5″, the surface density of HI is 4.5 $M_{\odot} pc^{-2}$, and the HI dispersion is 12.5km/s (it has to be also deprojected with inc = 39 degrees). Using the disk rotation in equation [1] we find that q changes from q = 0.1109 to 0.1216. The change in q is around 10%. Hence, our overall result will still hold.

4. Flaring of HI disks and its effect on q **:** In our study we have assumed that the disk thickness remains constant with radius. However, studies of edge on disk galaxies suggest that many HI disks start flaring i.e. the thickness increases beyond the R_{25} radius (O'Brien et al. 2010). Looking at equation [1] and [2], this suggests that the q increases with radius, or basically the halos become rounder with increasing radii in galaxies.

6. Conclusions

We present a method to determine the oblateness q of galaxy halos. Our method can be applied to gas-rich, face-on disk galaxies by applying the equation of hydrostatic equilibrium to the outer HI disks. We have used the HI velocity dispersion and HI surface densities of the disks to calculate q at the R_{25} and $1.5R_{25}$ radii. Our study shows that q remains fairly constant in the outer disk regions, beyond the R_{25} radius. Our results show that there is a significant correlation between the stellar mass and q, which suggests that galaxies with massive stellar disks have a higher probability of having spherical or prolate halos, whereas the low-mass dwarfs have oblate halos. Our most significant finding is that gas-rich galaxies with M(gas)/M(baryons)>0.5 have oblate halos (q<0.55), but stellar mass dominated galaxies have a range of q values. Also, the gas-dominated galaxies are dwarf galaxies. Hence, our results show that gas-rich dwarf galaxies have oblate halos whereas the larger galaxies have a range of halo shapes.

Acknowledgements. M.D. acknowledges the support of the Science and Engineering Research Board (SERB) MATRICS grant MTR/2020/000266 for this research. RI acknowledges financial support from the grant CEX2021-001131-S funded by MCIN/AEI/ 10.13039/501100011033, from the grant IAA4SKA (Ref. R18-RT-3082) from the Economic Transformation, Industry, Knowledge and Universities Council of the Regional Government of Andalusia and the European Regional Development Fund from the European Union and financial support from the grant PID2021-123930OB-C21 funded by MCIN/AEI/10.13039/501100011033, by "ERDF A way of making Europe" and by the "European Union" and the Spanish Prototype of an SRC (SPSRC) service and support funded by the Spanish Ministry of Science and Innovation (MCIN), by the Regional Government of Andalusia and by the European Regional Development Fund (ERDF).

References

Ansar, S., Kataria, S. K., & Das, M. 2023, MNRAS, 522, 2967.
Banerjee, A. & Jog, C. J. 2008, ApJ, 685, 254.
Banerjee, A. & Jog, C. J. 2011, ApJL, 732, L8.
Chua, K. T. E., Pillepich, A., Vogelsberger, M., et al. 2019, MNRAS, 484, 476.
Combes, F. 2014, Multi-Spin Galaxies, 486, 207.
Das, M., Sengupta, C., & Honey, M. 2019, ApJ, 871, 197.
Das, M., Ianjamasimanana, R., McGaugh, S. S., et al. 2023, ApJL, 946, L8.
Das, M., McGaugh, S. S., Ianjamasimanana, R., et al. 2020, ApJ, 889, 10.
Dubinski, J. 1994, ApJ, 431, 617.
Kent, S. M. 1987, AJ, 93, 816.
Kim, J.-. hoon. & Lee, J. 2013, MNRAS, 432, 1701.
Kreckel, K., Peebles, P. J. E., van Gorkom, J. H., et al. 2011, AJ, 141, 204.
Kregel, M., van der Kruit, P. C., & de Grijs, R. 2002, MNRAS, 334, 646.
Kumar, A., Das, M., & Kataria, S. K. 2022, MNRAS, 509, 1262.
O'Brien, J. C., Freeman, K. C., & van der Kruit, P. C. 2010, Astronomy & Astrophysics, 515, A62.
Olling, R. P. 1996, AJ, 112, 481.
Pearson, S., Price-Whelan, A. M., Hogg, D. W., et al. 2022, ApJ, 941, 19.
Posti, L. & Helmi, A. 2019, Astronomy & Astrophysics, 621, A56.
Ianjamasimanana, R., de Blok, W. J. G., & Heald, G. H. 2017, AJ, 153, 213.
Prada, J., Forero-Romero, J. E., Grand, R. J. J., et al. 2019, MNRAS, 490, 4877.
Sellwood, J. A. 2016, ApJ, 819, 92.
Yurin, D. & Springel, V. 2015, MNRAS, 452, 2367.

Discussion

F. Lelli: How will your results change if the HI disk is flared?

Answer: A flared HI disk will have a larger vertical scaleheight and so the parameter z_d will increase. If one examines the relation for q, it is clear that the denominator will decrease when z_d increases, thus making the q increase. So the halo will become rounder at larger and larger radii when the HI disk flares.

Gerhard Hensler: We know that galaxies are affected by their CGM, e.g. infall of HVCs. Could you include such perturbation into your modelling, i.e. relaxing the equlibrium assumption?

Answer: That is an interesting idea. The model in its present form assumes that the disk is in vertical hydrostatic equilibrium. So to include gas infall, such as that due to HVCs, we will have consider pertubation of the equilibrium equation. So it can be accommodated in the model but only for small rates of gas infall.

Dynamical Masses of Local Group Galaxies
Proceedings IAU Symposium No. 379, 2025
P. Bonifacio, M.-R. Cioni, F. Hammer, M. S. Pawlowski & S. Taibi, eds.
doi:10.1017/S1743921323002120

Planes of satellites in the nearby Universe

O. Müller

Institute of Physics, Laboratory of Astrophysics, Ecole Polytechnique Fédérale de Lausanne (EPFL), 1290 Sauverny, Switzerland
email: oliver.muller@epfl.ch

Abstract. Since the mid 70ies it is known that the dwarf galaxies around the Milky Way are arranged in a thin, polar structure. The arrangement and motion within this structure has been identified as a severe challenge to the standard model of cosmology, dubbed as the plane of satellites problem. More observational evidence for such structures has been put forward around other galaxies, such as the Andromeda galaxy, Cen A or NGC 253, among others, adding to the previously identified tensions. Solutions to the plane of satellite problem should therefore not only be tailored to the Milky Way, but need to explain all these different observed systems and environments.

Keywords. Dwarf galaxies

---

1. Introduction

The plane of satellites problem (Pawlowski 2018) is one of the most outstanding and heavily debated problems in near-field cosmology. It describes a discrepancy between the observed and predicted arrangement and motion of dwarf galaxy systems. In observations, dwarf galaxy systems around their host galaxies seem to be flattened and co-moving, while a more isotropic distribution and random motions is found in cosmological simulations. Together, these two feature – distribution and motion – constitutes the plane of satellite problem. It has been argued that this configuration of a satellite system is in severe tension with the standard model of cosmology based on statistical arguments, finding less than 1% of analogs in cosmological systems hosting such arrangements (Ibata et al. 2014a). Others argued that the frequency in cosmological simulations is much larger, at the 10% level (Cautun et al. 2015). It all depends on what is to be considered an analog of an observed system – no consensus has been established to date with different research teams using different definitions of what an analog of an observed satellite system is. For example, the counter-rotating dwarf galaxy Sculptor (Pawlowski et al. 2011) in the Milky Way plane may be included or excluded when studying the phase-space correlation.

While there is still a disagreement regarding the degree of actual tension within the cosmological paradigm, progress has been made on the observational side, with the plane of satellites problem not being constrained to the Local Group anymore. For example was it noted by Chiboucas et al. (2013) that the close-by M 81 satellite system exhibits some flattening, orientated within the Local Sheet. Tully et al. (2015) found that around Cen A, another neighbor of the Local Group, the dwarfs seem to be arranged in two parallel planar structures. Therefore, the plane of satellite problem must be considered in a wider context.

2. Observed planar structures

In the following I will review the observational evidence for planes of satellites in the nearby universe.

2.1. *The Local Group planes*

The best evidence for planes of satellites we have in the Local Group, namely for the Milky Way and the Andromeda galaxy. For the Milky Way, Lynden-Bell (1976) found that the then known dwarf galaxies (LMC, SMC, Draco, Ursa Minor, Sculptor), as well as the globular cluster Palomar 13 and the Magellanic stream, are arranged in a thin, polar structure. Independently, Kunkel & Demers (1976) found a slightly different plane (with an angle of more than 30 degrees from that of Lynden-Bell 1976), including Palomar 3 and 4 and the distant dwarf galaxy Leo-I. Their plane, however, did not include Palomar 13, nor the Magellanic stream. Lynden-Bell (1976, 1983) argued that the plane might of tidal origin and therefore should include the Magellanic stream. The discovery of dwarf galaxies around the Milky Way (e.g., Willman et al. 2005; Belokurov et al. 2007; Walsh et al. 2007) strengthened the picture of the existence of Lynden-Bell (1976)'s planar structure (Kroupa et al. 2005; Metz et al. 2007, 2009). Thanks to proper motion measurements of dwarf galaxies (e.g., Dinescu et al. 2004; Piatek et al. 2006, 2007; Kallivayalil et al. 2006), it became evident that the dwarfs are not only arranged in a thin plane, but that they also co-orbit within this plane (Metz et al. 2008; Pawlowski et al. 2012; Pawlowski & Kroupa 2020). This structure is now dubbed the Vast Polar Structure (VPOS).

For the Andromeda galaxy, Koch & Grebel (2006) found a planar alignment for a subsample of the dwarf galaxies, namely for the dwarf ellipticals. They note this arrangement seems aligned with the large-scale structure and polar to the disk of M 31. At the same time McConnachie & Irwin (2006) found a planar alignment of a subpopulation of dwarf galaxies with respect to the disk of M 31. New data from the PANDAS survey (Richardson et al. 2011; Martin et al. 2013) shed some light on the distribution of the system. Indeed, there seems to be a co-moving plane of satellites around M 31, which is neither aligned nor polar with its disk. The co-motion was established using line-of-sight velocities of 15 satellites, of which 13 seem to be consistent with a co-rotational signal (Ibata et al. 2013). Sohn et al. (2020) measured the proper motion of two satellites within this plane and found that they are aligned with it, which is further evidence for the co-rotating nature of the structure, however, observations of the remaining dwarfs are needed to further assess this situation. Santos-Santos et al. (2020) found another planar alignment of dwarf galaxies around M 31 using a 4-galaxy-normal density plot (Pawlowski et al. 2013). This plane is seen face-on from our point of view and consists of 18 dwarf galaxies. Based on its velocity dispersion, it should not be stable and disperse eventually.

2.2. *The Centaurus Group*

The Centaurus Group is a nearby galaxy aggregate consisting of two giant galaxies – the early-type Cen A galaxy at 3.8 Mpc (Harris et al. 2010) and the grand spiral M 83 at 4.8 Mpc (Karachentsev et al. 2007). Tully et al. (2015) argued based on the three dimensional positions of the dwarfs around Cen A that they are arranged in two thin planes. Discoveries of more dwarfs around Cen A (Crnojević et al. 2014, 2016; Müller et al. 2017) made this distinction between the two thin planes not as clear anymore (Müller et al. 2016). Müller et al. (2018a) studied the kinematics of the dwarf galaxy system and found that 14 out of 16 dwarf galaxies with available line-of-sight velocitieswere following a co-motion, consistent with a co-rotation within a plane. These numbers have been updated with recent MUSE observations (Müller et al. 2021a) to 21 out of 28 dwarf galaxies co-moving (Müller et al. 2021b). Comparing the Cen A satellite system to cosmological simulations finds similar disagreement between observations and predictions as for the Milky Way (Müller et al. 2021b). What is intriguing to note is that the flattened structure around Cen A – be it one thick or two thin planes – is aligned with the cosmic web (Libeskind et al. 2019), with its orbital pole pointing towards the Local Void (i.e. the plane is aligned with the Local Sheet).

Even though proper motion measurements will not be available for galaxies in the Centaurus Group, Kanehisa et al. (2023) analysed the Cen A satellite system under the assumption of a rotationally stabilized planar structure. This yields constraints for the transverse velocities of the dwarf galaxy satellites. Based on the predicted transvere velocities, five satellites are unlikely members of the plane. Disregarding these five off-plane dwarfs enhances the significance of the line-of-sight velocity trend from Müller et al. (2021b) by a factor of five. Interestingly, the line-of-sight velocity trend reported by Müller et al. (2021b) are consistent with either orbital direction, showing that a co-moving signal from line-of-sight velocities – i.e. red-shifted on one side and blue-shifted on the other – is not constraining.

Around M 83, Müller et al. (2018b) found evidence for another flattening of the satellite system. However, this was based on a smaller number of satellites, with larger uncertainties than that for Cen A. Because the distance uncertainties are as large as the putative planar structure, it is not clear whether it is physical or an artefact from the measurement errors.

2.3. *The Sculptor Group*

The Sculptor Group is another galaxy aggregate in our local neighborhood at 3.9 Mpc (Karachentsev et al. 2003). Compared to the Centaurus Group, it is however a very sparse environment. The dominant galaxy is the spiral galaxy NGC 253. Again, different surveys have targeted the vicinity of NGC 253 to search for dwarf galaxies and follow them up (Sand et al. 2014; Müller & Jerjen 2020; Martínez-Delgado et al. 2021; Mutlu-Pakdil et al. 2022). Martínez-Delgado et al. (2021) found a thin spatial alignment of nine dwarf galaxies based on their two and three dimensional positions. Only a handful (i.e. five) of these galaxies have velocities, but a co-moving trend is apparent, with four out of five being consistent with a co-rotational signal. Follow-up observations of their velocities are needed to put this on a more robust data basis.

2.4. *The NGC2683 group*

The NGC2683 galaxy is a field spiral galaxy at a distance of 9.4 Mpc (Karachentsev et al. 2015). Two teams recently surveyed the area and found 12 dwarf galaxies surrounding the galaxy (Carlsten et al. 2022; Crosby et al. 2023). A principle component analysis done by Crosby et al. (2023) of the on-sky distribution of the satellite system reveals an axis ratio $b/a = 0.23$ which is similar to that of other planar structures. Compared to an isotropic satellite system this is a 1 per cent outlier, but compared to the hydrodynamical Illustris-TNG100 simulation (Nelson et al. 2018) it is within one sigma of the predicted distribution (Crosby et al. 2023), i.e. not in tension. Only two satellites have velocity information though, so no robust assessment on co-rotation can be made. Again, follow-up observations of these dwarfs are needed to fully assess the plane-of-satellite problem for this galaxy group.

2.5. *The NGC2750 group*

The NGC 2750 group is at a larger distance (≈40 Mpc) than the previously discussed groups, so it represents quite the different environment than that of our own Local Group and its neigbors. Paudel et al. (2021) found evidence for a co-moving dwarf galaxy system, based on the motion of six satellites. Compared to the other systems presented before, the NGC 2750 dwarf galaxy system is not a flattened structure and more reminiscent of the findings by Ibata et al. (2014b) on the anti-correlation of diametrically opposed galaxy satellites. It certainly will be interesting to follow this group up and search for more dwarf galaxies, as well as get more velocities.

2.6. *The MATLAS survey*

A more statistical approach was done in the MATLAS survey (Habas et al. 2020; Poulain et al. 2021; Marleau et al. 2021). This MegaCam based survey targeted over 200 early-type galaxies between 10 to 45 Mpc, around which over 2000 dwarf galaxies were discovered (Habas et al. 2020). Heesters et al. (2021) used this catalog of dwarf galaxies to systematically measure the flattening of the satellite systems. Because only two dimensional information is available, the analysis was performed using the on-sky distribution of the dwarf galaxies. Heesters et al. (2021) found that around 30% of their satellite systems exhibit a significant flattening, more than what is expected from random distributions. However, none of the systems has enough line-of-sight velocities to study the kinematics of the dwarf galaxies, so it is not clear if there is co-motion of the dwarf galaxies as found for the e.g. Cen A. Following up the MATLAS dwarfs will be a major undertaking, which is currently underway (Müller et al. 2021c; Heesters et al. 2023).

3. A challenge or not?

The question about whether the plane of satellites problem is a challenge for ΛCDM cosmology was recently discussed in several papers. Pawlowski (2021) reviewed the situation for the three best studied systems – the Milky Way, the Andromeda galaxy, and Cen A – and concluded that there is currently no satisfying solution within standard cosmology to explain all the observations. Sales et al. (2022) classified the plane of satellites problem as a severe tension for cosmology. Shortly after this publication, Sales & Navarro (2023) commented that the plane of satellites problem is no longer in tension with ΛCDM based on the work of Sawala et al. (2022), which was, however, focusing only on the Milky Way satellite system. Xu et al. (2023) argued that for the right environment, the Milky Way satellite system may be explained, but also cautioned that to fully solve the plane of satellites problem the other observed planes of satellites also need to be taken into account.

Because the discussion is mainly focused on the Milky Way and its satellite system – which is indeed a well-studied system, but still consists of a sample size of one when comparing to cosmological simulations – it is imperative to study other systems as well. Fortunately, we live in an era of wide field surveys which will unlock the study of dwarf galaxy systems around hundreds of Milky Way analogs. However, we also need to be prepared for this incoming data and have the right tools to compare the observations to cosmology to make meaningful assessments of the status of this lively debated topic.

References

Belokurov, V., Zucker, D. B., Evans, N. W., et al. 2007, APJ, 654, 897.
Carlsten, S. G., Greene, J. E., Beaton, R. L., et al. 2022, APJ, 933, 47.
Cautun, M., Bose, S., Frenk, C. S., et al. 2015, MNRAS, 452, 3838.

Chiboucas, K., Jacobs, B. A., Tully, R. B., et al. 2013, AJ, 146, 126.
Crnojević, D., Sand, D. J., Caldwell, N., et al. 2014, APJL, 795, L35.
Crnojević, D., Sand, D. J., Spekkens, K., et al. 2016, APJ, 823, 19.
Crosby, E., Jerjen, H., Müller, O., et al. 2023, MNRAS, 521, 4009.
Dinescu, D. I., Keeney, B. A., Majewski, S. R., et al. 2004, AJ, 128, 687.
Habas, R., Marleau, F. R., Duc, P.-A., et al. 2020, MNRAS, 491, 1901.
Harris, G. L. H., Rejkuba, M., & Harris, W. E. 2010, PASA, 27, 457.
Heesters, N., Habas, R., Marleau, F. R., et al. 2021, A&A, 654, A161.
Heesters, N., Müller, O., Marleau, F. R., et al. 2023, arXiv:2305.04593.
Ibata, R. A., Lewis, G. F., Conn, A. R., et al. 2013, Nature, 493, 62.
Ibata, R. A., Ibata, N. G., Lewis, G. F., et al. 2014, APJL, 784, L6.
Ibata, N. G., Ibata, R. A., Famaey, B., et al. 2014, Nature, 511, 563.
Kallivayalil, N., van der Marel, R. P., Alcock, C., et al. 2006, APJ, 638, 772.
Kanehisa, K. J., Pawlowski, M. S., Müller, O., et al. 2023, MNRAS, 519, 6184.
Karachentsev, I. D., Grebel, E. K., Sharina, M. E., et al. 2003, A&A, 404, 93.
Karachentsev, I. D., Tully, R. B., Dolphin, A., et al. 2007, AJ, 133, 504.
Karachentsev, I. D., Tully, R. B., Makarova, L. N., et al. 2015, APJ, 805, 144.
Koch, A. & Grebel, E. K. 2006, AJ, 131, 1405.
Kroupa, P., Theis, C., & Boily, C. M. 2005, A&A, 431, 517.
Kunkel, W. E. & Demers, S. 1976, The Galaxy and the Local Group, 182, 241
Libeskind, N. I., Carlesi, E., Müller, O., et al. 2019, MNRAS, 490, 3786.
Lynden-Bell, D. 1976, MNRAS, 174, 695.
Lynden-Bell, D. 1983, Internal Kinematics and Dynamics of Galaxies, 100, 89
Marleau, F. R., Habas, R., Poulain, M., et al. 2021, A&A, 654, A105.
Martin, N. F., Ibata, R. A., McConnachie, A. W., et al. 2013, APJ, 776, 80.
Martínez-Delgado, D., Makarov, D., Javanmardi, B., et al. 2021, A&A, 652, A48.
McConnachie, A. W. & Irwin, M. J. 2006, MNRAS, 365, 902.
Metz, M., Kroupa, P., & Jerjen, H. 2007, MNRAS, 374, 1125.
Metz, M., Kroupa, P., & Libeskind, N. I. 2008, APJ, 680, 287.
Metz, M., Kroupa, P., & Jerjen, H. 2009, MNRAS, 394, 2223.
Mutlu-Pakdil, B., Sand, D. J., Crnojević, D., et al. 2022, APJ, 926, 77.
Müller, O., Jerjen, H., Pawlowski, M. S., et al. 2016, A&A, 595, A119.
Müller, O., Jerjen, H., & Binggeli, B. 2017, A&A, 597, A7.
Müller, O., Pawlowski, M. S., Jerjen, H., et al. 2018, Science, 359, 534.
Müller, O., Rejkuba, M., & Jerjen, H. 2018, A&A, 615, A96.
Müller, O. & Jerjen, H. 2020, A&A, 644, A91.
Müller, O., Fahrion, K., Rejkuba, M., et al. 2021, A&A, 645, A92.
Müller, O., Pawlowski, M. S., Lelli, F., et al. 2021, A&A, 645, L5.
Müller, O., Durrell, P. R., Marleau, F. R., et al. 2021, APJ, 923, 9.
Nelson, D., Pillepich, A., Springel, V., et al. 2018, MNRAS, 475, 624.
Paudel, S., Yoon, S.-J., & Smith, R. 2021, APJL, 917, L18.
Pawlowski, M. S., Kroupa, P., & de Boer, K. S. 2011, Astronomy & Astrophysics, 532, A118. doi:10.1051/0004-6361/201015021
Pawlowski, M. S., Pflamm-Altenburg, J., & Kroupa, P. 2012, MNRAS, 423, 1109.
Pawlowski, M. S., Kroupa, P., & Jerjen, H. 2013, MNRAS, 435, 1928.
Pawlowski, M. S. 2018, Modern Physics Letters A, 33, 1830004.
Pawlowski, M. S. & Kroupa, P. 2020, MNRAS, 491, 3042.
Pawlowski, M. S. 2021, Nature Astronomy, 5, 1185.
Piatek, S., Pryor, C., Bristow, P., et al. 2006, AJ, 131, 1445.
Piatek, S., Pryor, C., Bristow, P., et al. 2007, AJ, 133, 818.
Poulain, M., Marleau, F. R., Habas, R., et al. 2021, MNRAS, 506, 5494.
Richardson, J. C., Irwin, M. J., McConnachie, A. W., et al. 2011, APJ, 732, 76.
Sales, L. V., Wetzel, A., & Fattahi, A. 2022, Nature Astronomy, 6, 897.
Sales, L. V. & Navarro, J. F. 2023, Nature Astronomy.

Sand, D. J., Crnojević, D., Strader, J., et al. 2014, APJL, 793, L7.
Santos-Santos, I. M., Domínguez-Tenreiro, R., & Pawlowski, M. S. 2020, MNRAS, 499, 3755.
Sawala, T., Cautun, M., Frenk, C., et al. 2022, Nature Astronomy.
Sohn, S. T., Patel, E., Fardal, M. A., et al. 2020, APJ, 901, 43.
Tully, R. B., Libeskind, N. I., Karachentsev, I. D., et al. 2015, APJL, 802, L25.
Walsh, S. M., Jerjen, H., & Willman, B. 2007, APJL, 662, L83.
Willman, B., Dalcanton, J. J., Martinez-Delgado, D., et al. 2005, APJL, 626, L85.
Xu, Y., Kang, X., & Libeskind, N. I. 2023, arXiv:2303.00441.

Dynamical Masses of Local Group Galaxies
Proceedings IAU Symposium No. 379, 2025
P. Bonifacio, M.-R. Cioni, F. Hammer, M. S. Pawlowski & S. Taibi, eds.
doi:10.1017/S1743921323001424

The flattening of dark matter halo of Centaurus A galaxy (NGC 5128) out to 40 kpc

T. Veršič[1,2], M. Rejkuba[2], M. Arnaboldi[2], O. Gerhard[3], J. Hartke[4,5], C. Pulsoni[3] and G. van de Ven[1]

[1]Department of Astrophysics, University of Vienna, Türkenschanzstrasse 17, A-1180 Vienna, Austria
email: tadeja.versic@univie.ac.at

[2]European Southern Observatory, Karl-Schwarzschild-Str. 2, 85748 Garching, Germany

[3]Max-Planck-Institut für extraterrestrische Physik, Giessenbachstraße, 85748 Garching, Germany

[4]Finnish Centre for Astronomy with ESO (FINCA), University of Turku, FI-20014 Turku, Finland

[5]Tuorla Observatory, Department of Physics and Astronomy, FI-20014 University of Turku, Finland

Abstract. Cosmological simulations predict dark matter shapes that deviate from spherical symmetry. The exact shape depends on the prescription of the simulation and the interplay between dark matter and baryons. This signature is most pronounced in the diffuse galactic haloes that can be observationally probed with planetary nebulae and globular clusters (GCs). The kinematic observations of these halo tracers support intrinsic triaxial shape for the mass generating the gravitational potential. With discrete axisymmetric modelling of GCs as the halo tracers of NGC 5128 we investigate the overall mass distribution of this nearby giant elliptical galaxy. Our modelling approach constrains c_{200}, $(M/L)_{\star,B}$ and inclination. We derive a preliminary $M_{200} \sim 1 \times 10^{12}$ $M_{\odot}$ and flattening $q_{\rm DM} \sim 1.3$ indicative of prolate/triaxial halo for NGC 5128.

Keywords. (cosmology): dark matter; galaxies: halos; galaxies: elliptical and lenticular; evolution; individual: NGC 5128; galaxies: kinematics and dynamics

---

1. Introduction

Mass is the dominant driver of structure formation and evolution. In the concordance model of ΛCDM cosmology, dark matter is the dominant mass component in the Universe. ΛCDM has been very successful at explaining the observed properties of galaxy populations. However, there are still some predictions from the theory that have been challenging to constrain observationally. One such challenge is the intrinsic shape of the dark matter haloes. The shape is quantified with the axis ratios of a 3D ellipsoid: $p = b/a$ and $q = c/a$, where $a > b > c$ are semi-major, intermediate and minor axis. Halos with $a > b = c$ are considered prolate and $a = b > c$ are considered oblate, otherwise, the shape is triaxial.

NGC 5128 (CenA) is a nearby early-type galaxy (ETG), owing to its proximity and mass it is an excellent target for dynamical modelling of its halo tracers to constrain the flattening of the dark matter halo. At a distance of 3.8 Mpc (Harris et al. 2010) and with a stellar mass of $M_\star \sim 10^{11}\mathrm{M}_\odot$ this galaxy hosts 1700 discrete kinematic tracers, Globular Clusters (GCs) and Planetary Nebulae (PNe) with velocity measurements in the literature. Using PNe kinematics Peng et al. (2004a) found evidence for a triaxial potential of CenA.

In this proceeding, we outline the scientific rationale behind the dynamical studies of dark matter shapes, introduce NGC 5128 as a test case and discuss the first steps of the dynamical modelling of its halo tracers.

2. Dark matter halo shapes

Observationally, dark matter can only be probed through its gravitational influence on the baryons. We can use dynamical modelling to investigate the distribution of dark matter that best explains the motions of luminous matter in the dark matter-dominated galactic haloes. To carry out dynamical modelling and measure the shape of dark matter at these radii we need kinematic tracers of the galactic halo: GCs and PNe. In cosmological simulations, on the other hand, the distribution and shape of dark and baryonic matter can be determined directly through the positions of the simulated particles.

2.1. *Simulations*

Studies of N-body dissipationless simulations (Allgood et al. 2006) found a strong dependence of the halo shape with the different cosmological parameters and the dark matter fraction. These authors find dark matter haloes are prolate in the centres and become more spherical close to the virial radius, and similar conclusion was reached by Hayashi et al. (2007). In the Illustris simulations, Chua et al. (2019) found that baryonic physics acts to make the haloes rounder and more oblate. The impact varies with the halo mass and has strong radial variation, the strongest impact is for $M_{200} = 10^{12-13}\mathrm{M}_\odot$ at $r \sim 0.15 r_{200}$.

2.2. *Observational constraints*

The Milky Way provides a unique vantage point and the possibility to study a galaxy in the greatest detail. However, depending on the technique and data used the recovered shape of the dark matter halo of the Galaxy vary (Bland-Hawthorn and Gerhard 2016, and references therein).

Beyond the Milky Way isolated studies have found strongly prolate haloes for WLM, a gas-rich dwarf galaxy with the major axis of the dark and baryonic components perpendicular to each other (Leung et al. 2021). Khoperskov et al. (2014) found that polar ring galaxies are hosted in haloes that are more prolate in the outer regions and oblate in the centres. While dynamical modelling enables direct measurements of the shape, Pulsoni et al. (2018) find that ETGs have more diverse kinematic properties in their halos than in the central regions. With PNe they find all slow rotators and 40% of fast rotators show signatures of triaxial halos such as kinematic twists or misalignments.

3. The Centaurus A galaxy

NGC 5128 is the closest early-type galaxy to us classified as giant Elliptical or S0 and has clear evidence of a recent accretion history or possibly a major merger (Wang et al. 2020, and references therein). It hosts a radio source Cenutarus A and for simplicity, we will refer to the galaxy as CenA. Its proximity has enabled detailed studies of the resolved RGB (red giant branch) stars out to more than 25 $\mathrm{R_e}$(140 kpc) in Rejkuba et al. (2022) providing constraints on the stellar halo distribution. Spectroscopic observations of a large population of PNe (Peng et al. 2004a; Walsh et al. 2015) and

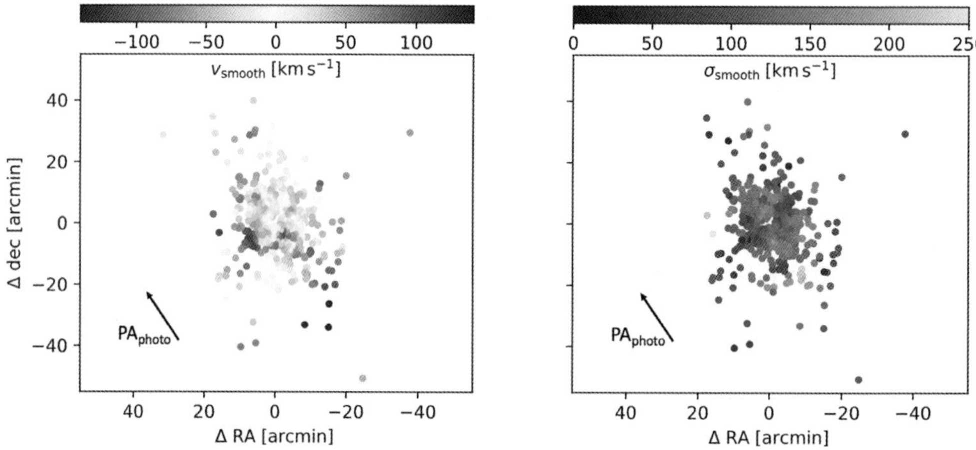

Figure 1. Smooth velocity and velocity dispersion field of GCs in CenA on the left and the right panel respectively. The photometric position angle of the stellar component is indicated with the arrow.

GCs (Hughes et al. 2023) over a vast halo area provide a necessary sample of kinematic tracers for the dynamical modelling that we are exploiting in this work.

3.1. *Kinematics of the globular clusters*

In this work, we use the kinematic sample compiled from the literature by Hughes et al. (2021a) of around 550 GCs out to a median distance of 40 kpc. To investigate the velocity field of the GCs used in this work we adopted the Gaussian smoothing approach from Pulsoni et al. (2018) tested on simulated galaxies for a similarly sized discrete sample of kinematic tracers. Figure 1 shows the smooth velocity on the left panel and velocity dispersion on the right of the GCs within 50 arcmin (~ 10 R$_e$). We can see signs of rotation both along the major and minor axis similar to the PNe velocity field. The velocity dispersion field shows a drop in the velocity dispersion along the minor compared with the major axis.

The GCs of CenA show bimodal colour distribution with red GCs exhibiting similar velocity properties as the field stars, traced by the PNe and blue GCs showing weaker rotation (e.g. Peng et al. 2004b). Woodley et al. (2010) confirmed the presence of metallicity bimodality first identified as a colour bimodality. Additionally, they found that $\sim 50\%$ of metal-rich GCs formed in the past 8Gyrs, supporting the continuous accretion and mergers in CenA.

3.2. *Kinematics of the planetary nebulae*

Peng et al. (2004a) found that the inner rotation of the PNe along the major axis of the galaxy is extended out to the limit of their data (80 kpc). They found evidence of a twist in the zero velocity curve which points to a triaxial or prolate potential of the galaxy. Using harmonic expansion Pulsoni et al. (2018) provided a systematic study of the PNe kinematics and computed the properties of a smooth velocity field from the discrete sample. The authors determined the radial variation of the kinematic position angle of the PNe and a drop in the velocity dispersion around 3 R$_e$($\sim$ 15 kpc).

3.3. *Comparison between GC and PNe velocity field*

To investigate the possible presence of substructures we use the smooth velocity field of the PNe from Pulsoni et al. (2018). We find the discrete sample of GCs is consistent with being drawn from the smooth velocity field of the PNe. The 1, 2, and 3σ outliers show no preferred position in the plane of the galaxy that would indicate a presence of a dynamically cold component that has recently been accreted. Such a component would not be dynamically relaxed and would bias dynamical modelling that is based on a steady-state assumption.

4. Dynamical modelling

We use CJAM (Watkins et al. 2013) implementation of Jeans Anisotropic MGE (Multi Gaussian Expansion), axisymmetric Jeans equations with the assumptions on the cylindrical alignment of the velocity anisotropy introduced by Cappellari (2008).

4.1. *Introduction to the discrete dynamical modelling*

We assume that a local line-of-sight velocity distribution (LOSVD) can be well described by a Gaussian function and use it as likelihood in our Bayesian analysis. Therefore, each kinematic tracer is used discretely to inform the likelihood to find the the best-fit parameters. This methodology has been used in the literature to investigate enclosed mass profiles, orbital properties of the kinematic tracers and flattening of the dark matter haloes on real and simulated galaxies (e.g. Zhu et al. 2016; Hughes et al. 2021b; Leung et al. 2021).

4.2. *Globular cluster tracer density profile*

The largest and most homogenous photometric catalogue of GC candidates is presented in Taylor et al. (2017) and we used it in this work to build the density profile of the kinematic tracers. To identify a sample of relaxed high-fidelity GCs from the GC candidates we analyzed the spatial symmetry of different magnitude selected populations. A dynamically relaxed population is expected to show a point-symmetric spatial distribution. To ensure well-characterized completeness we also require its magnitude distribution to be consistent with a theoretical Gaussian GC luminosity function.

We identified a population, which we called *intermediate population*: $-8.2 < R_0 \leq -7.6$ in the reddening corrected R-band magnitude. This population shows strong colour bimodality and we determine the colour separation of $(u-z)_0 \geq 2.7$ for the red population and $(u-z)_0 < 2.7$ for the blue, with approximately the same number of GCs in both. We computed the flattening of both red and blue GCs separately and found blue GCs are rounder. Finally, we binned them in elliptical radii and found the spatial distribution of GCs can be best described with a combination of an inner Sérsic and outer power-law profile.

Figure 2 shows the best-fit results for the blue GCs on the left and the red GCs on the right panel. The data is shown with colored points, the black line shows the best-fit model with dark and light grey 1 and 3σ uncertainties. To compare with the literature, the dotted black line shows the best-fit profiles for red and blue GCs from Taylor et al. (2017) work and the dashed orange and grey line show the golden and silver sample from Hughes et al. (2021a), where different datasets were used to analyze the contamination of the Taylor et al. (2017) GC candidates. Our profile is broadly consistent with the literature. In the inner regions, our profiles are steeper than the Hughes et al. (2021a) silver and Taylor et al. (2017) and slightly shallower than the Hughes et al. (2021a)

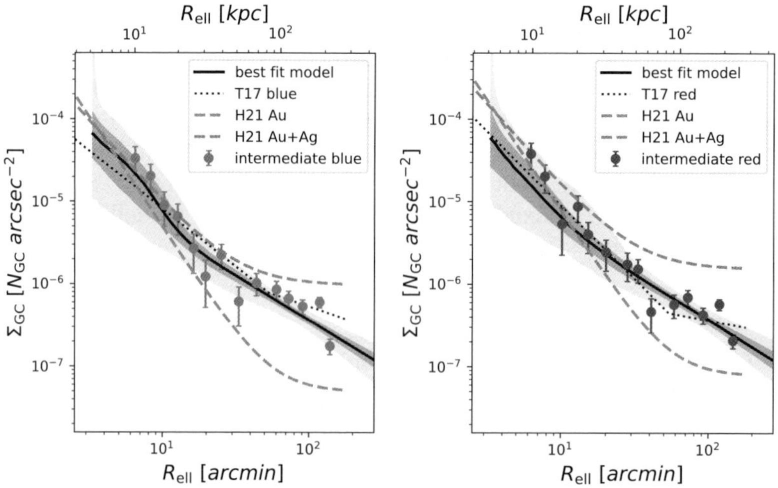

Figure 2. The tracer density profile for the red and blue globular clusters. The left panel shows the best-fit tracer density profile for the blue globular clusters and the right for the res globular clusters. Black solid lines show the best-fit models with dark and light grey showing 1 and 3 σ uncertainties. Dashed and dotted lines show literature results from Taylor et al. (2017) and Hughes et al. (2021a).

golden sample. In the outer regions, our tracer density drops faster than in all literature studies.

4.3. *Gravitational potential*

In our dynamical modelling, both the stars and dark matter contribute to the gravitational potential.

Using photographic plates, Dufour et al. (1979) found the de Vaucouleurs profile can best reproduce the stellar surface density on nested ellipses with a very mild flattening of 0.93 within 1R_e. Rejkuba et al. (2022) extended the stellar surface density profile out to 140kpc using resolved observations of RGB stars from HST observations. Halo stars were found to have a much flatter distribution with a flattening of 0.54. We used both results to create a 2-dimensional surface brightness profile in the range of $0.3<R/R_e<25$ accounting for the difference in the flattening of central and halo stellar components.

For the dark matter component, we assumed an NFW profile parametrised with the M_{200} and c_{200} parameters. For each set of M_{200} and c_{200} we decompose the profile with MGE and account for the assumed flattening of the dark matter and inclination. We define the flattening of the dark matter $q_{\rm DM} = c/a$, where a is aligned along the photometric major axis and c along the photometric minor axis.

4.4. *Likelihood analysis*

Our model has 7 free parameters: M_{200}, c_{200}, $q_{\rm DM}$, inclination, $(M/L)_{\star,\rm B}$, velocity anisotropy and rotation ($\beta_{z,\rm GC}$ and $\kappa_{\rm GC}$ as defined in Cappellari (2008)). For each combination of parameters, we use CJAM to compute the first and second velocity moments at the position of each GC and a Gaussian LOSVD as the likelihood to determine the best-fit parameters in a discrete approach. We model separately the red and blue GC populations.

Our modelling approach constrains c_{200}, $(M/L)_{\star,B}$ and inclination. We derive a preliminary dark matter mass $M_{200} \sim 1 \times 10^{12}$ $M_\odot$ and flattening $q_{\rm DM} \sim 1.3$. The resulting maximum likelihood solutions for the variables of the gravitational potential are consistent between the blue and red GCs. With κ_{GC} and $\beta_{z,GC}$ showing expected differences for both populations of GCs that reflect their distinct dynamic origin.

5. Conclusions

The advancement in our understanding of galaxy formation and evolution as well as the increase in the size and quality of the photometric and spectroscopic datasets has made it necessary to move away from spherical symmetry assumptions in dynamical modelling. With large spectroscopic datasets of halo tracers we can constrain the triaxiality of dark matter.

We find that the magnitude selected intermediate population from Taylor et al. (2017) sample of GCs shows point symmetric spatial distribution, evidence that this population is in approximate dynamical equilibrium. We find that both red and blue GCs have flattened distributions. We used this information to construct a tracer density profile that is broadly consistent with the literature but has a declining density profile in the outskirts of the galaxy. Discrete axisymmetric dynamical modelling can constrain the flattening of the dark matter potential out to 40 kpc in CenA. Detailed results will be presented in Veršič et al. 2023 in prep.

Questions

Jianling WANG: This is nice work. I am concerned about the systematics. The underlying assumption about the dynamics modelling is that the system is at equilibrium state and well relaxed, and single dynamical population tracers used. But it is well known that CenA has experienced a recent major merger, which could violate these assumptions, then what could be the systematic errors? And it would be useful to apply your modelling to a realistic model, for example (major merger models for CenA) to quantify the systematics.

Answer: Given that CenA shows evidence of recent accretion the presence of a non-relaxed component is of course a concern. We investigated the potential presence of a non-relaxed and kinematically cold component of our GCs sample with the smooth velocity field determined from the PNe. We found no evidence of a kinematically cold component in the spatially limited sample of GCs we used in this work. We, therefore, did not make further analysis of what the systematic errors might be in the presence of strong non-relaxed components. Literature studies have looked at how the enclosed mass measurements are biased in the presence of non-relaxed tracers (e.g. Alabi et al. 2016). But I am not aware of a study looking at the impact on the flattening of the potential. It would be very interesting to investigate and quantify this effect with the simulations as you mentioned. Thank you for the suggestion.

Acknowledgements

TV & GvdV acknowledge funding by ERC under EU H2020 grant No. 724857 (ArcheoDyn).

References

Alabi, A. B., Forbes, D. A., Romanowsky, A. J., Brodie, J. P., Strader, J., Janz, J., Pota, V., Pastorello, N., Usher, C., Spitler, L. R., Foster, C., Jennings, Z. G., Villaume, A., & Kartha, S. 2016, *mnras*, 460(4), 3838–3860.

Allgood, B., Flores, R. A., Primack, J. R., Kravtsov, A. V., Wechsler, R. H., Faltenbacher, A., & Bullock, J. S. 2006, *mnras*, 367(4), 1781–1796.

Bland-Hawthorn, J. & Gerhard, O. 2016, *araa*, 54, 529–596.

Cappellari, M. 2008, *mnras*, 390(1), 71–86.

Chua, K. T. E., Pillepich, A., Vogelsberger, M., & Hernquist, L. 2019, *mnras*, 484(1), 476–493.
Dufour, R. J., van den Bergh, S., Harvel, C. A., Martins, D. H., Schiffer, F. H., I., Talbot, R. J., J., Talent, D. L., & Wells, D. C. 1979, *aj*, 84, 284–301.
Harris, G. L. H., Rejkuba, M., & Harris, W. E. 2010, *pasa*, 27(4), 457–462.
Hayashi, E., Navarro, J. F., & Springel, V. 2007, *mnras*, 377(1), 50–62.
Hughes, A. K., Sand, D. J., Seth, A., Strader, J., Lidman, C., Voggel, K., Dumont, A., Crnojević, D., Mateo, M., Caldwell, N., Forbes, D. A., Pearson, S., Guhathakurta, P., & Toloba, E. 2023, *apj*, 947(1), 34.
Hughes, A. K., Sand, D. J., Seth, A., Strader, J., Voggel, K., Dumont, A., Crnojević, D., Caldwell, N., Forbes, D. A., Simon, J. D., Guhathakurta, P., & Toloba, E. 2021,a *apj*, 914a(1), 16.
Hughes, M. E., Jethwa, P., Hilker, M., van de Ven, G., Martig, M., Pfeffer, J. L., Bastian, N., Kruijssen, J. M. D., Trujillo-Gomez, S., Reina-Campos, M., & Crain, R. A. 2021,b *mnras*, 502b(2), 2828–2844.
Khoperskov, S. A., Moiseev, A. V., Khoperskov, A. V., & Saburova, A. S. 2014, *mnras*, 441(3), 2650–2662.
Leung, G. Y. C., Leaman, R., Battaglia, G., van de Ven, G., Brooks, A. M., Peñarrubia, J., & Venn, K. A. 2021, *mnras*, 500(1), 410–429.
Peng, E. W., Ford, H. C., & Freeman, K. C. 2004,a *apj*, 602a(2), 685–704.
Peng, E. W., Ford, H. C., & Freeman, K. C. 2004,b *apj*, 602b(2), 705–722.
Pulsoni, C., Gerhard, O., Arnaboldi, M., Coccato, L., Longobardi, A., Napolitano, N. R., Moylan, E., Narayan, C., Gupta, V., Burkert, A., Capaccioli, M., Chies-Santos, A. L., Cortesi, A., Freeman, K. C., Kuijken, K., Merrifield, M. R., Romanowsky, A. J., & Tortora, C. 2018, *aap*, 618, A94.
Rejkuba, M., Harris, W. E., Greggio, L., Crnojević, D., & Harris, G. L. H. 2022, *aap*, 657, A41.
Taylor, M. A., Puzia, T. H., Muñoz, R. P., Mieske, S., Lançon, A., Zhang, H., Eigenthaler, P., & Bovill, M. S. 2017, *mnras*, 469(3), 3444–3467.
Walsh, J. R., Rejkuba, M., & Walton, N. A. 2015, *aap*, 574, A109.
Wang, J., Hammer, F., Rejkuba, M., Crnojević, D., & Yang, Y. 2020, *mnras*, 498(2), 2766–2777.
Watkins, L. L., van de Ven, G., den Brok, M., & van den Bosch, R. C. E. 2013, *mnras*, 436(3), 2598–2615.
Woodley, K. A., Harris, W. E., Puzia, T. H., Gómez, M., Harris, G. L. H., & Geisler, D. 2010, *apj*, 708(2), 1335–1356.
Zhu, L., Romanowsky, A. J., van de Ven, G., Long, R. J., Watkins, L. L., Pota, V., Napolitano, N. R., Forbes, D. A., Brodie, J., & Foster, C. 2016, *mnras*, 462(4), 4001–4017.

Dynamical Masses of Local Group Galaxies
Proceedings IAU Symposium No. 379, 2025
P. Bonifacio, M.-R. Cioni, F. Hammer, M. S. Pawlowski & S. Taibi, eds.
doi:10.1017/S1743921323001813

New Satellite Dwarf Galaxies of NGC2683 and M104 and the Search for Satellite Planes

Ethan Crosby and Helmut Jerjen

Research School of Astronomy and Astrophysics, Australian National University, Canberra, ACT 2611, Australia
email: ethan.crosby@anu.edu.au

Abstract. The Satellite Plane Problem (SPP) has been a hotly discussed topic for the past two decades. During which, a diametric discussion has begun, between papers that suggest that satellite galaxy planes represent an exception of, and papers which suggest that they are consistent with ΛCDM simulations. However, this discussion has not moved far beyond analysis and re-analysis of galaxies in the Local Group which is a roadblock in producing a more complete and robust analysis. This is motivating an effort to characterise satellite galaxy systems in the wider local universe. We present here initial results of an extensive optical survey of NGC2683 and M104, with the purpose of identifying optical elusive satellite galaxy candidates for follow up observations. These systems are among the first that will allow us to control for mechanisms which are suggested to create or otherwise explain the significance of satellite planes.

Keywords. ΛCDM simulations, Satellite Plane Problem, dwarf galaxies

1. Introduction

Observations of satellites galaxies in the local universe reveal that they are generally distributed in confined 3D, rotating disks about their host galaxies. ΛCDM simulations however find that similar thin co-rotating disks are rare. This forms the basis of the Satellite Plane Problem (SPP) which refers to this tension between ΛCDM simulations and observations. The discussion surrounding the SPP begun when Kroupa *et al.* (2005) suggested that a thin disk of satellites around the Milky Way was inconsistent with ΛCDM simulations. Later that year a series of papers showed that this thin disk was more likely to be consistent with simulations (Zentner *et al.* 2005; Kang *et al.* 2005). This formed two groups of arguments, those who suggested that satellite planes are in contention with ΛCDM simulations and those who suggested they were consistent with ΛCDM simulations or that they don't exist. In the following years, the satellites in the thin disk of the Milky Way were found to be corotating (Pawlowski & Kroupa 2013) and similar co-rotating thin disks were found around M31 (Ibata *et al.* 2013; Conn *et al.* 2013) and Centaurus A (Tully *et al.* 2013; Müller *et al.* 2016, 2018) which were again suggested to be in contention with ΛCDM simulations. Around the same time, opposing suggested that the observed planes were unstable, transient features (Bahl & Baumgardt *et al.* 2014; Gillet *et al.* 2015; Buck *et al.* 2016) and that the original analysis may have been biased and exposed to the *"Look-Elsewhere"* effect (Cautun *et al.* 2015). The depth of information now known about the satellites of the Milky Way and Andromeda, particularly with recent *Gaia* and *HST* proper motions,has meant these two systems continued

Table 1. Details of the eight host galaxies imaged by the HSC.

Host Galaxy (1)	Morphology (2)	R.A. (J2000) (3)	DEC (J2000) (4)	$M_{V_T,0}$ (mag) (5)	D (TRGB - Mpc) (6)	M_* 10^{10} $M_\odot$ (7)	R_{200} kpc (8)
M104	SA(s)a	189.998	-11.623	-22.35	9.55	12.3	385
M66	SAB(s)b	170.062	+12.992	-21.65	9.6	6.4	273
M106	SAB(s)bc	184.740	+47.304	-21.34	7.31	4.7	246
NGC2903	SAB(rs)bc	143.042	+21.501	-21.29	9.33	4.6	244
NGC891	SA(s)b	35.639	+42.349	-20.98	9.12	5.2	255
NGC2903	SAB(rs)bc	143.042	+21.501	-21.29	9.33	4.6	203
NGC2683	SA(rs)b	133.172	+33.422	-20.89	9.36	2.1	190
NGC3521	SAB(rs)bc	166.452	-0.036	-20.91	8.03	2.3	194
UGCA127	Scd	133.362	+33.305	~ -18	10.6	0.4	110

Table 2: Basic properties of host galaxies included in the HSC imaging program. (1): Host Galaxy name. (2): de Vaucouleurs Galaxy Morphology (3): J2000 right ascension. (4): J2000 declination. (5): V-band absolute magnitude contained within sky-level isophote. (6) Tip of the Red Giant Branch distance estimate. (7): Stellar mass (Querejeta *et al.* 2015). (8): virial radius estimate. Data not referenced extracted from the NASA/IPAC Extragalactic Database.

to be the focus of the discussion. This has been used to argue that satellites planes continue to contend with simulations (Pawlowski & Kroupa 2019; Pawlowski & Toyn Sohn 2021) or that satellite planes are consistent with simulations (Sawalan *et al.* 2022; Santos-Santos *et al.* 2023).

Yet the universe is much more than just the Milky Way and M31, and it is quite possible that there is currently unquantifiable bias influencing our understanding. It is acknowledged the potential limitations and biases of our understanding, such as the *"Look-Elsewhere"* effect or any combination of potentially unique conditions present in the Local Group which may bias results, but have yet to fully account for them.

This motivates continued searches for satellite galaxies and testing the presence or absence of satellite planes outside of the Local Group. This way, we could begin to control the influence of particular galactic environments or galaxy formation scenarios and build a sample that is statistically robust.

2. The HSC Sample

To this end, in 2018 using the wide-field Hyper Suprime Camera (HSC) on the Subaru telescope on Maunakea, Hawaii, eight host galaxy environments were imaged, the details of which are described in Table 1. Seven pointings of the HSC were acquired for each environment, extending the field of view about each host galaxy environment to 3-5 degrees with exposures of 2×150 s per pointing. Originally g and r band images were requested, but bad weather limited the observations to g band only. These eight environments were imaged with the purpose of maximising the potential number of new satellite galaxy discoveries. As a consequence, these host galaxies reside near the edge of the Local Volume, a spherical volume of space about the Milky Way with diameter $\sim 10\,Mpc$, where galaxies here remain relatively under researched.

3. NGC2683

Our analysis of these environments begins with NGC2683, an isolated spiral galaxy about half the mass of the Milky Way in the Leo Spur, a spare filamentary structure outside of the Local Sheet, which hosts the Local Group and may have provided the conditions to form satellite planes. This offers an opportunity the test if large scale cosmological structures may influence the formation of satellites planes as some suggest (Pawlowski 2018; Santos-Santos *et al.* 2023).

3.1. *Completeness*

We have tested the completeness of our survey by simulating 4000 dwarf spheroidal galaxies and placing them in random locations throughout our images. We search for these simulated galaxies using the same process to identify dwarf galaxy candidates to emulate the real process for identifying candidates. We find that our extensive and deep HSC images, at the distance of NGC2683, extend to approximately $\sim 380\, kpc$ away from NGC2683, to a surface brightness limit of $\sim$ 26.5 mag arcsec^{-2}, with partial completeness down to $\sim$ 27.5 mag arcsec^{-2}. However, at the distance of 9.4 Mpc, compact dwarf galaxies with half-light radii < 300 pc can become difficult to discriminate from background galaxies, as the size of those galaxies at the distance of NGC2683 approaches the angular seeing limit, which smooths structures upon which this discrimination is primarily based. Given the results from the artificial galaxy simulation, our survey is 100 percent complete for galaxies with total absolute g magnitudes $M_{\rm g} < -10$, and 50 percent complete for $-10 < M_{\rm g} < -9$.

3.2. *New Dwarf Galaxies of NGC2683*

We searched the HSC images of NGC2683 by eye for extended objects with morphology resembling dwarf spheroidal (dSph) or dwarf irregular galaxies (dIrr). In this manner, dSph galaxies are inherently easier to identify their morphology and surface brightness are distinctly different from L_* galaxies. They are also more likely to be a satellite galaxy (as opposed to a field dwarf galaxy), given that these dwarf galaxies are generally found bound to nearby host galaxies (morphology-density relation). dIrr galaxies are however more difficult to identify, as the range of morphological traits can resemble a distant L_* galaxy due to PSF smoothing of granular features required to discriminate the two classes of galaxies and are additionally more likely to reside in the field than bound to a host.

With these considerations in mind, this survey found four satellite galaxy candidates in addition to eight satellite candidates previously reported. Pictures of the four new galaxy candidates are shown in Figure 1 and are predominantly dSph morphology as expected.

3.3. *The NGC2683 System*

It is of interest to see how the NGC2683 system of satellites compares to similar galaxy environments. Firstly, given the context this investigation is framed in, it would be useful to confirm whether NGC2683 has a plane of satellites or not. However, to unambiguously identify any plane of satellites, accurate 3D spatial coordinates and a line-of sight-velocity for each satellite candidate is required. With only 2D sky coordinates, we are limited to a simplified analysis instead. Using these sky coordinates, we take the 2D flattening of this system of satellites as the axis ratio of the ellipse formed from the eigenvectors of the covariance matrix from PCA dimensionality reduction (described in further detail in Crosby *et al.* (2023)). This 'PCA axis ratio' from this analysis is 0.26 ± 0.16 for NGC2683 and its system of satellites, the ellipse of which is shown in Figure 2. If we compare this to an underlying isotropic distribution of the same number of satellites, we find that the axis ratio in that case is $\approx 0.72 \pm 0.14$. This suggests that the satellites of NGC2683 are not isotropically distributed. We then compare NGC2683 to similar systems in the Illustris TNG100-1 ΛCDM simulation (Nelson *et al.* 2019). Similar systems are selected as having virial mass $5 - 50 \times 10^{11}\, M_\odot$ and those that had at least 12 sub-halos within two times the virial radius (R_{200}) with $M_g < -6$, to exclude dark halos and are separated by at least 1.5 Mpc from the nearest neighbouring galaxy group. In that case, the average

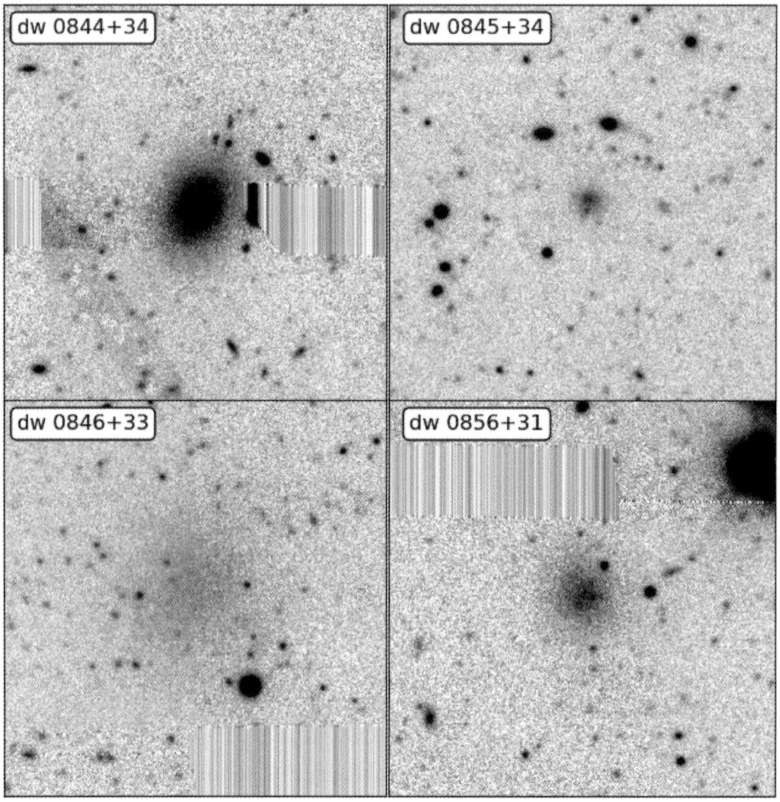

Figure 1. HSC-g images of the four newly detected satellite galaxy candidates within the NGC2683 group. North is up and east to the left. Each image has a side length of 100.8" or 4.6 kpc at the distance of NGC2683. 'Barcode' type anomalies are a result of missing information due to CCD gaps.

flattening for those 929 simulated systems is 0.44 ± 0.23, which on balance is consistent with the NGC2683 system. That is, the 2D distribution of satellites around NGC2683 is consistent with the Illustris TNG100-1 ΛCDM simulation. However, future spectroscopy and distance measurements for these candidates may yet reveal a thin co-rotating disk that is inconsistent with simulations.

In addition to the spatial distribution of these satellites, we can make useful comparisons of their photometry to simulations. In particular, there is a large 'magnitude gap', that is the difference in total absolute magnitudes between the brightest satellite of the NGC2683 system (which is KK69, $M_g = -14.5$) and NGC2683 itself ($M_g = -20.6$). Here we find that the magnitude gap is approximately 6.1 mag, which if we compare to the TNG100-1 simulation again, is larger than 95% of NGC2683-equivalent similar environments. This may be related to another small scale tension with ΛCDM simulations, the Too-Big-To-Fail problem.

4. New Satellites of M104

Now we move onto M104, which is an elliptical galaxy with a spiral-like dust lane situated in the sparse southern extension of the Virgo cluster with a stellar mass of $12.3 \times 10^{10}\, M_{\odot}$. At the distance of M104 (9.55 mpc), like NGC2683, our HSC images extend to around ~ 380 kpc once again, with similar detection completeness limits. However M104

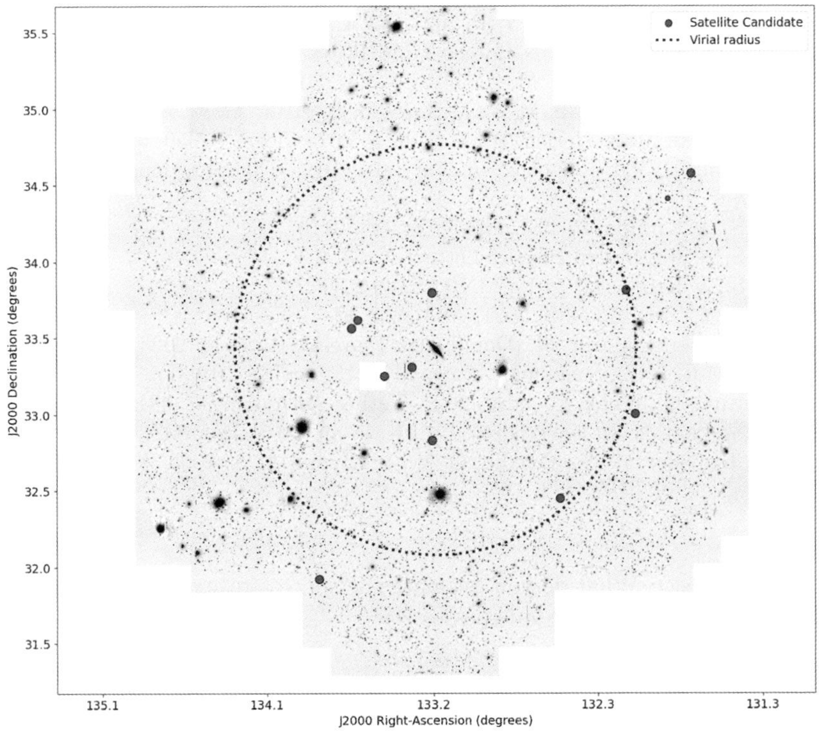

Figure 2. Map showing the HSC survey footprint of the NGC2683 environment. The locations of red circles represent those of all 12 satellites candidates. The black dashed line represents the approximate virial radius of NGC2683.

is a much more populated system with a total 75 dwarf galaxy candidates, where 40 of those are newly reported by us. This may still be a lower limit given M104 has an estimated virial radius of $\sim 380\ kpc$. There could be a number of satellites of M104 beyond the virial radius, as other surveys with a wider field of view have found dwarf galaxy candidates beyond this radius that have attributed them to M104 (Karachentsev *et al.* 2020).

We have yet to report detailed results for this survey (Crosby et al., in preparation), but with such a rich system of satellite galaxies, follow up observations for this system offer an exciting opportunity to explore the Satellite Plane Problem, or the Too-Big-To-Fail problem while minimising the impact of small-number statistics. We also meet our other scientific goals, given that M104 is relatively isolated and resides outside of local cosmic structures. To this end, we will follow up these satellite candidates with spectroscopic observations to extract line of sight velocities, which is vital to fully characterise the M104 satellite galaxy distribution.

References

Bahl, H., & Baumgardt, H. 2014, MNRAS, 438, 2916

Buck, T., Dutton, A. A., & Macciò, A. V. 2016, MNRAS, 460, 4348

Cautun, M., Bose, S., Frenk, C. S., et al. 2015, MNRAS, 452, 3838

Conn, A. R., Lewis, G. F., Ibata, R. A., et al. 2013, ApJ, 766, 120

Crosby, E., Jerjen, H., Müller, O., et al. 2023, MNRAS, 521, 4009

Gillet, N., Ocvirk, P., Aubert, D., et al. 2015, ApJ, 800, 34

Ibata, R. A., Lewis, G. F., Conn, A. R., et al. 2013, Nature, 493, 62

Kang, X., Mao, S., Gao, L., & Jing, Y. P. 2005, A&A, 437, 383
Karachentsev, I. D., Makarova, L. N., Brent Tully, R., et al. 2020, A&A, 643, A124
Kroupa, P., Theis, C., & Boily, C. M. 2005, A&A, 431, 517
Müller, O., Jerjen, H., Pawlowski, M. S., & Binggeli, B. 2016, A&A, 595, A119
Müller, O., Pawlowski, M. S., Jerjen, H., & Lelli, F. 2018, Science, 359, 534
Nelson, D., Springel, V., Pillepich, A., et al. 2019, Computational Astrophysics and Cosmology, 6
Pawlowski, M. S., & Kroupa, P. 2013, MNRAS, 435, 2116
Pawlowski, M. S. 2018, Modern Physics Letters A, 33, 6
Pawlowski, M. S., & Kroupa, P. 2019, MNRAS, 491, 3042
Pawlowski, M. S., & Tony Sohn, S. 2021, ApJ, 923, 42
Santos-Santos, I., Gámez-Marín, M., Domínguez-Tenreiro, R., et al. 2023, ApJ, 942, 78
Querejeta, M., Meidt, S. E., Schinnerer, E., et al. 2015, ApJSS, 219, 5
Sawala, T., Cautun, M., Frenk, C., et al. 2022, Nature Astronomy, 7, 481
Tully, R. B., Libeskind, N. I., Karachentsev, I. D., et al. 2015, ApJ, 802, L25
Zentner, A. R., Kravtsov, A. V., Gnedin, O. Y., & Klypin, A. A. 2005, ApJ, 629, 219

Discussion

Marcel Pawlowski: What do you estimate the success rate of the satellite candidates of M104 becoming real satellite galaxies to be?

Answer: The 75 reported candidates within the virial radius of M104 consist of what we consider 44 high probability or confirmed candidates (with spectroscopy or distance measurements) and 31 low probability candidates. These 31 low probability candidates include reported candidates from prior surveys which introduces problems; no two researchers have the same judgement or process when it comes to discriminating potential candidates. As a result these 31 'low probability' candidates includes objects we would not consider to be satellites of M104 but we have included nonetheless in acknowledgement of prior work and the literature. Approximately 40-50 of these candidates are likely to be real satellite galaxies, a number consistent with the success rates (80%) of other dwarf galaxy surveys our team has conduced.

Question: Would you expect to find a satellite plane around M104, given it resembles a miniaturised galaxy cluster with the number of satellites and their morphology?

Answer: We have no expectations as to the presence of a satellite plane around M104. But if this is indeed correct and M104 possesses no satellite plane due to its rich accretion history which built its extensive system of satellites, then this is an important find. It would show that satellite planes are not built, or are destroyed by, periods of high accretion of dwarf galaxies. It would help to confine hypotheses that preferential accretion has built the satellite planes we observe in the Local Group.

Marina Rejkuba: How do you characterise the completeness of the search for the satellites? In particular with respect to the gap in the luminosity distribution for NGC2683: how sure are you that this could not be the effect of the obscuration by the host galaxy (satellites behind the host)?

Answer: See Section 3.1 for an explanation of our characterisation of completeness. But obscurration of the brightest satellite would be extremely unlikely. Our spatial coverage of NGC2683 environment extends to 2 times the virial radii and NGC2683 itself covers less than 0.5% of that area, so it would be highly unlikely for a luminous to be completely obscured by NGC2683. In this context it is worth noticing that effective radius and total luminosity are correlated for dwarf galaxies. A luminous satellite would also have a large effective radius. We note that a luminous satellite could also be obscurred by a bright foreground star, but again we estimate the probability of this occuring to be $< 1\%$, given this field has also been scanned by blind 3D (α,δ,v_{rad}) neutral hydrogen surveys in the past, thereby excluding the possibility of a gas-rich satellite behind M104 or a bright foreground star.

Dynamical Masses of Local Group Galaxies
Proceedings IAU Symposium No. 379, 2025
P. Bonifacio, M.-R. Cioni, F. Hammer, M. S. Pawlowski & S. Taibi, eds.
doi:10.1017/S1743921323001485

A dynamical mass map of the nearby Fornax galaxy cluster

Avinash Chaturvedi[1,2], Nicola R. Napolitano[3,4] and Michael Hilker[2]

[1]Leibniz-Institut für Astrophysik Potsdam (AIP), An der Sternwarte 16, D-14482 Potsdam, Germany
email: avi.chaturvedi@aip.de

[2]European Southern Observatory, Karl-Schwarzschild-Straße 2, 85748 Garching, Germany

[3]School of Physics and Astronomy, Sun Yat-sen University, Zhuhai Campus, 2 Daxue Road, Xiangzhou District, Zhuhai, P. R. China

[4]INAF-Osservatorio Astronomico di Capodimonte, Salita Moiariello, 16, 80131-Napoli, Italy

Abstract. The nearby Fornax cluster (d $\sim$ 20 Mpc) provides an unparalleled opportunity to investigate the formation and evolution of early-type galaxies in a dense environment. Using the spectroscopic data from the ESO VLT/VIMOS spectrograph of the Fornax cluster, we have kinematically characterised the photometrically detected globular cluster (GC) candidates in the core of the cluster. We confirm a total of 777 GCs with new velocity measurements and compile the most extensive spectroscopic GC sample of 2341 objects in this environment. We used our GC radial velocity catalogue to perform dynamical mass modelling of NGC 1399, the central galaxy of the the Fornax cluster out to its 200 kpc. We find that both cusp (NFW) and core (Burkert) dark matter (DM) halo can produce the observed kinematics. Independent of the DM halo profile used, we find that inclusion of the intra-cluster GCs in mass-modelling can effect the mass-estimate.

Keywords. Fornax galaxy cluster– NGC1399 –globular clusters –kinematics –dynamical modelling

---

1. Introduction

Galaxy clusters are the largest gravitationally bound objects in the Universe, and studying their mass assembly provides important insights about galaxy formation and evolution in dense environments. In the current structure formation scenario, these massive structures grow by hierarchical merging (e.g. Kravtsov & Borgani 2012). Smaller low-mass systems form first and their subsequent mergers lead to massive structures. During the hierarchical formation, various physical processes are at work influencing their growth and evolution, such as ram pressure stripping and tidal interactions which can inflate galaxies. The nearby Fornax galaxy cluster is a very interesting and dynamically evolving environment (Kravtsov & Borgani 2012; Amorisco 2019). It is one of the most massive galaxy overdensities within 20 Mpc and is an ideal laboratory to learn about the mass-assembly process of galaxies.

Due to its proximity, Fornax provides a unique opportunity to map its complex kinematics from the core out to half of its virial radius ($R_{vir} \sim 0.7$ Mpc) using globular clusters (GCs) as kinematic tracers. This allows us to connect the large-scale kinematics of the major galaxies down to the scale of dwarfgalaxies, which constitute the

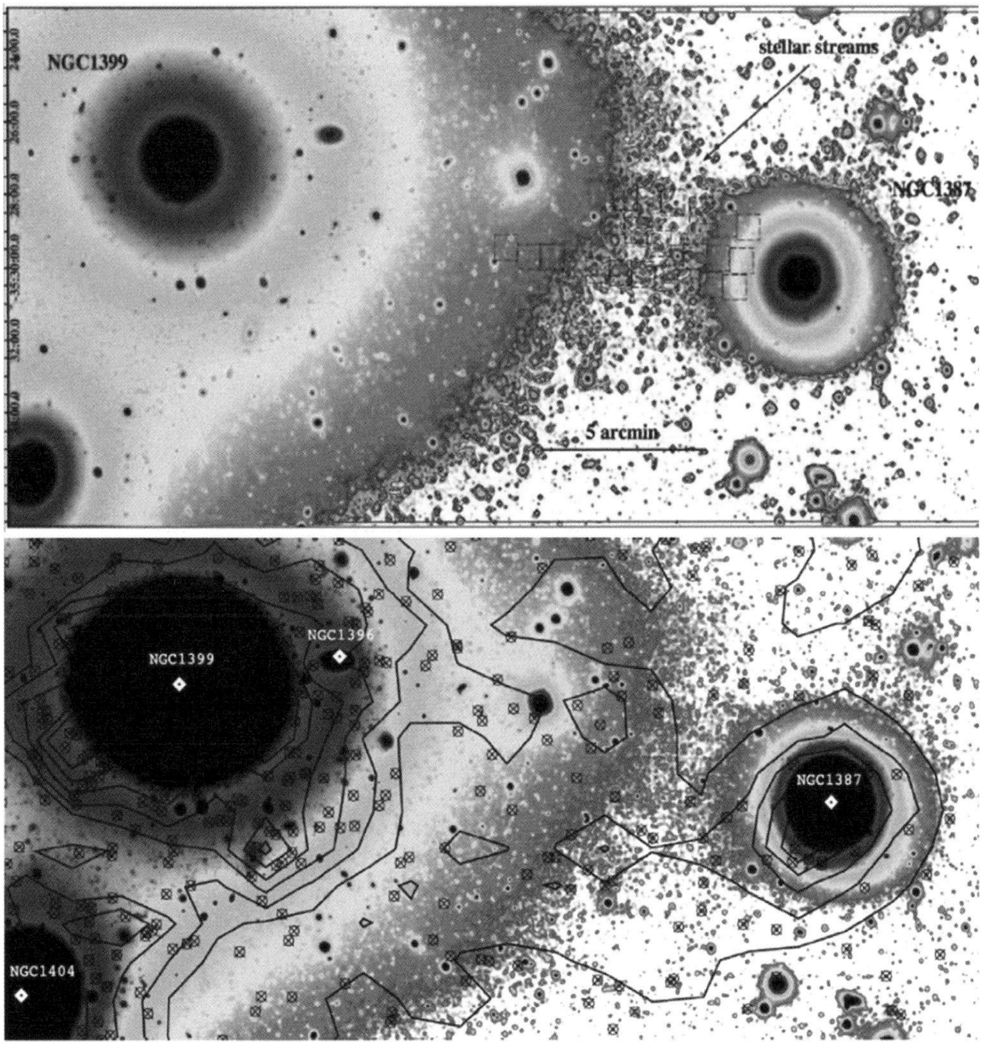

Figure 1. Top: g band image of the Fornax cluster core region, showing the presence of stellar streams westwards of NGC 1399 connecting to the neighbouring galaxy NGC 1387 (adopted from Iodice et al. 2016). **Bottom:** GCs density contours overplotted on the top image (adopted from D'Abrusco et al. 2016).

majority of kinematic sub-structures, and to build a detailed mass model of the cluster. Photometric studies from D'Abrusco et al. (2016) and Iodice et al. (2016) have revealed that, despite the regular appearance of the Fornax cluster, its assembly is still ongoing, as evidenced by the presence of stellar and GC tidal streams. In Figure 1, the left panel shows the stellar streams detected between NGC 1399, the central galaxy of Fornax, and its neighbouring galaxy NGC 1387. And the right panel shows the GCs overdensity in the same region.

To understand the mass-assembly of the Fornax cluster we are conducting the Fornax cluster VLT spectroscopic survey (Pota et al. 2018; Spiniello et al. 2018; Chaturvedi et al. 2022; Napolitano et al. 2022). In Chaturvedi et al. 2022, (hearafter FVSS-III), we produced the largest GCs radial velocity catalog using VIMOS observations and provided

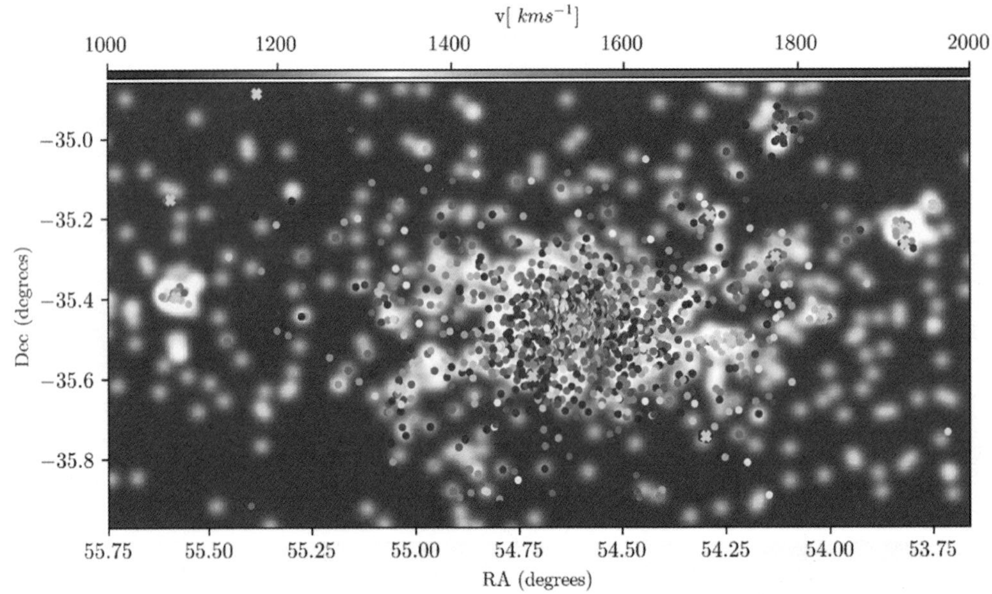

Figure 2. Full radial velocity map of GCs from FVSS-III within the Fornax cluster core region plotted over the surface density map of photometric GC candidates from the Fornax deep survey (Cantiello et al. 2020). Dots are colour coded according to the GCs line-of-sight velocity. Major galaxies are labelled in green. Background white color represent higher density of GCs from the photometric sample.

the spectroscopic confirmation and kinematical characterization of the GC substructures between the Fornax cluster galaxies. We also showed that the outer halo of NGC 1399 and intra-cluster substructures are dominated by blue GCs.

In this proceeding, we present results that will be expanded in a future work (Chaturvedi et al. in prep.) where we have performed the mass-modelling of NGC 1399 out to 200 kpc, with the main goal to learn about the total-mass estimate of NGC 1399 and to investigate the role of substructures in mass-modelling.

2. Kinematic sample of GCs for the dynamical modelling

For performing the mass-modelling work and to learn about the role of intra-cluster GCs populations on the mass profile of NGC 1399, we divided the full GC radial velocity catalog of Chaturvedi et al. (2022), into two samples. The first sample (hereafter referred to as *inner* sample) consists of the GCs within two effective radii of NGC 1399 where we exclude extreme radial velocity GCs that are probably not bound to the galaxy and just are chance-projections on its central region. The second sample (hereafter referred to as *full* sample) consists of the entire sample of GCs, including the ones in the outer halo of NGC 1399, extending out to five effective radii. Figure 2 shows the full radial velocity map of the *full* sample plotted over the surface density map of photometric GC candidates from the Fornax deep survey (Cantiello et al. 2020).

For the *inner* sample, we selected the GCs by performing cuts in the phase-space distribution. We first calculated the scatter in the radial velocities of GCs within two effective radii ($r_{\rm eff}$) of NGC 1399 and used $\pm 2\sigma$ of this velocity scatter as the lower and upper boundaries to select GCs belonging to the galaxy. The giant elliptical NGC 1404 lies at a projected distance of $2r_{\rm eff}$ from NGC 1399, and its outer halo GCs overlap in phase-space with the GCs of NGC 1399. Therefore, we removed the GCs belonging to

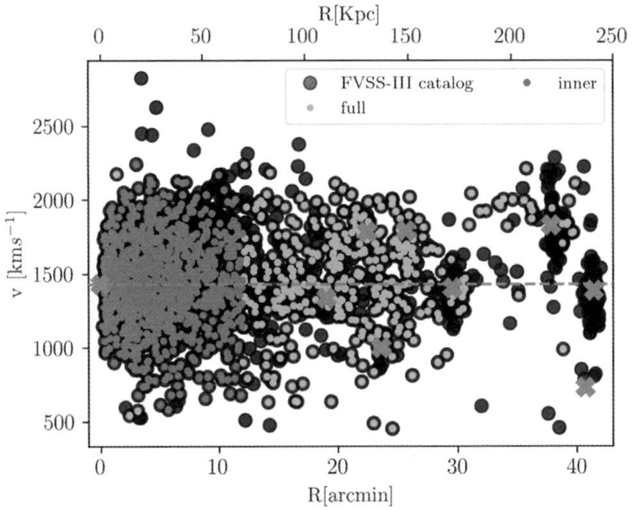

Figure 3. LOS velocities of our GC sample vs. cluster-centric distance. Black dots indicate the complete sample of GCs from paper FVSS-III. Red and orange dots indicate the GCs of the *inner* and *full* samples, respectively. Magenta crosses indicate major galaxies, and the horizontal dashed line marks the systemic velocity of NGC 1399.

NGC 1404 from the GCs sample selected within $2r_{\text{eff}}$. The GC selection around NGC 1404 was done in a similar way as for the *inner* sample but adopting the effective radius of NGC 1404.

For the *full* sample, we consider the entire FVSS-III catalog of GCs, although we limit it to 200 kpc cluster-centric distance due to spatial incompleteness of the spectroscopic samples further outside. Within 200 kpc, the GC systems of another four major galaxies, NGC 1387, NGC 1380B, NGC 1379 and NGC 1380, contribute to the sample. We remove GCs $2r_{\text{eff}}$ of these galaxies similarly as described above. Figure 3 shows the projected LOS velocity of the two different GCs samples. Red and orange dots denote GCs of the *inner* and *full* samples, respectively.

3. Dynamical Modelling

For the mass-modelling work, we used higher order Jeans moments, similar to what is described in Napolitano et al. (2009, 2011, 2014). The Jeans equation connects the tracer velocity distribution and its anisotropy to the total dynamical mass of the system. We assumed a non-rotating and spherical symmetry for the Jeans equations and solved it to obtain the second (velocity dispersion) and fourth order velocity moments (kurtosis). Inclusion of the fourth moment in our mass-modelling allows to alleviate and control the mass-anisotropy degeneracy (see Łokas & Mamon 2003; Łokas et al. 2005), faced in the Jeans modelling.

For the gravitational potential we include the two components stellar and dark matter (DM) mass. The stellar mass profile is obtained by integrating the deprojected g-band luminosity profile of NGC 1399. For the DM halo, we have considered two profiles describing the dark matter (DM) contribution to the gravitational potential, the cored Burkert profile (Burkert 1995) and the cuspy NFW profile (Navarro et al. 1997).

For determining the best fit parameters of the dynamical modelling we employed a maximum likelihood (ML) approach. We created likelihood functions corresponding to the observations, first corresponding to the velocity dispersion and second related to

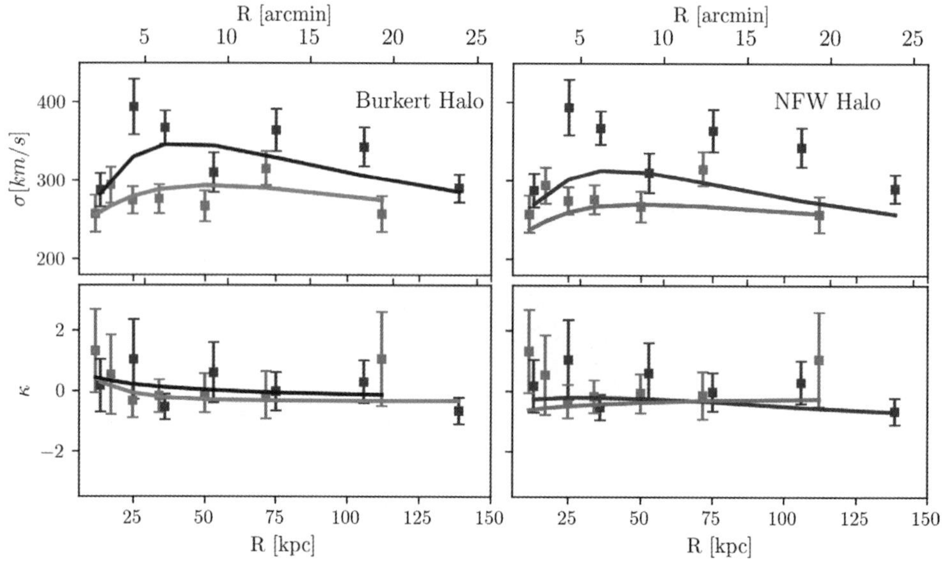

Figure 4. Velocity dispersion and kurtosis profiles of the *full* sample. Left: Best fit model for Burkert halo. Top and middle panels show the velocity dispersion and kurtosis profiles for the red and blue GCs (squares) and solid lines indicate the best fit models. Right: Same as left but for the NFW halo best parameters.

kurtosis of the GCs. Then we combined these two likelihood to determine the best set of parameters. Assuming a Gaussian line of sight velocity distribution (LOSVD), for a given population k of GCs (either red or blue) with a standard deviation equal to the uncertainty of the measured v_{rms}, the log-likelihood of observing a GC population having a velocity dispersion v_{rms}, given a set of model parameters is:

$$\ln p_{v_{rms}}(v_{rms_k}|\boldsymbol{M}_k) = -\frac{1}{2}\sum_{i=1}^{N}\left[\left(\frac{v_{rms,i}-\sigma_k}{\Delta v_{rms,k}}\right)^2 + \ln(2\pi\Delta v_{rms,i}^2)\right] \tag{1}$$

In our modelling the model parameters are constant mass-to-light ratio (Υ), DM scale radius (r_s) and density normalisation (ρ) of Burkert or NFW halo profile. In addition to these we have β_{in}, β_{out} anisotropy parameter which controls the shape of anisotropy profile in the Jeans equations.

To find the best-fit parameters, we utilize the EMCEE package (Foreman-Mackey et al. 2013), a python-based implementation of the affine-invariant Markov Chain Monte Carlo (MCMC) sampling algorithm. We performed the dynamical modelling for individual components, i.e. fitting the red and blue GCs kinematics independently, as well as for the combined two-component tracers, fitting both red and blue GCs kinematics simultaneously. In the next section, we present the results of two component modelling of the *full* sample and summarise the key findings.

4. Results

We focus on the results of the full sample. Figure 4 shows the best fit model to the observed kinematics of the full sample. Left panel in Figure 4, shows the best fit model

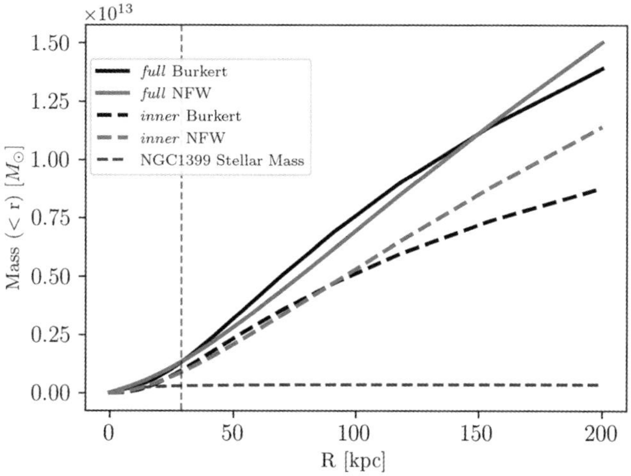

Figure 5. Enclosed mass profile of NGC 1399. The black and red solid lines indicate the enclosed mass for the *full* sample of Burkert and NFW halo, respectively. The dashed black and red lines denote the same but for the *inner* sample.

for the Burkert halo whereas the right panel shows the fit for the NFW halo. We find that both, NFW and Burkert halo, produce the observed kinematics quite well for both the *inner* and *full* sample.

Figure 5 shows the enclosed mass profiles obtained from our best-fit models for the *inner* and *full* sample. The black and red curves indicate the mass profiles for the Burkert and NFW profiles for the *full* sample, respectively, whereas the dashed curves show the same but for the *inner* sample. For the inner sample, we observe that the mass profiles of the Burkert (black dashed line) and NFW (red dashed line) halos agree well within 100 kpc, where the mass profiles are almost equal. Beyond that, the profiles start to diverge, and the NFW mass profile shows a mass ~1.5 times larger than the Burkert mass profile. In contrast to the *inner* sample, mass profile for the *full* sample, from both Burkert and NFW halos, are similar and are in excellent agreement with each other within the uncertainties. For the *full* sample, the mass profiles start deviating beyond 150 kpc.

5. Summary

We used the FVSS spectroscopic catalogue produced in FVSS-III to perform the mass-modelling of NGC 1399. Based on the phase-space and velocity space, we divided the GCs catalog in two kinematic samples around NGC 1399 to investigate the effect of intra-cluster GCs on the mass modelling. The *inner* sample of GCs is limited to $2r_{\mathrm{eff}}$ around NGC 1399, and the *full* sample extend to $\sim 6.0\ \mathrm{r_{eff}}$. We find that both DM halos, NFW (cusp) and Burkert (core), can reproduce the observed kinematics well, with the Burkert profile describing better the blue GCs. The total mass obtained from *full* sample for both the Burkert and NFW halo agrees well and is slightly higher then the mass obtained from the inner sample. Our work on mass-modelling of NGC 1399 shows that including the kinematics of GCs from the outer halos and intra-cluster region can change the total mass estimates. However to quantify and learn how significant is this mass change, requires a proper and detailed careful investigation of intra-cluster GCs role in the dynamical modelling.

References

Amorisco, N. C. 2019, MNRAS, 482, 2978

Burkert, A. 1995, ApJL, 447, L25.

Cantiello, M., Venhola, A., Grado, A., et al. 2020, Astronomy & Astrophysics, 639, A136. doi:10.1051/0004-6361/202038137

Chaturvedi, A., Hilker, M., Cantiello, M., et al. 2022, A&A, 657, A93.

Kravtsov, A. V. & Borgani, S. 2012, ARAA, 50, 353

D'Abrusco, R., Cantiello, M., Paolillo, M., et al. 2016, ApJL, 819, L31. doi:10.3847/2041-8205/819/2/L31

Iodice, E., Capaccioli, M., Grado, A., et al. 2016, ApJ, 820, 42. doi:10.3847/0004-637X/820/1/42

Pota, V., Napolitano, N. R., Hilker, M., et al. 2018, MNRAS, 481, 1744.

Napolitano, N. R., Romanowsky, A. J., Coccato, L., et al. 2009, MNRAS, 393, 329.

Napolitano, N. R., Romanowsky, A. J., Capaccioli, M., et al. 2011, MNRAS, 411, 2035.

Napolitano, N. R., Pota, V., Romanowsky, A. J., et al. 2014, MNRAS, 439, 659.

Napolitano, N. R., Gatto, M., Spiniello, C., et al. 2022, AA, 657, A94.

Navarro, J. F., Frenk, C. S., & White, S. D. M. 1997, ApJ, 490, 493.

Łokas, E. L. & Mamon, G. A. 2003, MNRAS, 343, 401.

Łokas, E. L., Mamon, G. A., & Prada, F. 2005, MNRAS, 363, 918.

Spiniello, C., Napolitano, N. R., Arnaboldi, M., et al. 2018, MNRAS, 477, 1880.

Foreman-Mackey, D., Hogg, D. W., Lang, D., et al. 2013, PASP, 125, 306. doi:10.1086/670067

Author index